BRITISH RUST FUNGI

BRITISH
RUST FUNGI

BY

THE LATE MALCOLM WILSON, D.Sc.

of the University of Edinburgh

AND

D.M.HENDERSON, B.Sc.

of the Royal Botanic Garden, Edinburgh

CAMBRIDGE

AT THE UNIVERSITY PRESS

1966

CAMBRIDGE UNIVERSITY PRESS
Cambridge, New York, Melbourne, Madrid, Cape Town,
Singapore, São Paulo, Delhi, Tokyo, Mexico City

Cambridge University Press
The Edinburgh Building, Cambridge CB2 8RU, UK

Published in the United States of America by Cambridge University Press, New York

www.cambridge.org
Information on this title: www.cambridge.org/9780521279260

First published 1966
First paperback edition 2011

A catalogue record for this publication is available from the British Library

Library of Congress Catalogue Card Number: 65-17202

ISBN 978-0-521-06839-0 Hardback
ISBN 978-0-521-27926-0 Paperback

PREFACE

This volume has had a somewhat chequered history. No major work on the rust flora of the British Isles, other than Wilson and Bisby's *Checklist*, has appeared since W. B. Grove's volume in 1913. Prior to his death in 1938, Grove intended to write an addendum to his *British Rust Fungi*. For several years before the Second World War, Dr Malcolm Wilson worked on a revision of Grove's book, but the stress of wartime and heavy teaching commitments stopped the project. After 1947 Dr Wilson decided that, in fact, a new book was required and after his retiral in 1951 he devoted much of his time to this task. He worked largely on a partially revised typescript of Grove's volume which he had prepared. At the time of his death in 1960 he had completed some two-thirds of the accounts of species and had collected notes for the rest of the work.

In 1960 I undertook to complete the text. In the course of this I have also revised some parts and rechecked all the references. I have also thought it valuable to provide keys to some major rust groups—the genus *Phragmidium*, rusts of *Compositae*, *Gramineae* and *Carex*.

The scope of introductory matter has been difficult to decide. Malcolm Wilson had already decided to leave out a discussion on phylogeny which was prominent in Grove's text. The Introductory Chapter is designed solely to enable the student to have at hand an outline of soral structure and life cycles. For more detailed studies and accounts of physiological races, cytology, etc., he must seek elsewhere; there are enough topics in these subjects to fill a companion volume at least as large as this present one.

A few of the illustrations (all reproduced ×660 unless otherwise stated) have been redrawn from other texts; the rest have been prepared from British collections in the herbarium of the Royal Botanic Garden, Edinburgh. The arrangement of genera follows our concept of a natural classification of these genera of temperate climate. Within large genera, e.g. *Puccinia* and *Uromyces*, the species are arranged according to the host family on which the teleutospore stage develops. This system is susceptible to criticism but for a flora of as restricted a geographical area as the British Isles it serves quite well; a rust flora of the north temperate hemisphere might be treated somewhat differently. The order of host families used is that in the second edition of Clapham, Tutin and Warburg's *Flora of the British Isles*. The nomenclature of native British hosts follows the same work except for a few more recent alterations.

The species concept we have tried to apply owes much to Ivar

Jørstad's work on rusts. We fully realise the extent to which rust taxonomy is susceptible to the taxonomy of their hosts, nevertheless the recognition at specific rank of morphologically distinct taxa, parasitising single host families, seems still to lead to the most workable species concept. Moreover, we recognise that, for example, the genera *Uromyces* and *Puccinia* are at many points unnatural but their proper revision cannot be undertaken in what is only a 'local' flora. It has also been necessary to restrict notes on subspecific categories to varieties and formae speciales, the intricacies of race structure and genetics of economically important rusts are outwith the content of this book.

In questions of nomenclature and synonymy *Enumeratio Uredinearum Scandinavicarum* by Hylander, Jørstad and Nannfeldt has been of great assistance. Only synonyms which have been used in works on British rusts have been listed. According to the latest *Code of Botanical Nomenclature* only names based on teleutospores are valid. Where I have felt it necessary to include an invalid name in synonymy I have enclosed it in square brackets or when a combination has been based on an invalid name the authority for the invalid name in square brackets. Also, when spore stages of a species have been well described abroad but have not been recorded in this country short descriptions have been included in square brackets with the hope of stimulating search for them by British mycologists. Although many ill-authenticated host records have been checked and discarded or definitely confirmed there is still a small hard core of doubtful host records. I have rather freely enclosed all host names for records, which I regard as not properly confirmed, in quotes.

This book could not have been written without the help and encouragement of many mycologists. In addition to a long succession of students and colleagues in Edinburgh, Malcolm Wilson was particularly indebted to E. A. Ellis, Esq., W. G. Bramley, Esq., Dr L. Ogilvie, the late G. R. Bisby, Kew, Mrs M. R. Gilson, Prof. E. A. Muskett, Belfast, Prof. J. Macdonald, St Andrews, and Dr Mary Noble, Edinburgh. My thanks are due to Dr R. W. G. Dennis, for advice and encouragement at Kew, to my assistant, Miss H. Prentice, for checking references and to many colleagues for advice and discussion. Dr C. E. Foister nobly undertook the preparation of the index. Dr H. R. Fletcher, Regius Keeper of the Royal Botanic Garden, Edinburgh, has encouraged the whole project. To the family of the late Malcom Wilson my deepest thanks are due for entrusting me with the completion of that fine mycologist's major work. The book will stand as a memorial to him.

D. M. H.

Royal Botanic Garden, Edinburgh
December 1963

CONTENTS

PUCCINIACEAE (*cont.*)

INTRODUCTION

Sori and spore stages of British rusts

A macrocyclic rust—that is one with a full complement of spore stages—develops five types of spore in the course of the life history, spermatium, aecidiospore, uredospore, teleutospore and basidiospore. The first four of these are produced in more or less organised groups or aggregations called sori, namely: spermogonia, aecidia, uredosori and teleutosori. The basidiospores are produced by the teleutospores so their aggregation or state of dispersal during formation is dependent upon the fate of the teleutospore when they reach maturity.

It should be noted that the application of these terms is not strictly according to their morphology but rather according to the position they occupy in the life cycle of the rust. Although this method may cause some confusion it has the merit of focusing attention on the biology of the fungi rather than on their morphology alone.

The following account deals with the structure and function of the soral and spore types and is restricted to the range displayed by members of the British rust flora.

Spermogonium (pl. spermogonia): the spermogonium bears spermatia (sing. spermatium). It develops on haploid mycelium and the spermatia are haploid spores with very limited powers of germination; their only known function is to fuse with receptive hyphae and initiate the dicaryon, the equivalent in the basidiomycetes of the diploid in other organisms. Spermogonia develop with and usually slightly before aecidia or teleutosori; they never accompany uredosori. In brachyforms spermogonia are associated with uredinoid aecidia—often also referred to as 'primary uredosori'.

In form, the spermogonia are more or less diffuse layers of fertile tissue without any clearly defined margin in the genera *Uredinopsis*, *Milesina* and *Pucciniastrum*, rather more clearly defined and lenticular in *Melampsorella*, *Melampsora*, *Coleosporium*, *Cronartium*, *Frommea*, *Kuhneola*, *Nyssopsora*, *Phragmidium*, *Tranzschelia*, *Triphragmium* and *Xenodochus* and more or less immersed, flask-shaped often with projecting paraphyses and receptive hyphae in the genera *Chrysomyxa*, *Cumminsiella*, *Gymnosporangium*, *Miyagia*, *Puccinia*, *Uromyces* and *Zaghouania*. In position, spermogonia may be subcuticular as in

Frommea, Kuhneola, Melampsora (on Dicotyledons), *Melampsorella, Melampsoridium, Milesina, Ochropsora, Pucciniastrum, Tranzschelia, Triphragmium, Uredinopsis* and *Xenodochus*, sub-epidermal in *Chrysomyxa, Coleosporium, Cumminsiella, Cronartium, Gymnosporangium, Hyalopsora, Melampsora* (on conifers), *Miyagia, Puccinia, Uromyces* and *Zaghouania* or intra-cortical in *Cronartium*. Spermatia are always unicellular and are usually enveloped in a sweetish secretion attractive to insects.

Aecidium (pl. aecidia). Aecidia are the structures in which the newly formed dicaryophase produces dicaryotic spores–aecidiospores. Aecidia are usually borne on a haploid mycelium in association with spermogonia. As regards their morphology aecidia can be grouped in five main types as follows:

(1) Aecidioid aecidia are the 'typical' aecidia of the genera *Uromyces* and *Puccinia*. A well-developed, but short, peridium is present, the peridial cells are rhomboidal or quadrate. The aecidiospores form in chains.

(2) Roestelioid aecidia are confined to the genus *Gymnosporangium*. The peridium is elongate and horn-like and the peridial cells long and narrow, usually at least twice as long as broad.

(3) Peridermioid aecidia occur in the genera *Coleosporium, Cronartium, Milesina* and *Pucciniastrum*—all on gymnospermous hosts. The peridium is strongly developed with elongate peridial cells. The aecidiospores are catenulate with coarsely verrucose walls.

(4) Caeomoid aecidia occur in the genera *Melampsora, Phragmidium* and *Xenodochus*. There is no well-defined peridium. The aecidiospores are in chains with finely ornamented walls.

(5) Uredinoid aecidia (often referred to as primary uredosori) occur in the genus *Trachyspora* and in certain species of the genera *Puccinia* and *Uromyces*. Uredinoid aecidia are usually accompanied by spermogonia and are grouped in the manner of aecidia. The aecidiospores resemble uredospores which may or may not be present but are formed following the initiation of the dicaryophase. Commonly occurring rusts with this type of aecidium are *Trachyspora intrusa, Puccinia hieracii* and *P. calcitrapae*.

Uredosorus (pl. uredosori). The uredosorus is borne on dicaryotic mycelium and bears dicaryotic spores, the uredospores. The uredospores arise singly by abstriction of the ends of unbranched hyphae often referred to as pedicels. A second generation of uredosori can result from infection by uredospores; they are 'repeating' sori. In the majority of genera of rusts the uredosori are not accompanied by paraphyses nor are they bounded by any peridial structure. In *Pucciniastrum* and *Milesina* a hemispheric cellular peridium occurs and the spores escape after release from their pedicels through a circular apical pore, the spores often adhering loosely in threads. In some rusts a few marginal non-sporogenous cells are present and these have been described as paraphyses. Old, often elongated uredospore pedicels have also been described and figured as paraphyses. Both of these 'false' paraphyses are quite different from the true uredosoral paraphyses which occur in the genera *Melampsora*, *Tranzschelia* and graminicolous species of *Puccinia* (*P. poae-nemoralis*, *P. brachypodii*). True paraphyses are clavate or capitate, often thick-walled at the apex, and occur scattered in the sori even before the first uredospores have developed.

Uredospores are usually capable of germination without any appreciable period of rest, given suitable conditions. Resting, thick-walled uredospores usually referred to as amphispores occur in the fern rust *Hyalopsora polypodii* and in *Puccinia microsora*. In both, teleutospores are rarely or imperfectly produced. Amphispores develop in otherwise typical uredosori and in general form and number of pores resemble uredospores.

Uredospores are always unicellular. In *Milesina*, *Hyalopsora*, *Coleosporium*, *Cronartium*, *Melampsora* and *Melampsoridium* pores in the spore wall are either absent or very difficult to observe; they are certainly not as clearly differentiated as in *Puccinia*, for example. The surface of uredospores is usually delicately echinulate, uniformly so in most rusts but in some the area about the pores is smooth and in *P. bullata* the lower half of the spore is more or less smooth. The position and number of pores in the walls are reasonably constant for any given species and are useful characters for identification. *Puccinia graminis* and *P. menthae* have four equatorial pores, equally distributed around the equator of the spore. The complex group of grass rusts referred to as *P. recondita* has between five and eleven pores scattered fairly regularly in the spore wall. *Puccinia hieracii* and *P. dioicae* have two pores placed in the upper half of the spore—referred to as supraequatorial pores.

Teleutosorus (pl. teleutosori). In most rust genera teleutospores are organised in well-differentiated teleutosori. But in some genera the organisation is irregular and the term is applied more for the sake of uniformity. By definition, teleutospore always produce basidia on germination.

The range of teleutosoral types is well represented in the British rust flora. In the genera *Melampsorella, Milesina, Pucciniastrum* and *Uredinopsis* the teleutospores develop in diffuse patches either within the epidermal cells or beneath the epidermis. In *Coleosporium, Melampsora, Melampsoridium* and *Ochropsora* the unicellular teleutospores form discrete layers one cell deep in a subcuticular or subepidermal position. In all the foregoing genera, spore production proceeds centrifugally in the sori and only one crop of spores is produced. In other genera spore maturation is much less regular within any one sorus, a well-developed sporogenous layer is developed and spore production continues over some time. This is so in the genera *Puccinia* and *Uromyces.* As a result of the continued activity and close cohesion of teleutospores (in *Cronartium*) or their involvement in a gelatinous matrix (*Gymnosporangium*) columns of spores are formed characteristic for the genus.

In the most frequent type of sorus in the genera *Puccinia* and *Uromyces* and their allies, the stalked teleutospores are borne on pulvinate sori without any peridium or paraphyses and the spores are shed when mature. There are many modifications of this type. In leptosporic sori, where the spores germinate without any appreciable period of dormancy (e.g. *P. arenariae* and *P. chrysosplenii*) the spores mature more or less simultaneously, are strongly coherent and are not shed. In the species *P. recondita* and *P. virgaureae*, the teleutosori have peripheral and partitioning, brown, cylindric paraphyses, the sori are deeply embedded in host tissue and only tardily erumpent. In *Miyagia* a ring of strongly thickened paraphyses bounds the teleutosorus.

It should be noted that although the teleutosoral structure of *Coleosporium* is very similar to that of *Melampsora* the teleutospores of *Coleosporium* become internally septate on germination. Each cell so formed produces a sterigma bearing a basidiospore so the whole structure may equally well be described as a thick-walled basidium.

The teleutospores of *Milesina, Hyalopsora, Melampsorella* and *Pucciniastrum* are simple, thin-walled, unpigmented cells often with an indistinct pore—a morphology presumably correlated with their protected position within the tissues of the host plant. Two-celled teleutospores are the rule in *Puccinia* but unicellular spores often occur, especially in some species. So frequently does this occur in *Puccinia allii* and *P. hordei* that the artificiality of separating the genera *Uromyces* and *Puccinia* on number of cells in the teleutospore is manifest. Certain races of *Miyagia pseudophaeria* also have predominantly unicellular teleutospores. In certain races of *P. recondita* alternating with aecidial hosts belonging to the *Ranunculaceae* three- or more-celled teleutospores are frequent. By some authors these have been segregated in the genus *Rostrupia*.

Summary of life cycles

The rust fungi are noteworthy for the diversity of spore stages in their individual life cycles. There is a maximum of five spore types, spermatium, aecidiospore, uredospore, teleutospore and basidiospore. They are usually formed in an orderly succession and are correlated with the alternation of haplophase and diplophase. Basidiospores, spermogonia, spermatia and aecidia are essentially haploid; aecidiospores, uredosori, uredospores, teleutosori and teleutospores are diploid. Furthermore, the alternation of generations is often correlated with an alternation between taxonomically diverse host plants. In *Puccinia graminis* the haploid sori, spermogonia and aecidia, develop on members of the *Berberidaceae* whereas the diploid uredosori and teleutosori occur on the *Gramineae*. Rusts which exhibit this host alternation are heteroecious; those in which the whole life cycle occurs on one host are auteocious.

Puccinia graminis the causal organism of the most intensively studied plant disease, Black Rust of Wheat, exemplifies the fullest type of life cycle; it is heteroecious with all five spore types. In spring germinating teleutospores release basidiospores which bring about infection of Barberry leaves. Germination of the basidiospores, penetration of the host and parasitic growth leads to the development of a haploid mycelium in the leaves. Under the epidermis on the upper surface of the leaf, flask-shaped spermogonia develop on yellow spots. The spermogonia rupture the epidermis and extrude spermatia in a sweet-smelling liquid, which is attractive to insects. Cross-fertilisation of infection centres is achieved by insect visitors. The spermatia are incapable of sustained growth, but diploidise receptive hyphae of the reciprocal heterothallic group with which they come in contact. Cross-fertilisation is followed by production of diploid, dicaryotic aecidiospores in the aecidia formed on the lower surface of the host leaves in association with the spermogonial groups. The aecidiospores discharged from the aecidia can infect only members of the *Gramineae*. Infection of a grass or cereal host leads to the establishment of a dicaryotic mycelium on which are developed first the uredosori, bearing dicaryotic uredospores, then the teleutosori with teleutospores which are at first dicaryotic. The uredospores are capable of spreading infection amongst the grass hosts and are essentially asexual conidia. The teleutospores are capable of germination only after a period of maturation, which corresponds in nature to winter dormancy. On germination, each of the two cells of the teleutospore germinates to produce a cylindrical, transversely septate, four-celled basidium. In the later stages of maturation of the teleutospores fusion of the pair of dicaryon nuclei takes place. During development of the basidium meiosis occurs and the resulting basidiospores are haploid. Basidiospores are forcibly discharged from the basidium; their role is reinfection of the barberry.

Thirty-eight native species out of a total of 218 recognised in this flora exhibit this heteroecious cycle with a full complement of spore stages at least occasionally in Britain. Commonly occurring species are *Coleosporium tussilaginis, Melampsoridium betulinum, Melampsora populnea* and *M. caprearum, Puccinia obscura, P. caricina, P. phragmitis* and *Uromyces dactylidis*. Those with all spore stages are known as eu-forms; heteroecious ones as hetereu-forms. It should be noted, however, that the aecidial host may be cut out of the cycle if the uredo stage can overwinter. Thus although it is known that many of the races of *Puccinia caricina* regularly form aecidia on one or other of the aecidial hosts, *Ribes, Parnassia, Pedicularis* or *Urtica*, it is quite clear that many races persist from year to year by uredo infections either as mycelium in leaves of the sedge host or by persistent uredosori. Those in which alternation takes place regularly are referred to as obligatorily heteroecious, examples are, *Uromyces dactylidis, Melampsora caprearum* and *Puccinia phragmitis*; those in which alternation is often avoided are referred to as facultatively heteroecious, *Melampsora caryophyllacearum*, fern rusts of the genus *Milesina*, many strains of *Puccinia recondita* and most of the *P. poae-nemoralis* complex are examples.

A second type of eu-form occurs where there is no alternation but all five spore forms occur on one host. These are the so-called auteu-forms of which good examples are the species of *Phragmidium* on *Rosa* and *Rubus* in which the aecidia are caeomoid and the common species *Puccinia lapsanae, P. violae, Uromyces armeriae* and *U. fabae* in which the aecidia are aecidioid, that is, with a well-developed peridium.

Brachy-forms are very similar to eu-forms as regards their life cycle, they have all five spore stages but the aecidia are morphologically similar to uredosori and lack peridia and the aecidiospores are exactly similar to uredospores. These uredinoid aecidia are usually accompanied by spermogonia and are most often large, often on petioles or stems and associated with some slight distortion of the host tissue. All the brachy-forms are autoecious and the aecidiospores infect the same host and uredosori result. The aecidia of brachy-forms have been called primary uredosori but this usage is probably inadvisable. Commonly occurring brachy-forms are *Kuhneola uredinis, Triphragmium ulmariae, Puccinia calcitrapae, P. hieracii, P. punctiformis* and *Trachyspora intrusa.*

Although there is still discussion on the direction and precise mechanism of other life cycles in the rust fungi there is no doubt that they can all be conveniently derived from the eu-forms by ommission of spore stages, in some accompanied by alternation of the host plant of one stage.

Opsis-forms are those which no longer form uredospores; all the British species of *Gymnosporangium* are heteropsis-forms forming spermogonia and aecidia on Juniper and teleutospores on dicotyledonous trees. Non-alternating autopsis-forms are represented by *Melampsora hypericorum, Xenodochus carbonarius, Puccinia smyrnii, P. hysterium* and *Uromyces scrophulariae.*

Micro-forms develop only teleutosori with or without spermogonia. It is generally agreed that micro-forms have evolved from eu-forms by loss of the other spore stages. However, there is an interesting relation between micro-forms and their most closely related eu-forms. This relation was particularly investigated by Tranzschel and Tranzschel's Law may be expressed, 'The most closely related species to a micro-form must be sought amongst heteroecious rusts which form their *aecidia* on the same or closely related host species, furthermore, the teleutosori of the micro-form usually have the same habit as the aecidia of the eu-form.' Thus the micro-forms on *Compositae* of the *Puccinia cnici-oleracei* group are related to the hetereu *P. dioicae* with aecidia on *Compositae* and teleutosori on *Cyperaceae* and the group of micro-forms on *Euphorbia* represented in Britain by *Uromyces tinctoriicola* are clearly derived from the heteroecious group of species represented in Britain by *U. pisi* or *U. anthyllidis* with aecidia on *Euphorbia* and teleutosori on members of the *Leguminosae.* Within the limits of the British rust flora, however, it is not possible to work out many of these correlations. The hetereu-forms most closely related to our micro-forms are frequently more southern in distribution and it has been held that the micro-forms and indeed other reductions in the length of life cycle have been evolved by selection in northern and montane climates where the growing season renders dependence on a heteroecious life cycle disadvantageous.

Lepto-forms are micro-forms in which the teleutospores germinate without a period of rest. The distinction, however, is not absolute, for a number of rusts with teleutospores only may have both resting teleutospores and leptospores which germinate immediately; indeed the two types may occur in the same sorus (*Puccinia clintonii* and *P. albulensis*). It is probably preferable to speak rather of teleutospores and leptospores. The leptospores are usually rather lighter in colour with smoother walls than the corresponding teleutospores and occur in pulvinate sori. Leptospores germinate *in situ* whereas the ordinary teleutospores are frequently pulverulent and dispersible.

Endo-forms have only sori which are morphologically aecidia but whose spores germinate like teleutospores to produce basidia. These forms have usually been segregated in the genus *Endophyllum*. There are only two species of this type in Britain, *E. sempervivi* and *E. euphorbiae-sylvaticae*.

Hemi-forms produce only uredosori and teleutosori, often these may be only the diplont stages of species of whose life cycle is unknown, but *Puccinia oxyriae* is a well-known species with a life cycle of this type for which no aecidial stage is known either on *Oxyria* or on an alternate host.

Summary of life cycles

	Spermogoria	Aecidia	Uredo	Teleuto
Hetereu-	±	+	+	+
Auteu-	±	+	+	+
Heteropsis-	±	+	−	+
Autopsis-	±	+	−	+
Brachy-	±	+[1]	+	+
Hemi-	−	−	+	+
Micro-	±	−	−	+
Lepto-	±	−	−	+
Endo-	±	−	−	+[2]

[1] Uredinial aecidia.   [2] Aecidioid teleutosori.

KEY TO THE GENERA OF RUSTS OCCURRING IN BRITAIN

1. Teleutosori not clearly defined; teleutospores within host cells or scattered in the mesophyll   **2**
Teleutosori distinct, either crustose or pulvinate   **5**

2. Teleutospores colourless   **3**
Teleutospores brown or yellowish; intra-epidermal   *Melampsorella*, p. 43

3. Teleutospores intra-epidermal   **4**
Teleutospores inter-cellular, in the mesophyll   *Uredinopsis*, p. 15

4. Uredospores colourless   *Milesina*, p. 17
Uredospores yellow-orange   *Hyalopsora*, p. 26

5. Teleutospores sessile, compressed, often angular, forming a subepidermal layer, usually not detached from the sorus   **6**
Teleutospores stalked, usually becoming detached and dispersed on maturity   **10**

6. Teleutospores producing an external basidium on germination; uredosori usually with capitate or clavate paraphyses or with hemispherical peridium   **7**
Teleutospores becoming internally septate on germination   **9**

7. Teleutospores in chains   *Chrysomyxa*, p. 58
Teleutospores in a single layer   **8**

8. Uredosori with hemispheric peridia; aparaphysate   *Melampsoridium*, p. 47
Uredosori with capitate paraphyses   *Melampsora*, p. 64

9. Teleutospores with a distinct apical thickening, gelatinising when moist   *Coleosporium*, p. 1
Teleutospores not thickened at apex and not gelatinising   *Ochropsora*, p. 10

10. Teleutospores triquetrously 3-celled   **11**
Teleutospores transversely septate or aseptate   **12**

11. Teleutospores with one pore per cell   *Triphragmium*, p. 111
Teleutospores with two pores per cell   *Nyssopsora*, p. 114

12. Teleutospores exclusively 1-celled   **13**
Teleutospores 2- or more-celled   **18**

13. Teleutospores united in columns   *Cronartium*, p. 51
Teleutospores in pulvinate or pulverulent sori   **14**

14. Teleutospores with basal pore   *Zaghouania*, p. 13
Teleutospores with median or superior pore   **15**

15. Teleutosori aecidioid or caeomoid (endo-forms)   **16**
Teleutosori not so   **17**

16. Teleutosori caeomoid   *Kunkelia*, p. 107
Teleutosori aecidioid   *Endophyllum*, p. 307

17. Teleutospore pedicel 1-septate *Trachyspora*, p. 363
    Teleutospore pedicel aseptate *Uromyces*, p. 310

18. Teleutospores united in waxy or gelatinous columns *Gymnosporangium*, p. 115
    Teleutospores in pulvinate or pulverulent sori 19

19. Teleutospores mostly more than 2-celled (wall of teleutospore often many-layered) 20
    Teleutospores mostly 2-celled (a proportion of 1-celled and 3- or more-celled spores may be present but then the wall of the spore is not obviously many-layered) 23

20. More than one pore in cells of teleutospore (except apical cell in *Xenodochus*) 21
    One pore in each cell of teleutospore 22

21. Apical cell with only one pore *Xenodochus*, p. 109
    All cells of teleutospores with two pores *Phragmidium*, p. 94

22. Teleutospore wall hyaline *Kuhneola*, p. 107
    Teleutospore wall brown *Frommea*, p. 110

23. Teleutospores with two pores in each cell *Cumminsiella*, p. 300
    Teleutospores with one pore in each cell 24

24. Cells of teleutospores readily seceding; pedicels often coherent at base; on *Prunus* or *Thalictrum* *Tranzschelia*, p. 302
    Not so 25

25. Uredosori and teleutosori with pigmented palisade-like marginal paraphyses (on *Sonchus*) *Miyagia*, p. 297
    Uredosori and teleutosori never both with conspicuous marginal paraphyses *Puccinia*, p. 122

NOTE

The illustrations are reproduced × 660 unless otherwise stated.

COLEOSPORIAECEAE

Coleosporium Lév.

Ann. Sci. Nat. Bot. Ser. 3, **8**, 373 (1847).

Spermogonia subepidermal with paraphyses and flexuous hyphae. **Aecidia** foliicolous, erumpent, with prominent, tongue-shaped peridia composed of a single layer of verrucose cells, dehiscing irregularly; aecidiospores ellipsoid or globoid with colourless, tesselate, superficially tuberculate wall. **Uredosori** erumpent, pulverulent, without peridia; uredospores globoid or oblong, catenulate, in wall structure resembling the aeciadiospores. **Teleutosori** subepidermal, indehiscent except through weathering, flattened, waxy, becoming gelatinous on germination; teleutospores sessile, in a single layer in lateral contact, unicellular, cylindroid, clavoid or prismatic with smooth, colourless walls, thin at the sides, strongly thickened and gelatinous above, becoming divided into four superposed cells, all of which, in autumn, can germinate as soon as mature, each producing a long sterigma bearing a basidiospore; basidiospores ovoid or ellipsoid, rather large, thin-walled. Predominantly heteroecious (two microforms on *Pinus* in North America and Eastern Asia); spermogonia and aecidia on the needles of *Pinus*; uredosori and teleutosori on various families of Angiosperms, especially *Compositae*.

Although Poirault & Raciborski (J. Bot. Fr. **9**, 330, 1895) concluded in 1895 that the structures which appear in the *Coleosporiaceae* in the position of teleutospores in other rust fungi can equally well be considered basidia, this conclusion has never gained wide acceptance, although it was recognised by Juel (Jahrb. Wiss. Bot. **32**, 361, 1898), by Maire (Bull. Soc. Myc. Fr. **18**, 28–9, 1902), by V. H. Blackman (Ann. Bot. Lond. **18**, p. 362, footnote, 1904) and by Raciborski (Bull. Acad. Sci. Cracovie, Math. Nat. 1909, p. 358, 1909).

The usual interpretation describes the 'teleutospores' of the *Coleosporiaceae* as germinating 'internally' directly to basidia. However it should be recognised that these 'teleutospores' are not units of dispersal, nor do they undergo a period of rest but germinate when they are mature. The teleutosorus might well be described as a basidiosorus.

There is some experimental evidence to suggest that the 'species', *C. campanulae*, *C. tussilaginis*, *C. senecionis*, discriminated solely by generic identity of the host, may not be strictly host-specialised.

Until recently many of the rusts of the genus *Coleosporium* which occur in central and northern Europe have been treated as species, but since it has been found impossible to distinguish them on morphological characters, it is now considered best to include them as races and race groups of one species with spermogonia and aecidia on *Pinus sylvestris*, and some closely related species of *Pinus* sub-genus *Diploxylon* and uredosori and teleutosori chiefly on *Compositae*, *Campanulaceae* and *Scrophulariaceae*. It is doubtful whether a similar conclusion can be drawn regarding the North American forms. Hedgecock (Mycol. **20**, 97, 1928) made a key for distinguishing species by their aecidial stages in which three British forms were included; Jørstad (1934, 30) stated 'even in Europe all races do not appear to possess exactly similar aecidia and very possibly the difference will prove constant'. Grove's suggestion that in some species the aecidia occur on only one of the two needles of the spur,while in other species they are found on both needles, does not appear to be tenable. Ludwig (Phytopath. **5**, 293, 1915) worked out a key for the determination of North American uredo- and teleuto-stages by morphological differences, but neither Ludwig's nor Hedgecock's methods have been widely accepted.

Klebahn (Z. Pfl.-Krankh. **34**, 289, 1924) showed that uredospores of *C. tussilaginis* can produce weak infection of *Senecio vulgaris* which is a common host for *C. senecionis*. Ludwig (*loc. cit.*) placed an American *Coleosporium* on *Senecio* in *C. campanulae* partially on account of morphological similarity.

In this and various other north-western European countries, forms of *Coleosporium* have also been found on hosts which are not closely related to any of the hosts of the native races; the infections are usually only in the uredospore condition but occasionally teleutospores are also present. Klebahn (Z. Pfl.-Krankh. **24**, 14, 1914) recorded the occurrence in Germany of a solanaceous plant, *Schizanthus grahamii* from South America, infected with *Coleosporium*, and, in the following two years, established by cultures that the infection was derived from various European races of the rust. Positive results were obtained by infection with uredospores of the 'species' *C. campanulae*, *C. euphrasiae*, *C. melampyri* and *C. tussilaginis*. In a similar manner another South American species, *Tropaeolum minus*, was shown by cultures to give positive results with *C. campanulae*, *C. tussilaginis* and *C. senecionis*. He also noted that various foreign species of *Senecio* and species of several other genera in the *Compositae*, such as Cineraria, in the Berlin Botanic Garden, had become infected. He considered it was very improbable that the rusts had been imported with these plants and concluded they had become infected from native races.

In Denmark, according to Rostrup (Overs., Biol. Medd. K. Danske Vidensk. Selsk.Forh. **1884**, p. 1, 1884), *C. senecionis* may infect *Crepis tectorum* and Sydow (Ann. Myc. **28**, 427, 1930) has stated that it infects *Chrysanthemum segetum* in Germany. Jørstad has found a *Coleosporium* on *Lactuca muralis* in Norway.

Jørstad (1934, p. 39, footnote) concluded from the investigations detailed in the previous paragraphs that 'This indicates that most, or all, European *Coleosporium* forms ought to be referred to one and the same plurivorous species, the specialization of which (in different physiological races) has begun, but is not yet quite completed.'

In this country the following introduced species have been attacked by *Coleosporium* forms and have probably been infected from native hosts; they have been recorded in the past under several specific names.

Under *C. senecionis*; *Senecio cruentus*, the cultivated Cineraria, *S. moorei* from Kenya, *S. smithii* from Patagonia, *S. fluviatilis* from central and south Europe, *Notonia grandiflora* from India, *Chrysanthemum carinatum*, *Calendula officinalis* (see p. 7).

Under *C. cacaliae*; *Cacalia hastata*, *C. suaveolens* (see p. 4).

Under *C. tussilaginis*; *Euryops evansii* from South Africa (see p. 10).

Under *C. petasitis*; *Petasites palmatus*, *P. japonicus* (see p. 6).

Under *C. narcissi*; '*Narcissus* sp.' (see p. 6).

Under *U. tropaeoli*; *Tropaeolum peregrinum* (see p. 9).

Coleosporium tussilaginis (Pers.) Lév.

Orbigny, Dict. Univ. Hist. Nat. **12**, 786 (1849).

Uredo tussilaginis Pers., Syn. Meth. Fung. p. 218 (1801).
[*Aecidium pini* Pers., Syn. Meth. Fung. p. 213 (1801), *p.p.*]

Spermogonia amphigenous, chiefly epiphyllous on pale or yellow spots, subepidermal or subcortical, scattered, or in two longitudinal rows, yellowish, becoming brown, conoid, flattened, 0·5–1 mm. long, 0·2–0·5 mm. wide. **Aecidia** amphigenous, laterally compressed, 1–3 mm. long, 1–5 mm. high, yellow becoming paler, dehiscing irregularly; peridial cells 35–70 μ long, 16–34 μ wide, walls equally thickened 3–5 μ thick or external wall thicker than internal, verrucose; aecidiospores globoid, ellipsoid, obovoid or angular, 20–40 × 16–27 μ; wall colourless 2–3 μ thick, densely verrucose. **Uredosori** hypophyllous, scattered, rounded or oblong, 0·4–0·7 mm. diam., soon naked, pulverulent, orange-yellow; uredospores globoid, ellipsoid or ovoid, 20–40 × 16–25 μ, wall colourless 1–1·5 μ thick, densely and finely verruculose. **Teleutosori** hypophyllous, rounded, scat-

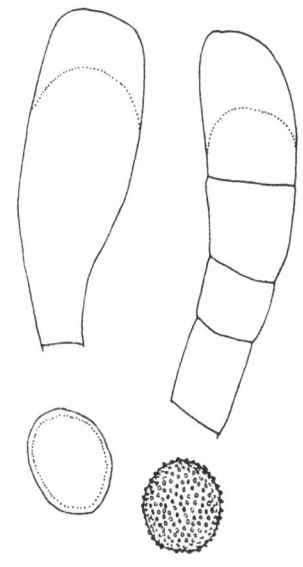

C. tussilaginis. Teleutospores and uredospores.

tered or confluent, forming waxy orange-red crusts, 0·4–0·8 mm. diam.; teleutospores clavoid to cylindric, rounded at the apex, attenuate or rounded at the base, 60–105 × 15–24 μ; wall thin at the sides, 12–30 μ thick at the apex, colourless, smooth; at first unicellular, then becoming 4-celled and greatly thickened and gelatinous at the apex. Hetereuform.
Spermogonia and aecidia on the needles of *Pinus sylvestris* and *P. nigra*; uredosori and teleutosori chiefly on the leaves and stems of *Compositae*, *Campanulaceae* and *Scrophulariaceae*. The aecidial stages collected in Britain cannot be closely correlated with the diplont races but there is observational evidence for connection between *Pinus sylvestris* and the races on *Senecio vulgaris* and *Tussilago farfara*.

The teleutospores are not shed but become exposed by weathering. From each cell a sterigma grows out bearing a basidiospore at its end; as the sterigmata from the lower cells are longer than those from the upper, the basidiospores are developed in an irregular layer just above the apices of the teleutospores. Infection takes place in the autumn on the current year's pine needles and the spermogonia and aecidia are produced in the following summer. On pine infection is important only when it is heavy on nursery stock.

The following races or race groups here arranged according to dicaryont host genera occur in this country:

Cacalia

Coleosporium cacaliae Otth, Mitt. Naturf. Ges. Bern, 1865, p. 179 (1866); Grove, Brit. Rust Fungi, p. 325; Wilson & Bisby, Brit. Ured. no. 11; Gäumann, Rostpilze, p. 117.

[Spermogonia and aecidia on *Pinus mugo* and on *P. sylvestris*]; uredospores and teleutospores on '*Cacalia hastata*' and '*C. suaveolens*'. Great Britain, rare.

This is an introduced form which has been recorded at Bath, Batheaston, Oxford and near Aberdeen. Spermogonia and aecidia have not been observed in Britain.

This rust was described by Otth growing on *Cacalia hastata* in the Bern Botanic Garden. Jørstad (1934, 42) suggested that the rust on *Cacalia* in European gardens is a native race from *Senecio* or other native hosts and listed it under *C. senecionis* (Hylander *et al.* 1953, 11).

Fischer (Bull. Soc. Bot. Fr. **41**, clxx, 1894) produced only spermogonia on *P. sylvestris* by infection with this race from *Adenostyles alpina*. Wagner (Z. Pfl.-Krankh. **6**, 11, 1896) found spermogonia and aecidia on a small tree of *P. mugo* which he had planted beneath infected *A. alpina*.

Calanthe

One collection of uredospores only of a *Coleosporium* on *Calanthe reflexa* made in Kew Gardens is in the Kew Herbarium.

Campanula

Coleosporium campanulae Lév., Ann. Sci. Nat. Bot. Ser. 3, **8**, 375 (1847). Valid specific name, *Coleosporium campanulae* Kickx, Fl. Crypt. Flandres, **2**, 54 (1867). Grove, Brit. Rust Fungi, p. 328; Wilson & Bisby, Brit. Ured. no. 12; Gäumann, Rostpilze, p. 113.

[*Uredo campanulae* Pers., Syn. Meth. Fung. p. 217 (1801).]

Spermogonia and aecidia on needles of *Pinus sylvestris*; uredospores and teleutospores on *Campanula carpatica* var. *turbinata*, *C. glomerata*, *C. isophylla*, *C. latifolia*, *C. planifera 'alba'*, *C. persicifolia*, *C. rapunculoides*, *C. rotundifolia*, *C. trachelium*, July–October. Great Britain and Ireland, frequent.

Klebahn (1904, 365) as the result of the work of Wagner and Fischer and his own investigations, distinguished three specialised forms, f. sp. *campanulae-rapunculoidis* Kleb., f. sp. *campanulae-trachelii* Kleb. and f. sp. *campanulae-rotundifoliae* Kleb.; judging from the list of British host species it appears that all these forms may occur in this country. The form on *Campanula persicifolia* is frequently found in gardens and it is noteworthy that this host was not infected by any of Klebahn's specialised forms (Klebahn, Z. Pfl.-Krankh. **5**, 82, 1905). Mains (Pap. Mich. Acad. Sci. **23**, 171, 1937) studied specialisation of this rust in North America. The host range of the various races is well tabulated by Gäumann (1959, 116).

Fischer (1904, 449) produced spermogonia and aecidia on *P. sylvestris* by sowing basidiospores from *C. trachelium* and Wagner (Z. Pfl.-Krankh. **8**, 257, 1898) infected the latter species with aecidiospores. No cultures on pine appear to have been carried out with other *Campanula* species.

This form is known to overwinter both in Europe and America by means of uredospores especially on rosette leaves; uredosori have been often observed, especially on *C. persicifolia*, during the winter in this country.

Magnus (Ber. D. Bot. Ges. **20**, 334, 1902) described and figured peripheral paraphyses in the uredosorus.

It was in this species, in 1880, that Schmitz (S.B. Niederrhein. Ges. Bonn, **37**, 159, 1880) first discovered dicaryons in mycelium and uredospores. Juel (Jahrb. Wiss. Bot. **32**, 361, 1898) described nuclear division in the basidium.

Melampyrum

Coleosporium melampyri (Rebent.) Karst., Bidr. Känned. Finl. Nat. Folk, **31**, 62 (1879); Grove, Brit. Rust Fungi, p. 327; Wilson & Bisby, Brit. Ured. no. 14.

Uredo melampyri Rebent., Prod. Fl. Neom. p. 355 (1804).

Coleosporium melampyri (Rebent.) Tul., Ann. Sci. Nat. Bot. Ser. 4, **2**, 136 (1854), *nomen fortuitum*.

Spermogonia and aecidia on *Pinus sylvestris*; uredospores and teleutospores on *Melampyrum arvense* and *M. pratense*. Great Britain and Ireland, frequent.

Klebahn (Z. Pfl.-Krankh. 5, 13 and 257, 1895) showed that aecidiospores of this race will infect *Melampyrum* but not *Euphrasia* or *Rhinanthus*. Also, by planting a young pot-plant of pine amongst a clump of strongly infected *Melampyrum* he obtained spermogonia in September and aecidia the following spring. Wagner (Z. Pfl.-Krankh. 8, 257, 1898) produced infection on *M. pratense* with aecidiospores from *P. montana* and Mayor (Bull. Soc. Neuchâtel. Sci. Nat. 48, 386, 1923) infected the same species with aecidiospores from *P. sylvestris*. Since the *Melampyrum* species are annual and die at the approach of winter it would seem probable that fresh infection must occur each year from aecidia; it is probable that the latter are more abundant than is supposed. In Finland the infection is heaviest on Scots pine up to 4 ft. high but it is not a serious disease of pine; infection of pine is unaffected by shading but *Melampyrum* is most heavily attacked in open habitats (Pohjakallio & Vaartaja, Acta For. Fenn. 55, 2, 1948).

Moreau (1914, 242) studied the details of nuclear division in the basidium.

Narcissus

[*Coleosporium narcissi* Grove, J. Bot. Lond. 60, 121 (1922)]; Wilson & Bisby, Brit. Ured. no. 15.

Spermogonia, aecidia and teleutosori unknown; uredosori on *Narcissus majalis*, June–July. England, very rare.

The only record of this rust is from Crown Colony, Holbeach, Lincs. where it was collected by F. Glover in 1920. This appears to be the first record of a *Coleosporium* on the *Amaryllidaceae*.

Petasites

[*Coleosporium petasitis* Cooke, Handb. Brit. Fungi, p. 521 (1871)]; Grove, Brit. Rust Fungi, p. 323; Wilson & Bisby, Brit. Ured. no. 16; Gäumann, Rostpilze, p. 121.

[*Uredo petasitis* DC., Fl. Fr. 2, 236 (1805).]
Coleosporium sonchi Plowr., Brit. Ured. Ustil. p. 250 (1889) *p.p.*

Spermogonia and aecidia on leaves of *Pinus sylvestris* and *P. nigra* ssp. *laricio*; uredospores and teleutospores on *Peta-sites albus, P. hybridus, P. japonicus* and *P. palmatus*, August–November. Great Britain and Ireland, frequent.

The records on *P. japonicus* and *P. palmatus* are from Peeblesshire (TBMS. 9, 142, 1924) and on *P. albus* from Norfolk (Ellis, Trans. Norf. Norw. Nat. Hist. Soc. 15, 372, 1943) and Perthshire.

The life cycle was demonstrated by Fischer (Bull. Soc. Bot. Fr. **41**, clxx, 1894) and Wagner (Z. Pfl.-Krankh. **6**, 10, 1896) for *P. hybridus* (as *P. vulgaris*) and by Mayor (Bull. Soc. Neuchâtel. Sci. Nat. **48**, 386, 1923) for *P. albus*.

Rhinanthus, Euphrasia, Odontites and Parentucellia

Coleosporium rhinanthacearum Lév., Ann. Sci. Nat. Bot. Ser. 3, **8**, 373 (1847); Grove, Brit. Rust Fungi, p. 326.
[*Uredo euphrasiae* Schum., Enum. Pl. Saell. **2**, 230 (1803).]
[*Uredo rhinanthacearum* DC. apud Poir., Lam. Encycl. Méth. Bot. **8**, 229 (1808).]
As *Coleosporium euphrasiae* Wint., Hedwigia, **19**, 54 (1880); Grove, Brit. Rust Fungi, p. 326; Wilson & Bisby, Brit. Ured. no. 13; Gäumann, Rostpilze, p. 110.

Spermogonia and aecidia on needles of *Pinus sylvestris*; uredospores and teleutospores on *Euphrasia officinalis* agg., *Odontites verna*, *Parentucellia viscosa*, *Rhinanthus minor* agg. July–September. Great Britain and Ireland, common.

It is not certain that *Odontites* can be infected from *Euphrasia* or *Rhinanthus*; no experiments on that point are recorded. Klebahn (Z. Pfl.-Krankh. **2**, 264, 1892; **4**, 9, 1894; **5**, 13, 1895) proved the relationship of the aecidium (*Peridermium stahlii*) to *C. euphrasiae*; he also proved that the parasite can be transferred from *Rhinanthus* to *Euphrasia* but not to *Senecio*, *Sonchus*, *Tussilago* or *Melampyrum*. Wagner (Z. Pfl.-Krankh. **8**, 261, 1898) infected *Euphrasia* with aecidiospores from *Pinus montana*. As all the dicaryont hosts of this rust are annual it appears that it is improbable that it overwinters in the uredospore stage, and in fact infections are rarely found far from Scots pine in Britain.

The records on *P. viscosa* are from Devon, Cornwall, the Channel Islands and from south-west Scotland.

Lindfors (1924, 5) described the development of the aecidium of this rust.

Senecio and allied genera

Coleosporium senecionis Kickx, Fl. Crypt. Flandres, **2**, 53 (1867); Grove, Brit. Rust Fungi, p. 320; Wilson & Bisby, Brit. Ured. no. 17; Gäumann, Rostpilze, p. 122.
[*Uredo farinosa* β *senecionis* Pers., Syn. Meth. Fung. p. 218 (1801).]
[*Uredo senecionis* Schum., Enum. Pl. Saell. **2**, 229 (1803).]

Spermogonia and aecidia on *Pinus sylvestris*, *P. nigra* ssp. *nigra* and ssp. *laricio*, May, June; uredospores and teleutospores on *Senecio congestus* var. *palustris*, '*S.jacobaea*', *S.squalidus*, *S.sylvaticus*, *S. viscosus*, *S. vulgaris* and cultivated or escaped *Calendula officinalis*, *S. mikanioides*, *S. candicans*, *S. cruentus*, *S. fluvi*atilis, *S. leucostachys*, *S. moorei*, *S. smithii* and perhaps this race on *Chrysanthemum carinatum* and on *Notonia grandiflora*; occurring all through the year. Great Britain and Ireland; very common on *S. vulgaris*, less so on other hosts.

This is the race of *Coleosporium* whose life history has been longest known; Wolff (Bot. Zeit. **32**, 184, 1874) first experimentally demonstrated it in 1872 and he was followed by Plowright (Grevillea, **11**, 52, 1882) as well as by Cornu, Hartig, Rathay, Thümen, Rostrup, Klebahn and Fischer. In an experiment carried out at Edinburgh in June 1929 aecidiospores from *Pinus sylvestris* placed on *Senecio vulgaris* produced uredospores in ten days. Aecidiospores from *P. sylvestris* placed on *S. vulgaris* in May, 1944, produced uredospores ten days later but gave negative results on *S. viscosus*, *Tussilago farfara*, *Campanula persicifolia* and cultivated *Narcissus pseudonarcissus*. The rust obviously overwinters on the common host, *S. vulgaris*, on which some can be found throughout the year, and probably the same is true of many other hosts. The rust was found on the island of Coll by Symington Grieve (TBMS. **9**, 41, 1924) in July 1922 on *S. smithii*, a South American species; there are no pines on this island and the nearest are found on the island of Mull, more than ten miles distant. The rust has also been found on *S. moorei*, a species from Kenya, cultivated in Edinburgh and on *Notonia grandiflora*, introduced into Kew Gardens from South Africa. It has been recorded on *Chrysanthemum carinatum* by Ellis (Trans. Norf. Norw. Nat. Hist. Soc. **16**, 174, 1946).

The uredospore stage of this rust on *Calendula officinalis* was collected by Beaumont in Devon (Rep. Seale Hayne Agric. Coll. **15**, 34, 1938) and by the late Alex Smith in Aberdeen in 1933 (*in litt.*). Experiments carried out late in September 1943 in Edinburgh showed that this host could be infected with uredospores from *Senecio vulgaris*; uredospores were produced 10 days after infection and soon afterwards considerable necrosis of the tissues round the sori took place. Terrier (Ber. Schweiz. Bot. Ges. **58**, 202, 1948) carried out this culture in Switzerland in late August and obtained uredospores in early September; he also found natural infections on *Calendula* when seedlings were planted among infected *S. vulgaris* in August, and noted that the infection soon disappeared.

From these cultures and observations, it appears that both in Scotland and in Switzerland infection of *Calendula* takes place most readily in the late summer and soon ceases. This is probably the rust described by Spegazzini as *Coleosporium calendulae* in Argentina, presumably on imported plants of *Calendula officinalis* (see Lindquist, Uredineana, **3**, 378, 1951).

Plowright (1889, 248) listed *S. jacobaea* as a host for this rust and Grove repeated the record but it has never been confirmed; Gäumann (*loc. cit.*) merely lists it as a recorded host.

Fischer (1898) showed that this race cannot be transferred to *Cacalia* or *Sonchus*, and on the basis of a series of infection experiments Wagner (Z. Pfl.-Krankh. **8**, 257, 1898) and Fischer (1898) described four formae speciales, f. sp. *nemorensis*, f. sp. *subalpini*, f. sp. *doronici* and f. sp. *silvatici*, of which only the last on *Senecio hybridus*, *S. sylvaticus*, *S. viscosus*, *S. vulgaris*, *Calendula officinalis*, *Schizanthus grahami* and *Tropaeolum minus* occurs in Britain.

Arnaud (Bull. Soc. Myc. Fr. **29**, 345, 1913) gave an account of the nuclear divisions in the teleutospore and Moreau (1914, 218) in addition described the divisions in the young aecidium. Sappin-Trouffy (1897, 188) described the development of the uredospores, spermogonia and aecidia as well as the germination of the aecidiospores. The development of the aecidium and the origin of the dicaryotic phase was described by Kursanov (1917, 21).

Sonchus

Coleosporium sonchi (Str.) Lév. & Tul., Ann. Sci. Nat. Bot. Ser. 4, **2**, 190, and tab. VIII (1854); Grove, Brit. Rust Fungi, p. 324; Wilson & Bisby, Brit. Ured. no. 18; Gäumann, Rostpilze, p. 126.

[*Uredo sonchi-arvensis* Pers., Syn. Meth. Fung. p. 217 (1801).]
Uredo tremellosa var. *sonchi* Str., Ann. Wetter. Ges. **2**, 90 (1810).

Spermogonia and aecidia on *Pinus sylvestris*; uredospores and teleutospores on *Sonchus arvensis*, *S. asper*, *S. oleraceus*, and *S. palustris*. August–November. Great Britain and Ireland, frequent.

Fischer (Bull. Soc. Bot. Fr. **41**, clxix, 1894), Klebahn (Z. Pfl.-Krankh. **5**, 69, 1895) and Wagner (Z. Pfl.-Krankh. **8**, 345, 1898) demonstrated the life cycle of this parasite. The two former also found that it could not be transferred to *Campanula*, *Senecio* or *Tussilago*.

It has been found on *S. palustris* near Ramsey, Huntingdon.

Sappin-Trouffy (1897, 169) and Moreau (1914, 202) described the development of basidia.

Tropaeolum

[*Uredo tropaeoli* Desm., Ann. Sci. Nat. Bot. Ser. 2, **6**, 243 (1836)]; Grove, Brit. Rust Fungi, p. 386.

Coleosporium tropaeoli Palm, Sv. Bot. Tidskr. **11**, 271 (1917), *nomen nudum*; Wilson & Bisby, Brit. Ured. no. 19.

Uredospores on *Tropaeolum peregrinum*. September–October. England and Channel Islands, very rare.

This rust was found by Capron at Shere, near Guildford in October, 1865 (see Cooke, J. Bot. Lond. **4**, 108, 1866) and by Rhodes at St Peter

Port, Guernsey, Channel Islands, in September 1931 (Grove, J. Bot. Lond. **70**, 112, 1932); it was recorded from Yorkshire as *Hemileia tropaeoli* Desm. (Mason & Grainger, Cat. Yorks. Fungi, 1937, 44). Klebahn (Z. Pfl.-Krankh. **24**, 14, 1914) produced uredospores on *T. minus* by infection with uredospores of *Coleosporium senecionis, C. tussilaginis* and *C. campanulae*. As there is no information on the other hosts of the *Tropaeolum* race in Britain it is not possible to place it with any of the other races.

Tussilago

Coleosporium tussilaginis (Pers.) Lév., Orbigny, Dict. Univ. Hist. Nat. **12**, 786 (1849); Grove, Brit. Rust Fungi, p. 322; Wilson & Bisby, Brit. Ured. no. 10; Gäumann, Rostpilze, p. 128.
Uredo tussilaginis Pers., Syn. Meth. Fung. p. 218 (1801).
Coleosporium sonchi Plowr., Brit. Ured. Ustil. p. 250 (1889) *p.p.*

Spermogonia and aecidia on the needles of *Pinus sylvestris*; uredospores and teleutospores on *Tussilago farfara* and on cultivated *Euryops evansii*. Great Britain and Ireland, very common.

The connection of the spore-forms on the alternate hosts was demonstrated by Klebahn (Z. Pfl.-Krankh. **2**, 268, 1892), Fischer (1898), Wagner (Z. Pfl.-Krankh. **8**, 258 and 345, 1898), Plowright (Gard. Chron. **25**, 415, 1899) and Mayor (Bull. Soc. Neuchâtel. Sci. Nat. **48**, 386, 1923). Klebahn (Z. Pfl.-Krankh. **34**, 294, 1924) stated that the mycelium may remain living in the needles for two years and produce a second crop of aecidia; this has also been observed near Edinburgh where, on young trees, aecidia were found on second- and third-year needles. Klebahn (*loc. cit.*) obtained a weak infection on *T. farfara* with uredospores from *Senecio vulgaris* and concluded that *C. tussilaginis* and *C. senecionis* were not so distinct as generally supposed.

Ashworth (Ann. Bot. Lond. **49**, 95, 1935) described scattered, emergent, receptive hyphae in this rust; she also carried out cultures on the pine.

Ochropsora Diet.

Ber. D. Bot. Ges. **13**, 401 (1895).

Spermogonia subcuticular with projecting paraphyses. **Aecidia** cupulate, then widely open with laciniate, revolute margin; aecidiospores catenulate, globoid, verrucose. **Uredosori** minute, subepidermal, surrounded with numerous paraphyses bound together at the base to form a peridium-like structure but with their upper ends free; uredospores globoid or ellipsoid, solitary, verrucose-echinulate, pores inconspicuous. **Teleutosori** minute, subepidermal, flat or slightly convex, forming waxy crusts; teleutospores in dense groups loosely laterally combined, cylindroid, clavoid or prismatic, at first continuous, then becoming divided into four superposed cells, the wall everywhere thin and smooth. Basidiospores on short sterig-mata, large, elongate-ovoid, or ellipsoid, rounded above. Heteroecious. A genus of three species, only one of which is British.

Ochropsora ariae (Fuck.) Ramsb.

Trans. Brit. Myc. Soc. 4, 337 (1914).
[*Aecidium leucospermum* DC., Fl. Fr. 2, 239 (1805).]
[*Uredo ariae* Schleich., Cat. Pl. Helv. Ed. 2, 38 (1807), *nomen nudum.*]
Melampsora ariae Fuck., Jahrb. Nass. Ver. Nat. 23–24, 45 (1870).
Caeoma sorbi Oud., Nederl. Kruidk. Arch. Ser. 2, 1, 177 (1872) and Arch. Néerland. 8, 383 (1873).
Ochropsora sorbi (Oud.) Diet., Ber. D. Bot. Ges. 13, 401 (1895); Grove, Brit. Rust Fungi, p. 329; Gäumann, Rostpilze, p. 1216.
Ochropsora ariae (Fuck.) P. & H. Sydow, Monogr. Ured. 3, 661 (1915); Wilson & Bisby, Brit. Ured. no. 65.

Spermogonia epiphyllous and on the tepals, scattered, subcuticular, obtusely cone-shaped, 100–125 μ wide, 60–70 μ high, whitish then brownish; spermatia ellipsoidal, 2 × 3 μ. **Aecidia** hypophyllous, scattered sparsely over the whole surface of the leaves, 0·1 mm. diam., cupulate, with torn revolute white peridium, peridial cells quadrate, external wall almost smooth, 6–10 μ thick, internal verrucose, 3–6 μ thick; aecidiospores globoid to ellipsoid, densely and delicately verruculose, 18–27 × 16–21 μ; wall hyaline, 1 μ thick. **Uredosori** hypophyllous, often on pale spots, scattered or in groups, small, rounded, 0·15–0·25 mm. diam., greyish or yellowish-white, surrounded by a circle of paraphyses forming a peridium, but the upper ends, when mature, are free and broadly cylindrical or clavate, slightly curved, up to 60 μ long and 14–18 μ wide, hyaline, wall about 1 μ thick; uredospores globoid, ellipsoid or ovoid, distantly verruculose or verruculose-echinulate, 19–28 × 16–22 μ; wall subhyaline or pale brownish about 1– 1·5 μ thick, with no perceptible pores. **Teleutosori** hypophyllous, on yellow or red spots, scattered or often in irregular groups, small, round or elongate, 0·25–

0·5 mm. diam., at first covered by the epidermis, translucent, pale flesh-colour, forming flat crustaceous pustules; teleutospores broadly cylindrical, apex rounded, wall smooth, colourless, scarcely 1 μ thick, at first unicellular then becoming

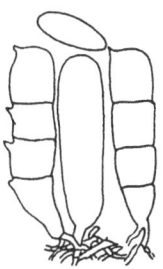

O. ariae. Germinating teleutospores and basidiospore (after Viennot-Bourgin).

4-celled, up to 70 μ long and 10–14 μ wide, contents opaque, grey, granular; basidiospores narrowly ellipsoid, 25 × 7–8 μ; wall thin, colourless. Hetereu-form.

Spermogonia and aecidia on leaves and tepals of *Anemone nemorosa*; uredospores and teleutospores on *Sorbus aucuparia*, Scotland, rare. Teleutospores once on crab apple.

The aecidia on *Anemone* are scarce in Great Britain. Immature teleutospores were recorded by J. Rees (*in litt.*) near Cardiff in September 1934;

they were on leaves of a crab apple immediately above a bed of infected anemones.

Uredospores and teleutospores were collected in Perthshire in 1964 on young trees of *Sorbus aucuparia*.

The teleutospores mature in the autumn and give rise to basidiospores at once. As shown by Fischer (1904, 455) and Dowson (Z. Pfl.-Krankh. **23**, 129, 1913) the mycelium of the aecidial stage is perennial in the rhizome of the anemone and infected leaves appear in the same spot year after year. Klebahn (Z. Pfl.-Krankh. **17**, 143, 1907) proved that the basidiospores infect the growing points of the rhizome in autumn, and produce the aecidia in the following spring. Soppitt stated (J. Bot. Lond. **31**, 273, 1893) that as the result of infecting seedling plants of the anemone in May 1892 with aecidiospores he found one leaf bearing aecidia in April 1893, and concluded that the aecidia could again produce the aecidial stage. This result has been regarded with doubt by subsequent investigators. In the anemone nearly every leaf of the infected plant is attacked as well as the flower shoots. The leaves become longer, narrower, paler green and are usually borne on longer petioles. They are often divided into fewer segments than the normal leaves. Fischer noted that when the fungus appears on the tepals adjoining cells develop chlorophyll.

Dowson showed that the mycelium occurs in the meristematic tissue of the growing point of the rhizome; it is intracellular in the buds and both intracellular and intercellular in the older parts. Its cells are uninucleate and, both in the rhizome and in the leaves, very complicated, irregular, coiled multinucleate haustoria are present. Although Sydow (*loc. cit.* p. 665) had emphasised in 1915 that the aecidial stage of this rust occurs only on *Anemone nemorosa*, Kursanov (1917, 17) described it on *A. ranunculoides* near Moscow. This anemone, he found, may be infected with the normal mycelium which produces binucleate aecidiospores or by a uninucleate form of the fungus in which the aecidiospores are uninucleate; in the aecidia producing uninucleate spores, however, a small percentage of binucleate spores is also found. A similar uninucleate form of the fungus has been found in plants of *A. nemorosa* growing in the Royal Botanic Garden, Edinburgh by Callen (TBMS. **24**, 109, 1940). The aecidiospores were generally uninucleate but a few chains of binucleate spores arising from binucleate basal cells also occurred.

Tranzschel first worked out the alternation of this species (Zbl. Bakt. II, **11**, 106, 1904) and it was confirmed by Klebahn (Z. Pfl.-Krankh. **15**, 80, 1905; **17**, 143, 1907) and Fischer (Ber. Schweiz. Bot. Ges. **15**, 18, 1905; Zbl. Bakt. II, **28**, 149, 1910). These investigators showed that aecidiospores produce infection on *Sorbus aucuparia*, *S. aria*, *S. torminalis*, *S.* '*scandica*', *Pyrus malus*, *P. communis* and *Aruncus silvestris* and other species; only young plants are susceptible to infection. The morphology and cytology of spore development was described by Soong (Flora, **133**, 345, 1939).

Zaghouania Pat.

Bull. Soc. Myc. Fr. **17**, 187 (1901).

Spermogonia flask-shaped, deeply sunk with well-developed projecting paraphyses. **Aecidia** subepidermal, rather deeply immersed, with or without a true peridium; aecidiospores catenulate, walls reticulate, intercalary cells evanescent. **Uredosori** subepidermal, uredospores borne singly, echinulate. **Teleutosori** developing within the uredosori or separately; teleutospores unicellular, ellipsoid, with one pore, basidia short, thick-walled, basidiospores sessile, not discharged.

The description is based upon that of Dumée & Maire (Bull. Soc. Myc. Fr. **18**, 17, 1902) who suggested that this rust should be included in a separate family of the Uredinales, the *Zaghouaniaceae*, and this view was supported by the Sydows (Monogr. Ured. **3**, 526) who considered the semi-internal basidium to be of great taxonomic importance. Dietel (Myk. Zbl. **5**, 68, 1914), however, did not recognise this family and considered this type of basidium to be merely an adaptation to the hot, dry conditions under which the rust develops.

Zaghouania is a monotypic genus with the species *Z. phillyreae* on species of *Phillyrea* in the Mediterranean area. Butler (Ann. Myc. **8**, 444, 1910) considered it to be closely allied to the genus *Chrysopsora* which he described in 1910 and which is also parasitic on the *Oleaceae*.

Zaghouania phillyreae Pat.

Bull. Soc. Myc. Fr. **17**, 187 (1901); Grove, Brit. Rust Fungi, p. 332; Wilson & Bisby, Brit. Ured. no. 321.

[*Uredo phillyreae* Cooke, Fungi Brit. Exs. 1, no. 592 (1871).]

Spermogonia amphigenous, chiefly epiphyllous, scattered among the aecidia, numerous, flask-shaped with well-developed ostiolar filaments, yellowish-orange; spermatia ovoid, hyaline, 4–5 × 2–3 μ. **Aecidia** amphigenous, chiefly hypophyllous and on the petioles and shoots; on the leaves, forming thickened swollen, circular areas, 2–10 mm. diam.; on the petioles and shoots in irregular groups which are often elongate, deforming the host, densely crowded, deeply immersed, small, long-closed and covered by the epidermis, orange; peridium slightly developed; aecidiospores globoid to ellipsoid, alveolate-reticulate, with orange contents, 20–28 × 12–17 μ, wall hyaline, 2–3 μ thick. **Uredosori** hypophyllous on small yellow spots, 1–6 mm. diam.; solitary or in groups, 0·3–1 mm. diam., rounded or irregular, flattened, at first covered by the epidermis, at length pale and pulverulent, orange-yellow; uredospores globoid to ovoid, echinulate, with orange contents, 16–28 × 12–16μ, wall hyaline, 2μ thick, with 4–6 minute, indistinct, scattered pores. [**Teleutosori** hypophyllous and on the stems, on yellow spots, scattered or loosely grouped, rounded or irregular, minute, up to 1 mm. long, yellowish; teleutospores ellipsoid or oblong, rounded at both ends, verrucose, before germination 28–35 × 14–21μ; wall hyaline, 2–3μ thick, with no obvious pore but germination always at the base; basidium 4-celled, semi-internal, 28–32μ long, 14–18μ wide; basidiospores subglobose, sessile, 12–15μ diam.] Auteu-form.

Spermogonia and aecidia on cultivated *Phillyrea latifolia*, Pevensey Churchyard, Sussex, August 1907 (Massee, J. Bot. Lond. **46**, 153, 1908); and on *P. latifolia* var. *media*, Chichester, Sussex, aecidia, 1869, uredospores, April 1874 (F. Paxton).

This species was probably introduced into the south of England together with one or both of its hosts; there has been no record of its recurrence since 1907. Teleutospores have not been detected on it in this country.

The fungus in the aecidial stage forms rounded, swollen pustules on the leaves and extensive patches on the stems. Every shoot of the year is usually attacked and contorted, and, in August, is covered and made conspicuous by a copious development of the orange aecidiospores. The attacked shoots remain short and are rapidly killed by the parasite; according to Maresquelle (Ann. Sci. Nat. Bot. Ser. 10, **12**, 93, 1930) they form an aborted witches' broom. The mycelium persists during the winter in the young shoots and uredospores are formed throughout the year. The teleutospores, which are mature in May, germinate immediately when placed in water and will infect young leaves, with production of aecidia in 15 days.

Rayss (Uredineana, **3**, 210, 1951) has recently recorded its occurrence in Palestine and has given a description of the spore forms.

MELAMPSORACEAE

Uredinopsis Magn.

Atti Congr. Bot. Internat. Genova, p. 167 (1893).

Spermogonia hypophyllous, minute, usually subcuticular, subglobose, without paraphyses. **Aecidia** subepidermal, peridermioid, peridium white, cylindric, fragile; aecidiospores globoid or ellipsoid, colourless, wall verrucose. **Uredosori** subepidermal, with an inconspicuous membranous peridium opening by apical pore; uredospores pedicellate, of two kinds: (1) the ordinary spores, almost sessile, fusiform with a slender beak, the wall smooth or with two longitudinal ridges; (2) amphispores with a long stalk, polyhedral and usually verrucose. **Teleutosori** diffuse; teleutospores intercellular, scattered singly or in irregular groups in the mesophyll, ellipsoid or globoid, 1- to several-celled, the wall colourless, thin and smooth. Heteroecious with aecidia on *Abies* and uredo- and teleutospores on ferns.

All the spores are colourless. The classification of the species depends largely on the uredospores, the other spore forms having much uniformity. The amphispores are usually produced late in the season and serve as a resting stage. The teleutospores overwinter in the dead fronds and the colourless 4-celled basidia, formed in the spring, pierce the epidermis; the subgloboid basidiospores are white, smooth and thin-walled. An account of the taxonomy and geographical distribution of the genus was published by Faull (Contr. Arn. Arb. **11**, 1938). Twenty-six species are known (Hiratsuka, Revision of the taxonomy of the Pucciniastreae, 1958, Tokyo), distributed in all the major regions of the world except Australasia. Three native and one introduced species occur in Europe and one native and one introduced species in Great Britain. Extensive cultural work on the genus has been carried out by Fraser (Mycol. **5**, 233, 1913, *ibid.* **6**, 25, 1914) and Faull (*loc. cit.*) in North America.

Uredinopsis americana Syd.

Ann. Myc. **1**, 325 (1903).

Uredinopsis mirabilis Magn., Hedwigia, **43**, 121 (1904).

[**Spermogonia** on needles of current season, hypophyllous, inconspicuous, round, colourless, subcuticular; spermatia elliptical. **Aecidia** hypophyllous, on needles of current season, in two rows, white, cylindrical, 0·2–0·3 mm. diam., 0·5–1 mm. high, rupturing at the apex; aecidiospores broadly ellipsoid or ovoid, white, 16–24 × 15–19 μ, closely and rather coarsely verrucose.] **Uredosori** hypophyllous on discoloured spots, scattered or aggregate, with a convex, colourless peridium, round, of two kinds: (1) 0·1–0·2 mm. diam.; uredospores hyaline, white in mass, abundant, extruded in tendrils, ellipsoid, obovoid or fusoid, 24–67 × 8–14μ, with a filamentous mucro about 4μ long; wall colourless, smooth, except for two opposing vertical rows of short, closely set cogs, walls 0·5–1μ thick with two pores near each end; pedicel; very short; [(2) 0·1–0·3 mm. diam. amphispores hyaline, white in mass, angularly obovoid or irregularly polyhedral, 19–39 × 11–22μ; walls hyaline, finely and closely verrucose, 0·5–1·5μ thick; pedicels long]. **Teleutosori** diffuse, amphigenous, mostly hypophyllous; teleutospores intercellular, scattered or loosely aggregate, in a single layer, colourless, ellipsoid or subspheroid, usually 4-celled, cruciate, 12–29 × 12–25μ with a single pore in the outer wall of each cell; walls hyaline, smooth, about 1μ thick. Hetereu-form.

[Aecidia on *Abies balsamea*.] Uredospores and teleutospores on *Onoclea sensibilis*. Introduced in a garden near London.

This American species was found in Britain by Hunter (see Faull, Contr. Arn. Arb. **11**, 67, 1938) in the uredospore and teleutospore condition; no amphispores were found. The aecidia are found in North America and have been produced by infection by Fraser (Mycol. **4**, 189, 1912; **5**, 236, 1913 and **6**, 25, 1914) and Faull. It is distinguished from *U. filicina* by the form of the ordinary uredospores.

An account of the teleutospores development in this and in other American species has been given by Pady (Can. J. Res. **9**, 458, 1933).

Uredinopsis filicina Magn.

Atti Congr. Bot. Internat. Genova, p. 167 (1893); Grove, Brit. Rust. Fungi, p. 379; Wilson & Bisby, Brit. Ured. no. 261; Gäumann, Rostpilze, p. 15.

[*Protomyces?filicina* Niessl, Rabh. Fung. Europ. no. 1659 (1873).]

[**Spermogonia** on needles of current season, amphigenous, minute, numerous, subcuticular; hemispherical in section, 82μ wide, 48μ high, apical pore slit-like; spermatia oblong or ellipsoidal, colourless, smooth, 5–6 × 2–2·5μ. **Aecidia** hypophyllous on needles of current season in two rows, white, cylindrical, 0·2–0·5 mm. diam. and up to 1·3 mm. high, rupturing at apex; aecidiospores broadly ellipsoid or ovoid, white, 19–24 × 16–22μ, closely and rather coarsely verrucose.] **Uredosori** hypophyllous on discoloured spots, thickly scattered, yellowish brown, round, with a convex, colourless peridium rupturing above, of two kinds: (1) large 0·1–0·3 mm. diam.; uredospores hyaline, ellipsoid, obovoid or fusoid, 24–46 × 8–13μ, with a conical, broad-based mucro about 12μ long; wall less than 1μ thick, colourless, smooth except

for a few low, scattered, hyaline warts; pedicels very short; (2) 0·1–0·4 mm. diam.; amphispores hyaline, obovoid to polyhedral, 14–30 × 8–22μ; wall 1–1·5μ

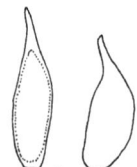

U. filicina. Amphispores.

thick, colourless, finely and closely verrucose, pedicels thin, as long as or longer than the spore. **Teleutosori** diffuse, amphigenous, mostly hypophyllous; teleutospores subepidermal, intercellular, scattered or loosely aggregate in a single layer, ellipsoid to roundish ovoid, usually

2- occasionally 1-celled, 18–24 × 15–16μ, wall hyaline, smooth, about 1μ thick; basidiospores 4–7 × 4·5–8μ. Hetereuform.
[Spermogonia and aecidia on *Abies*

mayriana]; uredospores and teleutospores on *Thelypteris phegopteris*. Uredospores, July; amphispores and teleutospores, August–September. England and Scotland, very rare.

This rust was first recorded in Great Britain from near Strachur, Argyllshire in 1929 (Wilson, 1934, 438) and subsequently from Yorkshire (Mason & Grainger, 1937) and from an early collection in Hertfordshire by Gregory (Trans. Hert. Nat. Hist. Soc. **23**, 136, 1950). The aecidial stage was first produced in Kamei's experiments (J. Facult. Agric. Hokkaido Imp. Univ. **24**, 364, 1933) in which he infected *Abies mayriana* from *Thelypteris* in Japan. Hiratsuka (1958) includes *A. homolepsis* and *A. veitchii* as hosts. It may be assumed that, if found in Europe, it will be on *A. alba*.

The peridium of the uredosori is pseudoparenchymatous, with cells isodiametric near the apex, becoming longer and prosenchymatous towards the periphery; the texture is rather tough and consistent, not friable. Fischer (1904, fig. 310) gives a beautiful and accurate drawing of the nature of the peridium and its relation to the epidermis. The ordinary uredosori are the first to appear and they dehisce promptly at maturity, their peridial cells and their spores are thin-walled and the spores germinate at once. The amphisporic sori appear later and are believed to dehisce only after overwintering; their peridial cells and spores are thick-walled, and the spores are said to germinate only after having passed the winter. The germination of overwintered amphispores and their infection of *T. phegopteris*, with the production of ordinary uredospores after about twelve days, was described by Kamei. Teleutospores were germinated by Dietel (Ber. D. Bot. Ges. **13**, 326, 1895). The teleutospores are to be found in large numbers borne on short lateral branches of the mycelium.

Milesina Magn.

Ber. D. Bot. Ges. 27, 325 (1909).

Spermogonia subcuticular or subepidermal, hemispheric, without paraphyses, but some with flexuous hyphae. **Aecidia** hypophyllous, subepidermal, erumpent, colourless, cylindric; aecidiospores ellipsoid or globoid, wall thin, closely and finely verrucose, or with short, irregularly disposed papillate ridges. **Uredosori** hypophyllous, subepidermal, pustular or immersed, dehiscent by a central pore;

peridium delicate, hemispheric, with cells often elongate at the sides and polygonal above; uredospores white in mass, shortly pedicellate, globoid or obovoid with colourless content, wall colourless, spinulose, echinulate, inconspicuously verrucose or smooth, the pores obscure. **Teleutosori** indefinite, teleutospores intraepidermal, one to few in each cell, few- to many-celled, vertically septate, forming a layer one cell thick, wall thin, colourless, smooth, one pore in upper wall of each cell. Heteroecious, with spermogonia and aecidia on *Abies* and uredospores and teleutospores on ferns. Thirty-four species are known.

The reasons for the use of the name *Milesia* F. B. White seem to us untenable although many have urged it (Rogers, Mycol. **45**, 250, 1948, Hylander *et al.* 1953). *Milesia* was described without mention of teleutospores by White whereas Magnus gave an excellent account of his *Milesina* fully satisfying the requirements of valid publication. Deighton (Taxon. **9**, 231, 1960) clearly states the case for rejection of *Milesia*.

Milesina blechni Syd.

Ann. Myc. **8**, 491 (1910); Grove, Brit. Rust Fungi, p. 377.
[*Aecidium pseudo-columnare* J. Kühn, Hedwigia, **23**, 168 (1884), *p.p.*]
[*Uredo scolopendrii* Schroet., Cohn, Krypt. Fl. Schles. 3 (3), 374 (1887), *p.p.*]
Melampsorella blechni Syd., Ann. Myc. **1**, 537 (1903).
Milesia blechni (Syd.) Arth., Bot. Gaz. **73**, 61 (1922); Wilson & Bisby, Brit. Ured. no. 56; Gäumann, Rostpilze, p. 26.

Spermogonia on needles of current season, amphigenous, mostly hypophyllous, immersed, flask-shaped in section. **Aecidia** hypophyllous on needles of current season, in two rows, one on each side of midrib, white, cylindrical, 0·3–0·4 mm. diam., with colourless, delicate peridium rupturing at the apex; aecidiospores ellipsoid, ovoid or globoid, white, 27–36 × 21–27 μ, densely and rather coarsely warted except on one side where the warts are minute, wall colourless, thin. **Uredosori** hypophyllous, scattered or grouped on greenish-brown areas, pustular, 0·2–0·4 mm. diam., yellowish, rupturing at a centrally placed stoma of the overlying epidermis; peridium hemispheric, hyaline, consisting towards the centre of isodiametric polygonal cells and laterally of elongated cells, walls of cells hyaline, thin, smooth; uredospores colourless, numerous, obovoid to ellipsoid, 26–45 × 15–23 μ, walls thin, 0·7–1 μ thick, with scattered rather coarse echinulations, pedicel about 12 μ long and 6 μ wide. **Teleutosori** on overwintered fronds on indefinite, extensive brown areas; teleutospores within the epidermal cells of the lower epidermis, occasionally in the upper epidermis, hyaline, rounded

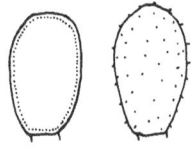

M. blechni. Uredospores (after Faull).

or irregular in outline, 1- to many-celled, cells of teleutospores with thin, smooth, colourless walls, irregularly polygonal, 8–16 × 6–11 μ, one pore in the outer wall of each cell. Hetereu-form.

Spermogonia and aecidia on *Abies alba* and *A. cephalonica*, uredospores and teleutospores on *Blechnum spicant*. Great Britain and Ireland, frequent and overlooked on *Blechnum*, rare on *Abies*.

Infection experiments were carried out in Germany by Klebahn (Z. Pfl.-Krankh. **26**, 257, 1916) who produced *Aecidium pseudo-columnare* on *Abies alba* and *A. cephalonica*, and this has been confirmed more recently by Mayor (Bull. Soc. Neuchâtel. Sci. Nat. **70**, 33, 1947).

The rust is widely distributed in Scotland and has been found up to 500 m. on the mountains. Uredospores may be found all the year round and the rust can evidently pass through the winter in this condition. The teleutospores may be found on the discoloured areas of leaves which have persisted in a green condition through the winter. Such leaves are borne especially on young plants which have not yet produced fertile leaves. The teleutospores germinate readily in April or May. There is little doubt that some of the specimens of *Aecidium pseudo-columnare* recorded in Scotland belong to this species.

Milesina scolopendrii (Faull) Henderson

Notes R. B. G. Edin. **23**, 504 (1961).

[*Ascospora scolopendrii* Fuck., Jahrb. Nass. Ver. Nat. **27–28**, 19 (1873).]

[*Uredo pteridum* White, Scot. Nat. **4**, 27 (1877–8).]

[*Uredo scolopendrii* Schroet., Krypt. Fl. Schles. 3 (3), 374 (1887)]; Grove, Brit. Rust Fungi, p. 378.

[*Milesina scolopendrii* [Fuck.] Jaap, Fungi Sel. Exs. no. 571 (1912).]

Milesia scolopendrii [Fuck.] Arth., Bull. Torr. Bot. Club, **51**, 52 (1924); Wilson & Bisby, Brit. Ured. no. 61; Gäumann, Rostpilze, p. 29.

Spermogonia on needles of current season, epiphyllous and hypophyllous, subcuticular, inconspicuous, colourless, plane, hemispherical to slightly flask-shaped in sectional view, 120–228μ broad by 100–188μ high; spermatia hyaline, narrowly elliptical, 4–5 × 1·5–2μ. **Aecidia** hypophyllous on needles of current season, in two irregular rows, one on each side of midrib, white, cylindrical, 0·4–0·5 mm. diam. by 0·7–1·5 mm. high; peridium colourless, delicate, rupturing at the apex; peridial cells polygonal, vertically elongate, overlapping, in a single layer, 28–56 × 20–36μ, with outer walls smooth, and inner walls finely and densely warted, the warts arranged in elevated, short lines; aecidiospores ellipsoid, ovoid or globoid, mostly elongated, white, 28–48 × 22–44μ, walls hyaline, thin, about 1μ thick, very densely and rather coarsely verrucose with warts irregular in outline, tapering to a blunt point and somewhat deciduous. **Uredosori** hypophyllous, pustular, scattered or loosely aggregate, frequently in rows between the lateral veins and parallel to

them, on greenish to brownish areas of indefinite extent, sometimes involving almost the entire frond, round, 0·1–0·3 mm. diam., covered by brownish epidermis, finally rupturing at a centrally placed stoma pore; peridium colourless, hemispheric, delicate, peridial cells elon-

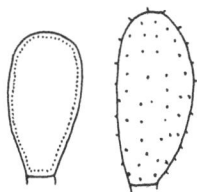

M. scolopendrii. Uredospores (after Faull).

gate and radially orientated at the sides of the peridium, irregularly polygonal, isodiametric to somewhat elongate in the upper part of the peridium, radially overlapping, 7–17μ across, with walls up to 1μ thick, smooth; uredospores colourless, pedicel up to 16μ long, obovoid or ellipsoid, 28–57 × 14–23μ, averaging about 37 × 19μ, spore wall hyaline, 0·5–1·5μ thick, quite strongly and rather sparsely echinulate. **Teleutosori** hypophyllous, occasionally also epiphyllous, on indefinite, brown areas on overwintered fronds; teleutospores intra-epidermal, sometimes within the guard cells, hyaline, rounded, or irregular in outline and conforming to the shape of the containing epidermal cell, often completely filling it, 1- to 40-celled, with anticlinal septa, at times several 1- to few-celled spores in a single epidermal cell, the cells of the teleutospores irregularly polygonal except along their free margins, 8–25 × 7–15 μ, wall thin, smooth, colourless, with a single pore; basidiospores globoid to subgloboid, 6–7·5 μ diam. Hetereu-form.

Spermogonia and aecidia on *Abies alba* and *A. concolor* by artificial inoculation; uredospores and teleutospores on *Phyllitis scolopendrium*. Great Britain and Ireland, scarce.

This rust has been found in a number of localities, expecially in the west of England and Scotland. The uredospores may be found during the whole of the year; the teleutospores are mature in the spring and germinate readily in April and May. Teleutospores were found in Scotland in 1933 and later in England. The rust occurs most frequently on sterile leaves especially on young plants. Successful infection experiments both on *Abies* and on the fern were carried out by Hunter (J. Arn. Arb. **17**, 34, 1936) who gave a detailed account of the development and structure of the spermogonium. Similar experiments were conducted by Mayor (Ber. Schweiz. Bot. Ges. **54**, 5, 1944). The description of the spermogonia and aecidia are from Hunter; that of the uredospores and teleutospores from Faull (Contr. Arn. Arb. **2**, 113, 1932).

Milesina murariae P. & H. Sydow

Monogr. Ured. **3**, 477 (1915).

[*Uredo murariae* Magn., Ber. D. Bot. Ges. **20**, 611 (1902).]
Milesina murariae Faull, Contr. Arn. Arb. **2**, 34 (1932); Wilson & Bisby, Brit. Ured. no. 59; Gäumann, Rostpilze, p. 24.

Spermogonia and **aecidia** unknown. **Uredosori** hypophyllous and petiolicolous, small, whitish, bullate, on yellowish fading areas of the leaf, covered by the light brown epidermis which finally ruptures at a centrally placed stomatic pore or by a slit, peridium colourless, hemispheric; peridial cells elongate and radially orientated at the sides, irregularly polygonal, isodiametric to somewhat elongate and irregularly disposed in the upper part of the peridium, 7–8 μ across, walls 1–2 μ thick, smooth; uredospores colourless, subgloboid, obovoid or ellipsoid, 23–37 × 14–23 μ, wall hyaline, 1·5–2·5 μ thick, strongly and rather sparsely echinulate; pedicel very short. [**Teleutosori** on overwintered fronds, on extensive brown areas often involving entire

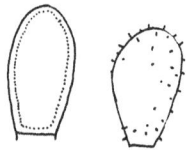

M. murariae. Uredospores (after Faull).

fronds; teleutospores intra-epidermal, hyaline, rounded or irregular in outline, 1- to many-celled, the cells irregularly polygonal, 10–25 × 7–16 μ with thin, smooth, colourless walls, with one pore in the outer wall of each cell.] Hetereuform?

On leaves and petioles of *Asplenium ruta-muraria*. Wales, Scotland, Ireland, rare, but easily overlooked.

This species has been found several times in Scotland (Wilson, 1934, 434), in Ireland (Hunter, TBMS. 1936) and Wales (Herb. Ellis). Teleutospores have been recorded only from Switzerland (Faull, *loc. cit.*). Although the aecidia are unknown it may be assumed that they occur on *Abies alba*.

Milesina kriegeriana (Magn.) Magn.

Ber. D. Bot. Ges. 27, 325 (1909); Grove, J. Bot. Lond. 59, 109 (1921).

[*Aecidium pseudo-columnare* Kühn, Hedwigia, 23, 168 (1884), *p.p.*]
Melampsorella kriegeriana Magn., Ber. D. Bot. Ges. 19, 581 (1901).
Hyalopsora kriegeriana (Magn.) Fischer, Ured. Schweiz, p. 583 (1904).
Milesia kriegeriana (Magn.) Arth., Mycol. 7, 176 (1915); Wilson & Bisby, Brit. Ured. no. 58; Gäumann, Rostpilze, p. 28.

Spermogonia on needles of current season, amphigenous mostly epiphyllous, numerous, irregularly scattered, inconspicuous, hemispherical in sectional view, subcuticular, about 0·1–0·15 mm. broad and high; spermatia narrowly elliptical, 3·5–5 × 1·5–2μ. **Aecidia** hypophyllous on needles of current season in two irregular rows on yellowish portions of the needles, cylindrical, 0·3–0·8 mm. diam., 0·5–1·3 mm. high; peridium colourless, rupturing at the apex, peridial cells polygonal, elongated vertically, overlapping, in a single layer, 32–68 × 16–25μ, with outer walls smooth and inner walls with fine, elevated ridges; aecidiospores ellipsoid, ovoid to globoid, white, 22–48 × 20–36μ, finely warted, walls hyaline, about 1μ thick. **Uredosori** hypophyllous, subepidermal, numerous, scattered or loosely grouped on greenish to brown areas of indefinite extent, pustular, round, 0·1–0·3 mm. in width, covered by brownish, discoloured epidermis with centrally placed stomatic pore; peridium hemispheric, delicate, peridial cells hyaline, isodiametrically to irregularly polygonal, overlapping, 7–14μ across, lateral ones radially elongated, walls of peridial cells hyaline, about 1μ thick, smooth; uredospores colourless, produced singly, with stalks 2–8μ long, obovoid to ellipsoid, 23–48 × 15–22μ, on an average about 33 × 18μ, walls of spores thin, 1μ or less, echinulate. **Teleutosori** formed in autumn on fronds of current season, hypophyllous, on brown areas of indefinite extent;

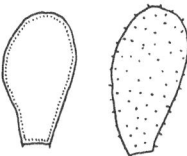

M. kriegeriana. Uredospores (after Faull).

teleutospores intra-epidermal, sometimes within the guard cells, hyaline, rounded, or irregular in outline and conforming to the shape of the containing epidermal cell, often completely filling it, 1- to 40-celled, with anticlinal septa; the cells of the teleutospores with thin, smooth, colourless walls, irregularly polygonal except along their free margins, 8–20 × 6–16μ; basidiospores 10–12 × 5–8μ, usually with a conspicuous oil drop. Hetereu-form.

Spermogonia and aecidia on *Abies alba, A. cephalonica, A. grandis, A. nordmanniana* and by inoculation on *A. concolor*; uredospores and teleutospores on

Dryopteris dilatata, *D. filix-mas*, and *D. carthusiana*. Spermogonia and aecidia, June–September, uredospores during the whole year, teleutospores from autumn to spring. Great Britain and Ireland, scarce but overlooked.

The first record of this rust on fern in Britain was by Boyd in Ayrshire (see Grove, J. Bot. Lond. **59**, 109, 1921). Aecidia were discovered on *A. alba* near Inverness in 1924 (TBMS. **9**, 142, 1924) but these may have belonged to some other species of *Milesina*. Teleutospores were found near Kelso, Scotland, in 1932 together with very abundant aecidia on *A. alba*. Aecidia on *A. grandis* were found in Northern Ireland in 1952. Uredospores have been found in many localities in Scotland, and in several places in England and Ireland (Hunter, TBMS. **20**, 116, 1936); the rust can evidently persist in the uredo stage throughout the year. It appears to be more widely distributed in the western parts of the country than in the east but it has been found in several localities in the east of Scotland and in Norfolk. At first sight this rust with whitish uredospores characteristic of all the fern rusts of the genus *Milesina* does not resemble other rusts. The uredosori sunk in the host tissue and discharging by a pore are also unlike the familiar aspect of a rust. Moreover, uredospore production is most abundant on moribund leaves lying in damp conditions near the ground rather than on the functional leaves freely exposed in the air.

The connection of the aecidium with the rust on the fern was first proved in Switzerland in 1931 by Mayor (Bull. Soc. Neuchâtel. Sci. Nat. **58**, 23, 1933) who infected *D. filix-mas* with aecidiospores from *A. alba* and also the latter with basidiospores. Many infection experiments were carried out by Hunter (J. Arn. Arb. **17**, 26, 1936) in England; she showed that young needles of *Abies* infected with basidiospores in May or June produced aecidia from 37–56 days after; ferns infected with aecidiospores produce uredosori within 32–64 days. The description of the spermogonia and aecidia given above is from Hunter's work.

In this species the teleutospores develop in the autumn and appear to be particularly prevalent on the barren fronds of young plants. According to Magnus (Ber. D. Bot. Ges. **19**, 581, 1901) they may germinate immediately. Germination takes place readily in May and June when fronds bearing them are kept damp. The teleutospores may be readily discovered by stripping off and examining portions of the epidermis from the brown areas of the infected leaf.

A detailed investigation of the development and structure of the spermogonium was made by Hunter (J. Arn. Arb. **17**, 127, 1936).

Milesina carpatorum Hyl., Jørst. & Nannf.

Opera Bot. Soc. Bot. Lundensis, **1**, 1 (1953).

[*Milesina carpatica* Wrobl., Sprawoz. Komisyi Fizyogr. Krakowie, **47**, 178 (1913).]
Milesia carpatica [Wrobl.] Faull, Contr. Arn. Arb. **2**, 55 (1932); Wilson & Bisby, Brit. Ured. no. 57.

Spermogonia and aecidia unknown. Uredosori hypophyllous, subepidermal, pustular, punctate, round, 0·08–0·2 mm. diam., covered by brownish-coloured epidermis with centrally placed substomatal pore, often occurring singly at the centre of small, sharply defined, dark brown spots, but also loosely aggregated on brownish areas of indefinite extent that appear to result from the extension and coalescence of the primary lesions; peridium hemispheric, delicate, peridial cells hyaline, isodiametrically to irregularly polygonal, 6–12μ across, walls of peridial cells hyaline, 0·5–1μ thick; uredospores colourless, short-stalked, obovoid, ellipsoid or subgloboid, 14–27 × 11–17μ, mean 20 × 14μ, walls thin, 0·5–0·7μ, with short, delicate echinulations. Teleutosori amphigenous, mostly hypophyllous, on overwintered fronds, on indefinite, extensive brown areas; teleutospores intra-epidermal, exceptionally in guard cells, hyaline, rounded, or

M. carpatorum. Uredospores (after Faull).

irregular in outline and conforming to the shape of the containing epidermal cell, often completely filling it, 1- to many-celled (up to 60 or more), with anticlinal septa; the cells of the teleutospores irregularly polygonal except along their free margins, 5–11 × 8–15μ, with thin, smooth, colourless walls. Hetereuform?

On *Dryopteris filix-mas*. England, rare.

This rust was found in 1936 by Hunter (TBMS. **20**, 116, 1936) at Newton Abbot, Devon, and it has not been recorded again. Until its discovery in England it was known only from Poland and Czechoslovakia. The description is taken from Faull (*loc. cit.*). It is distinguished from *M. kriegeriana* by its smaller uredospores and teleutospores.

Milesina vogesiaca Syd.

Ann. Myc. **8**, 491 (1910).

Milesia vogesiaca (Syd.) Faull, Contr. Arn. Arb. **2**, 103 (1932); Wilson & Bisby, Brit. Ured. no. 62; Gäumann, Rostpilze, p. 31.

Spermogonia on needles of current season, amphigenous, mostly epiphyllous, subcuticular, inconspicuous, colourless, 0·1–0·2 mm. broad by 0·1–0·2 mm. high, spherical to slightly flask-shaped in section, spermatia hyaline, narrowly elliptical, 4–5 × 1·5–2μ. Aecidia hypophyllous on needles of current season, in two irregular rows, one on each side of midrib on slightly yellowish discoloured portions of affected needles, white, cylindrical, 0·5–0·7 mm. diam. by 0·6–1 mm. high; peridium colourless, delicate, rupturing at the apex, peridial cells polygonal, elongated vertically, overlapping, in a single layer, 32–48 × 20–32μ, outer walls smooth, inner walls warted or with coarse, short, irregularly orientated ridges; aecidiospores ellipsoid, ovoid or globoid, mostly elongated, white, 32–46 × 24–30μ;

walls hyaline, about 1 μ thick with closely spaced, rather blunt, irregular, often deciduous verrucae. **Uredosori** hypophyllous, subepidermal, scattered or loosely grouped on greenish to brown areas of indefinite extent, pustular, round or slightly elongated, 0·1–0·3 mm. in length, covered by a slight brownish, discoloured epidermis with centrally placed stomatic pore; peridium hemispheric, delicate, peridial cells hyaline, isodiametrically to irregularly polygonal, 8–15 μ across, walls of peridial cells hyaline, 0·5–1·5 μ in thickness; uredospores colourless, white in mass, produced singly, short-stalked, obovoid to ellipsoid, 29–44 × 14–23 μ (on an average about 36 × 18 μ), walls less than 1 μ thick, smooth. **Teleutosori** hypophyllous and occasionally epiphyllous, forming indefinite extensive brown areas; teleutospores intra-epidermal sometimes within the guard cells, hyaline, rounded, or irregular in outline and conforming to

the shape of the containing epidermal cell, usually completely filling it, 1- to 50-celled, with vertical septa, the cells of the teleutospores with thin, smooth, colourless walls, irregularly polygonal

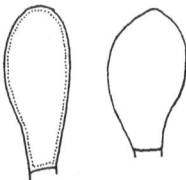

M. vogesiaca. Uredospores (after Faull).

except along their free margins, 9–17 × 8–14 μ, with a single pore in the outer wall of each cell. Hetereu-form.

Spermogonia and aecidia on *Abies alba*, by artificial inoculation; uredospores and teleutospores on *Polystichum setiferum*. Ireland, Killakee near Dublin, uncommon.

This species has been collected only by O'Connor from Ireland (Hunter 1936 c). Hunter (1936 a) has obtained spermogonia and aecidia on *A. alba* by artificial infection and (1936 b) has given a detailed account of the structure and development of the spermogonia. This is the only British species with perfectly smooth uredospores. The descriptions of the spermogonia and aecidia are from Hunter and those of the uredospores and teleutospores from Faull.

Milesina whitei (Faull) Hirats.

Monogr. Pucciniastreae, p. 124 (1936).

[*Milesina polystichi* Grove, J. Bot. Lond. **59**, 109 (1921).]
Milesia whitei Faull, Contr. Arn. Arb. **2**, 111 (1932); Wilson & Bisby, Brit. Ured. no. 63; Gäumann, Rostpilze, p. 33.

Spermogonia and **aecidia** unknown. **Uredosori** hypophyllous, subepidermal, scattered or loosely grouped on greenish to brownish areas of indefinite extent, inconspicuous, pustular, round, 0·15–0·3 mm. diam., covered by a buff, discoloured epidermis which finally ruptures at a central stoma pore; peridium very delicate, hemispheric, peridial cells hyaline, small, isodiametrically to irregularly polygonal, overlapping, at the base, more or less radially elongate, 8–15 μ

across; walls of peridial cells hyaline, less than 1 μ thick; uredospores colourless, white in mass, short-stalked, obovoid or ellipsoid, rarely subspherical, 22–40 × 17–22 μ, averaging about 30 ×

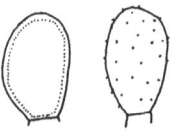

M. whitei. Uredospores (after Faull).

19μ, wall thin, less than 1μ thick, rather sparsely and finely echinulate. **Teleutosori** hypophyllous, on overwintered fronds, on indefinite, brown areas, at times involving entire pinnae; teleutospores intra-epidermal, hyaline, rounded, or irregular in outline and conforming to shape of the containing cell, often completely filling it, 1- to many-celled, with anticlinal septa; the cells of the teleutospores irregularly polygonal except along their free margins, 8–20 × 6–15μ, wall rather thin, smooth, colourless. Hetereuform?

Uredospores and teleutospores on *Polystichum setiferum*. Great Britain, Ireland, rare.

This species was recorded from Tintern (TBMS. **11**, 5, 1926), Ayrshire (J. Bot. Lond. **59**, 109, 1921) and Cornwall (Rilstone, J. Bot. Lond. **76**, 356, 1938) as *Milesina polystichi* and later from Devon and Wicklow (Hunter 1936c, 118) under its correct name.

Aggery (Bull. Soc. Hist. Nat. Toul. **68**, 15, 1935) appears to have described mycelium, haustoria, development of the uredosorus and method of infection of this rust under the name *Gloeosporium nicolai* (see Ramsbottom, TBMS. **25**, 334, 1942).

Milesina dieteliana (Syd.) Magn.

Ber. D. Bot. Ges. **27**, 325 (1909); Grove, Brit. Rust Fungi, p. 376.

[*Milesia polypodii* White, Scot. Nat. **4**, 162 (1877)]; Wilson & Bisby, Brit. Ured. no. 60; Gäumann, Rostpilze, p. 30.
[*Aecidium pseudo-columnare* Kühn, Hedwigia, **23**, 168 (1884), *p.p.*]
Melampsorella dieteliana Syd., Ann. Myc. **1**, 537 (1903).

Spermogonia on needles of current season, amphigenous, immersed and inconspicuous, colourless, plane, hemispherical to slightly flask-shaped in section, subcuticular, 120–225μ broad by 105–200μ high, usually broader than high; spermatia hyaline, narrowly elliptical, 4–5μ × 1·5–2μ. **Aecidia** hypophyllous on needles of current season, in two irregular rows, one on each side of the midrib, on slightly yellowish, discoloured portions of affected needles, white, cylindrical, 0·5–0·7 mm. diam. by 1·0–1·5 mm. high; peridium colourless, delicate, rupturing at the apex, peridial cells polygonal, elongate vertically overlapping, in a single layer, 22–42 × 28–60μ, with outer walls thin, smooth, inner walls 2·5–3·5μ thick with elevated, coarse, short, irregularly orientated ridges; aecidiospores ellipsoid, ovoid or globoid, hyaline, 28–54 × 20–36μ, with one side almost smooth, the other densely and rather coarsely warted, warts irregular in outline, tapering to a very blunt point and somewhat deciduous, walls hyaline, about 1μ thick. **Uredosori** hypophyllous, subepidermal, scattered or loosely grouped on greenish brown areas of indefinite

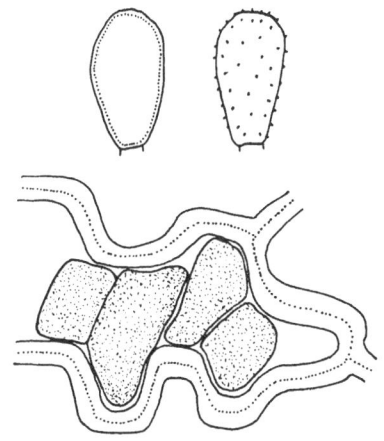

M. dieteliana. Uredospores and intraepidermal teleutospores (after Faull).

extent, pustular, round, 0·1–0·2 mm. diam., covered by a brownish discoloured epidermis with centrally placed stoma pore; peridium delicate but firm, peridial cells hyaline, isodiametrically to irregularly polygonal, 8–18 μ across, lateral ones more or less radially elongated, walls of peridial cells hyaline, 1 μ or less thick, smooth; uredospores colourless, short-stalked, obovoid to ellipsoid, sometimes subgloboid, 23–48 × 16–26 μ, on an average about 35 × 19 μ, walls 1·2 μ thick, diffusely and rather strongly echinulate. **Teleutosori** on overwintered fronds, hypophyllous, forming extensive indefinite brown areas; teleutospores intraepidermal, occasionally in the guard cells, hyaline, rounded, or irregular in outline and conforming to the shape of the containing epidermal cell, often completely filling it, 1- to many-celled (up to 50-celled), with anticlinal septa, a single pore in the outer wall of each cell; the cells of the teleutospores with thin, smooth, colourless walls, irregularly polygonal except along their free margins, 12–23 × 8–20 μ; basidiospores subglobular, 7–9 μ diam. Hetereu-form.

Spermogonia and aecidia on *Abies alba* and *A. concolor* by inoculation; uredospores and teleutospores on *Polypodium vulgare*. Great Britain and Ireland, frequent, but overlooked.

This species was first found in Perthshire by Buchanan White; he described only uredospores, but Faull (Contr. Arn. Arb. **2**, 82, 1932), who re-examined the type specimen, stated that teleutospores were also present.

The uredospores may be found during the whole of the year. Teleutospores were found in Scotland in 1932; they can usually be found in the spring on heavily infected leaves and germinate readily in April and May. Faull (*loc. cit.*) has pointed out that there is considerable variation in the size of the uredospores from different localities. The description of the spermogonia and aecidia is from Hunter (J. Arn. Arb. **17**, 34, 1936) who carried out numerous infection experiments on both *Abies* and *Polypodium*. She gave a detailed account of the structure and development of the spermogonia.

The life history of this rust was first experimentally determined by Mayor in Switzerland in 1936 (Bull. Soc. Neuchâtel. Sci. Nat. **61**, 117, 1936).

Aggery (1935) described the mycelium, development of the haustoria, development of the uredosorus and method of infection of this rust under the name *Gloeosporium polypodii* (see Ramsbottom, TBMS. **25**, 334, 1942).

Hyalopsora Magn.

Ber. D. Bot. Ges. **19**, 582 (1901).

Spermogonia subepidermal, oval in section, without paraphyses. **Aecidia** hypophyllous, erumpent, cylindric; aecidiospores globoid or ellipsoid, with bright yellow contents, wall thin, colourless, minutely verrucose or smooth. **Uredosori** subepidermal, peridium delicate, rarely observable when dry, or represented by paraphyses; uredospores pedicellate, obovoid, with orange-yellow contents, wall colourless, minutely verrucose or smooth, of two kinds: (1) thin-walled, usually with 4 equatorial pores; these are produced throughout the summer and germinate when mature; (2) thick-walled and larger, amphispores with 6–8 scattered pores which are formed usually in the later months and probably undergo a resting condition before germination. **Teleutospores** intra-epidermal, 2- to many-celled, wall colourless, thin, smooth, pores not observed, germinating immediately. Heteroecious with (as far as known) aecidia on *Abies* and uredospores and teleutospores on ferns.

Hyalopsora aspidiotus (Magn.) Magn.

Ber. D. Bot. Ges. **19**, 582 (1901); Grove, Brit. Rust Fungi, p. 374; Wilson & Bisby, Brit. Ured. no. 30; Gäumann, Rostpilze, p. 34.

[*Uredo polypodii* (Pers.) DC. β *polypodii-dryopteridis* Moug. & Nestl., Fl. Fr. **6**, 81 (1815).]

[*Uredo aspidiotus* Peck, Rep. N.Y. State Mus. **24**, 88 (1872).]

Uredo polypodii Schroet., Krypt. Fl. Schles. **3**, 374 (1887); Plowr., Brit. Ured. Ustilag. p. 256, *p.p.*

Melampsorella aspidiotus Magn., Ber. D. Bot. Ges. **13**, 288 (1895).

[**Spermogonia** hypophyllous, on one or on both sides of the midrib on circular yellow spots, usually elongated parallel to the long axis of the leaf, 1 × 0·5 mm. brown, oval, subepidermal, 0·3 mm. wide, 0·1–0·2 mm. deep; spermatia globoid, 3 μ diam., hyaline. **Aecidia** on yellowish leaves, from mycelium freely distributed through the mesophyll, hypophyllous in two rows one on either side of the midrib, 1–40 on each leaf, bladder-like, spherical, 0·5–0·7 mm. diam. and often only 0·2 mm. high, slightly elongated parallel to the long axis of the leaf, peridium delicate, rupturing at the apex, cells oblong, 18–32 × 15–20 μ, outer wall 4–5 μ in thickness, inner 6–8 μ, verrucose; aecidiospores globoid or broadly ellipsoid, angular, yellow, 21–24 × 16–19 μ; wall colourless, thin, 1·5 μ, verrucose.] **Uredosori** amphigenous, usually on yellowish areas, scattered, small, round, up to 0·5 mm. diam., golden-yellow, with a thin peridium, dehiscing irregularly, pulverulent; uredospores of two kinds: (1) with golden-yellow contents, ellipsoid or ovoid, covered uniformly with very faint scattered warts, 28–40 × 16–26 μ; wall about 1·5 μ thick with four indistinct equatorial pores; (2) amphispores, in smaller, mostly indehiscent sori, obovoid,

36–72 × 30–40 μ, wall 2·5–3·5 μ thick with very faint, hardly perceptible warts, and with 6–8 scattered pores. [**Teleutospores** intra-epidermal, often filling the cells completely, roundish or irregular, flattened where they are in contact, sometimes arranged in two layers, about 25 μ high, 21–35 μ or more wide, divided by

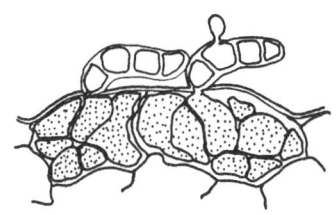

H. aspidiotus. Germinating teleutospores.

vertical septa into 3–5 (mostly 4) cells, wall thin, smooth, colourless; pore not perceptible.] Hetereu-form.

[Spermogonia and aecidia on *Abies alba*, the spermogonia in April on leaves of the previous season, the aecidia in May and June on leaves of the second previous season]; uredospores and teleutospores on *Thelypteris dryopteris*; uredospores June–August [teleutospores in May and June on the young leaves]. Scotland, very rare.

This rust has been recorded in Britain only from Aberdeenshire. The spermogonia and aecidia have not been found in this country and it is doubtful whether teleutospores have been discovered here either; the description of these is taken from Mayor (Bull. Soc. Neuchâtel. Sci. Nat. **47**, 67, 1921–2).

Usually the thin-walled uredospores appear early in the season and are followed by the sori containing the amphispores but Grove and Moss (Ann. Bot. Lond. **40**, 813, 1926) agreed that both kinds of spore may be developed in one sorus. Dietel (Oest. Bot. Zeit. **44**, 46, 1894) showed that the amphispores on germination develop a germ-tube. Pady (Ann. Bot. Lond. **49**, 71, 1935), working in Canada, found that the mycelium over-winters in the underground parts of its host and grows up with the young shoots in the spring.

The heteroecism of this rust was first proved in Switzerland by Mayor (Bull. Soc. Neuchâtel. Sci. Nat. **47**, 67, 1920–1 and **50**, 82, 1925) who obtained spermogonia and aecidia from teleutospore and also uredosori from aecidiospore inoculations; infections from teleutospores made in May on the leaves of *Abies alba* produced spermogonia in April in the following year and aecidia in April in the second year. Similar experiments were carried out in Canada by Bell and by Faull & Darker (Phytopath. **14**, 350, 1924) using aecidia from *A. balsamea* (*Peridermium pycnoconspicuum*) and producing uredospores on *T. dryopteris* in 22 days.

The cytology of uredospore development in this species was investigated by Kursanov (1917, 93), Lindfors (1924, 32), Bell and Moss; the account of the last investigator is the most complete and he concluded that a peridium is present and that the spores are stalked. Pady gave a description of teleutospore development and germination with the formation of the 4-celled basidium. The structure and development of the spermogonium was described in detail by Hunter (1927, *loc. cit.* 1936*b*).

Hyalopsora polypodii (Diet.) Magn.

Ber. D. Bot. Ges. **19**, 582 (1901); Grove, Brit. Rust Fungi, p. 375; Wilson & Bisby, Brit. Ured. no. 31; Gäumann, Rostpilze, p. 37.

[*Uredo linearis* var. *polypodii* Pers., Syn. Meth. Fung. p. 217 (1801).]
[*Uredo polypodii* [Pers.] DC., Fl. Fr. **6**, 81 (1815), *p.p.*]
Pucciniastrum polypodii Diet., Hedwigia, **38**, 260 (1899).

Spermogonia and **aecidia** unknown. **Uredosori** hypophyllous, minute, scattered, bullate, golden-yellow, with a thin peridium, rupturing irregularly; spores globoid, ellipsoid, with golden-yellow contents, of two kinds: (1) 22–35 × 13–20 μ, wall colourless, 1–1·5 μ thick, with faint, distant warts, with 4 equatorial pores; (2) amphispores, 26–38 × 18–29 μ, with colourless wall, 2–3 μ thick, with faint warts, with 6–8 scattered pores. [Teleutospores in the epidermal cells, often filling them completely, showing as yellowish-brown spots on the underside of the leaf, densely crowded, divided into 2–4 cells, each about 14–18μ diam., single cells subgloboid, wall thin, colourless, pore perceptible at the upper end, sometimes on the side.] Hetereu-form? On *Cystopteris fragilis*. Great Britain, scarce. Uredospores, June–October [teleutospores, June–September].

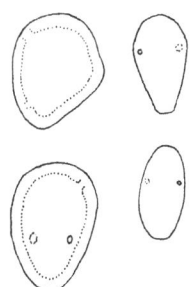

H. polypodii. Amphisphores (left) and uredospores.

This rust has been recorded from North Devon, Derbyshire, Norfolk, Yorkshire, several places in Wales and from central Perthshire where it was collected by Greville in 1823, Buchanan White in 1877 (Scot. Nat. **4**, 27, 1877) and where it is still quite plentiful. There is an early collection in the Persoon Herbarium, presumably from Britain, collected by Smith (Jørstad, Blumea, **9**, 1, 1958). It is doubtful whether teleutospores have been found in Britain. From observation of infected plants over several seasons they are certainly not regularly produced; from the same plants it is equally clear that uredo-perennation takes place. The thin-walled uredospores and occasionally the amphispores can germinate at once; the latter usually remain dormant until the following year. Dietel (Ann. Myc. **9**, 530, 1911) showed that infection by the uredospores can be performed easily, the time of incubation being 14 days; as the result of infection, thin-walled uredospores were at first formed but after a short time amphispores were produced; he did not obtain teleutospores. He also showed that the rust could overwinter by means of uredospores.

This species has a very wide host distribution and is known on at least 25 species of fern in 13 genera (Hiratsuka, 1958).

Bartholomew (Bull. Torr. Bot. Club, **43**, 195, 1916) showed that the cells of the mycelium are binucleate. The development of the uredosorus has been described by Moss (Ann. Bot. Lond. **40**, 813, 1926).

Hyalopsora adianti-capilli-veneris Syd.

Ann. Myc. **1**, 248 (1903); Grove, Brit. Rust Fungi, p. 378; Gäumann, Rostpilze, p. 40.

[*Uredo polypodii* γ *adianti-capilli-veneris* DC., Fl. Fr. **6**, 81 (1815).]

Spermogonia and **aecidia** unknown. **Uredosori** hypophyllous, producing spots on the upper surface, scattered, small, rounded or elliptical, 0·2–0·4 mm. diam., often on the nerves, covered by the epidermis, orange-yellow, surrounded by a few paraphyses; uredospores globoid or ellipsoid, minutely verrucose, of two kinds: (1) 20–34 × 16–25 μ, wall hyaline, 1–1·5 μ thick, with 4 equatorial pores; (2) amphispores, 23–31 × 14–21 μ, average 28 × 19 μ, wall 2·5–3 μ thick, pale brown, with 5–6 indistinct pores. [Teleutospores mostly hypophyllous, produced in the epidermal cells, smooth, hyaline, 1–4-celled, 12–20 μ high, up to 25 μ wide, single cells 9–13 μ diam., wall 1 μ thick.] Hetereu-form?

On the leaves of *Adiantum capillus-veneris*. Ireland, very rare.

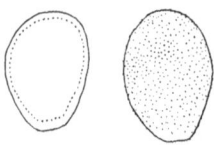

H. adianti-capilli-veneris. Amphispores.

Uredo polypodii on *Adiantum capillus-veneris* was recorded by Plowright (1889, 256) and Grove suggested that this was probably *H. adianti-capilli-veneris*. As this rust has been recorded from Austria, Italy, Spain and France it appeared possible that it might be found in British specimens of its host. Accordingly a search was made in the Edinburgh Herbarium and a few uredosori were found on a specimen of *Adiantum capillus-veneris* collected by Moore from limestone rocks, South Isle of Aran, Galway in 1843 (Wilson & Henderson, TBMS. **37**, 248, 1954). The uredosori are on leaves bearing mature sporangial sori and so the specimen must have been collected in the latter part of the year. The uredospores in this specimen are usually considerably thicker walled than those described by the Sydows (Monogr. Ured. **3**, 497) and the wall is slightly brownish; they must be regarded as amphispores and in consequence this species agrees with *H. aspidiotus* and *H. polypodii* in possessing two kinds of spores. It appears, however, that intermediate spore forms exist; the uredospore measurements given above are taken from Sydow (Monogr. Ured. **3**, 497) but in a specimen issued by Jaap, Fungi Sel. Exs. no. 570, the spores are 21–31 × 12–20·5 μ and the wall varies from 1·3–2·3 μ in thickness. The amphispore measurements given above are from the Irish specimen. There appears to be no definite peridium and according to Sydow its place is taken by a few paraphyses. The description of the teleutospores is taken from Sydow.

Pucciniastrum Otth

Mitt. Naturf. Ges. Bern, **1861**, 71 (1861).

Spermogonia subcuticular, flat, without paraphyses but some with flexuous hyphae. **Aecidia** with a thin cylindrical peridium made up of polygonal cells, verrucose on the inner wall; aecidiospores yellow, ellipsoid, verrucose except on one side which is usually thinner and smooth.

Uredosori with a delicate hemispherical peridium, opening at the summit with a pore, the cells around the orifice larger and sometimes echinulate; uredospores borne singly on short pedicels, yellow, echinulate, usually with four minute pores. **Teleutosori** subepidermal or intra-epidermal, forming indefinite crusts, or in small groups, or solitary in the mesophyll; teleutospores divided by vertical septa into 2–4 cells, walls brown. Heteroecious; the aecidia, when known, are found on *Abies*.

Following Fischer, Klebahn and Arthur the genera *Thecopsora* Magnus (1875) and *Calyptospora* Kühn (1869) are included here. Both of these genera form their teleutospores within the epidermal cells, the former without, and the latter with, great hypertrophy of the tissues of the host. The latter genus does not possess uredospores.

Pucciniastrum epilobii Otth

Mitt. Naturf. Ges. Bern, **1861**, 72 and 84 (1861); Wilson & Bisby, Brit. Ured. no. 247; Gäumann, Rostpilze, p. 41.

Uredo pustulata α *epilobii* Pers., Syn. Meth. Fung. p. 219 (1801).

Melampsora epilobii Fuck., Jahrb. Nass. Ver. Nat. **23–24**, 44 (1869).

Melampsora pustulata Schroet., Cohn, Krypt. Fl. Schles. 3 (1), 364 (1887).

Pucciniastrum pustulatum Diet., Engler & Prantl, Nat. Pflanzenfam. 1, 47 (1897); Grove, Brit. Rust Fungi, p. 366.

Pucciniastrum abieti-chamaenerii Kleb., Jahrb. Wiss. Bot. **34**, 387 (1900).

[*Uredo fuchsiae* Arth. & Holw. ex Arth., Amer. J. Bot. **5**, 538 (1918).]

[*Pucciniastrum fuchsiae* Hiratsuka, J. Facult. Agric. Hokkaido Imp. Univ. **21**, 98 (1927).]

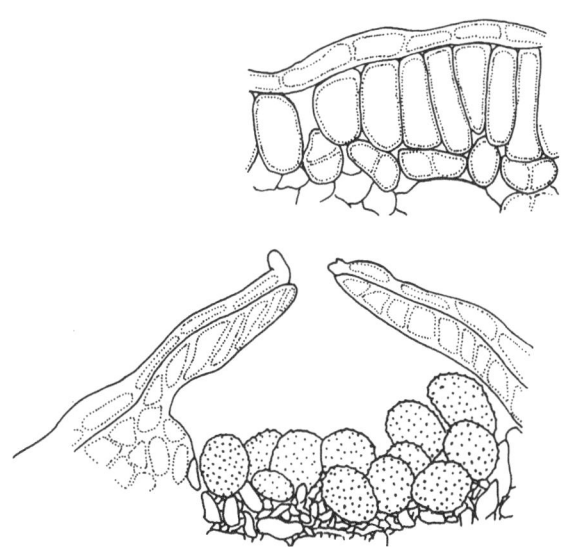

P. epilobii. Teleutosorus and uredosorus (after Savulescu).

Spermogonia hypophyllous, abundant, subcuticular, in section hemispherical, flattened, 102 μ wide, 23 μ high; spermatia ovoid, 1·5–3 μ. **Aecidia** hypophyllous, usually in two rows, cylindrical, about 1 mm. high and 0·25 mm. wide, whitish, splitting above or rarely at the sides, soon breaking down; cells of peridium polygonal, vertically elongate, 25–42 × 10–16 μ, with smooth outer walls and minutely verrucose inner walls; aecidiospores subgloboid, ovoid or ellipsoid, minutely verrucose with a smooth spot on one side, 14–21 × 10–14 μ, wall 1–1·5 μ thick. Uredosori hypophyllous, later also epiphyllous, generally seated on yellow or reddish spots, often on the stems, scattered or in groups, minute, rounded, 0·1–0·25 mm. diam., surrounded by the peridium and ruptured epidermis, at length opening with a central pore and then pulverulent, orange-yellow, peridium hemispherical, rather thin, made up of minute, cubical, smooth, thin-walled cells (wall about 2 μ thick), 10–18 μ long, cells near the ostiole rather thicker-walled (2–3 μ thick); uredospores ovoid or ellipsoid, 15–23 × 10–15 μ, wall 1–1·5 μ thick, hyaline, remotely and shortly echinulate. Teleutosori hypophyllous, rarely epiphyllous or on the stems, subepidermal, very minute, 0·1–0·2 mm. diam., mostly in groups, pale brown; teleutospores intercellular, cylindrical to globoid, 2–4-celled, 2-celled 18–28 × 7–15 μ, 4-celled 20–30 μ diam., lateral wall 1–1·5 μ, up to 3 μ thick at the apex, smooth, brown, with one pore in the upper part. Hetereu-form.

Spermogonia and aecidia on *Abies grandis*, June–July; uredospores and teleutospores on *Chamaenerion angustifolium, Epilobium anagallidifolium, E. montanum* and *E. palustre*, August–October. Great Britain and Ireland, frequent.

Before the publication of the Sydows' Monograph, Klebahn (Z. Pfl.-Krankh. **9**, 33, 1899), Fischer (Ber. Schweiz. Bot. Ges. **10**, 7, 1900), Tubeuf (Zbl. Bakt. II, **9**, 241, 1902) and Bubak (Zbl. Bakt. II, **16**, 155, 1906) demonstrated experimentally the connection of the rust on *C. angustifolium* with *Abies alba* and similar work was carried out in North America by Fraser (Mycol. **4**, 175, 1912) with *A. balsamea*. These investigators obtained successful results by use of the rust from *Chamaenerion* only. Klebahn, in the following year (Jahrb. Wiss. Bot. **34**, 386, 1900) sowed rust from *Chamaenerion* and from three species of *Epilobium* and produced infection only with the former; he confirmed this in the following year, when it was also confirmed by Fischer and Tubeuf. The Sydows, in consequence, distinguished two rusts, *P. abieti-chamaenerii*, confined to *Chamaenerion* which produced numerous teleutospores, the basidiospores infecting *A. alba* and *A. balsamea*, and *P. epilobii*, forming few teleutospores and many uredospores by means of which it persists through the winter, the aecidial host being unknown. Later, Weir & Hubert (Phytopath. **6**, 373, 1916, and **7**, 109, 1917) made reciprocal cultures in America using *C. angustifolium* and *A. lasiocarpa*. Cultures with teleutospores from *C. angustifolium* sown on *A. mayriana* were also made by Hiratsuka in 1926 (Jap. J. Bot. **6**, 23, 1932). Faull in 1938 (J. Arn. Arb. **19**, 163, 1938) confirmed the observations made by the Sydows and showed that the aecidial stage of the rusts on *Epilobium* was also borne on *A. balsamea*. He has pointed out that it differs from that on *Chamaenerion* and *A. alba* in a rather wider, more persistent peridium of the aecidium and rather more coarsely warted aecidiospores and slightly thicker walls of the cells of the peridium of the uredosorus and slightly narrower uredospores. This rust has been known as *P. pustulatum* Faull (non Diet.).

Hylander, Jørstad & Nannfeldt (1953) united the two races pointing out that Gäumann (Ber. Schweiz. Bot. Ges. **51**, 339, 1941) had shown that the race which infects *E. palustre* is able to infect numerous species of *Epilobium* and certain species of *Chamaenerion* and *Godetia*. For this reason the two races on *Chamaenerion* and *Epilobium* as well as that on *Fuchsia* are here united.

The aecidial stage was found on *A. grandis* near Lauder in Berwickshire recently. During the previous year this host species had been infected by basidiospores produced from overwintered teleutospores on *C. angustifolium*.

The rust was discovered on *E. anagallidifolium* by J. R. Matthews on Meall nan Tarmachan, Perthshire in 1922 at about an altitude of 3000 ft.; it has since been found on this host on several Scottish mountains, including Ben Loyal in Sutherland. Infected plants were grown for several months in Edinburgh and continued to form numerous uredospores but no teleutospores were produced. Healthy plants of *E. anagallidifolium* were readily infected by uredospores from this species (TBMS. **9**, 140, 1924) but failed to infect *E. obscurum*, *E. montanum* and *C. angustifolium*; it appears that the rust of *E. anagallidifolium* is a special form. This rust has been recorded in Norway north of 70° N. lat. by Jørstad (1940, 119) and in Iceland (1951, 53).

The uredo stage of this rust was found by J. Rees in January 1932 on *Fuchsia* cuttings (var. Golden Treasure) in a propagating house at Cardiff (Smith & Rees, TBMS. **16**, 308, 1932). Infection on *Fuchsia* was also recorded from Ayrshire in 1962. Affected leaves showed indefinite, pale areas on the upper surface and corresponding areas on the lower side bore numerous uredosori. The disease was most severe on the lower leaves and ultimately these became discoloured, dried up and fell off. The origins of the attacks are unknown. Previously the rust was known on *Fuchsia* from Guatemala and Costa Rica where it was described by Arthur & Holway as *Uredo fuchsiae*. Gäumann (Phytopath. Zeitschr. **14**, 189, 1942) has shown that the race of *P. epilobii* on *E. palustre* will attack *Fuchsia*, and uredospores from the latter infect *E. roseum*.

In the middle of April 1944 Alex. Smith (*in litt.*) described the occurrence of uredospores on shoots of *C. angustifolium*; he suggested that these shoots developed from buds which were initiated as the result of exposure or injury of the stolons and had become infected during the

3

previous autumn. In consequence, he concluded that the rust over-winters by means of mycelium in the buds. Jørstad (1953, 98) has suggested that probably in Norway uredo hibernation takes place in underground shoots and Weir & Hubert (Phytopath. **8**, 59, 1918) made similar suggestions for the fungus in North America.

The development of the spermogonia of this species was described by Hunter (J. Arn. Arb. **17**, 115, 1936). Pady (Can. J. Res. **9**, 458, 1933) gave an account of the development of the teleutospores.

Pucciniastrum circaeae (Wint.) De Toni

Sacc. Syll. Fung. **7**, 763 (1888); Grove, Brit. Rust Fungi, p. 365; Wilson & Bisby, Brit. Ured. no. 246; Gäumann, Rostpilze, p. 46.

[*Uredo circaeae* Schum., Enum. Pl. Saell. **2**, 228 (1803).]
[*Uredo circaeae* Alb. & Schw., Consp. Fung. Nisk. p. 124 (1805).]
[*Melampsora circaeae* [Alb. & Schw.] Thuem., Myc. Univ. no. 447 (1876).]
Phragmospora circaeae Wint., Hedwigia, **18**, 171 (1879).

[**Spermogonia** amphigenous, honey-coloured, subcuticular, 100–130μ wide, 25–35μ high. **Aecidia** amphigenous, generally hypophyllous, usually in two rows corresponding with the stomatal line, 1 mm. high, about 0·25 mm. diam., occasionally club-shaped and wider above; splitting irregularly at the apex; peridium of irregularly shaped cells in longitudinal rows, with finely punctate walls; spore mass pale yellow to bright orange; aeciodiospores globoid, polyhedroid, or ovoid 14–32 × 11–21μ; wall colourless, densely verruculose, about 2μ thick, often with a smooth strip where the wall is thinner.] **Uredosori** mostly hypophyllous, subepidermal, scattered or gregarious, often covering the whole surface of the leaf, small, round, 1–1·5 mm. diam., sometimes confluent, orange-yellow to pale yellow; peridia hemispherical, thin, dehiscent from a central pore, peridial cells small, irregularly polygonal, 8–18μ diam., walls 1·5–2μ thick, colourless, smooth, ostiolar cells larger, walls 1·8–

3μ, smooth; uredospores subgloboid, ovoid or ellipsoid, 16–24 × 12–16μ, wall 1–1·5μ, minutely echinulate, contents orange-yellow when fresh. **Teleutosori** hypophyllous, subepidermal, very small, scarcely visible to the naked eye; teleutospores in little groups, roundish or

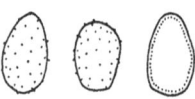

P. circaeae. Uredospores.

flattened at the sides, divided longitudinally into about 2–4 cells, 17–24 × 21–28μ; wall of uniform thickness (about 2μ), yellow, smooth. Hetereu-form.

[Spermogonia and aecidia on *Abies alba* by artificial culture]; uredospores and teleutospores (rare) on *Circaea* '*alpina*', *C. lutetiana* and *C. intermedia*, June–September. Great Britain and Ireland, frequent.

Both Bubak and Klebahn carried out infection experiments with basidiospores on *Abies alba* and other conifers without positive results but Fischer (Mitt. Naturf. Ges. Bern. **1916**, 134, 1917) was successful in obtaining spermogonia and aecidia on *A. alba*; the description of these is taken from his account. Teleutospores have been found only occasionally in this country. *Puccinia circaeae* has sometimes been confused with this species, but the *Pucciniastrum* has diffusely scattered, small, orange uredosori whereas the *Puccinia* has dark brown pulvinate teleutosori; both may occur on the same leaf.

Pucciniastrum agrimoniae (Diet.) Tranz.

Scripta Bot. Hort. Univ. Imp. Petrop. **4**, 301 (1895); Grove, Brit. Rust Fungi,
    p. 364; Wilson & Bisby, Brit. Ured. no. 245; Gäumann, Rostpilze, p. 48.
[*Uredo potentillarum* DC. var. *agrimoniae-eupatoriae*, Fl. Fr. **6**, 81 (1815).]
[*Caeoma (Uredo) agrimoniae* Schw., Trans. Amer. Phil. Soc. Ser. 2, **4**, 291 (1832).]
[*Coleosporium ochraceum* Bonord., Abh. Nat. Ges. Halle, **5**, 186 (1860).]
Thecopsora agrimoniae Diet., Hedwigia, **29**, 153 (1890).
Pucciniastrum agrimoniae-eupatoriae Lagh., Tromsö Mus. Aarsh. **17**, 92 (1895).

Spermogonia and **aecidia** unknown. **Uredosori** chiefly hypophyllous, pulvinate, small, 0·1–0·3 mm. diam., confluent, sometimes spread over the whole leaf, covered by the epidermis and surrounded by a hemispherical peridium, the cells of which are slightly thickened except round the pore where they are thick-walled and irregularly swollen, orange-yellow, fading to ochraceous; uredospores shortly ellipsoid or obovoid, 18–21 × 14μ, wall 1–1·5μ thick, echinulate, with indistinct pores. [**Teleutosori** indefinite, brown; teleutospores subepidermal, intercellular, cuneate, smooth, each divided into 4 cells by two anticlinal walls, 30 × 21–30μ.] Hetereu-form?

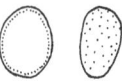

P. agrimoniae. Uredospores.

On *Agrimonia eupatoria*, uredospores, July–September. England, southern Scotland and Ireland, frequent in the south, absent in north-west Scotland.

Teleutospores of this species have not been found in Britain and are rare everywhere; they were described by Tranzschel & Dietel (Hedwigia, **29**, 152, 1890). Klebahn (Z. Pfl.-Krankh. **17**, 149, 1907) proved that the fungus could maintain itself by overwintered uredospores.

The development of the uredosorus has been described by Ludwig & Rees (Amer. J. Bot. **5**, 55, 1918), Colley (J. Agric. Res. **15**, 52, 1918), Dodge (J. Agric. Res. **24**, 885, 1923) and discussed by Moss (Ann. Bot. Lond. **40**, 813, 1926) who summarised the work of these investigators.

Pucciniastrum areolatum (Fr.) Otth

Mitt. Naturf. Ges. Bern, **1863**, 85 (1863).
[*Licea strobilina* Alb. & Schw., Consp. Fung. Nisk. p. 109 (1805).]
[*Uredo padi* Schum. & Kunze, Deutschl. Schwamme, **8**, 4 (1817).]
Xyloma areolatum Fr., Obs. Myc. **2**, 258 (1817).
Thekopsora areolata (Fr.) P. Magn., S.B. Ges. Naturf. Fr. Berlin, p. 58 (1875);
    Wilson & Bisby, Brit. Ured. no. 252; Gäumann, Rostpilze, p. 53.

[*Aecidium strobilinum* [Alb. & Schw.] Wint., Rabh. Krypt. Fl. Ed. 2, **1** (1), 260 (1882).]
Pucciniastrum padi (Schum. & Kunze) Engler & Prantl, Nat. Pflanzenfam. **1**, 47 (1897).
Thecopsora padi (Schum. & Kunze) Kleb., Jahrb. Wiss. Bot. **34**, 378 (1900); Grove, Brit. Rust Fungi, p. 368.

Spermogonia on the underside of cone scales, forming a flat crust of irregular form, up to 4 mm. across, subcuticular, surrounded by the ruptured cuticle, whitish, exuding a sugary liquid with a strong smell. **Aecidia** crowded, covering the upper or sometimes the lower side of the scales of cones, hemispherical or angular through mutual pressure, about 1–1·25 mm. wide, 0·75–1 mm. high, reddish or pale brown, at length opening by a slit and cupulate; peridium firm, hard, composed of cells 23–26 μ long and about 22 μ wide, external wall of cells sparingly verruculose or punctate, strongly thickened (up to 15–20 μ) internal wall thinner (2·5–3·5 μ) verruculose, almost without lumen; aecidiospores angular-globoid, ovoid or ellipsoid, 20–30 × 16–22 μ, densely verrucose on wall 4–6 μ thick, hyaline, with a narrow smooth stripe where the wall is 3 μ thick. **Uredosori** hypophyllous, clustered on spots, 1–5 mm. wide, brownish above, reddish or purplish below, and usually bordered by the veins, covered by the epidermis and by a hemispherical peridium which opens at the summit by a pore; peridial cells minute, cubical, thin-walled, thicker towards the pore, hyaline or yellowish, 8 μ long, 8–12 μ wide, smooth, ostiolar

cells up to 22 μ long and very thick-walled, with no or almost no lumen; uredospores subgloboid, ovoid, or ellipsoid, 15–21 × 11–15 μ, wall hyaline, shortly echinulate, 1·5–2 μ thick. **Teleutosori** epiphyllous or occasionally hypophyllous, minute or largely confluent, forming red then brown, shining crusts which are bounded by the veins, up to 1 cm.

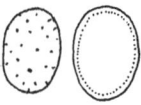

P. areolatum. Uredospores.

diam. Teleutospores intra-epidermal, densely crowded, subspherical, ovoid or oblong prismatic 22–30 × 8–14 μ divided by anticlinal walls into 2, 3 or 4 cells, walls thin, slightly thickened above, brown, smooth, with a pore in the upper and inner corner of each cell. Hetereuform.

Spermogonia and aecidia on cone scales of *Picea abies*, August–November; uredospores and teleutospores on *Prunus padus* August, September. England, Scotland, uncommon.

This rust has been recorded from two localities in Yorkshire (Mason & Grainger, 1937, 43) and from several places in Scotland: Dumfriesshire (A. Lorrain Smith), near Edinburgh, Drumnadrochit (TBMS. **3**, 57, 1908), Inverary, Aviemore, and from the Tay, Dee and Moray areas (Trail, 1890, 324).

The connection of the spore forms on the two hosts has been experimentally demonstrated by Klebahn (Jahrb. Wiss. Bot. **34**, 347, 1900, *idem*, **35**, 660, 1900 and Z. Pfl.-Krankh, **17**, 150, 1907), Tubeuf (Arb. Biol. Abt. Land-u. Forstw. Kais. Gesundh. **2**, 1, 164, 1901) and Fischer (Ber. Schweiz. Bot. Ges. **12**, 8, 1902).

The basidiospores infect the cone scales of the spruce in spring, about the time of pollination. The mycelium spreads through the axis of the cone from the upper scales to the lower. The aecidia develop in summer and mature on the fallen cones in the same year; their spores germinate in the following May and then infect the leaves of the bird cherry, on which they produce uredospores in the summer and teleutospores in the autumn.

According to Roll-Hansen (Medd. Norske Skogforsøksv. 9(4), 504, 1947) Norway spruce shoots, especially terminal ones, often become infected with the mycelium of this rust and assume an S-shape and undergo necrosis. On the shoot infections aecidia are few or more often absent. This shoot infection has not been observed in Britain. On cones a large number of the aecidia do not open before the second spring after ripening but the spores are still able to germinate. The rust attacks several *Prunus* species in Scandinavia and causes a serious shot-hole disease of plums. It has probably been overlooked on these hosts in Britain.

The development of the uredospores, teleutospores and basidia has been described by Sappin-Trouffy (1897, 174).

Pucciniastrum pyrolae Diet. ex Arth.

N. Amer. Fl. 7, 108 (1907); Grove, Brit. Rust Fungi, p. 367; Wilson & Bisby, Brit. Ured. no. 249; Gäumann, Rostpilze, p. 50.

[*Uredo pyrolae* Mart., Prod. Fl. Mosq. Ed. 2, 229 (1817).]
[*Melampsora pyrolae* Schroet., Cohn, Krypt. Fl. Schles. 3(1), 366 (1887).]
[*Pucciniastrum pirolae* Diet., Engler & Prantl, Nat. Pflanzenfam. 1, 47 (1897).]

Spermogonia and aecidia unknown. Uredosori amphigenous, mostly hypophyllous, sometimes on petioles, scattered or in small groups, causing reddish or reddish-brown spots on the upper surface of the leaf, subepidermal, minute, round, 0·1–0·4 mm. diam., long-covered by the epidermis, brownish-yellow; peridia hemispherical, firm; upper peridial cells small, isodiametrically to irregularly polygonal, 9–18μ across, lateral ones somewhat radially elongate, walls of peridial cells thin, about 2μ thick, gradually thickened below toward the orifice, smooth, colourless, somewhat overlapping, ostiolar cells large, 32–46μ long, coarsely to sparsely aculeate above, greatly thickened below; uredospores broadly ellipsoid, 23–42 × 10–19μ; wall 1·5–2·5μ thick, with pointed warts, nearly colourless; contents orange-yellow when fresh. [Teleutosori hypophyllous, subepidermal, inconspicuous, flat, forming an even layer of laterally united cells; teleutospores oblong or

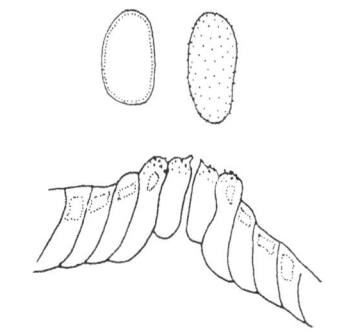

P. pyrolae. Uredospores and apex of peridium (after Savulescu).

columnar, 24–28 × 10–12 μ; wall uni-
formly thin, about 1 μ, colourless.]
Hetereu-form?
Uredospores and teleutospores on

Pyrola media, P. minor, 'P. rotundifolia'
and Orthilia secunda. England and Scotland, rare.

The structure of the peridium of the uredosorus distinguishes this from other species of Pucciniastrum. The uredospores are often more closely warted at one end, though this is not invariably so.

Teleutospores are infrequent and have not been found in Britain; previous to their initial discovery in North America it was uncertain in what genus the rust should be placed. The rust overwinters in this country by means of uredospores; this also occurs in Norway (Jørstad, 1940, 119), North America (Weir & Hubert, Phytopath. 8, 57, 1918), Russia (Treboux, Myk. Zbl. 5, 120, 1914), Siberia and Japan.

It has recently been recorded in Scotland on Orthilia secunda by Henderson (TBMS. 37, 248, 1954). The record on Pyrola rotundifolia is extremely doubtful.

The development of the uredosorus has been investigated by Kursanov (1923, 92) and Moss (Ann. Bot. Lond. 40, 813, 1926).

Pucciniastrum vaccinii (Wint.) Jørst.

Skr. Norske Vidensk.-Akad. Oslo, I, 1951, 2, 55 (1952).
[Aecidium ? myrtilli Schum., Enum. Pl. Saell. 2, 227 (1803).]
[Uredo vacciniorum DC., Fl. Fr. 6, 85 (1815).]
[Thecopsora myrtillina Karst., Bidr. Känned. Finl. Nat. Folk, 31, 59 (1879)]; Gäumann, Rostpilze, p. 58.
[Thecopsora ? vacciniorum [DC.] Karst., Bidr. Känned. Finl. Nat. Folk, 31, 58 (1879)]; Grove, Brit. Rust Fungi, p. 371; Wilson & Bisby, Brit. Ured. no. 254; Gäumann, Rostpilze, p. 60.
Melampsora vaccinii Wint., Rabh. Krypt. Fl. Ed. 2, 1 (1), 244 (1882).
[Pucciniastrum myrtilli [Schum.] Arth., Res. Sci. Congr. Intern. Bot. Vienne, 337 (1906).]

[Spermogonia hypophyllous, occasionally epiphyllous, numerous, inconspicuous, subcuticular, in section hemispherical to conoidal, 41–105 μ wide, 15–23 μ high. Aecidia hypophyllous in two rows on yellow spots and on cones, cylindric, small, 0·2–0·3 mm. diam., 0·5–1 mm. high; peridial cells delicate, readily falling apart, slightly overlapping, the outer wall very thin, smooth, the inner wall 4–5 μ thick, moderately verrucose; aecidiospores globoid or broadly ellipsoid, 18–27 × 15–21 μ; wall colourless, thin, 1–1·5 μ, finely and evenly verru-

cose.] Uredosori hypophyllous, scattered or somewhat gregarious, subepidermal, small, round, 0·08–0·2 mm.

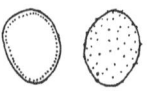

P. vaccinii. Uredospores.

diam., bullate, dehiscent by a small central pore, yellowish-red fading to pale yellow in colour, long-covered by the overarching epidermis, peridia hemispherical, firm, peridial cells small, iso-diametrically to irregularly polygonal, 6–18μ across, somewhat overlapping, walls uniformly thin, 1–2μ, smooth, nearly colourless, ostiolar cells ovate or oblong, 20–35 × 7–15μ, walls smooth, nearly colourless, rather thick, 2·5–6μ, often nearly obliterating the lumen; uredospores subgloboid, broadly obovoid or ellipsoid, 18–31 × 13–21μ; wall colourless, 1–2μ thick, minutely echinulate, contents orange-yellow when fresh, often with rudimentary paraphyses inter-mixed. **Teleutosori** hypophyllous, forming small, brown, indehiscent crusts, teleutospores intra-epidermal, prismatic or cylindric, 14–17 × 7–10μ, wall nearly colourless, uniformly 1μ thick, smooth. Hetereu- or hemi-form?

[Spermogonia and aecidia on leaves and cones of *Tsuga canadensis* and *T. caroliniana*]; uredospores and teleutospores on *Vaccinium myrtillus*, *V. oxycoccos*, *V. uliginosum* and *V. vitis-idaea*; uredospores only, May–October. Great Britain and Ireland, scarce.

The spermogonia and aecidia have not been found in Europe or Asia, only in North America (Hiratsuka, 1958).

The rust is widespread on *V. myrtillus*, occasionally found on *V. vitis-idaea*, infrequent on *V. uliginosum* (Trail, 1890, 324 and Dennis & Gray, Trans. Bot. Soc. Edin. **36**, 220, 1954 and on Ben Lui, Perthshire) and has been found once on *V. oxycoccos* in Northumberland (Wilson & Henderson, TBMS. **37**, 248, 1955). The species has been recorded from Orkney (Trail, *loc. cit.*) and Shetland Islands (Dennis & Gray, *loc. cit.*) and South Uist (Heslop Harrison, *in litt.*).

The teleutospores are very rare, and are most readily found on the dead or fallen leaves; in Europe they appear to have been found on *V. myrtillus* and *V. uliginosum*. Teleutospores have not been found on *V. vitis-idaea* (Hiratsuka, 1936, 325). The rust on this latter host has been regarded as a separate species by Hiratsuka and Gäumann on account of its larger uredosori but the reason for separation appears to be insufficient as pointed out by Jørstad (1934, 25). Uredosori on *V. vitis-idaea* from Aviemore measure 0·2–0·5 mm., those on *V. uliginosum* (Ben Lui, Perthshire) 0·2–0·5 mm., while those on *V. myrtillus* are smaller, 0·15–0·25 mm.

In North America *Rhododendron* (*Azalea*), *Gaylussacia*, *Lyonia*, *Menziesia* and *Pernettya* are attacked and in Japan *Hugeria* and *Lyonia* in addition to species of *Vaccinium* (Hiratsuka, 1958).

Clinton (Rep. Conn. Agric. Exp. Sta. **1909–10**, 719, 1911 and **47**, 499, 1924) in eastern North America infected *Gaylussacia baccata* by sowing aecidiospores from *Tsuga canadensis*. Fraser (Mycol. **5**, 237, 1913 and **6**, 27, 1914) obtained aecidia by sowing teleutospores from *Vaccinium canadensis*.

In north-western Europe it is clear that the rust does not alternate, for it occurs whether conifers are present or not. No culture work has been carried out in Europe or with European host species.

The development of the spermogonia has been described by Adams (Bull. Penn. State Coll. Agr. Exp. Sta. **160**, 3, 1919), and Hunter (1936*b*), and that of the aecidia by Adams and of the uredospores by Moss (Ann. Bot. Lond. **40**, 813, 1926). Pady (Can. J. Res. **9**, 458, 1933) has investigated the development of the teleutospores.

Pucciniastrum goeppertianum (Kühn) Kleb.

Wirtswechselnden Rostpilze, p. 391 (1904).

[*Aecidium columnare* Alb. & Schw., Consp. Fung. Nisk. p. 121 (1805).]
Calyptospora goeppertiana Kühn, Hedwigia, **8**, 81 (1869); Grove, Brit. Rust Fungi, pp. 59 and 372; Wilson & Bisby, Brit. Ured. no. 5; Gäumann, Rostpilze, p. 65.
Melampsora goeppertiana (Kühn) Wint., Rabh. Krypt. Fl. Ed. 2, **1** (1), 245 (1882).
C. columnare [Alb. & Schw.] Kühn, Rabh. Fungi Europ. no. 3521 (1886).

[**Spermogonia** hypophyllous, inconspicuous, flattened, convex, subcuticular, in section hemispherical, 42–137μ wide, 13–30μ high. **Aecidia** hypophyllous, arranged in two long rows parallel to the midrib, on yellowish spots, cylindrical or sac-like, white, with torn or slit margin; peridial cells thin-walled, inner wall verrucose; aecidiospores broadly ellipsoid, uniformly and finely verrucose, orange-red, 21–30 × 14–18μ.] **Uredosori** absent. **Teleutosori** caulicolous, systemic, forming a continuous layer around the swollen stems, at first clear-pink, then becoming reddish-brown and forming a polished crust, later becoming dull; teleutospores up to 12 in each epidermal cell, densely crowded, prismatic, occasionally unicellular but generally divided by longitudinal walls into 2, 3 or 4 cells, up to 42μ high, 12–22μ wide, wall yellowish-brown, smooth, at sides 0·5–0·8μ thick, up to 3μ thick at the summit, with a pore at the upper and inner corner of each cell. Heteropsis-form?

[**Spermogonia** and **aecidia** on the first year needles of *Abies alba*]; teleutospores on the stems of *Vaccinium vitis-idaea*, doubtfully native, and on imported *V. corymbosum*.

The inclusion of this rust as a British species rests almost entirely on the statements made by Grove that it has been found in England, Wales and Scotland and a specimen in his collection in the British Museum Herbarium labelled 'Scotland'. His figure 278 shows an affected branch of *V. vitis-idaea* which is stated to have been collected in Scotland. Special search for this rust has been carried out by several mycologists in Scotland for many years without success. Plowright (1889, 271) stated that teleutospores had not been found in England up to 1889. Specimens of an aecidium on several species of *Abies* collected at Lyme Regis were sent by Plowright to J. Kühn who identified them as his *Aecidium pseudo-columnare*, now known to be the aecidial stage of a *Milesina*. Similarly the specimens of aecidia figured by Grove from near Torquay, collected by Parfitt in 1867 and recorded by Cooke as *Peridermum columnare* (Cooke, 1871, 535 and Micro. Fungi, Ed. 4, p. 194) prove on examination to be the aecidial stage of a *Milesina*.

Massee published a description of *C. goeppertiana* in 1907 (Kew Bull. 1907, 1) but gave no reference to any discovery of the teleutospore stage on *V. vitis-idaea* in Britain; he stated that leaves of *Abies nordmanniana* bearing peridermia had been received from Wales. Later Massee (Mildews, Rusts and Smuts, 1913, 177) stated that this rust grows on *V. vitis-idaea* in this country but mentioned no localities where it had been found. The rust was recorded from Ulster (district Antrim, Derry and Tyrone) by Adams & Pethybridge (Proc. Roy. Irish Acad. **28**, 120, 1910) but no specimens can be traced.

As *P. goeppertianum* has been recorded from France, Holland, Belgium, Denmark, Norway, Sweden, Spain, Germany and Switzerland, its occurrence in this country would not be unexpected, but until well-authenticated specimens are found here its status remains doubtful.

In the spring of 1958 the rust was found by M. Perrin in a nursery in Dorset, on a bush of *Vaccinium corymbosum* sent from the United States in 1953, as a rooted cutting. This discovery was confirmed by L. Ogilvie (*in litt.*). The infected bush was burnt.

The life history of *P. goeppertianum* was experimentally demonstrated in Europe by Hartig (Allg. Forst-Jagdzeit. p. 289, 1880), Kühn (Hedwigia, **26**, 28, 1887) and Bubak (Ann. Myc. **2**, 361, 1904; Zbl. Bakt. ii, **16**, 154, 1906) and in North America by Arthur in 1909 (Mycol. **2**, 231, 1910) and Fraser (Science, **30**, 814, 1909) using the rust on different species of *Abies* and *Vaccinium*.

Accounts of the life history of the rust in Europe were given by Hartig (1894, 161) and Tubeuf & Smith (1897, 370). Maresquelle (Ann. Sci. Nat. Bot. Ser. 10, **12**, 83, 1930) described the deformations caused by the rust on *V. vitis-idaea*. Pady (Can. J. Res. **9**, 458, 1933) has given a very complete account of the development of the teleutospores.

Pucciniastrum guttatum (Schroet.)
Hyl., Jørst. & Nannf.

Opera Bot. Soc. Bot. Lundensis, **1**, 1, 81 (1953).
[*Caeoma galii* Link, L. Sp. Pl. Ed. 4, 6 (2), 21 (1825).]
Melampsora guttata Schroet., Abh. Schles. Ges. Vaterl. Cult. Nat. Abth. 1869–72, 26 (1870).
Melampsora galii Wint., Rabh. Krypt. Fl. Ed. 2, 1 (1), 244 (1882).
Thecopsora galii De Toni, Sacc. Syll. Fung. 7, 765 (1888); Grove, Brit. Rust Fungi, p. 370; Wilson & Bisby, Brit. Ured. no. 253; Gäumman, Rostpilze, p. 64.
Pucciniastrum galii Fisch., Beitr. Kryptogamenfl. Schweiz. Ser. 2, **2**, 471 (1904).
Thekopsora guttata (Schroet.) P. & H. Syd., Monogr. Ured. 3, 467 (1915).

Spermogonia and aecidia unknown. Uredosori hypophyllous, occasionally on the petioles or stems, scattered or gregarious, at times thickly scattered over the whole surface of the leaf, subepidermal, minute, round, 0·1–0·25 mm. diam., pulvinate, yellowish-orange, peridia hemispherical, firm, delicate, upper peridial cells small, irregularly polygonal, 8–15µ diam., walls smooth, thin, less than 1·5µ, colourless, ostiolar cells rounded 10–17µ high, walls 1·5–2·5µ thick, smooth, nearly colourless; uredospores subgloboid, ovoid or ellipsoidal, 13–24 × 10–17µ, wall minutely echinulate, 1–1·5µ thick, colourless, contents orange-yellow when fresh. Teleutosori amphigenous, inconspicuous, on dark brown discoloured areas of indefinite extent, at times involving entire leaves, teleutospores intra-epidermal, globoid, subgloboid or somewhat angular, 20–

30µ diam. divided vertically into 2–4 cells (usually 4), wall 1·2–2µ thick, slightly thicker at the apex, yellowish-brown. Hetereu-form?

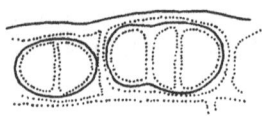

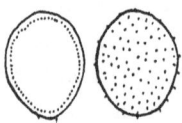

P. guttatum. Teleutospores and uredospores (after Savulescu).

Uredospores and teleutospores on *Galium odoratum, G. palustre, G. saxatile, G. uliginosum, G. verum* and *Sherardia arvensis.* June–September. England and Scotland, rare.

This rust was found by Soppit in 1889 (see Plowright, TBMS. **1**, 59, 1899) and subsequently by Boyd (Grove, J. Bot. Lond. **59**, 314, 1921) in south-west Scotland; by Ellis in Norfolk (1934, 503); it has also been found near Aberdeen by Alex. Smith and in Inverness (*in litt.*). According to Gäumann (1959, 65) it uredo-perennates probably by infected basal rosettes of the hosts and field observations on the rust on *Galium odoratum* in Scotland confirm this. Uredosori can be found on overwintered leaves in March.

Uredo goodyerae Tranz.

Trav. Soc. Nat. St Petersburg Sect. Bot. **23**, 27 (1893). [Teleutospores unknown.]

[*Pucciniastrum goodyerae* [Tranz.] Arth., N. Amer. Fl. **7**, 105 (1907)]; Wilson & Bisby, Brit. Ured. no. 248; Gäumann, Rostpilze, p. 52.

[Spermogonia, aecidia and teleutosori unknown.] Uredosori amphigenous, scattered or crowded in small groups, subepidermal, minute, round, 0·1–0·3 mm. across, orange-yellow, then pale yellow, long-covered by the epidermis, peridia hemispherical, delicate, firm, dehiscent by an apical pore, peridial cells small, isodiametrically to irregularly polygonal, 7·5–15µ across, lateral ones

radially elongate, walls of the peridial cells thin, smooth, colourless or sub-hyaline, ostiolar cells rather large, 32–42µ high, finely echinulate above; uredospores ovoid, 23–34 × 16–20µ, wall finely echinulate, uniformly thick, 1·5–2µ thick, colourless, contents pale yellow. Imperfect uredo-form.

On *Goodyera repens,* April–October. Scotland, rare.

This rust has been found only in Inverness, Moray and Banff in Scotland. It was mentioned by Grove (1913, 344) under *Melampsora orchidi-*

repentis and also recorded under this name at Forres in 1912 (TBMS. **4**, 33, 1913). Repeated attempts to discover teleutospores in this country have failed, nor have they been discovered elsewhere.

Melampsorella Schroet.

Hedwigia, **13**, 85 (1874).

Spermogonia subcuticular, minute, without ostiolar filaments but some with flexuous hyphae. Aecidia minute, subepidermal, erumpent, with shortly cylindrical peridium which ruptures at the apex and falls soon after dehiscence, peridial cells small, elongate-polygonal; aecidiospores globoid or ellipsoid, wall hyaline, contents orange-yellow. Uredosori subepidermal, minute, aparaphysate; peridia hemispherical, made up of small irregularly polygonal cells; uredospores pedicellate, ellipsoid or obovate, echinulate, with orange-yellow contents and thin hyaline walls, usually with four minute pores. Teleutosori hypophyllous, forming irregular discoloured areas, the teleutospores loosely grouped within the epidermal cells, globoid, 1-celled and producing a 4-celled basidium; basidiospores globoid or ovoid, hyaline.

Heteroecious with aecidia on *Abies* and *Picea* and uredosori and teleutosori on dicotyledonous hosts.

Melampsorella caryophyllacearum Schroet.

Hedwigia, **13**, 85 (1874); Grove, Brit. Rust Fungi, p. 360; Wilson & Bisby, Brit. Ured. no. 52; Gäumann, Rostpilze, p. 74.

[*Uredo pustulata cerastii* Pers., Syn. Meth. Fung. p. 219 (1801).]
[*Uredo caryophyllacearum* Chev., Fl. Envir. Paris, **1**, 405 (1826).]
Melampsorella cerastii [Pers.] Wint., Hedwigia, **19**, 56 (1880).
Melampsora cerastii Wint., Rabh. Krypt. Fl. Ed. 2, **1** (1), 242 (1882).

Spermogonia amphigenous, mostly epiphyllous, subcuticular, scattered, honey-coloured, minute, depressed hemispherical, 100–320μ broad, 25–60μ high; spermatia ellipsoid, 3·5–5 × 1·5–3·5μ, colourless. Aecidia hypophyllous, in two rows, one on each side of the midrib, erumpent, hemispherical or shortly cylindrical, 0·4–1 mm. long, 0·2–0·8 mm. diam., reddish-yellow, peridia colourless, rupturing irregularly at the apex, peridial cells irregularly polygonal, 25–55 × 15–30μ, elongate vertically, inner walls thin, verruculose, outer thin, smooth; aecidiospores subgloboid, ellipsoid or polygonoid, with orange-yellow contents, 16–30 × 14–20μ, wall 1–2μ thick, densely verruculose, colourless. Uredosori hypophyllous, rarely epiphyllous or on petioles, subepidermal, usually arising below a stoma, in groups or scattered, sometimes over the whole surface, minute, circular, 0·1–0·4 mm. diam., orange-yellow, peridia hemispherical, delicate, with a small apical pore, upper peridial

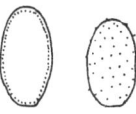

M. caryophyllacearum. Uredospores.

cells small, irregularly polygonal, 10–20μ across, lateral elongate, wall 1·5–3μ thick, smooth, nearly colourless; uredospores ellipsoid or ovoid, 16–30 × 12–21μ, wall 1–1·5μ thick, sparsely echinulate, colourless, contents orange-yellow, pores minute, generally two towards each end. **Teleutosori** hypophyllous, rarely epiphyllous, often covering the whole under surface of the leaf on whitish or pale reddish areas; teleutospores 1- rarely 2-celled, within the epidermal cells, solitary or in groups, globoid or ellipsoid, sometimes angular, 12–25μ across, wall smooth, thin; basidiospores globoid, 7–10μ diam., nearly colourless, with thin, smooth wall. Hetereu-form. Spermogonia and aecidia on leaves of *Abies alba, A. cephalonica, A. lowiana, A. nordmanniana* and *A. pinsapo*; uredospores and teleutospores on *Cerastium arcticum, C. arvense, C. glomeratum, C. semidecandrum, C. tomentosum, C. holosteoides, Stellaria graminea, S. holostea* and *S. media*; spermogonia and aecidia from June to August; uredospores from May onwards; teleutospores, May. Great Britain and Ireland, uncommon.

In this rust the mycelium of both stages is perennial. The teleutospores develop on the leaves of the dicaryont host as they appear in the spring and germinate immediately and the basidiospores infect the fir. On the fir the mycelium spreads slowly and infected spots become recognisable in the following autumn as elongate swellings. Later on, these show a rough, cracked surface as the result of infection by various other fungi and the stem may be easily broken across in this region. Early in the following summer buds give rise to infected shoots which are thicker and rather shorter than ordinary shoots and grow vertically upwards; they bear short, thickened needles arranged spirally on them. Spermogonia and aecidia soon develop on these leaves which fall in the following August, soon after the aecidiospores have been shed. Little chlorophyll is present in the infected shoots and they are very conspicuous during the summer on account of their pale yellow colour. They branch freely and thus increase the size of the broom. Infected shoots appear to be very susceptible to frost and may become killed back in the winter almost to the base; some buds however usually remain alive and give rise to infected shoots in the following year. There appears to be no record of the production of cones on the brooms but male flowers were found on brooms in Perthshire by R. M. Adam. Brooms usually continue to grow for a number of years and may attain a large size.

Most of the records of the aecidial stage appear to be from Scotland and Ireland and especially from the western regions. The most frequent aecidial host is *A. alba*; the record on *A. nordmanniana* is from Perthshire (Wilson, TBMS. **9**, 140, 1924) and those on *A. cephalonica* and *A. lowiana* from near Dublin (TBMS. **11**, 16, 1926). Burrell (Trans. Norf. Norw. Nat. Hist. Soc. **7**, 255, 1901) has given an account of the distribution on *A. alba* in Norfolk.

Changes in the structure of the shoots of *Abies* resulting from infection were studied by de Bary (Bot. Zeit. p. 257, 1867), Mer (Rev Gén. Bot. p. 153, 1894) and others.

There is some evidence that a specialisation into biological races has begun on different species of the Alsineae.

On the diplont host the mycelium grows up with the young shoots and gradually extends into all the leaves, making them in some cases smaller and giving them a slightly yellowish appearance. Uredospores are produced on the infected leaves and sometimes also on the sepals. As this diplont is entirely independent it is often found at a considerable distance from any *Abies* and in quite different plant communities. The specimen on *Cerastium arcticum* was found at an altitude of about 3000 ft. on Ben Nevis and that on *C. semidecandrum* on a salt marsh on the coast of Fife; the sepals on the latter were heavily infected with uredosori.

Fischer (Z. Pfl.-Krankh. **11**, 321, 1901) was the first to detect the alternate hosts of this species by infection experiments. He sowed aecidiospores from *A. alba* on *Stellaria nemorum* and in the following year (*ibid.* **12**, 193, 1902) performed the reverse infection. His results have been confirmed repeatedly on other hosts by Tubeuf, Klebahn and Bubak in Europe, by Arthur and by Weir and Hubert and others in America and by Hiratsuka (Jap. J. Bot. **6**, 1, 1932) in Japan.

A rust which produces similar witches' brooms caused by *Melampsorella* on both *Picea* and *Abies* species occurs in North America and the fungi responsible have been regarded as one species, but recently Pady (Trans. Kansas Acad. Sci. **43**, 147, 1940 and **44**, 190, 1941 and Mycol. **38**, 477, 1946), who has examined the spermogonia, aecidia and haustoria, has tabulated the differences of the forms on the two aecidial hosts and has suggested that two distinct species are involved, indistinguishable in the dicaryont stages. Boyce (Trans. Conn. Acad. Arts Sci. **35**, 329, 1943) considered that the brooms on *Picea* result from infection with *Peridermium coloradense*. Nevertheless, Hiratsuka (1958) places the *Picea* races in *Melampsorella caryophyllacearum*.

The development of the uredosorus has been investigated by Moss (Ann. Bot. Lond. **40**, 813, 1926) and the morphology and ontogeny of the spermogonium has been described by Hunter (Bot. Gaz. **83**, 1, 1927). The intra-epidermal development of the teleutospores and basidial formation is described by Pady (1946, *loc. cit.*).

Melampsorella symphyti Bub.

Zbl. Bakt. ɪɪ, **12**, 423 (1904); Grove, Brit. Rust Fungi, p. 363; Wilson & Bisby, Brit. Ured. no. 53; Gäumann, Rostpilze, p. 78.

[*Uredo symphyti* DC., Lam. Encycl. Meth. Bot. **8**, 232 (1808).]

[**Spermogonia** on needles of current season, chiefly hypophyllous, crowded, often spread over whole surface of leaf, minute, orange-yellow. **Aecidia** on needles of current season, hypophyllous, in two rows parallel to the midrib, not crowded, cylindrical, 0·5–0·8 mm. high; peridia colourless, firm, opening at the summit by a cleft, at length torn to the base into 3–5 segments, peridial cells irregularly polygonal, 25–55 μ long, 17–22 μ wide, walls 2–2·5 μ thick, colourless; aecidiospores globoid, subgloboid or ovoid, 20–40 × 18–29 μ, wall 1·5–2 μ thick, verrucose, contents orange-yellow.] **Uredosori** hypophyllous, subepidermal, scattered or gregarious, often covering the whole surface of the leaf, small, round, 0·1–0·3 mm. across, covered by the epidermis which finally ruptures at a centrally placed pore, often somewhat pulverulent, rich golden-yellow, peridia hemispherical, firm, peridial cells small, isodiametrically or irregularly polygonal, 8–20 μ across, walls thin, smooth, colourless or subhyaline; uredospores obovoid, broadly ellipsoid or ellipsoid, 22–33 × 18–25 μ, wall sparsely and coarsely echinulate, 1–1·5 μ thick. **Teleutosori** hypophyllous, developed within the epidermal cells, forming large, whitish or pinkish patches, many crowded in each cell, pale yellow; teleutospores 1-celled, subgloboid or ellipsoid, occasionally somewhat angu-

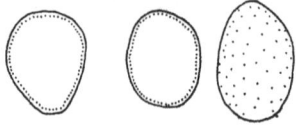

M. symphyti. Uredospores.

lar, 8–20 μ across, wall uniformly thin, less than 1 μ, smooth, nearly colourless. Basidiospores globoid or ovoid, 4–4·5 × 3·5–4 μ. Hetereu-form.

[*Aecidia* on *Abies alba*, May–June]; uredospores on *Symphytum asperum, S. officinale, S. tuberosum*; May–September; uredospores and teleutospores (May) on *S. asperum × officinale* (= *S. × peregrinum*). Great Britain and Ireland, scarce.

The aecidial stage of this rust has not been found in Britain and is known only in Czechoslovakia and Bulgaria.

The mycelium is perennial in the rhizome of *Symphytum* and usually all the leaves on an infected plant produce uredospores in the earlier part of the summer; these are generally smaller than healthy leaves and are paler green. Leaves produced in the later part of the year may not produce spores.

Teleutospores were discovered by Alex. Smith on *S. × peregrinum* in May, 1932, near Ludlow (see Grove & Chesters, TBMS. **18**, 274, 1934); only the older leaves were infected with this stage. On germination, the apical end of the teleutospore bursts through the papillate projection which has pierced the outer epidermal cell wall. The basidia are rather thick and short and give rise to four basidiospores on long, tapering sterigmata. In moist weather the teleutospore crusts are covered by masses of intertwined basidia and become whitish. The account of the aecidium is based on that of Bubak (*loc. cit.*) who worked out the life cycle in 1903. He (Bubak, Zbl. Bakt. II, **16**, 155, 1906) was not able to infect leaves of *S. officinale* with uredospores from *S. tuberosum* and also failed to bring about infection with aecidiospores on leaves of either of these species. He suggested that infection might take place in some other way (e.g. on the rhizome) or that it only became obvious in the following year.

Melampsoridium Kleb.

Z. Pfl.-Krankh. **9**, 21 (1899).

Spermogonia subcuticular, inconspicuous, without paraphyses but sometimes with flexuous hyphae. **Aecidia** erumpent, minute, cylindrical, rupturing at the apex; peridial cells small, irregularly polygonal; aecidiospores ellipsoid, with reddish-yellow contents and thin walls, verrucose except for a smooth spot. **Uredosori** subepidermal, minute, peridium hemispherical, peridial cells minute, irregularly polygonal, the ostiolar cells ovoid-conical, extending into an acute or acuminate, conical, spine-like apex; uredospores pedicellate, echinulate, usually smooth above. **Teleutosori** hypophyllous, subepidermal, indehiscent; teleutospores compacted laterally in a single layer, unicellular, clavoid or prismatic, wall smooth, thin, coloured, with a single, often indistinct pore; basidium 4-celled with globoid basidiospores.

Heteroecious with aecidia on *Larix* and uredosori and teleutosori on dicotyledonous trees.

Melampsoridium hiratsukanum Ito

J. Facult. Agric. Hokkaido Imp. Univ. **21**, 9 (1927); Wilson & Bisby, Brit. Ured. no. 55.

[**Spermogonia** amphigenous mostly hypophyllous, subcuticular, minute, lenticular, 90–126 × 30–55 μ, honey-yellow; spermatia oblong, 5–9 × 2–4 μ, hyaline. **Aecidia** hypophyllous, cylindrical, 0·5–2 mm. across, up to 1·4 mm. high, rupturing at the apex, peridial wall colourless, cells quadrilateral or hexagonal in face view, 23–36 × 14–20 μ, overlapping, inner wall thick, closely verrucose, with uniform papillae; aecidiospores ellipsoid, 18–26 × 15–20 μ, wall verruculose, 1·8–2 μ thick, with a smooth area on some spores, hyaline, contents orange when fresh.] **Uredosori** hypophyllous, subepidermal, producing on the upper surface reddish-yellow spots which are limited by the veins, scattered or in groups, circular or elliptical, 0·2–0·4 mm. diam., covered by the epidermis which finally ruptures at a centrally placed pore, orange-yellow; peridium hemispherical, strongly developed, upper

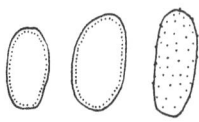

M. hiratsukanum. Uredospores.

peridial cells small, irregularly polygonal 8–18 μ across, lateral ones somewhat radially elongate, walls of peridial cells

1–2·5 μ thick, the inner rather thicker, colourless, smooth, ostiolar cells extending into long, sharp spines, 32–60 μ long; uredospores ovoid or ellipsoid, 21–34 × 10–18 μ, wall colourless, thin, uniformly sparsely and finely echinulate. **Teleutosori** hypophyllous, covered by the epidermis, scattered or grouped, minute, up to 0·5 mm. diam., at first yellow, then becoming brown or purple-brown, at length blackish-brown; teleutospores prismatic or clavate, smooth, yellow, 32–45 × 10–16 μ, wall about 1 μ thick. Hetereu-form.

[Spermogonia and aecidia on *Larix leptolepis*, and *L. dahurica* and *L. europaea* by innoculation]; uredospores and teleutospores on *Alnus cordata*, *A. glutinosa* and *A. incana*. Scotland and Ireland, scarce.

Only the uredospore and teleutospore stages of this rust have been found in Britain. The specimens found in this country, have often been regarded as *Melampsoridium alni* but on examination have proved to be *M. hiratsukanum*; similar confusion of the two species has occurred on the continent.

The rust was first recorded in this country by Farquharson (Ann. Scot. Nat. Hist. 1911, 240) from Aberdeen and was described from the same place, where it occurred in a forest nursery on two-year seedlings of *A. glutinosa* and *A. incana*, by Wilson (TBMS. **9**, 140, 1924). Specimens from these collections were submitted to Hiratsuka for identification. It was also found by Boyd in Ayrshire and recorded by Grove (J. Bot. Lond. **59**, 315, 1921). Since that time it has been found in Angus, Argyllshire, Dumfriesshire and in West Ross-shire; it has also been found on *Alnus cordata* in Aberdeen. The record of *M. alni* from Ireland by O'Connor (1936) is probably also this species. Infection is heaviest on young or coppiced trees.

M. hiratsukanum differs from *M. alni* in its shorter and broader uredospores (21–34 × 10–15 μ as against 32–46 × 8–15 μ) with a less marked smooth area at the apex. It closely resembles *M. betulinum* but may be distinguished by its smaller uredospores and much longer ostiolar cells of the peridia in the uredosori; judging from the description of Hiratsuka the peridium of the aecidial stage is also larger.

Hiratsuka (Jap. J. Bot. **6**, 1, 1932) by infection experiments proved the connection with the aecidium on various species of *Larix*.

Melampsoridium betulinum (Fr.) Kleb.

Z. Pfl.-Krankh. **9**, 17 (1899); Grove, Brit. Rust Fungi, p. 358; Wilson & Bisby, Brit. Ured. no. 54; Gäumann, Rostpilze, p. 68.
[*Uredo populina* var. *betulina* Pers., Syn. Meth. Fung. p. 219 (1801).]
Sclerotium (*Xyloma*) *betulinum* Fr., Syst. Myc. **2**, 262 (1822).

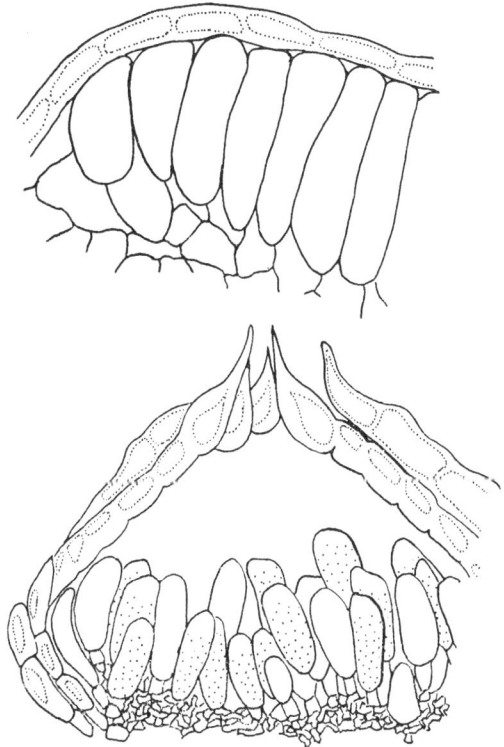

M. betulinum. Teleutosorus and uredosorus.

Spermogonia amphigenous, rather numerous, scattered, subcuticular, pale yellow, flattened-conical, $50-65 \times 20-30\mu$; spermatia ovoid, $1 \cdot 5-2\mu$. Aecidia hypophyllous, solitary or in longitudinal rows on one or both sides of the midrib, subepidermal, $0 \cdot 1-1 \cdot 5$ mm. wide, $0 \cdot 3-1$ mm. long, $0 \cdot 3-0 \cdot 5$ mm. high, peridia light reddish-orange fading to white, rupturing at apex with irregularly torn margin, peridial cells small, rhomboidal, $25-30\mu$ long, $12-20\mu$ wide, inner walls finely verrucose, $2-3\mu$ thick, outer of equal thickness, smooth; aecidiospores globoid or ellipsoid, $16-24 \times 12-18\mu$, wall $1-1 \cdot 5\mu$ thick, finely and closely verrucose except a small smooth and slightly thinner area on one side. Uredosori hypophyllous, with yellow spots showing on upper side, subepidermal, scattered, $0 \cdot 1-0 \cdot 5$ mm. diam., yellow, peridia hemispherical, firm, at length opening at the summit, peridial cells small, polygonal, $8-18\mu$ across, those at the sides radially elongate, ostiolar cells ovate-conical tapering into an acute apex; uredospores ellipsoid or subclavate, $22-38 \times 9-15\mu$, wall colourless, $1-1 \cdot 5\mu$ thick,

4

with distant spines but smooth at apex. Teleutosori hypophyllous, subepidermal, scattered, often covering the whole leaf, orange becoming yellowish-brown; teleutospores prismatic, rounded at both ends, in a palisade-like layer, 30–52 × 8–16μ, wall nearly colourless, smooth, laterally 1μ thick up to 2μ at the apex. Hetereu-form.

Spermogonia and aecidia on *Larix decidua* and *L. leptolepis*, April–June; uredospores and teleutospores on *Betula pubescens* and *B. pendula*. Great Britain and Ireland, very common.

Plowright was the first to discover that the rust on the birch was connected with an aecidium on the larch (Gard. Chron. **8**, 41, 1890 and Z. Pfl.-Krankh. **1**, 130, 1891). He performed the experiment in both directions and his conclusions were confirmed by Klebahn and by Ashworth (La Cellule, **43**, 189–200, 1934) who infected *L. decidua* in England and by Hunter (1936*b*, 120) who infected *L. decidua* and *L. laricina* in the United States. The aecidial stage is often supposed to be rare in this country but this can be explained by its inconspicuous appearance and short duration. It was first discovered by Keith (Scot. Nat. **7**, 271, 1883–4) near Inverness, then by Plowright at King's Lynn (*loc. cit.*), and later by Borthwick & Wilson (Notes R. B. G. Edin. **8**, 79, 1913) and since then has been found in several places in Scotland; it was discovered on *L. leptolepis* in the Clyde area in 1933, and it appears that in Scotland it may generally be found, if searched for carefully in mid-May, wherever birches and larches grow together. Liro (Acta Soc. Fauna Fl. Fenn. **29**, no. 7, 1907) in Finland failed to bring about infection of *L. decidua* and *L. sibirica* with basidiospores of *M. betulinum*; he believed that the aecidial stage does not develop in northern regions. He showed that the uredospores will not remain viable during the winter. He also showed (Ann. Myc. **6**, 580, 1908) that the rust persists, probably in the form of mycelium, in the leaves which sometimes remain attached throughout the winter to infected seedlings; further, he suggested that mycelium can overwinter in the buds of such seedlings. Klebahn in Germany (Z. Pfl.-Krankh. **22**, 342, 1912) failed to bring about infection with overwintered uredospores. D'Oliveira & Pimentel claimed that overwintering takes place in the buds (For. Abstr. **15**, 2646, 1953).

On several occasions extending over a number of years heavily infected seedlings have been removed in autumn to a garden in Edinburgh, but have always failed to reproduce the rust in the following year; these, however, did not retain any leaves throughout the winter.

The presence of completely infected shoots on some young birches in West Ross-shire early in 1958 which had been heavily infested with uredosori in 1957, suggested that occasional bud perennation either as spores or mycelium can take place in Britain. The rust spreads so rapidly that only a few overwintered foci are necessary to produce widespread infection, given the right climatic conditions. The aecidial stage has not been found in North America (Hunter, J. Arn. Arb. **17**, 120, 1936). Weir & Hubert (Phytopath. **8**, 55, 1918) failed to bring about infection with overwintered teleutospores and considered that the rust overwinters on birch, producing uredosori in the next season, so that larch is not essential for its perpetuation.

As the result of culture experiments Klebahn (1904, 402 and Z. Pfl.-Krankh. **15**, 100, 1905) suggested the existence of specialised forms which are not sharply differentiated; these are f. sp. *betulae-verrucosae* on *B. verrucosa* and f. sp. *betulae-pubescentis* on *B. pubescens* and *B. nana*.

The development and cytology of the uredospores, teleutospores and basidia were described by Sappin-Trouffy (1897, 165) and Kursanov (1922, 96) referred to the development of the peridium of the uredosorus. Ashworth (Ann. Bot. Lond. **49**, 95, 1935) described the occurrence of substomatal and emergent hyphae in infected leaves of *L. decidua*. Hunter (1936*b*) gave an account of the spermogonia.

Cronartium Fries

Obs. Myc. **1**, 220 (1815).

Spermogonia caulicolous, flat, large, spreading between the cork layer and the cortex without definite limit, dehiscing by longitudinal slits in the bark, without paraphyses but some with flexuous hyphae. **Aecidia** caulicolous, erumpent, hemispherical or sac-like, with peridium generally several cells in thickness, usually with irregular or circumscissile dehiscence; aecidiospores ellipsoid with tesselated, coarsely verrucose walls with sometimes a smooth spot on one side. **Uredosori** hypophyllous, with a delicate hemispherical peridium dehiscing at the summit with a narrow pore; uredospores borne singly on pedicels, echinulate, without obvious pores. **Teleutosori** erumpent, often arising out of the uredosori, the catenulate 1-celled teleutospores united into a cylindrical column, horny when dry, germinating as soon as mature with a 4-celled basidium. Basidiospores minute, spherical.

Heteroecious, with spermogonia and aecidia on the stems of *Pinus*. Several are known to produce secondary or repeating aecidia.

Cronartium flaccidum (Alb. & Schw.) Wint.

Hedwigia, **19**, 55 (1880); Rabh. Krypt. Fl. Ed. 2, **1** (1), 236 (1882); Wilson & Bisby, Brit. Ured. no. 20.
Sphaeria flaccida Alb. & Schw., Consp. Fung. Nisk. p. 31 (1805).
Erineum asclepiadeum Willd., Funck, Cryptogam. Gewächse Fichtelgeb. p. 145 (1806).
Cronartium asclepiadeum (Willd.) Fr. Obs. Myc. **1**, 220 (1815); Grove, Brit. Rust Fungi, p. 313; Gäumann, Rostpilze, p. 81.

Cronartium paeoniae Cast., Cat. Pl. Marseill. p. 217 (1845).
[*Peridermium cornui* Rostr. ex Kleb., Hedwigia, **29**, 29 (1890).]

This rust species includes one host alternating and one non-alternating race, the latter consisting of the aecidial stage only, viz.:

Peridermium pini [Pers.] Lév., Mem. Soc. Linn. Paris, 212 (1826), p.p., emend.; Kleb. Hedwigia, **29**, 28 (1890); Wilson & Bisby, Brit. Ured. no. 66.
[*Aecidium pini* Pers., Syn. Meth. Fung. p. 213 (1801), p.p.]

Spermogonia rather flat, irregular, forming blisters up to 3 mm. diam. Aecidia erumpent from the bark forming large orange-yellow bladders, grouped or spread more or less equally over a large part of the branch, 2–7 mm. long, 2–3 mm. diam., peridium minutely verrucose on the exterior, more strongly verrucose on the inner side, firm, consisting of 2–3 layers of cells, with stiff, hair-like outgrowths from the inner surface expanded at the base and up to 1 mm. long, cells of the peridium up to 80μ long, 16–36μ wide, walls verrucose 4–6μ thick; aecidiospores globoid-ellipsoid or polyhedroid, 24–31 × 16–23μ, walls hyaline, verrucose, 3–4μ thick. Uredosori hypophyllous, scattered or in groups, 0·15–0·25 mm. diam., pustular, at length opening with round pore, peridium thin, made up of cells 25μ long 15μ wide with walls everywhere 2–3μ thick; uredospores ovoid or ellipsoid, 18–30 × 14–20 μ, wall hyaline, sparsely and shortly echinulate, 1·5–2·5μ thick, with indis-tinct pores. Teleutosori hypophyllous, usually arising in the uredosori, the spore mass cylindrical, emerging as a straight or curved, waxy mass, 1–2 mm. long, 50–120μ wide, yellowish or reddish-brown; teleutospores ellipsoid, yellow or pale yellowish-brown, 20–60 × 10–16

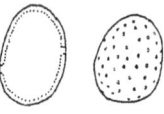

C. flaccidum. Uredospores.

μ, wall smooth, up to 1μ thick, slightly thicker at the apex. Hetereu- and repeating aecidial-forms.

Spermogonia and aecidia on *Pinus sylvestris*, May, June; uredospores and teleutospores on *Paeonia mascula* and *Tropaeolum majus*, July–October. The alternating race, southern England, rare; the repeating race on Pine, Great Britain and Ireland, scarce.

This species is plurivorous in its diplont stages (on various species of *Gentiana*, *Impatiens*, *Nemesia*, *Paeonia*, *Pedicularis*, *Schizanthus*, *Tropaeolum*, *Verbena* and *Vincetoxicum* in Europe) but not in its aecidial stage. The relationship of the aecidial stage on *Pinus* with the rust on *Vincetoxicum officinale* was first proved by Cornu (C.R. Acad. Sci. Paris, **102**, 930, 1886) while Geneau de Lamalière (Assoc. Fr. Avanc. Sci. **23**, 628, 1895) first showed the connection with the rust on *Paeonia mascula*. Klebahn, Fischer and many others confirmed these results and proved the relation between aecidia and other spore forms on species of *Cynanchum*, *Verbena*, *Pedicularis*, *Impatiens*, *Tropaeolum* (Klebahn, Z. Pfl.-Krankh. **24**, 10, 1924) and other genera. Hiratsuka (Jap. J. Bot. **6**, 24, 1932) and others in Japan showed that the aecidiospores of this species on *P. densiflora* infect various species of *Paeonia*. Fragoso (1925, 374) mentions a *Peridermium* on *P. halepensis* in Spain, which he suggests belongs to *C. flaccidum*. Klingstrom has also carried out extensive inoculation of pines with basidiospores (personal communication).

In this country the rust on *Paeonia* was found by Plowright in Norfolk in 1889 and by others in the south of England (on *P. mascula*, Lackham, Wiltshire, Bond, *in litt.*) and on *Tropaeolum majus* near Norwich (Ellis TBMS. **20**, 10, 1935).

No successful experimental cultures have been carried out in this country with the non-alternating race known as *Peridermium pini* and, in consequence, its distribution here is not definitely known. It has been reported from Norfolk by Ellis (*loc. cit.*) and by Alex. Smith (*in litt.*) from Suffolk but as *C. flaccidum* has been recorded in Norfolk, these records may belong to it.

In Scotland Stevenson (1879, p. 256) recorded *P. pini* from the Tweed, Forth, Tay and Moray areas; there is a specimen in the Johnson Herbarium at Edinburgh collected in Berwickshire. Alex. Smith (*in litt.*) reported it as abundant in Banffshire and it has been collected recently in Morayshire and near Kelso. It is a troublesome forest disease on Scots pine in north-east Scotland. There is no record of the alternating race from Scotland. In Ireland Pethybridge (J. Dept. Agric. Tech. Instruct. Ireland, **2**, 500, 1911) described a serious outbreak of disease of Scots pine in Waterford caused by this rust. He assumed that it was due to *C. peridermium-pini* (Willd.) Liro which is now regarded as *C. flaccidum* on *Pedicularis*, but he was unable to find any spore stages on *Pedicularis*. As the alternating race of *C. flaccidum* has not otherwise been recorded from Ireland it appears probable that Pethybridge's record refers to *P. pini*.

The aecidial mycelium perennates in the cortex producing a canker of the stem; according to Haack (Z. Forst-Jagdwesen, **46**, 3, 1914) it may persist in a living condition for many years. Large cankers have been found on mature trees of *P. sylvestris* in Morayshire and near Kelso. As the result of infection experiments, Haack discovered that the aecidiospores could again infect the pine and produce aecidia; he was able to infect unwounded branches 2–6 years old which were still bearing needles but found that the presence of wounds increased the amount of infection. These results were confirmed by Klebahn (Flora, **111**, 194, 1918) and by Liese (Mitt. Deutsch. Dend. Ges. 1928, 165), both of whom concluded that only certain forms of *P. sylvestris* were susceptible. Klebahn (Z. Pfl.-Krankh. **48**, 369, 1938), definitely showed that infection takes place through wounds.

There are records of an aecidial perennating rust of this type on

Pinus mugo (as *P. montana,* Fischer, 1904, 436) and *P. pinaster,* Gäumann, 1959, 82) in Central Europe.

The development and cytology of the uredospores and teleutospores of *C. flaccidum* were described by Sappin-Trouffy (1897, 179).

Peridermium pini has been found parasitised by *Tuberculina maxima* in Inverness and Roxburghshire. Liro has also described this parasite in Finland.

The cytology of *P. pini* was investigated by Klebahn (*loc. cit.* 1938) who found that the hyphae are uninucleate and the aecidiospores binucleate; on germination some aecidiospores give rise to uninucleate, others to binucleate germ-tubes.

A specimen in Berkeley's Herbarium labelled 'Overton Longeuville, May 1858 [Hunts.] on *Pinus ponderosa*' in Kew Herbarium is mentioned here tentatively. Only a cankered stem is present; the host is certainly not of the *P. sylvestris* group and is probably correctly named. The aecidia present could well be any of several species recorded on *P. ponderosa* in North America. The infection has produced slight fusiform swelling of the branch. The aecidiospores are broadly ellipsoid 28–36 × 16–22 μ, closely verrucose.

Cronartium ribicola J. C. Fischer

Hedwigia, **11**, 182 (1872); Grove, Brit. Rust Fungi, p. 316; Wilson & Bisby, Brit.
  Ured. no. 21; Gäumann, Rostpilze, p. 85.
Cronartium ribicola H. A. Dietr., Pl. Fl. Balt. Crypt. IV, 21, (1854) and Arch. Naturk.
  Liv. Ehst. Kurl. II, **1**, 287 (1856) *nomen nudum.*]
[*Peridermium strobi* Kleb., Abh. Nat. Ver. Bremen, **10**, 153 (1887).]

Spermogonia caulicolous, subcortical, forming blisters 2–3 mm. diam., yellow, with a small ostiole, exuding a sugary fluid; spermatia ovoid to ellipsoid 2 × 4·5 μ. **Aecidia** caulicolous, erumpent, rounded or oblong, 2–6 × 1–3 mm. and up to 3 mm. high, rupturing above or along the sides, orange-yellow, peridium of 2–5 layers of cells up to 60 μ long, 15–30 μ wide, smooth or minutely verrucose externally, more strongly verrucose internally; aecidiospores globoid, ellipsoid or polyhedroid, orange, 22–29 × 18–20 μ, wall coarsely verrucose except on an elliptic area, 2–2·5 μ thick except on the smooth spot where it is 3–3·5 μ thick. **Uredosori** hypophyllous, on pale spots, forming groups 1–5 mm. diam.,

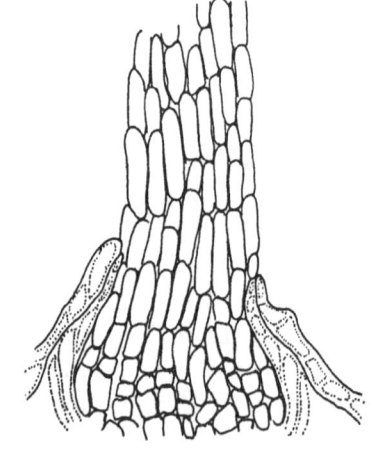

C. ribicola. Teleutosorus (after Savulescu).

pustular, 0·1–0·3 mm. diam., yellow, surrounded by a delicate peridium opening by a central pore; uredospores ellipsoid to obovoid, distantly and sharply echinulate, orange, 21–25 × 13–18 μ; wall colourless, without distinct pores, 2–3 μ thick. Teleutosori hypophyllous, crowded, especially along the veins of the leaf, sometimes covering the whole blade, arising in the uredosorus, orange to brownish-yellow, producing columns of teleutospores up to 2 mm. long and 120–150 μ thick, teleutospores ellipsoid cylindric, 30–70 × 10–21 μ; wall colourless, smooth, 2–3 μ thick. Hetereuform.

Spermogonia and aecidia on *Pinus aristata*, *P. ayacahuite*, *P. cembra*, *P. flexilis*, *P. griffithii*, *P. lambertiana*, *P. monticola*, *P. parviflora* and *P. strobus*, March–June; uredospores and teleutospores on *Ribes nigrum*, *R. sanguineum*, *R. sylvestre* and *R. uva-crispa*, July–October. Great Britain and Ireland, frequent.

This is the well-known Blister Rust of the Weymouth or white pine which has caused such extensive damage of the five-needled pines in Europe and North America. It appears that all five-needled pines may become infected although it is known that the species vary in susceptibility; the Balkan pine, *P. peuce*, was said to be immune but it is now known to be slightly susceptible (Tubeuf, Z. Pfl.-Krankh. 41, 369, 1931). *Pinus strobus* is moderately susceptible while *P. albicaulis*, *P. flexilis* and *P. monticola* are highly susceptible (Spaulding, Phytopath. 15, 591, 1925). A large number of species of *Ribes* has been shown to be attacked and of these *R. nigrum* is probably the most susceptible (Taylor, Phytopath. 12, 298, 1922). Some forms of *R. sylvestre* are highly resistant especially some strains of 'Rote Hollander' (Tubeuf, Z. Pfl.-Krankh. 43, 433, 1933) and the Norwegian variety 'Viking' (Hahn,, Trans. Bot. Soc. Edin. 30, 137, 1929; Phytopath. 26, 860, 1936). Almost all seedlings obtained from 'Viking' are resistant. Germination and penetration of young leaves of a resistant variety takes place but the hyphae soon necrose. On older leaves more extensive growth takes place but no symptoms develop. Riker & Kouba (Phytopath. 30, 20, 1940) have brought forward evidence to show that young rust-resistant trees of *P. strobus* occur in the United States. Patton & Riker (For. Abstr. 20, 1463, 1958) have selected resistant strains of *P. strobus* and *P. monticola*. Infection of the pine by basidiospores takes place on stem or on needles of twigs not more than three years old; on the latter the germ-tube grows through a stoma-pore. A yellow spot is produced where the tissue contains abundant mycelium. When the needle is infected the mycelium grows down into the short shoot and then into the stem where it spreads in the cortex. After some time the cortex swells, spermogonia are produced and then aecidia. Aecidia are formed from two to four years after infection, spermogonia usually one year earlier. Infected branches may produce aecidia for many years but at length the mycelium usually rings the branch which then dies.

The infection of *Ribes* takes place by the growth of the germ-tubes of the aecidiospores through the stomatal openings. The uredosori are generally produced on the under surface of the leaf but occasionally are also found on the petiole, bud scales or young stems. Overwintering of the rust on *Ribes* has been established but the exact method by which this takes place is doubtful and it probably only occurs under special conditions.

Cronartium ribicola is generally considered to be an Asiatic species, the original host being *P. cembra* although Gäumann (1959, 89) regards it as a relict species with disjunct distribution between Central Europe and Eastern Asia. It was first discovered in Europe by Dietrich (*loc. cit.*) in 1854 in Estonia on *Ribes*; this investigator named the fungus *C. ribicola* but gave no description of it. He also found the fungus on *P. strobus* and named it *Peridermium pini* f. *corticola* in ignorance of its identity with *C. ribicola*. Several investigators (Woronin, Oersted, Rostrup) mentioned the fungus on *Ribes* in 1871 but did not describe it. The fungus on *P. strobus* was named *Peridermium strobi* by Klebahn (*loc. cit.*) in 1887 and in the following year (Ber. D. Bot. Ges. **6**, xlv, 1888) he experimentally proved its connection with *C. ribicola*; he also carried out the infection of *P. strobus* by means of teleutospores in 1903–4 (Z. Pfl.-Krankh. **15**, 86, 1905).

Pinus strobus was introduced into Europe from North America in 1705. The fact that the rust was not found on the tree until nearly 150 years after its introduction has been brought forward as additional evidence that *C. ribicola* is not a native of Europe but was introduced from Asia on the comparatively resistant *P. cembra* and then spread rapidly on the more susceptible *P. strobus* (see Lepik, Festi Metsand, Aastar, **8**, 177, 1937). Spaulding (U.S. D. A. Tech. Bull. **87**, 26, 1929) has suggested that *Pinus pumila* may prove to be the original aecidial host of the rust.

Successful infection experiments have been frequently repeated by many investigators using various species of *Ribes* and several five-needled pines.

Repeated attempts to infect *P. strobus* by means of aecidiospores have failed (see Spaulding, U.S. D. A. Bull. **957**, 37, 1922) and it may be concluded that host alternation is obligatory.

The species was first discovered in this country at King's Lynn by Plowright, who found it on the leaves of black, white and red currant in

1892; he discovered the aecidial stage on *P. strobus* in the following year and in 1899 infected the currant with aecidiospores from the pine (Gard. Chron. **26**, 94, 1899). Moore (1959, 119) records it as occasional on gooseberry.

The first discovery of the species in North America was made in the State of New York on *Ribes* in 1906 and it was found on *P. strobus* in 1909; later findings have indicated that it was present in the north-eastern United States as early as 1898. It is generally agreed that it was introduced from Germany on young plants of *P. strobus*. It is now widely distributed in the north-eastern states and in eastern Canada. It was found at Vancouver, British Columbia, in 1921 and it has been ascertained that it was introduced there in 1910 on a single shipment of 1000 young plants of *P. strobus* from France. It has since spread throughout the range of *P. monticola* in British Columbia, south through Washington and Oregon into northern California, where it has appeared on *P. lambertiana* and *P. albicaulis*.

Very complete accounts have been given of the morphology of this species by Spaulding (U.S. D. A. Bur. Pl. Indust. Bull. **206**, 1–88, 1911; U.S. D. A. Bull. **957**, 1–100, 1922; *idem*, Tech. Bull. **87**, 1–58, 1929), Colley (J. Agric. Res. **15**, 619, 1930) and Tubeuf (Z. Pfl.-Krankh. **46**, 49 and 113, 1936) while Moir (U.S. D. A. Bull. **1186**, 1–31, 1924) has given an account of its distribution in western Europe.

A detailed investigation of the cytology of this species was made by Colley (*loc. cit.*) and Pierson (Nature, **131**, 728, 1933) described the fusion of spermatia with the receptive or flexuous hyphae of the spermogonium.

The white pine blister rust has caused extensive damage to plantations of *P. strobus* in this country and on the continent and the few plantations of *P. monticola* which exist in Scotland have been very seriously attacked. In consequence, in Europe the planting of five-needled pines as forest trees has been almost entirely discontinued. The disease on the black currant is serious and, when heavily infected, the bushes lose their leaves soon after midsummer and fruiting is prevented; the disease on red currants is not so serious and the gooseberry is rarely infected. Spraying of the currant has little or no effect on the disease. The best method of control is the removal of one of the host plants and this has proved to be impracticable in Europe. In certain areas in the eastern United States control of the disease has been obtained by the complete eradication of all species of *Ribes*. Treatment of excised cankers with acti-dione has given promising results on Pine (Moss, Plant Dis. Reptr, **42**, 703, 1958).

Certain biological agents are unfavourable to the development of this fungus. Rodents commonly gnaw the bark of the lesions on the pines and so reduce the number of aecidiospores produced (Mielke, J. For. **33**, 994, 1935). *Tuberculina maxima* a parasitic fungus, known as purple mould, often attacks the aecidia in this country and prevents the development of the aecidiospores; this fungus is known from other parts of Europe (see Lechmere, Naturw. Z. Forst-Landw. **12**, 491, 1914) and western North America (Mielke, Phytopath. **23**, 299, 1933; Hubert, Phytopath. **25**, 253, 1935). In cities and industrial centres in this country, in Europe generally (see Tubeuf, Z. Pfl.-Krankh. **46**, 507, 1936) and in North America (Spaulding, U.S. D. A. Tech. Bull. **87**, 21, 1929) where five-needled pines and *Ribes* occur together this rust is conspicuously absent, the smoke and fumes in the air apparently being unfavourable to the fungus.

Chrysomyxa Unger

Beitr. Vergl. Path. p. 24 (1840).

Spermogonia amphigenous or episquamous, subepidermal, paraphysate. **Aecidia** with a well-developed membranous peridium, consisting of a single layer of cells and dehiscing irregularly at the apex; aecidiospores with a hyaline, coarsely warted wall without pores. **Uredosori** subepidermal, erumpent, pulverulent, with or without a very delicate peridium; uredospores produced in rows by basipetal abstriction, resembling aecidiospores. **Teleutosori** pulvinate, waxy; teleutospores in simple or branching chains, 1-celled with thin, colourless walls, germinating without a resting period.

Nearly all are heteroecious species with aecidia on *Picea* and uredospores and teleutospores on the *Ericaceae*. *C. abietis* is micro-cyclic. Dietel suggested that the genus is related to *Thecopsora* (i.e. *Pucciniastrum*), because the host plants of both genera are found usually in the *Ericaceae*.

Savile (Can. J. Res. **28**, 318, 1950) has described the North American species and has emphasised the importance of the size, dimensions and spacing of the warts on the aecidiospores and uredospores in distinguishing the species.

Chrysomyxa abietis Unger

Beitr. Vergl. Path. p. 24 (1840); Wilson & Bisby, Brit. Ured., no. 6; Gäumann, Rostpilze, p. 101.
Blennoria abietis Wallr., Allg. Forst-Jagdzeit. **17**, 65 (1834).

Spermogonia, aecidia and uredosori absent. Teleutosori hypophyllous, on yellow or orange spots, elongate, 0·5–10 mm. long, 0·3–0·5 mm. broad, 0·5 mm. high, orange to reddish-brown; teleutospores in chains 70–100μ long, oblong, 20–30 × 10–15μ, wall, hyaline, smooth, 1μ thick. Micro-form. On needles of *Picea abies*, *P. rubens* and *P. sitchensis*, March–May. England, Scotland, Northern Ireland, scarce.

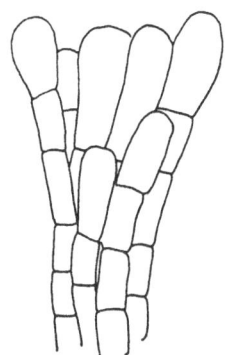

C. abietis. Teleutospores.

This rust was first recorded by Somerville from Durris, near Aberdeen, in 1911 (Quart. Journ. For. **5**, 277), and in 1915 (*ibid*. **9**, 68) the same author stated that he had received no further records of its occurrence. In 1915 Trail stated: 'It is not more than six or seven years ago that I first observed *Chrysomyxa abietis* in Aberdeenshire, and it is probable that the fungus has only recently made its appearance in the north of Scotland' (Trans. Roy. Scot. Arbor. Soc. **29**, 187, 1915). In 1915 it was stated to be quite common at Novar in Ross-shire. It appears that the rust has spread rapidly in Scotland; in England it was recorded from Northumberland, and from Devon and Cornwall in 1953 (*in litt.*). In 1934 it was found near Kelso, sparingly on *P. sitchensis* in the vicinity of infected *P. abies* and in 1952 on *P. rubens* from Argyll. In some seasons it may cause considerable defoliation of spruce.

Infection is heaviest when a cold spring delays shoot development so as to coincide with maturation of the teleutospores (Murray, For. Comm. Booklet, **4**, 1953).

In 1938 it was recorded in Norway by Jørstad (Nytt Mag. Naturv. **78**, 153, 1938) on *P. sitchensis*, *P. pungens* and *P. engelmannii*.

Very complete accounts of the development of the sorus were given by both Kursanov (1917) and Lindfors (1924). The general mycelium is uninucleate and the binucleate condition arises by fusion of cells at the base of the sorus. Usually only the spores in the upper part of the chains function and these, on germination, give rise to a 4-celled basidium. The teleutospores germinate without any period of rest. In the early stages of development when only the striking transverse yellow bands on the needles are visible the fungus may be confused with *Chrysomyxa rhododendri* but in the latter spermogonia soon develop and in the later stages the aecidia are readily distinguished.

Chrysomyxa empetri Cummins

Mycol. **48**, 602 (1956).

[*Chrysomyxa empetri* [Pers.] Schroet., Cohn, Krypt. Fl. Schles. **3** (1), 372 (1887)]; Grove, Brit. Rust Fungi, p. 311; Wilson & Bisby, Brit. Ured. no. 7; Gäumann, Rostpilze, p. 99.
[*Uredo empetri* Pers. ex DC., Fl. Fr. **6**, 87 (1815).]

[**Spermogonia** amphigenous on needles of current season, uniseriate, conspicuous, yellowish then reddish-brown, aparaphysate, subepidermal, 140–160μ broad, 100–135μ deep; spermatia subgloboid to ellipsoid 5–7 × 5·5–10μ, extruded in a sticky liquid. **Aecidia** amphigenous on needles of current season, uniseriate, on pale-yellowish portions, elliptical to subcircular in transverse section, 0·5–1·5 mm. wide, 0·5–2 mm. high; peridium colourless, rupturing at apex; peridial cells polygonal, elongate vertically, 19–54 × 32–76μ, outer walls smooth, about 1μ thick, inner walls coarsely verrucose, 4–5μ thick; aecidiospores yellow, ellipsoid or ovoid, 21–34 × 30–47(55)μ, wall closely and coarsely verrucose, 0·3–1·5μ thick excluding the warts.] **Uredosori** epiphyllous, one or few on a leaf, pustular, subepidermal, circular or elliptical to linear, 0·2–2 mm. long; peridium distinct, adhering to the epidermis which ruptures at maturity, 15–17μ thick; peridial cells in a single layer, angular, 10–20μ diam., wall 3–4μ thick; uredospores orange, catenulate, pulverulent, ellipsoid, ovoid or subgloboid, 25–49 × 20–31μ; wall hyaline, closely and coarsely verrucose, 0·2–1·0μ

thick, excluding the warts, warts cylindrical to slightly stellate or irregular, 0·7–2·2μ high × 0·3–1·0μ wide, 0·7–2·5(3·0)μ spacing. [**Teleutosori** epiphyllous on overwintered leaves, one or few on a leaf, yellow, cushion-shaped, waxy, subepidermal, subcircular to elongate, often nearly as long as the leaf; teleutospores catenulate, 3–6 in a chain, smooth, thin-

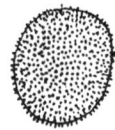

C. empetri. Uredospore.

walled, contents yellow, 19–24 × 18–21μ; basidia 4-celled, pale yellow, up to 65μ long, 7–8μ diam.; basidiospores with yellow contents, very thin-walled, subgloboid to ellipsoid, 10–15μ diam., usually about 12μ.] Hetereu-form.
[Spermogonia and aecidia on several species of *Picea* in North America]; uredospores throughout the summer and autumn on *Empetrum hermaphroditum* and *E. nigrum*; [teleutospores in spring and early summer]. North Wales, Scotland and Ireland, usually scarce but locally abundant in some seasons.

Only the uredospore stage has been found in this country. The specimen on *E. hermaphroditum* was collected by M. R. Gibson on Cairngorm in 1938.

Teleutospores have been recorded from Greenland (Rostrup, Medd. Grøen. **3**, 536, 1888; Lind, 1913, 280), from Norway by Lagerheim (Tromsö Mus. Aarsh. **16**, 107 and 119, 1893) and by Jørstad (1935, 50) and from North America by Faull (J. Arn. Arb. **18**, 141, 1937). Faull (*loc. cit.*) also described the aecidial stage and carried out successful infections on *Picea glauca* and *P. rubens* with teleutospores and on *E.*

nigrum with aecidiospores. Savile (Can. J. Res. **28**, 322, 1950) has given a more recent account of its occurrence in northern Canada. The leaves of *Empetrum nigrum* are recurved forming a cavity on the under side into which the stomata open. The sori are borne on the morphological upper surface and as this bears no stomata, their position is exceptional (Fischer, Mitt. Naturf. Ges. Bern. **1915**, 231, 1916). According to Lindfors (1924) the mycelium in *Empetrum* is binucleate. The rust is widely distributed in the north temperate region and occurs on *Empetrum rubrum* in the Falkland Islands (Arwidsson, Sv. Bot. Tidskr, **30**, 401, 1936).

Chrysomyxa pirolata Wint.

Rabh. Krypt. Fl. Ed. 2, **1** (1), 250 (1882); Gäumann, Rostpilze, p. 103.

[*Aecidium ? pyrolae* DC., Fl. Fr. **6**, 99, (1815).]
[*Caeoma pyrolatum* Schw., Trans. Amer. Phil. Soc. N.S. **4**, 294 (1834).]
[*Uredo pirata* Körnicke, Hedwigia, **16**, 28 (1877).]
Chrysomyxa pyrolae [DC.], Rostr. Bot. Z bl. **5**, 127 (1881); Grove, Brit. Rust Fungi, p. 312; Wilson & Bisby, Brit. Ured. no. 8.

[**Spermogonia** epiphyllous on cone scales, numerous, subepidermal, flat, 0·5–1 mm. wide, 50–100 µ high, inconspicuous. **Aecidia** amphigenous on cone scales, large, oblong or irregular in shape, often confluent and then 0·5–1 cm. diam., one to few on each scale, bullate, with a convex, evanescent peridium, white or yellowish, pulverulent; aecidiospores ellipsoid, orange, 22–37(–46) × 17–35 µ, wall hyaline, 2·0–4·7 µ thick, including warts, the numerous, large, crowded warts, polygonal to elongate, 1·5–4·7 µ wide × 2·0–8·0(–10·0) µ long]. **Uredosori** hypophyllous, often covering the whole surface uniformly, circular, 0·5–1 mm. diam., soon naked, surrounded by the torn epidermis and a very delicate evanescent peridium, yellow to orange; uredospores ellipsoid or obovoid, 19–33 (–36) × 13–24 µ, wall excluding warts 0·4–1·0 µ thick, hyaline, warts circular to elongate, entire, 0·5–1·8 µ high × 0·5– 2·0 µ wide to 4 µ long, spacing very vari-

able, about 1·0–3·2 µ. **Teleutosori** hypophyllous, prominent, flat, covering the whole leaf surface uniformly, waxy,

C. pirolata. Uredospores.

yellowish-red, turning to blood-red, brown when dry; teleutospores irregularly oblong or ellipsoid, 7–10 × 14–26 µ, in a series 100–130 µ long, wall colourless, uniformly 1 µ thick or less, smooth. Hetereu-form.

[Spermogonia and aecidia on the cone scales of *Picea abies* in Europe and on several species in North America]; uredosori on the leaves, petioles and petals of *Pyrola minor*, *P. rotundifolia* and its subsp. *maritima*, April–August; teleutosori, June. Great Britain, rare.

C. pirolata has been recorded in the uredospore stage from several localities in north-east Scotland and is abundant on *P. rotundifolia* subsp. *maritima* near Southport, Lancashire. Teleutospores are rarely formed in this country but were found by Alex. Smith near Cullen, Banffshire, in 1922 (Wilson, 1934, 410). The aecidial stage has not been found in this country but occurs in Central Europe on the cone scales of *Picea abies* and in North America on *P. glauca* and *P. mariana*.

Rostrup, Arthur and Kern expressed the view that this species was heteroecious and the connection was demonstrated by Fraser who sowed basidiospores on the cones of *Picea mariana* in Nova Scotia (Mycol. 4, 183, 1912) and on *P. glauca* (*ibid.* 17, 84, 1925) and obtained the aecidia.

The mycelium is systemic in both phases; it is perennial only in *Pyrola* where it maintains itself through the winter in the buds and leaves, and so is independent of alternation (Liro, Acta Soc. Fauna Fl. Fenn. 29 (6), 20, 1906).

Kursanov (1922) pointed out that the uredosorus closely resembles an aecidium, the uredospores being developed in chains with intercalary cells; it differs, however, from a typical aecidium in being covered by two layers of thin-walled cells, interpreted by Kursanov as peridium.

There has been some confusion between this species and *Pucciniastrum pyrolae* (see p. 37) but in the latter the uredosori are scattered and there is no distinct peridium. The record of *C. pirolata* for Edinburgh quoted by Grove is probably based on that given in the English Flora, 5 (9), p. 378, by Berkeley on specimens collected by Greville; these are in the Edinburgh Herbarium and are *Pucciniastrum pyrolae*.

Jørstad (K. Norske Vidensk. Selsk. Skr. 38, 1935) described it as epidemic in some seasons in Norway, where it may substantially reduce the yield of spruce seed, he had earlier suggested that the aecidium may be able to repeat itself.

The rust on *Ramischia secunda* described by Lagerheim as *Chrysomyxa ramischiae* differs in having primary and secondary uredosori. It has not been recorded in Britain.

Chrysomyxa rhododendri de Bary

Bot. Zeit. 37, 809 (1879); Grove, Brit. Ured. p. 384; Wilson & Bisby, Brit. Ured. no. 9; Gäumann, Rostpilze, p. 94.
[*Uredo rhododendri* DC., Fl. Fr. 6, 86 (1815).]
Chrysomyxa ledi (Alb. & Schw.) de Bary, var. *rhododendri* (de Bary) Savile, Can. J. Bot. 33, 491 (1955).

Spermogonia amphigenous, numerous, inconspicuous, 110–115μ diam. 100–125μ high, honey-coloured then reddish-brown. **Aecidia** amphigenous, erumpent on transverse, yellowed zones of the leaves, elongate parallel to the midrib, up to 3 mm. long; peridium membranaceous, at length irregularly torn, white; aecidiospores more or less ellipsoid, yellow, 24–28 × 20–22μ, wall 1μ thick, densely verruculose, except for a longitudinal smooth zone. Uredosori usually hypophyllous, often on yellowish spots, scattered or in small groups, confluent in heavy infections, 0·5–2 mm. diam., pulverulent, spore mass orange; uredospores 22–27 × 16–22 μ, wall 1·5 μ thick, finely verruculose except for longitudinal smooth patch, pores indistinct. Teleutosori hypophyllous forming as laccate brownish-red areas 0·5–2 mm. diam.; teleutospores unicellular, superposed 6–10 one above the other, cylindric, 20–30 × 10–14 μ, wall, thin, smooth, colourless. Hetereu-form.

Aecidia on *Picea abies* and less commonly on *P. sitchensis*, July to September; uredospores on *Rhododendron ciliatum × glaucophyllum, R. cinnarbarinum × maddeni, R. concatenans, R. ferrugineum, R. hippopheaeoides × racemosum, R. hirsutum, R. keleticum, R. oolotrichum, R. ponticum* (II rare), *R. spiciferum, R. yunnanense × roylei* and *Azaleodendron*, 'Dr Masters' (*R. japonicum* hybrid), throughout the summer; teleutospores abundant on *R. ponticum*

and less common on other species, October–July. In scattered localities in England, Scotland and Ireland. On *R. ponticum* in Scotland and Ireland only.

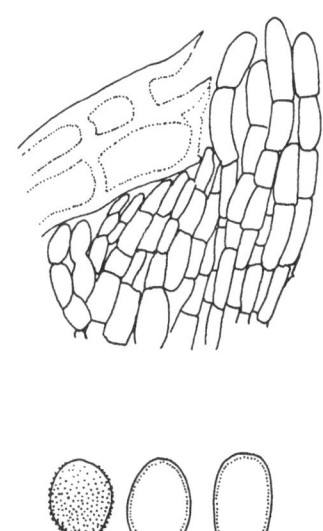

C. rhododendri. Teleutosorus and uredospores.

This is an introduced species first recorded by Boyd who found the uredospore and teleutospore stages on *Rhododendron hirsutum* at Douglas Castle, Lanarkshire in 1913, the aecidial stage was found in the Solway area (Trans. Bot. Soc. Edin. **26**, xxxiii, 1914) and an account of the rust was published in 1915 (Trans. Roy. Scot. Arbor. Soc. **29**, 187, 1915). Specimens of the aecidial stage were collected by Trail in Aberdeenshire in September 1916 (J. Bot. Lond. **55**, 135, 1917). Specimens of *Picea abies* heavily infected with the aecidial stage were received from Northern Ireland in 1924. It was found on *Rhododendron* in Cheshire in 1930 and on *R. roylei* in Cornwall in 1937 (Ashworth, J. Roy. Hort. Soc. **63**, 487, 1938) and has been recorded on many species since then.

At the only known locality for the aecidia on *Picea sitchensis*, *P. abies* is also infected and the diplont host is *Rhododendron ponticum*.

On *R. ponticum* the teleutospores germinate in June or July just as the flush of young spruce leaves appear and the spermogonia and aecidia appear on the spruce from that time onwards. The aecidiospores can at once infect the Rhododendron leaves, where the mycelium winters, producing its spores in the following spring. Infection is particularly heavy on the leaves of the young soft shoots growing up from cut-back plants of *R. ponticum*. Teleutosori occur on leaves of *R. ponticum* several years old.

On Rhododendrons, other than *R. ponticum*, uredospores predominate and on many teleutospores have not been observed; infected spruce are not usually associated and the rust is clearly uredo-perennating and independent of alternation. On *R. ponticum*, on the contrary, uredosori often are not produced or are sparse and the rust is obviously obligatorily heteroecious. Furthermore, *R. ponticum* has been found infected only in close proximity to infected spruce. Natural stands of *R. ponticum* are not recorded as being attacked by any *Chrysomyxa* in Asia Minor nor are there records of infection elsewhere.

Chrysomyxa rhododendri is very common in Europe wherever spruce and *R. hirsutum* or *R. ferrugineum* grow in proximity. Its host relations were first investigated by de Bary (*loc. cit.*).

Melampsora Cast.

Obs. Pl. Acotyl. **2**, 18 (1843).

Spermogonia subcuticular or subepidermal, conical or hemispherical, without paraphyses but sometimes with flexuous hyphae. **Aecidia** caeomoid usually foliicolous, without peridium or sometimes with peripheral, paraphysis-like hyphae which may unite to form a rudimentary peridium, aecidiospores catenulate, globoid or ellipsoid, with verrucose walls. **Uredosori** subepidermal, pulverulent, with a thin, evanescent peridium; uredospores borne singly on pedicels, globoid or ellipsoid, with indistinct pores; with capitate or clavate paraphyses. **Teleutosori** subcuticular or subepidermal, forming crusts consisting of a single layer of spores; teleutospores 1-celled, adhering laterally, with coloured walls, with one indistinct apical pore, germinating in spring; basidia typically 4-celled, producing globoid, colourless or yellowish basidiospores.

Melampsora is often regarded as intermediate in character between the *Melampsoraceae* and *Pucciniaceae*. All British species are macrocyclic (except *M. hypericorum*); heteroecious when the teleutospores are on woody plants (with one exception, *M. amydalinae*) and autoecious when on herbaceous plants.

Melampsora lini (Ehrenb.) Lév.

Ann. Sci. Nat. Bot. Ser. 3, **8**, 376 (1847); Wilson & Bisby, Brit. Ured. no. 46; Gäumann, Rostpilze, p. 193.

[*Uredo miniata* var. *lini* Pers., Syn. Meth. Fung. p. 216 (1801).]
[*Uredo lini* Schum., Enum. Pl. Saell. **2**, 230 (1803).]
Xyloma lini Ehrenb., Sylvae Myc. Berol. p. 27 (1818).
Melampsora lini Desm., Pl. Crypt. no. 2049 (1850); Grove, Brit. Rust Fungi, p. 355.

Var. *lini*

Uredosori amphigenous and on the stems, subepidermal, scattered or grouped, often crowded, rounded, 0·3–0·75 mm. diam. or on the stems, up to 1·5 mm. long, at first covered with a parenchymatous peridium but soon naked, surrounded by the ruptured epidermis, orange: paraphyses capitate, 40–64 μ long, 15–25 μ wide, wall much thickened above; uredospores globoid or ellipsoid, contents orange-yellow, 16–25 × 13–20 μ; wall about 2 μ thick, finely echinulate-verruculose, colourless, pores scattered. **Teleutosori** amphigenous or on the stems, scattered or grouped, confluent and elongate, subepidermal, at first reddish-brown, then black and shining; teleutospores cylindric-prismatic, 35–55 × 10–20 μ, wall about 1 μ thick throughout, brown. Hemi-form.

On *Linum catharticum*, May–October. Great Britain and Ireland, frequent.

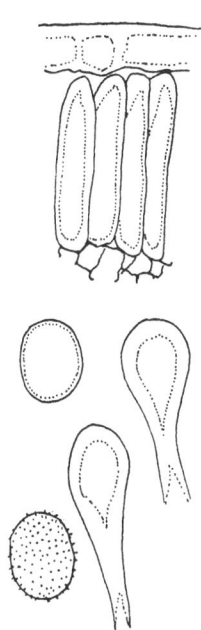

M. lini. Teleutosorus, uredo-paraphyses and uredospores.

The type variety occurs almost everywhere the host is to be found. It is strongly specialised to *L. catharticum* (Buchheim, Ber. D. Bot. Ges. **33**, 73, 1915).

There appears to be no record of spermogonia or aecidia in this country. In Britain it overwinters by uredosori as in Norway (Jørstad, 1960, 126).

var. *liniperda* Körnicke

Land.-Forstwirtsch. Z. Prov. Preussen, p. 10 (1865).
M. lini var. *major* Fuck., Jahrb. Nass. Ver. Nat. **23-4**, 44 (1869).
M. liniperda (Körnicke) Palm, Sv. Bot. Tidskr. **4**, 4 (1910).

Spermogonia amphigenous, often enclosed within the annular aecidial groups, subepidermal, numerous, scattered, inconspicuous, pale yellow or orange, flattened globoid and bulging from the leaf as small papillae; spermatia 3–5 × 2–4 μ, extruded in the form of a tendril, without paraphyses but with flexuous hyphae, 0·7 mm. long and about 2 μ wide. **Aecidia** amphigenous, usually hypophyllous, prominent, in crescentic, annular or rounded groups, 0·3–0·5 mm. diam., orange, with a rudimentary peridium, soon becoming naked and surrounded by ruptured epidermis, aecidiospores globoid or subgloboid, contents pale orange, 21–28 × 10–27 μ, wall finely verruculose, about 1 μ thick. **Uredosori** and **teleutosori** as in var. *lini* but teleutospores up to 60–80 μ long. Auteu-form.

On cultivated *Linum usitatissimum*, June–October. Great Britain and Ireland.

5

Uredospores collected in Scotland in 1946 on self-sown seedlings in a field which had borne a diseased crop in the previous year, are rather larger than those of var. *lini* averaging $27 \cdot 5 \times 15\mu$, with some up to 32μ long.

Arthur (J. Mycol. **13**, 201, 1907) was the first to establish the full life history of this variety when he produced spermogonia and aecidia on *Linum usitatissimum* and *L. lewsii* by infection with teleutospores from the former species. Spermogonia and aecidia were produced in Ireland in 1921 by infection of cultivated flax with basidiospores from over-wintered teleutospores (Pethybridge, Lafferty & Rhynehart, J. Dept. Agric. Ireland, **21**, 175, 1921). Palm (*loc. cit.*) showed that uredospores from *L. catharticum* would not infect *L. usitatissimum*; this was confirmed in Ireland in 1920 (Pethybridge & Lafferty, J. Dept. Agric. Ireland, **20**, 334, 1920 and **22**, 103, 1922) but it was found that uredospores from *L. usitatissimum* produced slight infection of *L. angustifolium* (= *L. bienne*). Voglino (Bull. Soc. Bot. Ital. **1896**, 38, 1896) distinguished var. *viscosi* Vogl. on *L. viscosum* which possesses teleutospores $55-70 \times 20-22\mu$.

The wild species of *Linum* occurring in Europe do not appear to act as hosts for the rust occurring on cultivated flax; specialised forms have been distinguished by Buchheim (Ber. D. Bot. Ges. **33**, 73, 1915) on *L. alpinum*, *L. catharticum*, *L. tenuifolium* and *L. strictum*. In Australia a wild species, *L. marginale*, is susceptible to the rust from some cultivated varieties and acts as a perennial host (Waterhouse & Watson, J. Roy. Soc. N.S.W., **77**, 138, 1943).

Considerable work has been done on the genetics both of this fungus and of its hosts. In a series of recent papers Flor has shown that hybridisation in the aecidium produced new races and that the host-parasite relation can be interpreted on a gene-for-gene model. In Britain, Colhoun (Ann. Appl. Biol. **35**, 582, 1948) has recorded the resistance or susceptibility of fibre and of linseed varieties in Northern Ireland and Arif (TBMS. **37**, 353, 1954) isolated sixteen physiologic races from British collections.

Fromme (Bull. Torr. Bot. Club, **39**, 113, 1912) has made detailed cytological study of this rust and Moss (Mycol. **21**, 79, 1929) has described the development of the uredosorus: Allen (J. Agric. Res. **49**, 765, 1934) has given an account of the spermogonia and aecidia and studied its heterothallism.

Melampsora vernalis Wint.

Rabh. Krypt. Fl. Ed. 2, **1** (1), 237 (1882); Grove, Brit. Rust Fungi, p. 357; Wilson & Bisby, Brit. Ured. no. 51; Gäumann, Rostpilze, p. 188.
[*Caeoma saxifragarum* Schlecht., Fl. Berol. **2**, 121 (1824).]

Spermogonia scattered, subepidermal, honey-coloured then brownish, up to 120 μ high, 140–220 μ wide. **Aecidia** hypophyllous or on the calyx, scattered or spread equally over the leaf surface, subepidermal, minute, round, soon naked, pulverulent, yellowish-orange; aecidiospores angular-globoid to ellipsoid, with orange contents, 16–26 × 12–22 μ, wall densely and minutely verrucose, 2–2·5 μ thick. **Teleutosori** amphigenous, mostly hypophyllous and on the stems, irregularly or densely grouped, subepidermal, minute, punctiform, at

first yellowish then brown or blackish-brown; teleutospores irregular, ellipsoid, irregularly prismatic or clavoid, brown,

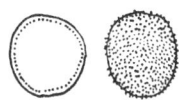

M. vernalis. Uredospores.

35–55 × 13–18 μ, wall 1·5–2 μ, at apex up to 3 μ thick. Opsis-form.
On *Saxifraga granulata*, June–September. Scotland, rare.

This was recorded by Plowright (Gard. Chron. **8**, 41, 1890; J. Roy. Hort. Soc. **12**, cxi, 1890; TBMS. **1**, 59, 1898) from specimens collected at Portlethen, Kincardine, Scotland, by James Taylor in June 1890. Teleutospores were present on the lower leaves and stems. It had been previously reported in Scotland by Greville and Trail. Grove doubtfully included uredospores in the life history of this species after the account of Voglino (Bull. Soc. Bot. Ital. **1896**, 39, 1896) and Fischer (1904, 511). Uredospores have not been found in this country nor does Jørstad (1960, 126) record them and it is doubtful if they ever occur. Plowright, Dietel (Forstl. Naturw. Z. **4**, 374, 1895) and Klebahn (Z. Pfl.-Krankh. **22**, 339, 1912) obtained teleutospores without the formation of uredospores by infecting *S. granulata* with aecidiospores.

Paraphyses-like cells, which closely resemble those present in some races of *Melampsora epitea* (see p. 80) project from the marginal pseudoparenchyma of the aecidium; they have been described and figured by Lindfors (Sv. Bot. Tidskr. **4**, 201, 1910).

Melampsora euphorbiae Cast.

Obs. Pl. Acotyl. **2**, 18 (1843); Grove, Brit. Rust Fungi, p. 353; Wilson & Bisby, Brit. Ured. no. 40; Gäumann, Rostpilze, p. 180.
Uredo euphorbiae-helioscopiae Pers., Syn. Meth. Fung. p. 215 (1801).
Xyloma (Placuntium) euphorbiae Schub., Fic. Fl. Dres. **2**, 31 (1823).
Melampsora euphorbiae-dulcis Otth, Mitt. Naturf. Ges. Bern, **1868**, 70 (1868); Wilson & Bisby, Brit. Ured. no. 41; Gäumann, Rostpilze, p. 183.
Melampsora helioscopiae Wint., Rabh. Krypt. Fl. Ed. 2, **1** (1), 240 (1882); Gäumann, Rostpilze, p. 179.

Spermogonia subepidermal, flattened-hemispherical, orange. **Aecidia** caeomoid, on leaves, amphigenous, up to 0·5 mm. diam., on the stems, up to 4 mm. long, orange, without paraphyses; aecidiospores globoid to ellipsoid, 21–28 × 19–24 μ, wall densely verrucose, 1·5 μ thick. **Uredosori** amphigenous, generally hypophyllous, rarely on the stems, scattered, minute, up to 500 μ diam., yellow, paraphyses, numerous, capitate, 14–16 μ wide at the apex; uredospores globoid to ellipsoid, with yellow contents 15–22 × 12–20 μ, wall more or less densely echinulate, colourless, without perceptible pores. **Teleutosori** subepidermal, amphigenous and on the stems, minute, often confluent and elongate on the stems, reddish-brown then black; teleutospores cylindrical-prismatic, not thickened at the apex, 18–60 × 7–15 μ, wall reddish-brown or brown, 1·5–3 μ thick. Auteu-form.

On *Euphorbia amygdaloides*, *E. cyparissias*, *E. dulcis*, *E. exigua*, *E. helioscopia*, *E. hyberna*, *E. paralias* and *E. peplus*.

Spermogonia and aecidia doubtfully recorded; uredospores and teleutospores

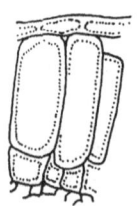

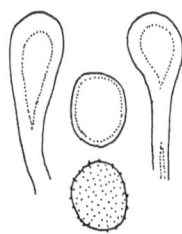

M. euphorbiae. Teleutosorus on *E. peplus*; uredo-paraphyses and uredospores.

common on *E. helioscopia* and *E. peplus* otherwise scarce, often uredospores only.

Melampsora euphorbiae sensu lato is regarded as a collective species made up of a number of races or special forms which, in part, show slight morphological differences; some of these have been regarded as species by previous investigators. Müller (Zbl. Bakt. II, **19**, 441 and 544, 1907) after a detailed morphological investigation of these forms and after carrying out a number of cultures concluded that it was impossible to establish strictly limited, morphological species from these forms and created instead five collective types. The Sydows, in their Monograph (375, **3**, 1915), recognised these types to a considerable extent as provisional species and this procedure has been adopted by other investigators. In most of these forms the aecidial stage has been found only rarely and its occurrence in this country is doubtful.

The following forms have been recorded in Great Britain and Ireland.

On *E. amygdaloides* found once, at Haslemere in 1932 (TBMS. **18**, 13, 1933) and assigned to *M. euphorbiae*. Müller found only uredospores on this species and consequently did not name this rust.

On *E. cyparissias* in the uredospore condition only, in Cumberland, at St Andrews and in Norfolk. This has been described by Müller as *M.*

euphorbiae f. sp. *euphorbiae-cyparissiae* and he stated that it occurs only on *E. cyparissias*. Dietel (Forstl. Naturw. Z. **4**, 374, 1895) produced spermogonia and aecidia on *E. cyparissias* by infecting with teleutospores and Jacky (Ber. Schweiz. Bot. Ges. **9**, 27, 1899) stated that he obtained the uredospores by infection with the basidiospores without the intervention of the aecidial stage.

On *E. dulcis* as *M. euphorbiae-dulcis* found by J. C. Arthur at Glengariff, Eire, and recorded by Grove and Chesters (TBMS. **18**, 265, 1933). Only uredospores and teleutospores were found. This form is distinguished by its small, ovoid to cylindrical teleutospores (18–40 × 7–15 μ). The connection of the different spore stages on this host was first proved by Dietel (Oest. Bot. Zeit. **39**, 356, 1889).

On *E. exigua* as *M. euphorbiae* f. sp. *euphorbiae-exiguae* which occurs only on this species. Müller has described the aecidial stage. It was recorded by Purton (Mid. Flora, **3**, 298, 1821) in 1821 and occurs occasionally in the Midlands and south of England.

On *E. helioscopia* as *M. helioscopiae* is commonly found in Great Britain and Ireland. The Sydows (*loc. cit.*) separated it on account of its larger teleutospores (40–60 × 7–12 μ). It has been shown by Müller that it can attack only *E. helioscopia*. According to Klebahn (Z. Pfl.-Krankh. **17**, 153, 1907) teleutospores sown on the same host produced spermogonia and aecidia.

The form on *E. hyberna* which has been recorded from Devon by Hadden (J. Bot. Lond. **54**, 54, 1916) and from Killarney (TBMS. **22**, 10, 1936) has not been studied by any worker and its relations are uncertain.

A form on *E. paralias* has been collected on the coast of Cardigan, Wales and near Bristol (TBMS. **39**, 391, 1956); only uredospores have been found. It has also been found on the coast of Belgium (Bommer & Rousseau, Bull. Soc. Bot. Belg. **25**, 166, 1886) in the uredospore stage only; nothing appears to be known of this form.

On *E. peplus* f. sp. *euphorbiae-pepli* has been described by Müller (*loc. cit.*); this has been recorded frequently in England and Ireland. Pieschel (Phytopath Zeitschr. **7**, 393, 1934) described the occurrence of an albino of this form in Germany in which the uredospores, germ-tubes and paraphyses are completely destitute of orange colouring matter; the teleutospores possessed the normal dark-brown wall.

Treboux (Ann. Myc. **10**, 306, 1912) found that uredospores on *E. glareosa* germinated readily in early spring and concludes that the rust can persist through the winter by this method; it is probable that many of the British forms perennate in the same way.

The development of the uredosori and teleutosori on *E. dulcis* has been described by Sappin-Trouffy (1897, 160) and on *E. uralensis* and *E. lucida* by Kursanov (1922, 94).

Melampsora hypericorum Wint.

Rabh. Krypt. Fl. Ed. 2, **1** (1), 241 (1882); Grove, Brit. Rust Fungi, p. 354; Wilson & Bisby, Brit. Ured. no. 42; Gäumann, Rostpilze, p. 185.

Caeoma hypericorum Schlecht., Fl. Berol. **2**, 122 (1804).

[*Uredo hypericorum* DC., Mem. Soc. Agric. Paris, **10**, 235 (1807) and Fl. Fr. **6**, 81 (1815).]

Mesopsora hypericorum [DC.] Diet., Ann. Myc. **20**, 30 (1922).

Spermogonia and **uredosori** unknown. **Aecidia** hypophyllous, caeomoid, with yellow or orange spots on upper surface, rounded or oblong, flatly pulvinate, 0·3–0·5 mm. diam., soon naked, orange; aecidiospores in short chains, without paraphyses, ellipsoid to polygonal or subclavate, contents pale orange, 18–28 × 10–18 μ; wall colourless, tesselate, warted, about 2 μ thick with no perceptible pores. **Teleutosori** hypophyllous, subepidermal, rounded, about 0·1 mm. diam., reddish-brown then dark brown; teleutospores cylindric-prismatic, more or less rounded above, 28–40 × 10–17 μ, wall thickened above up to 3 μ. Opsis-form.

On *Hypericum androsaemum*, *H. hirsutum*, *H. maculatum*, *H. perforatum*, *H. pulchrum*, *H. tetrapterum* and on cultivated *H. calycinum*, *H. cernuum*, *H. elatum* and *H. patulum*. Great Britain and Ireland, May–October. Frequent on *H. androsaemum*, scarce on *H. pulchrum* and *H. perforatum* and occasionally found on the other host species. Both in England and Ireland the records are generally from the south-western counties.

In this rust Plowright (1889, 243) considered that the first formed spores were uredospores but Gobi and Tranzschel (Scripta Bot. Hort. Univ. Imp. Petrop. **3**, 103, 1891) regarded them as aecidiospores, since they are in chains with intercalary cells and without paraphyses and in this they were followed by Fischer (1904, 506) Grove and Sydow (Monogr. Ured. **3**, 384). Dietel (*loc. cit.*) considered that they are uredospores since they are in chains, without paraphyses and possess tesselated, warted walls and he removed the species to the new genus *Mesopsora*; he also suggested that the aecidial stage should occur on one of the *Abietineae*. Later (Ann. Myc. **39**, 156, 1941) he reverted to the opinion that the first-formed spores are aecidiospores and suggested that they are borne in repeating aecidia; he pointed out that the spores resemble the aecidiospores of other species of *Melampsora* on *Larix*. He has suggested that this is a heteroecious species which produced its primary aecidia with spermogonia on one host (possibly *Larix*) and its secondary aecidia (without spermogonia) and subsequent spore stages on *Hypericum*. Jørsted (1953, 115) considered it to be an opsis-form with repeating aecidia with spermogonia suppressed. In north-west Scotland viable sori can be found on *H. androsaemum* throughout the year.

Grove recorded *H. humifusum* as a host but gave no information regarding specimens in this country; none is present in the Plowright or Grove herbaria in the British Museum. In the Kew Herbarium there is an unnamed specimen collected in 'Edinburgh' by Leighton which might possibly have been on this host but this has been recently determined to be on a prostrate form of *H. pulchrum*; specimens on this form have recently been found in Scotland, in Sutherland, and near Peebles. There appears to be no evidence of infection of *H. humifusum* in this country.

Klebahn (Z. Pfl.-Krankh. **15**, 106, 1905) failed to obtain infection on various species of *Hypericum* and on *Abies alba*, *Picea excelsa* and *Larix decidua* with overwintered teleutospores from *H. hirsutum*. Müller (Zbl. Bakt. **17**, 210, 1907) considered the form on *H. montanum* to be a biological race since he could not infect other species of *Hypericum* with spores from it.

In Edinburgh infected plants of *H. androsaemum* and *H. elatum* have been grown for several years in close proximity to *H. perforatum*, *H. pulchrum*, *H. humifusum* and *H. calycinum* but none of these species has become infected; aecidiospores from *H. elatum* also failed to infect *H. androsaemum*. Moore (Bull. Min. Agric. **139**, 80, 1947) noted that rust on *H. androsaemum* from Ross, Hereford did not attack *H. calycinum*, *H. patulum*, *H. henryi* and *H.* × *moserianum* in the same shrubbery.

Melampsora allii-populina Kleb.

Z. Pfl.-Krankh. **12**, 25 (1902); Grove, Brit. Rust Fungi, p. 347; Wilson & Bisby, Brit. Ured. no. 34; Gäumann, Rostpilze, p. 137.
[*Caeoma alliorum* Link, Sp. Pl. **2**, 7 (1825), *p.p.*]
[*Uredo ari-italici* Duby, Bot. Gall. **2**, 899 (1830).]
[*Caeoma ari-italici* [Duby] Wint., Rabh. Krypt. Fl. Ed. 2, **1** (1), 256 (1882)]; Grove, Brit. Rust Fungi, p. 388.

Spermogonia yellowish, pulvinate, about 140μ wide and 100μ high. **Aecidia**, caeomoid, in groups on yellowish spots, about 1 mm. wide, surrounded by the epidermis and a rudimentary peridium, bright orange-red; aecidiospores globoid, ovoid or angular-globoid, 17–23 × 14–19μ, wall about 2μ thick, but sometimes thicker and then obviously thin at certain spots, verruculose.] **Uredosori** hypophyllous, rarely epiphyllous, on yellowish spots, round, scarcely 1 mm. wide, bright reddish-orange, paraphyses mostly capitate with thin stalks, 50–60 × 14–22μ, wall uniformly 2–3μ thick; uredospores oblong or clavoid, rarely ovoid, 24–38 × 11–18μ, wall 2–4μ thick, sometimes thicker at one end but without equatorial thickening, with indistinct scattered pores, distantly echinulate, but smooth at apex. **Teleutosori** hypophyllous, subepidermal, scattered over the leaf singly and in groups, 0·25–1 mm. diam., pulvinate, blackish-brown, not shining, teleutospores prismatic, rounded at both ends, 35–60 × 6–10μ, wall

light brown, 1–1·5μ thick, scarcely
thickened above. Hetereu-form.
['Spermogonia and aecidia on *Allium
ursinum* and *Arum maculatum*', May];
uredospores and teleutospores on *Populus nigra* and *P. trichocarpa* and possibly
other *Populus* species, cf. list under
Melampsora larici-populina, p. 74.

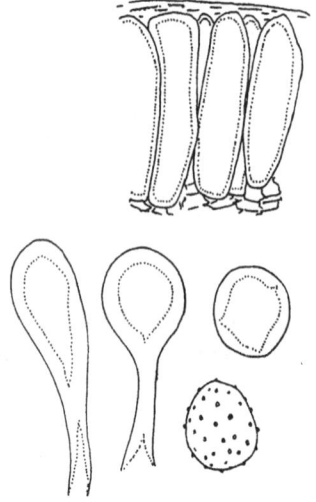

M. allii-populina. Teleutosorus, uredo-
paraphyses and uredospores.

Klebahn (Z. Pfl.-Krankh. **12**, 22, 1902) first showed that the aecidium
of this rust is produced on species of *Allium*. Liro (Acta Soc. Fauna Fl.
Fenn. **29** (7) 9, 1907) has reported the presence of a rudimentary
peridium around the aecidiospores.

The aecidial stage of Melampsoras appear to be rare in this country
on *Allium* hosts, and there are no specimens which can be unequivocably referred to any of the three species concerned, *M. allii-populina*,
M. allii-fragilis and *M. salicis-albae*. There are only a few unplaceable
specimens and many unsatisfactory or incorrect records. The status of
all three is doubtful in Britain and requires investigation. In the Kew
Herbarium there is a specimen of a caeomoid aecidium on *A. ursinum*
collected by J. W. Ellis in Merioneth which belongs to one of this group.
Cooke (Micro. Fungi, Ed. 4, p. 217), stated that *Caeoma alliorum* was
common but his specimen of *Trichobasis alliorum* from Edinburgh (No.
425, Fung. Brit. Exs. Ed. 2) is the uredospore stage of *Puccinia allii* on
leek. Stevenson (Myc. Scot. No. 1292) under *Uredo alliorum* DC. gives
'Scotland, Dr M. C. Cooke' but as pointed out by Trail (Scot. Nat.
N.S. **4**, 326, 1889–90) this also may be *P. allii*. Grove's only reference on
Allium is to Plowright who in turn quoted Cooke whose records of this
species are quite clearly not trustworthy. The specimen of *C. alliorum*
from the Hennedy Herbarium recorded in the British Association List
of Fauna and Flora of the Clyde Area (1901, 66) is no longer in existence
and has never been checked. *Caeoma ari-italici* was shown to be an

aecidial stage of *M. allii-populina* by Cruchet (Bull. Soc. Vaud. Sci. Nat. **56**, 485, 1927) who infected *Arum maculatum* and various species of *Allium* from teleutospores; it was found near Salisbury by F. J. Tatum in 1897 and this appears to be the only, but unconfirmed, British record although it has been found in Germany, France, Switzerland, Spain and Russia. Tranzschel (Trans. Russ. Acad. Sci. **2**, 99, 1927; Abs. in RAM. **6**, 755, 1927) in Russia established a connection between *Caeoma ari* and a species of *Melampsora* on *Populus nigra* var. *pyramidalis* to which he gave the name of *M. ari-populina*.

The aecidia of *M. salicis-albae* and *M. allii-fragilis* are also borne on *Allium* and are morphologically indistinguishable from those of *M. allii-populina*; the aecidia of *Puccinia sessilis* which are also produced on *Allium* are easily distinguished by the presence of a well-developed peridium.

Sappin-Trouffy (1897, 170) described the development of the aecidium.

Melampsora larici-populina Kleb.

Z. Pfl.-Krankh. **12**, 32, 1902; Grove, Brit. Rust Fungi, p. 348; Wilson & Bisby, Brit. Ured. no. 45; Gäumann, Rostpilze, p. 132.

[*Caeoma laricis* Hart., Wichtige Krankh. Waldbäume, p. 93 (1847), *p.p.*]

[**Spermogonia** amphigenous, subepidermal, prominent, up to 150μ diam. and 100μ high, yellow. **Aecidia** caeomoid on yellow spots, 0·5–1 mm. diam., orange, with a rudimentary peridium; aecidiospores globoid or ovoid, 17–22 × 14–19μ, wall 1–1·5μ thick, colourless, finely verruculose.] **Uredosori** mostly hypophyllous, in little groups, causing yellowish angular spots on the upper side, rarely solitary and epiphyllous, up to 1 mm. wide, at first covered by the raised epidermis and a peridium of colourless cells, 10–12 × 5–6μ, distributed over the whole leaf-surface; paraphyses clavate to capitate, 40–70 × 14–18μ, wall strongly thickened (up to 10μ above); uredospores broadly ellipsoid 30–40 × 13–17μ, wall about 2μ thick at the equator, echinulate except at the smooth apex. **Teleutosori** epiphyllous, minute, but united in groups and confluent, distributed over nearly the whole leaf, subepidermal, red-brown, at length black; teleutospores prismatic, rounded above and less so below, 40–50 × 7–10μ, wall pale, scarcely 1μ thick, but reaching

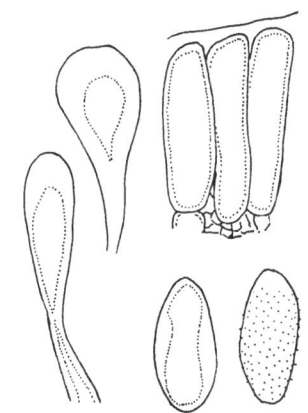

M. larici-populina. Teleutosorus, uredoparaphyses and uredospores.

2·5–3μ at the faintly coloured apex; without an evident pore. Hetereu-form. [Spermogonia and aecidia on *Larix decidua* and *L. leptolepis*, May, June];

uredospores and teleutospores on *Populus* × *canadensis*, *P.* × *generosa*, *P. laurifolia*, *P. nigra*, *P. nigra* var. *italica*, *P.* × robusta and *P. trichocarpa*, July–October. Great Britain and Ireland, frequent.

The description of the aecidia, which have not with certainty been found in Britain, is after Klebahn (Z. Pfl.-Krankh. **9**, 141, 1899). This species is distinguished from *M. allii-populina* by the elongate uredospores with distinct equatorial thickening, the epiphyllous teleutosori and the slightly thickened teleutospore apex.

Hersperger (Mitt. Naturf. Ges. Bern. 1928, xxvii) showed that the aecidial stage possesses a rudimentary peridium made up of loosely connected cells, 29–39 × 19–25 μ, of which the outer wall is thin and the inner 2–4 μ thick.

The connection of the spore stages by culture experiments was proved by Klebahn (*loc. cit.*) and confirmed by Mayor (Bull. Soc. Neuchâtel. Sci. Nat. **46**, 38, 1920 and **58**, 30, 1933) in Switzerland. Hiratsuka (Jap. J. Bot. **6**, 1, 1932) also carried out infection experiments in Japan. Ashworth (Ann. Bot. Lond. **49**, 95, 1935) described 'receptive hyphae' in this species. Hybridisation between different strains of this rust were carried out by Van Vloten (Tijdschr. Plantenz. **55**, 196, 1949). Peace (For. Comm. Bull. **19**, 1952) summarised the susceptibilities of poplars to Melampsoras in some sites in England. Although he did not attempt to identify the species more closely, most of his information probably refers to *M. larici-populina*, although *M. allii-populina* may also be involved. He classed as susceptible, *P. candicans*, *P. tacamahaca*, *P. maximowiczii*, *P. yunnanensis*, *P.* × *berolinensis* group, *P.* × *generosa;* as intermediate in susceptibility *P. trichocarpa*, *P. nigra* var. *italica*, *P.* × *eugenii*, *P.* × *robusta*, *P.* × *serotina*; and as resistant *P. deltoides*, *P. nigra* var. *betulifolia*, *P.* × *laevigata*, *P.* × *marilandica*, *P.* × *regenerata* and *P.* × *gelrica*. Somewhat similar susceptibilities were earlier noted by Cameron (Scot. For. Jour. **50**, 146, 1936).

Melampsora populnea (Pers.) Karst.

Bidr. Känned. Finl. Nat. Folk, **31**, 53 (1879).

Sclerotium populneum Pers., Syn. Meth. Fung. p. 125 (1801).
Melampsora tremulae Tul., Ann. Sci. Nat. Bot. Ser. 4, **2**, 95 (1854); Grove, Brit. Rust Fungi, p. 349; Wilson & Bisby, Brit. Ured. no. 50.

Spermogonia and **aecidia** described on succeeding pages. **Uredosori** hypophyllous, minute, 0·5 mm. diam., pulvinate, pulverulent, yellowish-orange; para- physes distributed throughout the sorus, 40–60 × 8–23 μ, clavoid or somewhat capitate, with wall 3–9 μ thick; uredospores ovoid, broadly ellipsoid or obovoid, 15–25 × 11–18 μ, wall 2–3 μ thick with stout, rather distant spines about 2 μ apart. **Teleutosori** hypophyllous, subepidermal, scattered over the whole leaf, 0·5–1 mm. diam., dark brown; teleutospores prismatic, rounded at both ends, pale brown, 22–60 × 7–12 μ; wall 1–2 μ thick, not thickened above, smooth, pores not evident. Hetereuforms.

Spermogonia and aecidia on *Larix*, *Pinus* and *Mercurialis*; uredospores and teleutospores on *Populus alba* and *P. tremula*. Great Britain and Ireland, frequent.

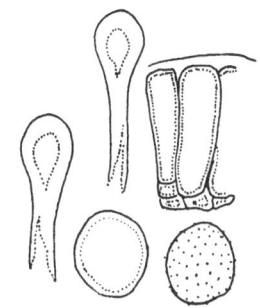

M. populnea. Teleutosorus, uredo-paraphyses and uredospores.

As used here *M. populnea* includes various races, usually distinguished as species, which do not differ essentially in morphological characters but possess different aecidial hosts under which they are discussed in the following accounts.

Larix

Melampsora larici-tremulae Kleb., Forstl. Naturw. Z. 6, 470 (1897); Grove, Brit. Rust Fungi, p. 349; Gäumann, Rostpilze, p. 134.
[*Caeoma laricis* Hartig, Wichtige Krankh. Waldbaume, p. 93 (1874), *p.p*]
Melampsora laricis Hartig, Allg. Forst-Jagdzeit, **61**, 326 (1885).
Melampsora tremulae f. *laricis* Hartig, Lehrb. Baumkrankh. Ed. 2, p. 14 (1889).

Spermogonia amphigenous, grouped or slightly scattered, 70–80 μ diam., 30–45 μ high, pale yellow. **Aecidia** caeomoid, generally hypophyllous, scattered or in small groups, on yellowish spots, minute, rounded or oblong and then up to 0·75 mm. long, pale yellow, soon naked, pulverulent; aecidiospores globoid, ovoid or angular, 14–17 × 12–16 μ, wall 1 μ thick, finely verruculose.

Spermogonia and aecidia on *Larix decidua*; uredospores and teleutospores on *Populus alba* and *P. tremula*. Great Britain, scarce(?).

The aecidial stage, *Caeoma laricis*, has been found in the Clyde, Dee and Moray areas in Scotland; Trail (1890, 323) found it growing in close proximity to infected *P. tremula* near Aberdeen and the two stages have been seen closely associated in the Clyde area.

The connection between the different spore stages was experimentally proved by Hartig (Bot. Zbl. **23**, 362, 1885), Klebahn (Z. Pfl.-Krankh. **4**, 12, 1894), Fischer (1898, 90) and more recently by Mayor (Bull. Soc. Neuchâtel. Sci. Nat. **46**, 38, 1920 and **54**, 56, 1929) who infected *Larix leptolepis* as well as *L. decidua*.

The teleutospores germinate after the winter's rest and the conditions under which this takes place have been studied by Dietel (Zbl. Bakt. II, **35**, 272, 1912).

The rust can overwinter as mycelium in poplar buds in Central Europe (Klebahn, Z. Pfl.-Krankh. **22**, 342, 1912 and Van Vloten, Tijdschr. Plantenz. **50**, 49, 1944) and is usually independent of alteration.

Mercurialis

Melampsora aecidioides Plowr., Brit. Ured. Ustil. p. 241 (1889).
[*Melampsora rostrupii* Wagner, Oest. Bot. Zeit. 46, 274 (1896) *nomen nudum*; Wagner, ex Kleb., Z. Pfl.-Krankh. 7, 342 (1897) *nomen prov.*]; Grove, Brit. Rust Fungi, p. 351; Gäumann, Rostpilze, p. 139.
[*Uredo confluens* var. *mercurialis-perennis* Pers., Syn. Meth. Fung. p. 214 (1801).]
[*Melampsora aecidioides* (DC.) Schroet., Cohn, Krypt. Fl. Schles. 3 (1), 362 (1889).]

Spermogonia epiphyllous or a few hypophyllous, in small clusters, minute, lenticular, up to 200μ wide and 90μ high, honey-coloured. Aecidia hypophyllous and on the petioles and stems in clusters, 0·5–2 cm. wide, on pale yellow spots, often circinate around the spermogonia, about 1–1·5 mm. diam., often confluent and elongate, up to 5 mm. long on the leaves and on the stems up to 3 cm. long, pulverulent, bright orange; aecidiospores globoid, globoid-angular or ovoid, 13–24 × 11–17μ; wall 1–1·5μ thick, finely and densely verruculose.

Spermogonia and aecidia on *Mercurialis perennis*, April–June. Great Britain, frequent.

The connection of the two forms was first shown by Rostrup (Overs. K. Danske Vidensk. Selsk. Forh. **1884**, 14, 1884), and confirmed by Plowright (1889, 241) in England; Klebahn (Z. Pfl.-Krankh. **6**, 337, 1896), Wagner and others confirmed these results by culture on the continent. Germinating teleutospores are easily found on fallen leaves of *P. tremula* or *P. alba* in spring where *Mercurialis* is infected. The large yellow spots show conspicuously on the upper surface of the leaves. Peace (Rep. For. Comm. **1959**, 68) recorded slight infection of *P. grandidentata* by this race.

Mercurialis magnusiana Wagner, which has its aecidium on *Chelidonium* and *Corydalis*, closely resembles *M. rostrupii* and is included under *M. populnea*; it has not been recorded for Britain but Plowright (1889, 241) mentions a *Melampsora* on *P. tremula* from which he could not obtain aecidia on *Larix*, *Pinus* or *Mercurialis*—this might be *M. magnusiana*.

Accounts of development of the aecidium have been given by Blackman & Fraser (Ann. Bot. Lond. **20**, 35, 1906) and Kursanov (1922).

Pinus

Melampsora pinitorqua Rostr., de farligste Snyltere i Danmarks Skove, p. 10 (1889); Grove, Brit. Rust Fungi, p. 350; Wilson & Bisby, Brit. Ured. no. 50; Gäumann, Rostpilze, p. 136.
[*Caeoma pinitorquum* de Bary, Monatsber. K. Akad. Wiss. Berlin, p. 624 (1863).]

Spermogonia subcuticular or partially subepidermal, punctiform, yellow, up to 130μ wide, 15μ high. **Aecidia** caeomoid, erumpent through the cortex of young shoots and occasionally on the leaves, solitary, linear, reaching 20 × 3 mm., reddish-orange; aecidiospores globoid or ovoid, pale reddish-yellow, 14–20 × 13–17μ, wall about 2μ thick, finely verruculose.

Spermogonia and aecidia on *Pinus nigra*, *P. nigra* subsp. *laricio*, *P. pinaster* and *P. sylvestris*, May and June. Southeastern England, local.

This form was recorded (Massee & Crossland, 1905, 183 and Mason & Grainger, 1937, 44) on *Pinus sylvestris* and *Populus tremula* from Yorkshire. It was found in Surrey in 1930 on *P. sylvestris* and there is a specimen in the Kew Herbarium collected in the Isle of Wight by Crossland on *P. tremula*. It has been reported by Peace (Forestry, **18**, 47, 1944) from Kent and Sussex in 1942–4. The Forestry Commission have recently reported it from Hampshire, Northamptonshire and Lincolnshire; the records on *P. pinaster* and *P. nigra* subsp. *laricio*, both of which are rather resistant (Peace, For. Comm. Bull. **19**, 1952), are from Kent. It causes severe distortion and die-back of Scots Pine when attack is heavy. Peace (Rep. For. Comm. 1959, 68) records the hybrid *P. tremula × tremuloides* as an additional host. One of the parents, *P. tremuloides*, and *P. grandidentata* are resistant.

In this race the uredospore wall is for the most part 2μ thick but the side walls may be thickened up to 5–6μ with thinner spots (pores?) in them; the teleutospores are shorter than usual, 20–35μ long.

The connection of the spore forms on the two hosts was first demonstrated by Rostrup and was confirmed by Hartig (Lehrb. Baumkrankh. 1882, 72) and Klebahn (Z. Pfl.-Krankh. **12**, 39, 1902 and **17**, 154, 1907). Detailed accounts of the rust on *P. sylvestris* have been given by Hartig (Wichtige Krankh. Waldbaume, 1874, 83) in Germany and Sylven (Medd. Statens Skogsförsöksanst. **13–14**, 1077, 1917) in Sweden. Hartig was of the opinion that the mycelium perennates in the pine shoots but Sylven considered this improbable. Jørstad (Medd. Norske Skogforss. **6**, 19, 1923) stated that the mycelium may persist during the winter in the buds of the aspen but later (Jørstad, 1960, 150) suggested that it is obligatorily heteroecious. Klingstrom (Studia For. Suecica, **6**, 1963) lists nine species of *Pinus* susceptible to this rust and has carried out extensive infection experiments.

Melampsora capraearum Thüm.

Mitt. Forstl. Versuchsw. Oest. **2**, 34 and 36 (1879).
[*Uredo capraearum* DC., Lam. & DC. Syn. Pl. Gall. p. 48 (1806).]
[*Caeoma laricis* Hart. Wichtige Krankh. Waldbaume, p. 93 (1874), *p.p.*]

Melampsora farinosa (Pers.) Schroet., Cohn Krypt. Fl. Schles. 3 (1), 360 (1887).
Melampsora larici-capraearum Kleb., Z. Pfl.-Krankh. 7, 326 (1897); Grove, Brit. Rust
Fungi, p. 338; Wilson & Bisby, Brit. Ured. no. 43; Gäumann, Rostpilze, p. 149.

Spermogonia amphigenous, subcuticular, pustulate, occasionally confluent, hemispherical in section, 30–100μ high, 18–38μ wide, spermatia 3–4 × 1·5–2·8μ. Aecidia caeomoid, minute, scattered, pale orange, soon naked, pulverulent; aecidiospores globoid, angular-globoid or ellipsoid, 15–25 × 12–17μ, wall up to 2μ thick, finely verruculose, with many, scattered, indistinct pores. Uredosori hypophyllous, showing as yellow spots on the upper side, variable in size and arrangement, 1–3 mm. wide; paraphyses capitate, 50–70 × 18–26μ, wall 5–6μ thick above, stalk 5–6μ wide; uredospores ovoid or ellipsoid, 14–21 × 13–15μ, wall 2–2·5μ thick, firm, distantly echinulate, distance between warts 2–2·5μ, with scattered pores. Teleutosori epiphyllous, subcuticular, 1 mm. or more wide, dark reddish-brown, frequently confluent in extensive crusts; teleutospores prismatic, rounded below, 30–45 × 7–14μ, wall light brown, thin (1μ) but up to 10μ thick at the apex

which is pierced by a slightly excentric pore. Hetereu-form.
Spermogonia and aecidia on *Larix decidua* and *L. leptolepis*, April–May;

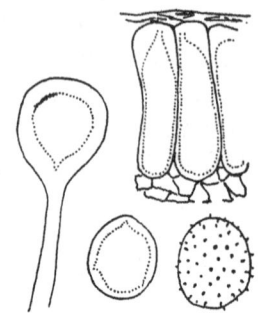

M. capraearum. Teleutosorus, uredoparaphysis and uredospores.

uredospores and teleutospores on *Salix aurita, S. caprea, S. caprea × viminalis* and *S. cinerea* subsp. *atrocinerea* and subsp. *cinerea*; teleutospores from September onwards. Great Britain, common.

The teleutospores are distinguished from those of all other *Melampsora* species on *Salix* by being thickened above, the reddish teleutosori on the upper side of the leaf are also distinctive. Plowright remarked that the aecidium is not uncommon early in the year on larch foliage but it is very inconspicuous and easily overlooked; he found aecidiospores in company with the uredospores on *S. caprea* at West Malvern in June 1900. They occur in Scotland in May and June both on *Larix decidua* and *L. leptolepis*; but it is impossible to say, without experiment, to which species of *Melampsora* any given aecidium on larch is to be assigned, although *M. capraearum* is undoubtedly the commonest. It is probable that this rust is obligatorily heteroecious in this country as in Norway (Jørstad, 1940, 36).

The connection of the different spore stages was first proved by Klebahn (Z. Pfl.-Krankh. 7, 325, 1897) and was frequently confirmed by this investigator, by Jacky (Ber. Schweiz. Bot. Ges. 9, 25, 1899), Liro (Acta Soc. Fauna Fl. Fenn. 29, no. 6, 1906), Mayor (Bull. Soc. Neuchâtel. Sci. Nat. 46, 36, 1920), who infected both *L. decidua* and *L.*

leptolepis, and Wilson (1934, 416) in Scotland. Klebahn (Z. Pfl.-Krankh. **12**, 39, 1902) also infected *L. occidentalis* and Hunter (1936*b*, 116) *L. laricina*. Klebahn was unable to infect *Salix cinerea* but Schneider (Zbl. Bakt. II, **16**, 1906) infected this and several other species of *Salix*. Hiratsuka (Jap. J. Bot. **6**, 11, 1932) in Japan has infected *L. leptolepis* and various Japanese species of *Salix*.

'Receptive hyphae' on infected larch leaves are described by Ashworth (Ann. Bot. Lond. **49**, 95, 1935) and Hunter (1936*b*, 119) has given an account of the development of the spermogonia. Dietel (Zbl. Bakt. II, **31**, 97, 1911) gave an account of the conditions required for the germination of the teleutospores.

Melampsora larici-pentandrae Kleb.

Forstl. Naturw. Z. **6**, 470 (1897); Grove, Brit. Rust Fungi, 345; Wilson & Bisby, Brit. Ured. no. 44; Gäumann, Rostpilze, p. 148.

[**Spermogonia** epiphyllous, subcuticular, 60–100μ wide, 30–50μ high. **Aecidia** caeomoid, bright orange; aecidiospores angular-globoid or ovoid, 18–26 × 13–10μ, wall faintly verruculose, 1·5–2μ thick.] **Uredosori** mostly hypophyllous but single ones epiphyllous, up to 1 mm. diam., bright orange; uredospores ovoid or broadly ellipsoid, smooth at the apex, 26–44 × 12–16μ, wall about 2μ thick, echinulate; paraphyses clavate to capitate, 40–60μ long, apex up to 22μ broad. **Teleutosori** hypophyllous, subepidermal, scattered or grouped and often covering the whole surface of the leaf, minute, about 0·5 mm. diam., often confluent, yellowish-brown, finally brownish-black, teleutospores prismatic or cylindrical, rounded at both ends, yellowish-brown, 28–38 × 6–11μ, wall 1μ thick. Hetereuform.

[Spermogonia and aecidia on *Larix decidua*, April–May]; uredospores and teleutospores on *Salix pentandra*, *S. fragilis × pentandra*. Scotland and Ireland, scarce.

This species was first satisfactorily recorded as British by Grove (J. Bot. Lond. **59**, 314, 1921) who described a specimen on *S. pentandra*, collected by Boyd in Ayrshire. It is not known whether the aecidia exist in this country. According to Jørstad (1960, 141) the rust is often independent of host alternation in Norway.

The connection between the spore forms was first proved by Klebahn (Z. Pfl.-Krankh. **7**, 330, 1897) and this investigator mentions the spermogonia and states that the aecidia differ by their deep-orange colour from all others occurring on *Larix*; the uredospores are larger than those of *M. amygdalinae* with a thicker wall, and the echinulations are slightly further apart.

Melampsora ribesii-viminalis Kleb.

Jahrb. Wiss. Bot. **34**, 363 (1900); Grove, Brit. Rust Fungi, 342; Wilson & Bisby, Brit. Ured. no. 49.

[**Spermogonia** generally epiphyllous, grouped, 65–80μ high, 150–190μ wide. **Aecidia** caeomoid generally hypophyllous, on yellow spots which show on both sides of the leaf, scattered or arranged in circular groups, 0·5–1·5μ diam. often confluent, orange, surrounded by the ruptured epidermis; aecidiospores globoid to ovoid, rarely angular, 16–23 × 14–18μ, wall 2–4μ thick, finely and densely verruculose.] **Uredosori** hypophyllous, minute, about 0·25 mm. diam. scattered or in groups, pale orange-yellow; uredospores more or less rounded, 15–19 × 14–16μ, wall 2μ thick, uniformly echinulate; paraphyses capitate or more often clavate, hardly thickened above, 50–70 × 18–25μ. **Teleutosori** epiphyllous, developed between the cuticle and the epidermis, 0·25–0·5 mm. diam., scattered or in groups over the whole leaf, dark brown, shining; teleutospores prismatic, rounded at both ends, more or less irregular, 25–40 × 7–14μ; wall thin, clear brown, not thickened above, with no evident pore. Hetereu-form.

[Spermogonia and aecidia on *Ribes uva-crispa*]; uredospores and teleutospores on *Salix viminalis*. Great Britain, rare.

There is considerable doubt if the aecidial stage has been found in this country; an aecidium was found by E. J. Tatum near Salisbury (TBMS. **1**, 97, 1899), but this may equally well be a stage in the life history of *M. ribesii-purpureae* (see p. 89). The uredospore stage on *S. viminalis* was recorded by Sprague (1952, 5) from Gloucestershire and doubtfully from Scotland (Wilson, 1934, 417). Klebahn (*loc. cit.*) and Mayor (Bull. Soc. Neuchâtel. Sci. Nat. **51**, 75, 1926 and **54**, 56, 1929) demonstrated the alternation between *Ribes* and *Salix*. The teleutospores can be distinguished from those of all other British species except *M. laricicapraearum* by their subcuticular position. A race of *M. epitea* is also recorded on *S. viminalis* but that species has subepidermal teleutospores and the paraphyses of the uredosori strongly thickened at the apex. Scaramella (Chanousia, **11**, 72, 1932) reported overwintering of mycelium on the willow.

Melampsora epitea Thüm.

Mitt. Forstl. Versuchsw. Oest. **2**, 38 and 40 (1879); Wilson & Bisby, Brit. Ured. no. 37.

Melampsora hartigii Thüm., Mitt. Forstl. Versuchsw. Oest. **2**, 41 (1879).
[*Uredo epitea* Kunze & Schum., Myk. Hefte, **1**, 68 (1817), *p.p.*]

Spermogonia amphigenous, epiphyllous or hypophyllous, subcuticular, variable in shape and size, globose, lenticular, or slightly convex, 60–120μ high, 150–200μ wide. **Aecidia** caeomoid, amphigenous or hypophyllous, scattered or in groups, sometimes with a rudimentary peridium, orange; aecidiospores globoid, angular-

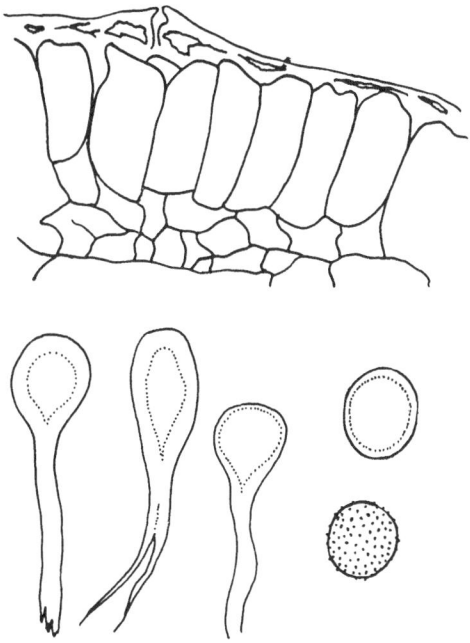

M. epitea. Teleutosorus, uredo-paraphyses and uredospores.

globoid or ovoid, 15–25 × 10–21 μ, wall finely verruculose 1·5–2·5 μ, occasionally up to 5 μ thick. **Uredosori** amphigenous, orange-yellow, at first covered then surrounded by the ruptured epidermis, rounded oval or elongated, 0·5–1·5 mm. wide, orange-yellow, with numerous capitate paraphyses, 35–80 μ, occasionally up to 90 μ long, with heads 15–24 μ diam. occasionally up to 35 μ wide, usually thick-walled but some with thin walls; uredospores ellipsoid or globoid, 12–25 × 10–18 μ, wall aculeato-verrucose, 1·5–3 μ thick, with very indistinct pores. **Teleutosori** subepidermal, amphigenous

but mostly hypophyllous, scattered or in groups, minute, 0·25–1 mm. diam. but often confluent and larger, yellowish-brown, at length blackish-brown; teleutospores prismatic, rarely clavoid, rounded at both ends or slightly tapering towards the apex, pale brown, 20–50 × 7–14 μ, wall 2 μ or slightly thicker at the apex. Hetereu-form.

Spermogonia and aecidia on *Euonymus, Larix, Ribes, Saxifraga* and several genera of the *Orchidaceae*; uredospores and teleutospores on *Salix*. Great Britain and Ireland, frequent.

This is regarded as a collective species made up of various races and race groups. Usually they have been looked upon as distinct species but they are not separable by clear morphological characters but only by the host genus or family of the haplont. *M. epitea* is characterised by possessing subepidermal teleutospores which are not thickened above and have no distinguishable pore; the uredospores are roundish or oval with uniformly echinulate walls. The heads of the inner uredosoral paraphyses are comparatively small and mostly thick-walled; the peripheral paraphyses are normally thin-walled more clavate and generally larger than the more distinctly capitate inner ones (see Jørstad, 1940, 18).

Within this group two varieties are recognised.

var. *reticulatae* (A. Blytt) Jørst.

Skr. Norske Vidensk.-Akad. Oslo Math. Nat. Kl. **1940** (6), 31 (1940). Wilson & Bisby, Brit. Ured. no. 38.

[*Caeoma saxifragae* Wint. Rabh. Krypt. Fl. Ed. 2, **1** (1), 258 (1882), *p.p.*]
Melampsora reticulatae Blytt, Chria. Vidensk. Selsk. Forh. **1896**, 65 (1896); Gäumann, Rostpilze, p. 172.

Spermogonia and aecidia on *Saxifraga aizoides*; uredospores and teleutospores on *Salix arbuscula*, *S. herbacea*, *S. lanata* and *S. reticulata*. Locally frequent on the mountains of central and northern Scotland.

The aecidial stage on *Saxifraga aizoides* has been associated with the diplont on *Salix reticulata* since 1904 (Fischer, 1904, 492 and others) but successful cultures were not made until 1951 when Henderson (TBMS. **36**, 315, 1953) by sowing aecidiospores obtained uredospores on *S. reticulata* after 14 days; *S. herbacea*, *S. lanata* and other species of *Salix* were not infected.

It differs from other races formerly placed under *M. arctica* in its larger uredospores ($24–32 \times 16–22\cdot5\,\mu$) with larger paraphyses, up to $90\,\mu$ long, with width of head $25–35\,\mu$ and thickness of wall up to $10\,\mu$.

The aecidial stage was collected by Trail in Aberdeenshire in 1890. Henderson reported it as quite common on certain of the higher hills in Perthshire. The diplont was probably first recorded by Gardiner (Fl. Forfar, p. 298) as *M. mixta* on *S. reticulata* in 1848; it was collected by Trail at Braemar in 1882.

Lind (Biol. Medd. K. Danske Vid. Selsk. **6**, 5, 1927 and Jørstad (1940, 126, footnote) have stated that overwintering may take place by mycelium in the haplont host but neither gave any evidence for this statement. Henderson (*loc. cit.*) has discussed its survival and dispersion; he has found that very soon after the snow has melted, groups of interconnected infected shoots of *S. aizoides* may be present among a mat of healthy plants. A group was kept growing in isolation at Edinburgh during the winter; in the following year spermogonia appeared on 20 March and mycelium was present in the pith of the current and of the previous year's growth. He noted that infected portions or even whole plants of the saxifrage may be swept down mountain streams and take root at lower levels. Although the lower altitudinal limit of *S. reticulata* in Scotland is about 600 m. groups of infected saxifrage plants were found at an altitude of 200 m. on Ben Lui. In consequence of these observations carriage by spring floods seems an important mode of dispersal in this country.

Jørstad (1951, 16) has stated that the haplont of this rust is fairly common in Iceland although *S. reticulata* is not present in the island and he has suggested that *S. lanata* or *S. herbacea* may serve as hosts for the diplont; he has also noted that, in all arctic willows, overwintering by uredo takes place in the buds.

Lindfors (1924, 4) described the fusion of basal cells in the aecidium and nuclear division in the spermogonium and aecidium of this rust.

var. *epitea*

The remainder of the races included in *M. epitea* are grouped below according to their aecidial hosts. They are morphologically indistinguishable but each has a distinctive host specialisation pattern in both diplont and haplont stages.

Saxifraga

> Alternating between montane willows and *Saxifraga hypnoides*,
> *S. oppositifolia* or independent of aecidial host.

Melampsora arctica Rostr., Medd. Grønl. 3, 535 (1888); Grove, Brit. Rust Fungi,
  p. 346; Wilson & Bisby, Brit. Ured. no. 37, *p.p.*
Melampsora alpina Juel, Övers. K. Vet. Akad. Föhr. **1894** (8), 417 (1894); Gäumann,
  Rostpilze, p. 170; Wilson & Bisby, Brit. Ured. no. 37, *p.p.*

On *Saxifraga hypnoides* the rust has been rarely found but is locally frequent. Unnamed specimens collected by Greville probably about 1822 on Ben Venue, Perthshire are in the Edinburgh Herbarium; it was found by Wilson in 1934 on Ben Lui and has been recorded recently by Henderson (TBMS. **36**, 315, 1953) from several localities in Perthshire. It was first recorded on this host by Dietel (Sitz. Ber. Naturf. Ges. Leipzig, **37**, nos. 15 and 16, 1890) near Leipzig, and recently by Jørstad (1951, 17) from Iceland.

Henderson carried out inoculations on *Salix* with aecidiospores and found that uredospores were produced within three weeks on *S. herbacea*, *S. sadleri* (= *herbacea* × *lanata*) and *S. herbacea* × *myrsinites*; uredospores produced on *S. sadleri* in these cultures were transferred to other plants of *S. herbacea* and formed uredospores and teleutospores in the autumn. *S. lanata* remained uninfected in these cultures.

Although no direct proof has been brought forward there appears to be no doubt that in the rust on this saxifrage overwintering takes place by means of persistent mycelium. Plants bearing numerous rust sori were found in June, 1934 at about 800 m. on Ben Lui where no *S. herbacea* appeared to be present in the vicinity and Henderson has noted a similar occurrence in Perthshire in June, where spermogonia and aecidia were well developed on plants at about 1000 m. which had just been released from their snow covering and where, although *S. herbacea* was present in the neighbouring turf no rust sori could be found on the leaves.

The aecidia on *Saxifraga oppositifolia* have been found only once in Britain, on Ben Lui, Perthshire at an altitude of about 1000 m. The saxifrage host was growing in the vicinity of *Salix herbacea* which had borne uredospores and teleutospores in the previous year; only a very few aecidia were discovered and there appeared to be no indication of systemic infection (Wilson, J. Bot. Lond. **57**, 161, 1919). In the account of these specimens it was stated that paraphyses with thin, colourless walls and yellow granular contents, 'ending in a swollen head which is always smaller than the aecidiospores' were present. Although these paraphyses were not mentioned by Juel (Övers. K. Vet. Akad. Förh. **8**, 417, 1894) in his account of *M. alpina*, somewhat similar structures were described by Klebahn (Z. Pfl.-Krankh. **17**, 156, 1907) in a caeoma which appeared in the Botanic Garden in Hamburg on an unnamed species of *Saxifraga* collected in Spitzbergen: he assigned this to *M. alpina* on account of its structure and because in culture it readily infected *Salix herbacea*. According to this account the aecidial sorus is surrounded by a small amount of pseudoparenchyma from which paraphysis-like hyphae grow up and project at the periphery of the spore chains. These marginal paraphyses have also been described and figured on *M. alpina* by Lindfors (Sv. Bot. Tidskr. **4**, 197, 1910); he described them as a ring of sterile hyphae of which the end cells enlarge to form definite heads; the latter being smaller in width than the aecidiospores beside which they are found. The structure to which they give rise has been regarded as a rudimentary peridium by Hersperger (Mitt. Naturf. Ges. Bern, 1928, xxvii, 1928) in *Melampsora larici-populina* (see p. 74). The relation between aecidia on *Saxifraga oppositifolia* and diplont stages on *Salix herbacea* (and their specialisation to these hosts) was proved experimentally by Jacky (Ber. Schweiz. Bot. Ges. **9**, 49, 1899) and confirmed by Klebahn (Z. Pfl.-Krankh. **17**, 156, 1907). There is no experimental evidence on this fungus for Britain.

On *Salix arbuscula* collections have been made by Henderson on Ben Lui and Creag an Lochan, Perthshire. On four of these the uredospores are amphigenous, the paraphyses heads measuring $17–25(21·5)\mu$ in width, the wall 3–8 μ thick; on one with hypophyllous uredosori the paraphyses approach var. *reticulatae*, being 20–30(26)μ in width with wall 5–8(7)μ thick. Amphigenous teleutospores were found on two specimens. In cultures made on two occasions from the specimens with the smaller paraphyses the uredospores infected *S. reticulata* but not *S. repens*, *S. lanata* nor *S. herbacea*.

Schneider (Zbl. Bakt. II, **13**, 223, 1904, and **16**, 77, 1906) showed that both *S. reticulata* and *S. arbuscula* could be infected weakly by f. sp. *nigricantis* of *M. larici-epitea* but it appears unlikely that rusts collected on *S. arbuscula* at between 400–600 m. in Scotland would have their aecidial stage on *Larix* sp.

Uredospores and teleutospores on *S. herbacea* were first collected in Britain by Wheldon & Wilson (Ann. Scot. Nat. Hist. 1911, 37) in 1911 and the same rust has since been collected on a number of mountains in central and northern Scotland at altitudes usually over 800 m. (Wilson, J. Bot. Lond. **53**, 48, 1915, and **57**, 161, 1919). In the Shetland Islands it has been found by Spence (*in litt.*) at a low altitude. It is generally quite common and not obviously associated with aecidial stages in the field. Jørstad (1940, 31) considered that in Norway and in Iceland (1951, 20), the mycelium of this rust hibernates in the buds during the winter in this host and this, from field observations, certainly takes place in Scotland.

The hybrids, *S. herbacea* × *lanata* and *S. herbacea* × *myrsinites*, have been infected from the aecidia on *Saxifraga hypnoides* by Henderson (*loc. cit.*) the heads of the inner paraphyses produced in the uredosori measure 15–19·6 μ diam. with wall 2–4·6 μ thick. These hybrids have not been found infected in the field.

On *S. lanata* two collections of rust have been made in Glen Fee, Angus, and there is also a rusted specimen from Glen Callater in the Edinburgh Herbarium. All these specimens agree with var. *reticulatae*. As already stated, cultures carried out on this species with aecidiospores from *Saxifraga aizoides* and from *S. hypnoides* gave negative results; also cultures with uredospores from *Salix arbuscula* failed to produce infection. The aecidial host of the rust on *S. lanata* in Scotland is, in consequence, unknown. In Iceland, Jørstad (1951, 20) has suggested it is obligatorily heteroecious as primary uredospores (i.e. those developing from a perennating mycelium) have not been found upon it; he added 'Observations in the field indicate that *Saxifraga aizoides*, *S. cespitosa* and *S. hypnoides* serve as caeoma (aecidial) hosts.' No rust is known on *S. cespitosa* in Scotland.

Rust on *Salix lapponum* has been found recently on two specimens in the Edinburgh Herbarium, collected in this country: one by W. Gardiner, by sides of streams, Glen Clova, Angus, the other by E. S. Marshall, Coire Chaim, near Dalwhinnie, Inverness. In both specimens only uredosori are present, these are hypophyllous, partially covered by the leaf indumentum and rather inconspicuous; the paraphyses are 17–25(21)μ diam., with wall 5–8(6·5)μ thick.

Melampsora lapponum was described by Lindfors in Sweden in 1913 (Sv. Bot. Tidskr. **7**, 48, 1913) and the aecidial stage, *Caeoma violae*, in 1910 (*idem*, **4**, 198); he proved the connection of the spore stages by culture from *Viola epipsila*. According to Lindfors *M. lapponum* is distinguished from typical *M. epitea* by its wider paraphyses, up to 30μ across and the head which is thin-walled, up to 4μ thick.

Jørstad stated (1953, 116) that two rusts occur on *S. lapponum*, (1) typical *M. epitea*, (2) *M. lapponum*, with aecidia on *Viola epipsila* and *V. palustris*, but the two are not well defined; he added that possibly some of the rusts under *M. epitea* which are approaching the var. *reticulatae* in structure may really belong to *M. lapponum*.

It appears from their measurements that the Scottish specimens on *S. lapponum* belong to *M. epitea* not to *M. lapponum*.

There is a specimen of *M. epitea* on *Salix myrsinites* in the Edinburgh Herbarium collected in Glen Clova in 1847, bearing uredosori in which the paraphyses measure 16–22μ across the head; the rust has also been recently collected on this host by N. F. Robertson in Inverness-shire.

In this account of the rust forms previously included in *M. arctica* it has been shown that some can almost certainly persist throughout the winter in the mycelial condition both in their haplont and in their diplont hosts, e.g. var. *reticulatae* on *Saxifraga aizoides* and on *Salix reticulata*; in consequence this parasite can maintain itself independently on either of its hosts. A quite similar example is the race on *Saxifraga hypnoides* and *Salix herbacea*.

Both the arctic willows and saxifrages can multiply under natural conditions by the gradual separation of their shoots which then grow into new individal plants. When an infected plant reproduces itself vegetatively in this way the resultant plants are also infected. As a result the independence of the haplont and diplont stages referred to in the last paragraph may be continued indefinitely.

It has been shown that some of the forms placed under *M. arctica* can, under certain circumstances, be obligatorily heteroecious (see Schneider-Orelli, Zbl. Bakt. II, **25**, 436, 1910) but as pointed out by Arthur

(Mycol. **20**, 41, 1928) it is probable that dissemination of spores through the air plays only a small part in geographical distribution. In the north, however, high winds during the cold season may carry infected fragments of the hosts over the surface of ice for long distances. This method of distribution was suggested by Sernander (Den skandinavska vegetationens spridningsbiologi, Uppsala, 1901) who described it as 'winter scattering' and it has been discussed later by Lind (Biol. Medd. K. Danske Vid. Selsk. **6**, 1, 1927), who referred particularly to the distribution of fragments of *Saxifraga* sp. containing perennial mycelium which, after coming to rest, might produce roots in the following spring and develop into infected plants; he considered that in this way spread of rust fungi might take place across large bodies of water and even between continents. Arthur (1929, 169) suggested that by this method fragments of hosts with adherent spores could be transported; these 'may lodge against similar hosts or be massed in favourable spots for the seeds to grow and obtain a footing, and infection be thus established when the next warm season opens'. He concluded that these methods may, in part, account for the circumpolar distribution of some species.

Euonymus

Melampsora euonymi-capraearum Kleb., Jahrb. Wiss. Bot. **34**, 358 (1900); Grove, Brit. Rust Fungi, p. 339; Wilson & Bisby, Brit. Ured. no. 39; Gäumann, Rostpilze, p. 175.

Aecidia generally hypophyllous, bright-orange, in elongated clusters, 1–1·5 mm. long, on yellow spots, 0·5–1 cm. long; aecidiospores 18–23 × 14–19 µ, wall up to 5 µ thick.

Spermogonia and aecidia on *Euonymus europaeus*, April–June; uredospores and teleutospores on *Salix atrocinerea*, *S. aurita*, *S. caprea* and *S. cinerea*. England and Scotland, scarce

The aecidial stage has been found by Plowright near Salisbury (TBMS. **1**, 97, 1898) in Worcestershire, Gloucestershire (Sprague, p. 5), Suffolk (Mayfield, 1935, 10) and Cornwall (Rilstone, *in litt.*).

The diplont stage appears to be scarce in England; it has been found twice in Scotland; near Troon, Ayrshire (Scot. Crypt. Soc. Rep. 1928) and New Galloway, Kircudbrightshire, in 1931; both uredosori and teleutosori were present in the two localities.

Plowright (TBMS. **1**, 97, 1898) using aecidiospores from specimens near Salisbury obtained uredospores on *Salix cinerea*; he thus confirmed the cultures made by Rostrup (Overs K. Danske Vidensk. Selsk. Forh. **1884**, 13, 1884). Cultures with positive results were also made by Klebahn (Z. Pfl.-Krank. **7**, 329, 1897). The mycelium of this race may overwinter in the willow buds according to Scaramella (Chanousia, **11**, 72, 1932).

Larix

Melampsora larici-epitea Kleb., Z. Pfl.-Krankh. **9**, 88 (1899); Grove, Brit. Rust
   Fungi, p. 340; Wilson & Bisby, Brit. Ured. no. 37, *p.p.*; Gäumann, Rostpilze,
   p. 152.
[*Caeoma laricis* Hart., Wichtige Krankh. Waldbaume, p. 93 (1874), *p.p.*]

Spermogonia and aecidia on *Larix deci-* *aurita* × *repens*, *S. calodendron*, *S. cap-*
dua and *L. leptolepsis*; uredospores and *rea*, *S. caprea* × *viminalis*, *S. daphnoides*,
teleutospores on *Salix* spp. The records *S. fragilis*, *S. nigricans*, *S. phylicifolia*,
on all willows require confirmation; the *S. phylicifolia* × *repens*, *S. purpurea*, *S.*
following have been recorded: *Salix alba*, *triandra*, *S. triandra* × *viminalis* and *S.*
S. atrocinerea, *S. atrocinerea* × *triandra*, *viminalis*.
S. atrocinerea × *viminalis*, *S. aurita*, *S.*

Klebahn showed in 1899 (*loc. cit.*) that this rust produces its aecidia
on *Larix decidua* and according to Grove (1913, 341) this was confirmed
by Plowright in this country in May 1900 using teleutospores from *S.*
atrocinerea; no other reference to this culture has, however, been found.

Mayor has infected *L. leptolepis* using teleutospores from *S. viminalis*
(Bull. Soc. Neuchâtel. Sci. Nat. **46**, 37, 1920) and from *S. nigricans*
(*idem*, **58**, 30, 1933). An aecidial stage on *L. leptolepis* was also found in
Scotland in 1934 but this may belong to other species of *Melampsora*.

In the aecidial stage on *Larix* the caeomata are hypophyllous, scattered or in series on one or both sides of the midrib with corresponding
yellow spots on the upper side of the leaf; they are roundish or oblong,
0·5–1·5 mm. diam. and pale orange-yellow. They have been found on
the same leaf as the aecidial stage of *Melampsoridium betulinum* (see
p. 49).

Numerous infection experiments have been carried out by Hiratsuka
(Jap. J. Bot. **6**, 1, 1932) with this species in Japan where he has proved
that several specialised forms of the rust exist on *L. leptolepis* and on
Japanese species of *Salix*.

Many specialised forms of this rust have been described on the continent by Fischer, Klebahn, Schneider, Jørstad (1940, 25 and 1953, 103)
and others and this work has been summarised by Gäumann (1959, 154)
who recognises six *formae speciales* in this group. There is insufficient
evidence on the behaviour of this group of rusts in Britain to correlate
them with the extensive experimental results of these workers.

Orchidaceae

Melampsora repentis Plowr., Z. Pfl.-Krankh. **1**, 131 (1891); Wilson & Bisby, Brit. Ured. no. 47; Gäumann, Rostpilze, p. 162.

Melampsora orchidi-repentis Kleb., Jarhb. Wiss. Bot. **34**, 369 (1900); Grove, Brit. Rust Fungi, p. 343.

[*Caeoma orchidis* (Alb. & Schw.) Wint. Rabh. Krypt. Fl. Ed. 2, **1** (1), 256 (1882).]

[*Uredo confluens* var. *orchidis* Alb. & Schw. Consp. Fung. Nisk. p. 122 (1805).]

Spermogonia hardly projecting, lenticular, mostly under the stomata, about 80 μ high, 170 μ wide. **Aecidia** caeomoid, hypophyllous, on large pale yellow spots, 1–2 cm. long, groups rounded or elongated, up to 2 cm. long, often confluent, bright orange; aecidiospores angular-globoid to ovoid, 15–25 × 10–20 μ, wall 1–1·5 μ thick, densely verruculose.

Spermogonia and aecidia on *Gymnadenia conopsea*, *Listera ovata*, *Dactylorchis maculata* agg., *D. praetermissa* and *D. incarnata*, May–June; uredospores and teleutospores on '*Salix aurita*' and *S. repens*. Great Britain and Ireland, scarce.

Plowright (*loc. cit.*), and Gard. Chron. **8**, 41, 1890, and J. Roy. Hort. Soc. **12**, cxi, 1890) produced uredospores on *S. repens* by infection with aecidiospores. Klebahn confirmed this, using '*O. latifolia*', *S. repens* and *S. aurita*. Soppitt found the two stages growing together at Southport on *D. maculata* and *S. repens*; the rust has been found recently in Fife growing on *D. incarnata* and *S. repens* in close proximity. The aecidium on *G. conopsea* has been recorded in Norfolk (Ellis) and in Inverness-shire and on *O. praetermissa* from Norfolk (Ellis).

The caeoma on *Goodyera repens* mentioned by Grove (1913, 344) is now recognised as the uredosorus of *Uredo goodyerae* (see p. 42).

Plants of *S. repens* were found in May 1931 at Tentsmoor, Fife, bearing uredospores on the leaves, young shoots, and female catkins; the infection of the latter was particularly heavy, the young capsules being covered almost completely by rust; there were no orchids in the the immediate vicinity, and it appears improbable that they had been infected by aecidiospores at this early date. It is probable that the rust can exist through the winter as mycelium in the stems and buds.

Ribes

Melampsora ribesii-purpureae Kleb., Jahrb. Wiss. Bot. **35**, 667 (1901); Grove, Brit. Rust Fungi, p. 342; Wilson & Bisby, Brit. Ured. no. 48.

[**Spermogonia** chiefly epiphyllous, subepidermal, crowded in small groups, subconical, punctiform; 60–80 μ high, 150–200 μ wide. **Aecidia** hypophyllous, in small groups on discoloured spots, round, bright yellow; aecidiospores globoid or broadly ellipsoid, 18–26 × 15–20 μ, wall colourless, irregularly 1·5–3 μ thick, finely verrucose.]

[Spermogonia and aecidia on *Ribes alpinum* and *R. uva-crispa*]; uredospores and teleutospores on *Salix purpurea*. England and Scotland, rare.

British records of the aecidial stage are doubtful; the only reference to it is that by Plowright (TBMS. **1**, pt. 3, 97, 1899) who stated that E. J. Tatum of Salisbury had found *Caeoma confluens* on *Ribes uva-crispa* in the neighbourhood of a number of Salices. This may, however, have been the aecidial stage of *M. ribesii-viminalis*.

Melampsora amygdalinae Kleb.

Jahrb. Wiss. Bot. **34**, 352 (1900); Wilson & Bisby, Brit. Ured. no. 36; Gäumann, Rostpilze, p. 144.

Spermogonia subcuticular, slightly projecting, orange, becoming darker, 100 μ wide, 50 μ high. **Aecidia** caeomoid, produced on the stems and young leaves, usually hypophyllous, occasionally epiphyllous, up to 1 mm. long but usually in groups 3–10 mm. long on the leaves and up to 1 cm. on the stems, confluent, bright orange, aecidiospores globoid, angular-globoid or ovoid, 18–23 × 14–19 μ, wall 2 μ in thickness, minutely verruculose. **Uredosori** hypophyllous, producing a pale spot on the upper surface, occasionally epiphyllous and also on the stems, scattered, minute, about 0·5 mm. diam. rounded, orange; paraphyses capitate, 35–80 μ long, 10–18 μ wide above, stalk 4–6 μ wide, or clavate, 10–15 μ wide above, stalk 4–10 μ wide, wall 2–5 μ thick; uredospores ovoid or oblong, 20–40 × 11–18 μ, wall 1·5 μ thick, smooth at apex, aculeate-verruculose at base, warts about 2 μ apart. **Teleutosori** hypophyllous, occasionally epiphyllous, subepidermal, in small groups spread over the whole leaf, minute, 0·3–0·5 mm. diam., at first reddish-brown, later brownish-black; teleutospores cylindric or prismatic often irregular, rounded at both ends, yellowish-brown, 8–15 × 7–14 μ; wall 1 μ thick, not thickened above. Auteu-form.

On *Salix triandra*, spermogonia and aecidia, April; uredospores, March–October, teleutospores, July–October. England, local.

The description is taken partly from Klebahn (*loc. cit.*) who showed that this is the only autoecious *Melampsora* species on *Salix*. The uredospores resemble those of *M. larici-pentandrae* but are shorter, the wall is thicker and the warts are more distant.

Specimens of this species, sent to Greville in 1823 by Baxter from Oxfordshire, are in the Edinburgh Herbarium. It was first recorded from this country by Plowright in 1902 (Gard. Chron. **32**, 55, 1902) from specimens collected by E. J. Tatum near Salisbury. An account of this rust from osier beds in Somerset was published by Ogilvie (Ann. Rep. Agric. Hort. Res. Sta. Long Ashton, 1931, 133). Cankers are produced on the young stems by the aecidial stage and later by the uredospore stage, the latter being perennial; the uredospore stage also causes distortion of the leaves, considerable leaf fall and stunting of growth. Infected rods are worthless for basket making. Klebahn (Z. Pfl.-Krank. **48**, 384, 1938) observed uredosori on leaves of freshly opened buds of *S. triandra* in April, thus confirming the overwintering by mycelium in the buds. Overwintered infected leaves bearing teleuto-spores are the chief source of infection in Britain. Uredosori on uncut rods of the willow can also serve as sources of infection in the following season.

Ogilvie (*loc. cit.*) produced the aecidial stage by infection by means of germinating teleutospores but was unable to infect *S. alba* and *S. fragilis*. Mayor (Bull. Soc. Neuchâtel. Sci. Nat. **58**, 29, 1933) produced spermogonia and aecidia on both *S. triandra* and *S. pentandra* by means of teleutospores from the former, but a large range of other species of *Salix* was not susceptible to the rust.

Further observations on the rust, particularly on the conditions favouring spore germination and the prevention of disease have been made by Ogilvie & Hutchinson (Ann. Rep. Agric. Hort. Res. Sta. Long Ashton, 1932, 125), and Ogilvie (*idem*, 1932, 131).

Melampsora salicis-albae Kleb.

Jahrb. Wiss. Bot. **35**, 679 (1901).

Melampsora allii-salicis-albae Kleb., Z. Pfl.-Krankh. **12**, 19 (1902); Grove, Brit. Rust Fungi, p. 345; Wilson & Bisby, Brit. Ured. no. 35; Gäumann, Rostpilze, p. 159.

[*Caeoma allii-ursini* (DC.) Wint., Rabh. Krypt. Fl. Ed. 2, 1(1), 255 (1882).]

[**Spermogonia** slightly prominent, lenticular, about 120μ high and 210μ wide. **Aecidia** caeomoid, hypophyllous and on the stems, in groups on yellowish spots about 1 mm. diam., surrounded by the ruptured epidermis, bright orange-yellow; aecidiospores irregular, generally angular, rarely oblong, 17–26 × 15–18μ, wall 1–1·5μ thick, densely verrucose.] **Uredosori** of two kinds: (1) in summer and autumn on the leaves, hypophyllous, 0·5 mm. wide, on inconspicuous discoloured spots; (2) in spring, erumpent from the bark of young twigs and as much as 5 mm. long, afterwards on the young leaves, up to 2 mm. long and densely crowded; paraphyses capitate, 50–70 × 15–20μ, not thickened above, absent from the cortical sori; uredospores all similar, distinctly oblong, sometimes clavate or pyriform, 20–36 × 11–17μ; wall 2μ thick, smooth above, distinctly echinulate below. **Teleutosori** amphigenous, subepidermal, scattered thinly over the leaf surface singly or in groups, dark brown; teleutospores prismatic, rounded at both ends, brown, 24–45 × 7–10μ; wall scarcely 1μ thick, not thickened above, without evident pore. Hetereu-form.

[Spermogonia and aecidia on *Allium ursinum*]; uredospores and teleutospores on *Salix alba* and its var. *vitellina*. Great Britain, probably frequent.

Klebahn (Z. Pfl.-Krankh. **12**, 17, 1902, and Jahrb. Hamburg. Wiss. Anst. **1902**, 1, 1903) first proved the connection of the spore stages and this has been confirmed by Mayor (Bull. Soc. Neuchâtel. Sci. Nat. **54**, 55, 1929 and **58**, 28, 1933, and **61**, 116, 1936). Various species of *Allium* have been infected and Mayor has infected several species of *Salix* which possess pale catkin bracts while those with discoloured bracts remained uninfected.

The aecidium on *Allium* is indistinguishable from that of *M. allii-fragilis* or *M. allii-populina*.

This species can winter by its teleutospores or by the perennial mycelium in the cortex of the branches on which the uredospores appear in spring before the aecidium is produced; these sori are without paraphyses. The branches on which they appear have not developed a cork layer and were probably infected in the previous autumn through the young epidermis (Klebahn, Z. Pfl.-Krankh. **48**, 369, 1938). This investigator also described uredosori in April on small leaves and on the brown scales of a recently opened bud, suggesting the presence of overwintered mycelium in the buds. Infected shoots of *S. alba* var. *vitellina* are rendered too brittle for basket making by the attack of this rust. Grove (*loc. cit.*) mentioned a specimen collected in Yorkshire by Crossland which he refers doubtfully to this species; it is also recorded in the Catalogue of Yorkshire Fungi (Mason & Grainger, 1937, 43) and from Oxford. *Uredo vitellinae* on *S. vitellina* is recorded by Greville (Fl. Edin. 1824, 437) and this record probably refers to *M. salicis-albae*. Ellis (Trans. Norf. Norw. Nat. Hist. Soc. **14**, 106, 1936) recorded its occurrence in Norfolk.

Melampsora allii-fragilis Kleb.

Jahrb. Wiss. Bot. **35**, 674 (1901); Grove, Brit. Rust Fungi, p. 344; Wilson & Bisby, Brit. Ured. no. 33; Gäumann, Rostpilze, p. 157.

[**Spermogonia** scarcely projecting, lenticular, about 200 μ wide. **Aecidia** caeomoid, amphigenous and on the stems and bulbils, clustered on discoloured spots, surrounded by the ruptured epidermis, up to 2 mm. long and 0·5–1 mm. wide, bright orange-yellow; aecidiospores irregularly angular or ellipsoid, rarely globoid, 18–25 × 12–19 μ, wall 1–2 μ thick, finely verrucose.] **Uredosori** hypophyllous or partly epiphyllous, minute, 0·5 mm., circular, surrounded by the torn epidermis, reddish-orange, causing reddish-yellow spots on the upper side; paraphyses capitate or clavate, 50–70 × 15–20 μ, with thin pedicel and uniformly thickened membrane (3–5 μ); uredospores obovoid, 22–33 × 13–15 μ, wall up to 3 μ thick, distantly echinulate, but smooth and somewhat thinner above. **Teleutosori** chiefly epiphyllous, subcuticular, scattered or in groups, pulvinate, 0·25–1·5 mm. broad, dark brown, shining; teleutospores prismatic, rounded at both ends, 30–48 × 7–14 μ, wall pale brown, 1 μ thick, without evident pores; basidiospores orange. Hetereu-form.

[Spermogonia and aecidia on *A. ursinum*]; uredospores and teleutospores on *Salix fragilis* and *S. pentandra* and the hybrid between them. Great Britain, probably scarce.

The possible aecidial records for this species are dealt with under *M. allii-populina*. There is no good evidence that the aecidia of either have been found in Britain.

There is a specimen which appears to be the diplont of this species collected at Southampton (Kew Herbarium) and another in Edinburgh Herbarium from Forfar collected by U. K. Duncan.

The connection of the spore stages was first demonstrated by Klebahn and has been confirmed by Mayor (Bull. Soc. Neuchâtel. Sci. Nat. **58**, 27, 1933, and **61**, 115, 1936).

This species is distinguished from *M. larici-pentandrae* by its aecidial host and by its teleutosori which are mostly epiphyllous and sub-cuticular. The many species of *Allium* on which this rust forms aecidia are summarised by Gäumann (1959, 156).

PUCCINIACEAE

Phragmidium Link

Magn. Ges. Naturf. Fr. Berlin, 7, 30 (1815).

Spermogonia subcuticular, conical or flattened, without paraphyses but with flexuous hyphae. **Aecidia** caeomoid, usually encircled by incurved paraphyses; spores in chains, each with numerous pores. **Uredosori** definite, usually encircled by paraphyses; uredospores borne singly on pedicels, often with paraphyses intermixed, pores numerous, scattered. **Teleutospores** 2- to several-celled by transverse septa; wall thick, laminate, usually coarsely verrucose, the middle layer dark and rigid; pores two or more in each cell; pedicels often swollen below; basidiospores globoid.

All the species are autoecious and confined to the family *Rosaceae*.

The genus was divided into two sections by Arthur (1934, 78), *Earlea* with smooth or almost smooth teleutospores with firm non-hygroscopic pedicels and *Euphragmidium* with verrucose teleutospores and hygroscopic pedicels, swelling in the lower portion.

KEY TO BRITISH SPECIES OF PHRAGMIDIUM

1. Teleutospores smooth or almost smooth, pedicel non-hygroscopic   2
   Teleutospores verrucose, pedicel hygroscopic, swelling below   4

2. Teleutospores with short apiculus; on *Sanguisorba*   *P. sanguisorbae*, p. 102
   Teleutospores without apiculus; on *Potentilleae*   3

3. Uredospores verrucose   *P. fragariae*, p. 100
   Uredospores echinulate   *P. potentillae*, p. 101

4. On *Rosa*   5
   On *Rubus*   7

5. Teleutosori brown; teleutospores 6–8-celled; on *Rosa* sect. *Pimpinellifoliae*
      P. rosae-pimpinellifoliae, p. 103
   Teleutosori black   6

6. Pore membrane of uredospores and aecidiospores hemispheric projecting far into lumen of spores; teleutospores abruptly apiculate
      P. tuberculatum, p. 106
   Pore membrane of uredospores and aecidiospores less conspicuous; teleutospores gradually apiculate   *P. mucronatum*, p. 104

7. Uredospores aculeate or verrucose   8
   Uredospores finely echinulate   9

8. Teleutospores 3–5-(mostly 4-) celled   *P. violaceum*, p. 98
   Teleutospores 4–7- (mostly 6-) celled   *P. bulbosum*, p. 95

9. Teleutospores 4–8- (mostly 6–7-) celled; uredospore wall 1·5–2·5 μ thick
P. acuminatum, p. 98

Teleutospores 5–10- (mostly 7–8-) celled; uredospore wall 2–3 μ thick
P. rubi-idaei, p. 96

Phragmidium bulbosum (Str.) Schlecht.

Fl. Berol. **2**, 156 (1824); Gäumann, Rostpilze, p. 1201.
Puccinia mucronata β Puccinia rubi Pers., Syn. Meth. Fung. p. 230 (1801).
Puccinia rubi DC., Fl. Fr. **2**, 218 (1805), (non P. rubi Schum. (1803) = Phr. rubi-idaei
(DC.) Karst.).
Uredo bulbosum Str., Ann. Wetter. Ges. **2**, 108, (1810).
Phragmidium rubi (Pers.) Wint., Rabh. Krypt. Fl. Ed. 2, **1** (1), 230 (1882), p.p.;
Grove, Brit. Rust. Fungi, p. 297; Wilson & Bisby, Brit. Ured. no. 71.

Spermogonia epiphyllous, in minute clusters, hemispherical or conical. **Aecidia** hypophyllous, solitary or in minute groups, rounded, 0·5–1 mm. diam., or elongated on the nerves, orange, surrounded by clavate, curved, hyaline paraphyses 45–75 μ long, 7–12 μ thick, scarcely or not thickened at the apex; aecidiospores subgloboid or ellipsoid, yellow, 18–28 × 14–22 μ, wall 1–2 μ thick with few, rather large and irregular, flat, crowded warts, with 1–4 pores. **Uredosori** hypophyllous, scattered, minute or confluent, yellow, surrounded by paraphyses; uredospores globoid or ellipsoid, 20–28 × 14–21 μ, wall sparingly and rather distantly echinulate, yellow, 1–2 μ thick, with 2–4 pores. **Teleutosori** hypophyllous, small, on brownish spots, scattered, circular, up to 0·5 mm. diam., pulverulent, black; teleutospores cylindric, of 4–7 (mostly 6) cells, not constricted, rounded above with a hyaline apiculus up to 12 μ long, mostly 60–75 × 25–29 μ, wall brown, 5–7 μ thick, coarsely warted, with 2–4 pores to each cell, pedicel persistent, hyaline, up to 140 μ long, at the base up to 22 μ wide. Auteuform.

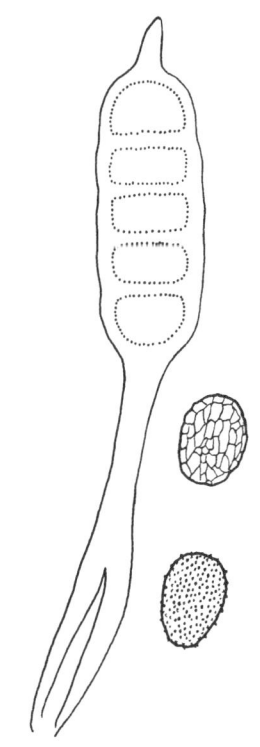

On Rubus fruticosus, section Corylifolii and on R. caesius. July–September. Great Britain and Ireland, common.

P. bulbosum. Teleutospore aecidiospore and uredospore (after Jørstad).

The occurrence on Rubus caesius is recorded by Plowright (1889, 224) and Grove; the rust on R. saxatilis also recorded by the latter has now been transferred to Phragmidium acuminatum (see p. 98).

This species forms the same type of purplish or reddish spots on the upper leaf surface above the sori, as *P. violaceum*. It is rather less common than the latter, possibly on account of its restriction to a smaller number of *Rubi*.

Klebahn (Z. Pfl.-Krankh. **17**, 139, 1907, and **22**, 321–50, 1912) showed experimentally that *R. caesius* and species of section *Corylifolii* may serve as hosts for the same race of *P. bulbosum*; he also succeeded in transferring this race to four species in other sections although the resulting infections were only weak. By infection with aecidiospores he was successful in producing uredospores.

Jørstad (1953*a*, 9) found that in Norway only species of *Corylifolii* were infected but that in Sweden and Denmark the rust occurred also on *R. caesius*.

The germination of the teleutospores was first described by Tulasne (Ann. Sci. Nat. Bot. Ser. 4, **2**, 147, 1854). Sappin-Trouffy (1897) gave an account of the life history and cytology.

Phragmidium rubi-idaei (DC.) Karst.

Bidr. Känned. Finl. Nat. Folk, **31**, 52 (1879); Grove, Brit. Rust Fungi, p. 298; Wilson & Bisby, Brit. Ured. no. 72; Gäumann, Rostpilze, p. 1205.

[*Uredo rubi-idaei* Pers., Syn. Meth. Fungi. p. 218 (1801).]
Puccinia rubi Schum., Enum. Pl. Saell, **2**, 235 (1803). (non *Phr. rubi* Wint. (1882) = *Phragmidium bulbosum* (Str.) Schlecht.)
Puccinia rubi-idaei DC., Fl. Fr. **6**, 54 (1815).
Puccinia gracilis Grev., Fl. Edin. p. 428 (1824).

Spermogonia epiphyllous, few, in groups surrounded by the aecidia, inconspicuous, conical, yellow, 45–90μ diam., 30–35μ high. **Aecidia** epiphyllous, in annular groups, pulvinate, orange-yellow fading to pale yellow with peripheral incurved, clavate, paraphyses 40–70μ long, 14–18μ thick, with uniformly thin walls; aecidiospores globoid or broadly ellipsoid, 16–24 × 14–18μ, wall pale yellow, 1·5–2·5μ thick, sparsely echinulate. **Uredosorus** hypophyllous, very small, 0·1–0·2 mm. diam., scattered, pale-orange, surrounded by numerous, incurved, thin-walled, clavate paraphyses, 40–70μ long, apex 14–24μ broad; uredospores broadly ellipsoid, yellow, 18–25 × 14–18μ, wall 1·5–2·5μ thick, sparsely echinulate, pores obscure. **Teleutosorus** hypophyllous, very small, 0·3–0·7 mm. diam., black; teleutospores cylindric, 6–10-celled (mostly 7–8), rounded below, tapering at the apex to a hyaline apiculus, 3–13μ long, chocolate-brown, coarsely verrucose, 80–135 × 28–35μ, wall 5–6μ thick with 3 pores in each cell, pedicel 80–120μ long, colourless except at the apex, swelling in water to lanceolate and 14–27μ broad in lower part; basidia 4-celled, basidiospores subgloboid, 4–6μ diam. Auteu-form.

On *Rubus idaeus*, spermogonia and aecidia, April–July; uredospores, May–October; teleutospores, July–October. Great Britain and Ireland, frequent.

The spermogonia and aecidia are much rarer than the other spore-forms. Klebahn (Z. Pfl.-Krankh. **17**, 142, 1907) by sowing teleutospores produced aecidiospores, and by sowing the latter, uredospores; the teleutospores germinate in spring. Zeller (J. Agric. Res. **34**, 857, 1927)

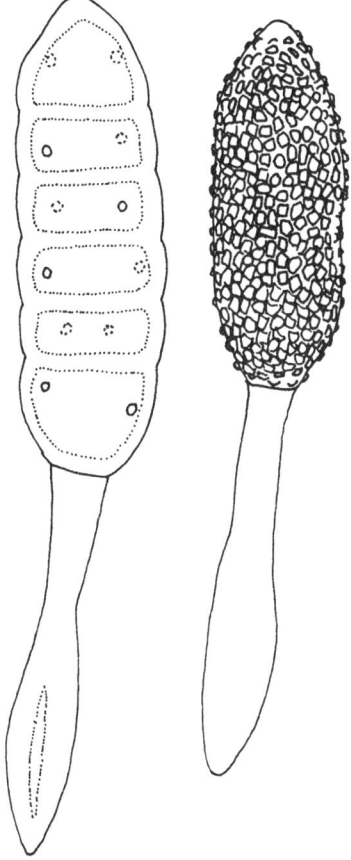

P. rubi-idaei. Teleutospores.

described the occurrence of uredosori and teleutosori on the stems in North America, but these do not appear to have been observed in this position in this country. Zeller & Lund (Phytopath. **24**, 257, 1934) studied the life history and conditions of infection in North America and investigated the susceptibility of various species and varieties of *Rubus*; they also gave notes on the cytology of the rust.

Phragmidium acuminatum (Fr.) Cooke

Handb. Brit. Fungi, p. 490 (1871).
Aregma acuminata Fr., Obs. Mycol. 1, 226 (1815).
Phragmidium rubi-saxatilis Liro, Bidr. Känned. Finl. Nat. Folk, 65, 421 (1908);
Grove, Brit. Rust Fungi, p. 297; Wilson & Bisby, Brit. Ured. no. 73.

Spermogonia absent? Aecidia amphigenous, generally hypophyllous, scattered or grouped, rounded, minute, about 0·3–0·5 mm. diam. elongated on the nerves, up to 5 mm. long, yellow, surrounded by numerous hyaline, cylindrical paraphyses, rounded at the apex, 50–75 μ long, 5–8 μ thick; aecidiospores globoid, subgloboid or ellipsoid, sparsely aculeate, yellow, 18–28 × 16–22 μ, wall 2–3 μ thick, with 3–7 pores. Uredosori hypophyllous, scattered, minute, rounded, yellow, surrounded by cylindrical-clavate, hyaline, paraphyses 50–80 μ long and, at the apex, up to 16 μ wide; uredospores globoid, subgloboid, ellipsoid or ovoid, yellow, 18–28 × 16–23 μ, wall 1·5–2·5 μ thick, sparsely echinulate, pores indistinct. Teleutosori hypophyllous, scattered or grouped, minute, sometimes confluent, pulverulent, black; teleutospores cylindrical or fusoid-cylindrical 4–8 generally 6–7-celled, not constricted at the septa, with a hyaline or pale-coloured apical papilla 7–18 μ long, 52–110 × 25–32 μ, wall 3–4 μ thick, densely verrucose, chestnut-brown, each cell with 2–4 pores; pedicels persistent, hyaline, up to 130 μ long, and up to 11 μ wide in the upper part, dilated to 18 μ near the base. Auteu-form.

On *Rubus saxatilis*. Scotland, very rare.

This species is distinguished from *Phragmidium bulbosum* by the larger number of cells in the teleutospore. It has been found in Britain only by Keith at Forres in 1879.

Phragmidium violaceum (C. F. Schultz) Wint.

Hedwigia, 19, 54 (1880) and Rabh. Krypt. Fl. Ed. 2, 1 (1), 231, (1882); Grove, Brit. Rust Fungi, p. 295; Wilson & Bisby, Brit. Ured. no. 75; Gäumann, Rostpilze, p. 1196.
Puccinia violacea C. F. Schultz, Prod. Flor. Starg. p. 459 (1806).

Spermogonia epiphyllous, usually in the lighter centre of a conspicuous dark violet or reddish leaf spot, densely crowded, hemispherical; spermatia oval. Aecidia caeomoid, hypophyllous or on the stems on conspicuous reddish spots with violet-red margins, scattered or confluent, circular or elongate up to 1 cm. when on the nerves and stems, orange-yellow, surrounded by straight or slightly curved, hyaline paraphyses up to 60 μ long and 18 μ wide; aecidiospores ellipsoid or oblong, 19–30 × 17–24 μ, wall yellow, 3–3·5 μ thick, distantly and strongly aculeate-verrucose, spines about 4–5 μ apart. Uredosori hypophyllous, on similar spots, scattered or confluent, minute, orange-yellow, surrounded by numerous, hyaline, curved, clavate or capitate paraphyses, 45–60 μ long, 14–22 μ wide; uredospores globoid or ellipsoid, yellow, 19–30 × 18–25 μ; wall 3–4·5 μ thick with indistinct pores, distantly and strongly aculeate-verrucose. Teleutosori hypophyllous, on similar spots, scattered or confluent, up to 1 mm. or more in diameter, pulvinate, black; teleutospores chestnut-brown, oblong or widely cylindrical, 1–5-(mostly 4-) celled, not or slightly constricted, rounded at both ends, with a short yellowish apiculus, 3–4μ long, covered by numerous hyaline warts, 65–100 × 30–36μ, wall with 2–4 pores in each cell, 6–8μ thick, pedicel persistent, hyaline, up to 190μ long, swollen at the base up to 18μ wide. Auteu-form.

On *Rubus fruticosus* agg. belonging to several sections, on cultivated blackberries, loganberries and *R. laciniatus*; spermogonia and aecidia, May; uredospores, June–September; teleutospores, July until the following spring. Great Britain and Ireland, very common.

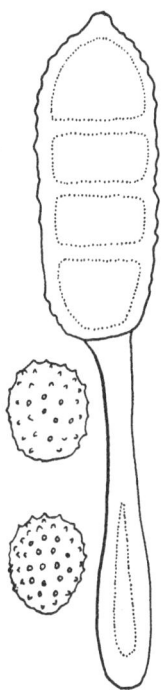

P. violaceum. Teleutospore, aecidiospore and uredospore.

This species is recognised by its large, conspicuous, red and purple spots on the upper surface above the sori and, on microscopical examination, by the predominance of 4-celled teleutospores. They pass the winter on the leaves which often remain green on the plant, and germinate in early spring. Leaves may become very heavily infected with uredospores and when this occurs the edges frequently turn up; peduncles and sepals may also be infected. Spermogonia and aecidia are found 12–14 days after infection with teleutospores.

According to Vleugel (Sv. Bot. Tidskr. **2**, 123, 1908) and Klebahn (Z. Pfl.-Krank. **22**, 333, 1912) *P. violaceum* infects neither *R. caesius* nor most of the species of section *Corylifolii* but it is found on many species of the other sections. Eriksson (Arkiv Bot. **18**, no. 18, 1923) also carried out cultures on several species of *Rubus*; he obtained infections with teleutospores and with aecidiospores. Jørstad (1953*a*) has concluded that species from several sections are attacked in Norway.

This rust was recorded by Rilstone (J. Bot. Lond. **70**, 319, 1932) as very common on *R. adscitus* Genev. in Cornwall; it has been found on *R. laciniatus* Willd. at two localities in Scotland and is also recorded from Kent on this species.

Teleutospore germination was described by Blackman (New Phytol. **2**, 10, 1903) who also described the cytology and life history (Ann. Bot. Lond. **18**, 323, 1904). His work on nuclear migration was repeated and confirmed by Welsford (Ann. Bot. Lond. **29**, 293, 1915). Ashworth (Ann. Bot. Lond. **49**, 95, 1935) described emergent hyphae in the spermogonia.

It should be noted that *Phragmidium candicantium* recorded by Jørstad (1953*a*) in Norway has not been found in Britain. It appears to favour *Rubi* of section *Candicantes* as hosts and can be differentiated by the spacing of the echinulations on the uredospore wall (2·5–3 μ apart in *P. candicantium*, 4–5 μ in *P. bulbosum* and 1·5–2 μ in *P. violaceum*). Morever *P. candicantium* tends to have a proportion of teleutospores with 6–7, often even 8, cells per spore.

Phragmidium fragariae (DC.) Rabh.

Herb. Myc. no. 1987 (1855).
Puccinia fragariae DC., Lam. Encycl. Méth. Bot. **8**, 244 (1808).
Puccinia fragariastri DC., Fl. Fr. **6**, 55 (1815).
Phragmidium granulatum Fuck., Symb. Myc. p. 46 (1869); Gäumann, Rostpilze, p. 1179.
Phragmidium fragariastri (DC.) Schroet., Cohn Krypt. Fl. Schles. **3** (1), 351 (1887); Grove, Brit. Rust Fungi, p. 290; Wilson & Bisby, Brit. Ured. no. 67.

Spermogonia in small clusters, honey-coloured, often surrounding the aecidia. **Aecidia** caeomoid, mostly hypophyllous or on the veins and petioles, scattered, rounded or oblong, 0·5–2 mm. long, often confluent, orange, surrounded by clavate paraphyses; aecidiospores sub-globoid, ovoid or ellipsoid, sometimes angular, orange-yellow, 17–28 × 14–21 μ, wall about 2 μ thick, densely verrucose. **Uredosori** hypophyllous, on small yellow spots, scattered, minute, soon naked, orange-yellow, surrounded by and including numerous, hyaline, capitate, thin-walled paraphyses up to 80 μ long and 10–20 μ wide; uredospores globoid or ovoid, orange-yellow, 18–25 × 16–22 μ, wall densely verrucose, 2 μ thick, with indistinct pores. **Teleutosori** hypophyllous, scattered, minute, punctiform, soon naked, pulverulent, brown or blackish-brown; teleutospores oblong or widely cylindric, 2–5- (mostly 4-) celled,

rounded at both ends, very slightly constricted 40–70 × 25–28 μ, wall pale brown, sometimes slightly thickened and paler

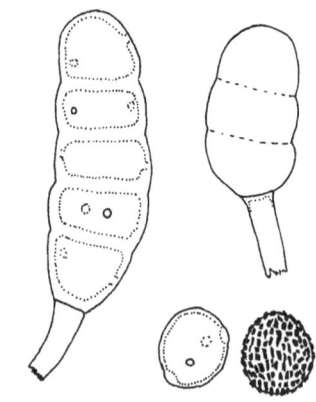

P. fragariae. Teleutospores and aecidiospores.

at the summit but never papillate, sometimes with a few rather delicate warts which are more abundant towards the apex but generally quite smooth, with usually 3 pores per cell, pedicel colourless, persistent, up to 35 μ long. Auteu-form.

On *Potentilla sterilis*, March–October. Great Britain and Ireland, very common.

The uredospores of this species are distinguished from those of its allies by being densely and rather coarsely verruculose and very similar to the aecidiospores, from which, in fact, they differ almost solely in being abstricted singly and not in chains. The aecidium is one of the earliest rusts of spring, showing on the leaves as soon as they are well developed, and often attacking the calyx. Gäumann (1959, 1181) stated that the surface of the teleutospore may vary considerably, even in one infection, from coarsely warted to almost smooth.

The rust is very common on *P. sterilis*; the records on *P. reptans* from Dublin (TBMS. **11**, 16, 1926) and *P. erecta* from Windsor (*ibid.* **10**, 6, 1928) made on forays of the British Mycological Society are almost certainly incorrect.

The specific epithet *fragariae* DC. (1808) has priority, although it was an error for *fragariastri* and later rejected by De Candolle himself in 1815.

Phragmidium potentillae (Pers.) Karst.

Bidr. Känned. Finl. Nat. Folk, **31**, 49 (1879); Grove, Brit. Rust Fungi, p. 291; Wilson & Bisby, Brit. Ured. no. 70; Gäumann, Rostpilze, p. 1181.
Puccinia potentillae Pers., Syn. Meth. Fung. p. 229 (1801).

Spermogonia, amphigenous, few, surrounded by the aecidia, 100–160 μ diam., 25–40 μ high, yellow. **Aecidia** amphigenous, scattered or aggregated, round, 0·5–1·5 mm. diam., surrounded by the ruptured epidermis and by clavate or cylindrical paraphyses, up to 80 μ long, and 6–10 μ wide, orange; aecidiospores subgloboid, ovoid or ellipsoid, yellow, 18–28 × 14–21 μ, wall densely verrucose, about 2 μ thick. **Uredosori** hypophyllous, scattered, rounded, minute, 0·5–1 mm. diam., often confluent, at first covered by the swollen epidermis, surrounded by abundant, clavate, curved paraphyses up to 80 μ long and 10–20 μ wide, orange-yellow; uredospores globoid to ovoid, yellow, 21–24 × 16–19μ, wall finely echinulate, 1·5–2 mm. thick. with scattered, indistinct pores. **Teleutosori** hypophyllous, scattered or grouped, rounded, soon naked, black; teleuto-

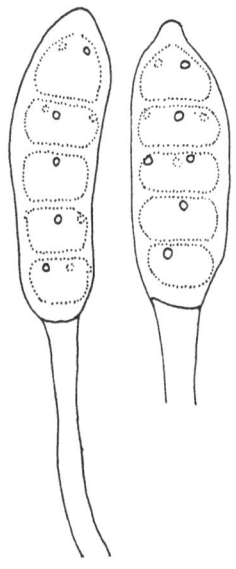

P. potentillae. Teleutospores.

spores cylindric or clavoid, 3–6-celled (occasionally 1–2), hardly constricted at the septa, rounded or bluntly papillate at the apex, base rounded, 42–80 × 20–28 μ, cells generally equal in size, except for the larger uppermost one, wall, smooth, brown, 3–4 μ thick, with 2–3 pores in the upper part of each cell, pedicel persistent, hyaline, of equal thickness, up to 175 μ long, 8–12 μ thick. Auteu-form.

On *Potentilla anglica*, '*P. argentea*', *P. tabernaemontani* and on cultivated '*Potentilla* sp.'. Great Britain and Ireland, scarce.

This species is more closely allied to *Phragmidium sanguisorbae* than to *P. fragariae*. The finely echinulate uredospores and the papillate teleutospores distinguish it from the latter. In the uredospores stage it appears to be indistinguishable from *Frommea obtusa*.

Greville (Fl. Edin. p. 428, 1824) recorded the species on *Potentilla argentea* near Edinburgh but there is no material to substantiate this. On *P. tabernaemontani* it occurs at a few localities in East Lothian and O'Connor (1936, 391) recorded it on *P. anglica* from Killarney but only uredosori are present in his material.

Eriksson (Arkiv Bot. **18**, 28, 1923) carried out cultures on several species of *Potentilla* and found no evidence for specialisation; infections were carried out with teleutospores and aecidiospores.

Hiratsuka (Jap. J. Bot. **7**, 227, 1935) confirmed the relationship of the spore forms experimentally.

Phragmidium sanguisorbae (DC.) Schroet.

Cohn, Krypt. Fl. Schles. 3 (1), 352 (1887); Grove, Brit. Rust Fungi, p. 292; Wilson & Bisby, Brit. Ured. no. 74.

Puccinia sanguisorbae DC., Fl. Fr. **6**, 54 (1815).

Phragmidium poterii [Schlecht.] Fuck., Symb. Myc. p. 46 (1870); Gäumann, Rostpilze, p. 1183.

Spermogonia amphigenous, on purple spots, in rounded groups, flat, honey-coloured. **Aecidia** caeomoid, amphigenous, often circinate around the spermogonia, minute, larger and elongated on the petioles and nerves, orange, surrounded by curved, clavoid, paraphyses; aecidiospores globoid or ellipsoid, yellow, 18–25 × 16–22 μ, wall densely verrucose with 6–8 indistinct pores. **Uredosori** mostly hypophyllous, scattered, very small, 0·25 mm. or less diam., orange-yellow, surrounded by clavoid, curved paraphyses, 30–50 μ long, 10–17 μ wide; uredospores globoid to ellipsoid, orange-yellow, 17–24 × 16–20 μ, wall echinulate 1–1·5 μ thick with 6–8 indis-

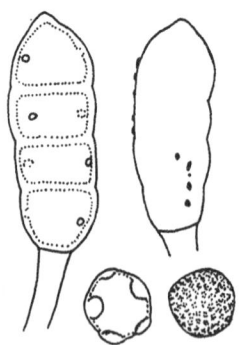

P. sanguisorbae. Teleutospores and uredospores.

tinct pores. **Teleutosori** hypophyllous, scattered, rounded, minute, 0·25–1 mm.

diam., soon naked, surrounded by numerous clavoid, curved paraphyses up to 60μ long; teleutospores ellipsoid-oblong or widely cylindrical, 2–5-(mostly 4-) celled, slightly constricted at septa, apex rounded with a hyaline papilla up to 5μ long, base rounded, 40–70 × 20–26μ, wall with minute, scattered warts, brown, with 2–3 pores in each cell; pedicels hyaline, 20–25μ long and up to 14μ thick. Auteu-form.

On *Poterium sanguisorba* and *P. polygamum*, aecidia, April–June; teleutospores, July–November. Great Britain, frequent in England and rare in southern Scotland, Ireland.

The collections on *Poterium polygamum* were made in Yorkshire and Oxford.

Phragmidium rosae-pimpinellifoliae Diet.

Hedwigia, **44**, 339 (1905); Gäumann, Rostpilze, p. 1186.

Phragmidium rosarum f. *rosae-pimpinellifoliae* Rabh., Fungi Europ. no. 1671 (1873) [*nomen nudum*].

As *Phragmidium disciflorum* James, Grove, Brit. Rust Fungi, p. 293 *p.p.*

[**Spermogonia** chiefly caulicolous, irregularly scattered, often in groups, inconspicuous, 15–30μ high, honey-coloured.] **Aecidia** caeomoid, on branches, petioles and veins of the leaves and fruits, confluent in large pustules up to 10 cm. long, orange-yellow when fresh, surrounded by cylindric-clavate paraphyses sometimes abundant but usually few or none, 30–50μ long, 10–15μ wide: aecidiospores ellipsoid or globoid 18–27 × 15–20μ, wall nearly or quite colourless, 2–2·5μ thick, finely verrucose with 6–8 indistinct pores. **Uredosori** hypophyllous, scattered, minute, 0·1 mm. diam., orange, surrounded by numerous strongly incurved cylindrical paraphyses, 30–50μ long, 8–12μ wide, sometimes slightly dilated above; uredospores obovoid-globoid, 18–25 × 16–20μ, wall nearly or quite colourless, 2–2·5μ thick, closely echinulate, pores scattered, indistinct. **Teleutosori** hypophyllous, scattered, very small, 0·1 mm. diam., chestnut-brown; teleutospores oblong-cylindric to obovoid-cylindric, 6–8-celled, the uppermost cell rather longer, not constricted, rounded above, the apiculus prominent, 14–16μ long, coloured and roughened except at the apex, rounded below, chestnut-brown, 70–115 × 25–34μ, wall very opaque, 3–5μ thick, finely verrucose especially in the upper part, usually nearly or quite smooth below; pedicel one and a half times length of spore, colourless except near the spore, swelling in water in lower part to up to 25μ thick. Auteu-form.

On *Rosa pimpinellifolia* and on cultivated roses especially (perhaps exclusively) those with *pimpinellifoliae* parentage. Great Britain, uncommon.

This rust was included by Grove under *P. disciflorum* (= *P. mucronatum*) but has been shown in Switzerland by Bandi (Hedwigia, **42**, 118, 1903) who made cultures with aecidiospores, to be biologically distinct. It differs from *P. mucronatum* in its smaller aecidiospores which are surrounded by few or no paraphyses, its brown teleutosori and much more finely verrucose teleutospores which are usually almost or quite smooth below. Bandi found that repeating aecidia occur in this species.

Phragmidium mucronatum (Pers.) Schlecht.

Fl. Berol. **2**, 156 (1824); Gäumann, Rostpilze, p. 1190.

Puccinia mucronata α *Puccinia rosae*
Pers., Syn. Meth. Fung. p. 230 (1801).
Phragmidium subcorticium Wint., Rabh.
Krypt. Fl. Ed. 2, **1** (1), 228 (1882).
Phragmidium disciflorum James, Contr.
U.S. Nat. Herb. **3**, 276 (1895); Grove,
Brit. Rust Fungi, p. 293; Wilson &
Bisby, Brit. Ured. no. 69.

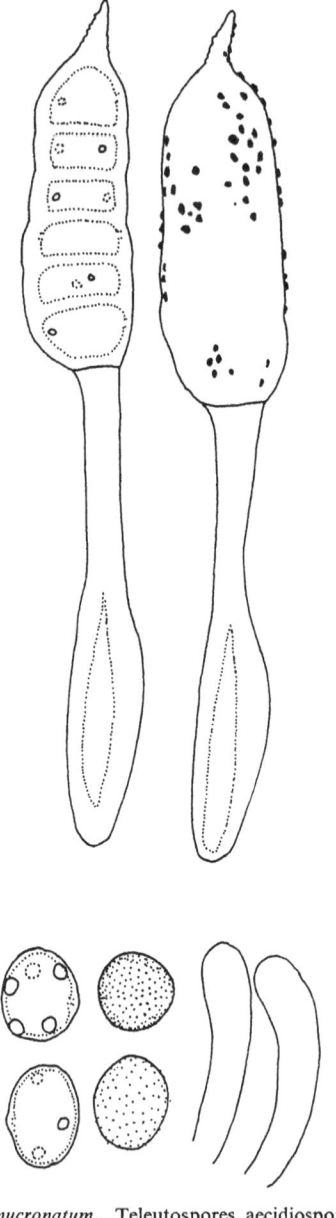

Spermogonia epiphyllous, in small groups, flat, 110–115 μ diam. 35–40 μ high, pale honey-yellow. **Aecidia** caeomoid, on branches and petioles, hypophyllous on the leaf veins, and on the fruits, often confluent on branches, on the leaves mostly roundish, bright orange, surrounded by a circle of upright clavate or clavate-capitate, hyaline, thin-walled paraphyses up to 80 μ long, 8–18 μ wide; aecidiospores globoid to ellipsoid, 20–28 × 17–21 μ, wall finely and distantly verruculose-echinulate, pale yellow, 1–2 μ thick, pores 2–2·5 μ in diameter the inner surface of the membrane almost flush with the inner surface of the spore wall. **Uredosori** hypophyllous, scattered or in groups, very small, 0·1–0·2 mm. diam., rounded, soon naked, pale orange, surrounded by a circle of clavoid, curved colourless paraphyses, up to 70 μ long, and 7–8 μ wide; uredospores globoid, ellipsoid or ovoid, 20–28 × 14–20 μ, wall closely echinulate, pale yellow, about 2 μ thick, with 8 or more scattered pores, 2–2·5 μ diam., the membrane almost flush with the inner surface of the spore wall. **Teleutosori** hypophyllous, scattered or in groups, rounded, minute or confluent, black; teleutospores ellipsoid to subfusoid, 6–8- rarely 5–9-celled, not constricted, rounded at the base with an almost hyaline, distinctly roughened, apical papilla 7–13 μ long gradually expanding into the spore apex, 64–90 × 22–23 μ, wall blackish-brown, 5–7 μ thick, coarsely verrucose with almost hyaline tubercles, with 2–3 pores in each

P. mucronatum. Teleutospores, aecidiospores, uredo-paraphyses and uredospores.

cell; pedicels one and a half times length of spore, upper half colourless, lower half coloured, swelling greatly in water. Auteu-form.
On '*Rosa arvensis*', *R. canina*, '*R. centifolia*', '*R. involuta*', '*R. laxa*', '*R. rubiginosa*', '*R. rugosa*', '*R. tomentosa*', '*R. villosa*' and many cultivated forms, spermogonia and aecidia, May–June; uredospores and teleutospores, July–October. Great Britain and Ireland, very common.

The teleutospores can germinate in the spring and produce the aecidial stage (Jacky, Zbl. Bakt. II, **18**, 91, 1907; Eriksson, Arkiv. Bot. **18**, no. 18, 1924) but they germinate with difficulty (Müller, Landw. Jahrb. **15**, 719, 1886; Bandi, Hedwigia, **42**, 118, 1903; Bewley, Scient. Hort. **6**, 97, 1938; Williams, Ann. Appl. Biol. **25**, 730, 1938). It also overwinters by the mycelium of the aecidial stage which is possibly perennial. Eriksson (*loc. cit.*) stated that aecidia on the stems are not accompanied by spermogonia. The structure of the deformed and swollen stems has been investigated by Eriksson (*loc. cit.*) and Wenzl (Z. Pfl.-Krankh. **46**, 204, 1936).

Bewley and Williams investigated the rusts found on cultivated roses, on briars (forms of *R. canina*), on *R. laxa* and on *R. rubiginosa*. They found considerable variation in the number of cells and in the length of the teleutospores and stated that strains on cultivated roses, briars and *R. laxa* and certain varieties of *R. canina* show constant morphological differences. *Rosa laxa* and certain varieties of *R. canina* were not attacked by the rust from cultivated roses and the latter were resistant to certain inoculates from *R. canina*. It is almost certain that they used both *Phragmidium mucronatum* and *P. tuberculatum* in their experiments. Severity of attack by the rust was found to depend on dew formation and the presence of a high water table in the soil; resistance was increased when transpiration was not excessive and root action sufficiently active to supply the water required.

This rust was figured by Hooke in his Micrographia in 1665 (p. 121, table 12, fig. 2), this being the first published enlarged drawing of a microfungus. Its cytology was investigated by Sappin-Trouffy (1897, 158) and Moreau (Le Botaniste, **13**, 162, 1914).

A general account of the life history and the disease on the rose was published by Cochrane (Cornell Univ. Agric. Exp. Sta. Mem. 1944, 268); this includes an account of the effect of temperature and humidity on the germination of the spores. The same author has given an account of the effect of artificial light on the germination of the uredospores (Phytopath. **35**, 458, 1945).

Phragmidium tuberculatum J. Müller

Ber. D. Bot. Ges. 3, 391 (1885); Grove, Brit. Rust Fungi, p. 388; Gäumann, Rost-pilze, p. 1188.

Spermogonia epiphyllous in minute groups, honey-coloured, 90–100μ diam., 20–40μ high; spermatia ovoid, 2·5–3·3 × 1·6μ. **Aecidia** on branches, petioles and nerves of the leaves in elongate pustules, on the leaves in smaller, rounded pustules, surrounded by clavate, hyaline paraphases; aecidiospores globoid, subgloboid or ellipsoid, densely verrucose, hyaline, 20–30 × 18–24μ, wall rather thick with 6–8 scattered pores 4·5μ diam. with membranes apparently hemispheric in optical section. **Uredosori** hypophyllous, scattered or in groups, very small, pale yellow, surrounded by inwardly curved, clavate paraphyses, up to 60μ long and 6–18μ wide, uredospores globoid, subgloboid, ellipsoid or ovoid, verrucose-echinulate, yellow, 20–25 × 16–24μ, wall 1·5μ thick with 6–8 scattered pores, 4·5μ, pore membrane in optical section hemispheric, intruding into the lumen of the spore. **Teleutosori** hypophyllous, scattered or in groups, small, black; teleutospores ellipsoid-oblong to cylindroid, 4–6-celled, not constricted at the septa, base rounded, the uppermost cell longer than the rest, with a pale or hyaline apical papilla up to 22μ long abruptly passing into the apical spore membrane, verrucose, chestnut-brown, 55–110 × 30–36μ, wall 6–7μ thick with 2–3 pores in each cell; pedicel hyaline, persistent, equal to the spore, base enlarged up to 30μ diam. Auteu-form.

Spermogonia, aecidia, uredospores and teleutospores on *Rosa rubiginosa*, *R. rugosa* and various undetermined cultivars. Probably common, but not hitherto adequately differentiated.

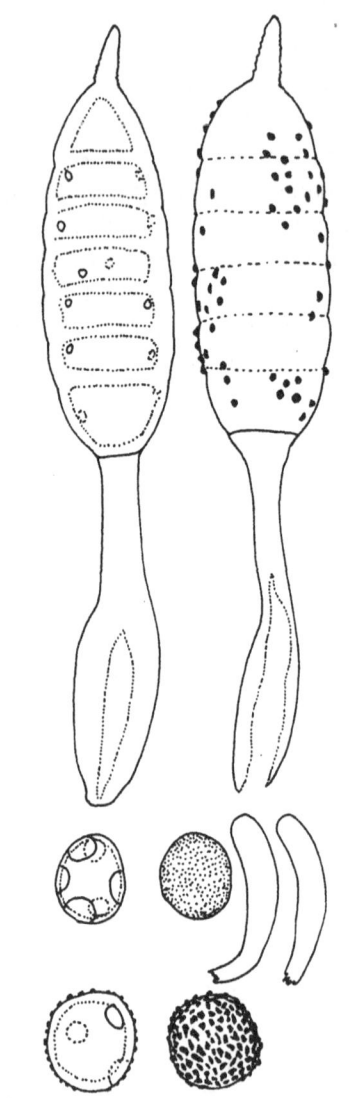

P. tuberculatum. Teleutospores, aecidiospores, uredo-paraphyses and uredospores.

This species was recorded by Ramsbottom (TBMS. **4**, 103, 1913) but dropped from the Wilson & Bisby list. It has been misunderstood due to attempts to differentiate it from *Phragmidium mucronatum* by the number of cells in the teleutospore. That method is quite inadequate. The abrupt contraction of the teleutospore apex into the apiculus and the large uredospores pores in *P. tuberculatum* readily distinguish it from *P. mucronatum*. Information on host range is inadequate to make any detailed statements.

As noted under *P. mucronatum*, William's (Ann. Appl. Biol. **25**, 730, 1938) experiments probably involved a mixture of the two species.

Kunkelia nitens (Schw.) Arth.

Bot. Gaz. **63**, 504 (1917).

This rust was recorded by Moore (1959, 193) on an importation of dewberry in 1931. The infection was completely eradicated by destroying the infected plants.

Kuhneola Magn.

Bot. Zbl. **74**, 169, (1898).

Spermogonia epiphyllous, subcuticular, with a flat hymenium, without paraphyses. **Aecidia** uredinoid, epiphyllous, without peridium or paraphyses; aecidiospores solitary, wall pale yellow or colourless, verrucose, with obscure pores. **Uredosori** usually hypophyllous with or without peripheral paraphyses; uredospores solitary, wall pale yellow, echinulate. **Teleutosori** hypophyllous, without paraphyses; teleutospores unicellular, hyaline or coloured, firmly connected in more or less elongate chains, with colourless or faintly coloured wall, smooth, with one apical pore, germinating immediately when mature; pedicel very short or wanting.

The genus is largely confined to the *Rosaceae*, but occurs on the *Malvaceae* in America. The only British species occurs on *Rubus*.

Arthur (1934, 93) considered that each chain represents a many-celled teleutospore but Dietel (Ann. Myc. **10**, 205, 1912) held that the teleutospores are unicellular and develop from the consecutive cells of a hypha; they are not held together by an external membrane as teleutospores of *Phragmidium*. The Sydows (Monogr. Ured. **3**, 313) agreed with this view of their structure but Cummins (1959) regards them as multicellular structures.

Kuhneola uredinis (Link) Arth.

Res. Sci. Congr. Bot. Vienne, p. 342 (1905); Wilson & Bisby, Brit. Ured. no. 32.
Oidium uredinis Link, Sp. Pl. Ed. 4, 6 (1), 123 (1824).
Chrysomyxa albida Kühn, Bot. Zbl. 16, 154 (1883).
Uredo muelleri Schroet., Cohn, Krypt. Fl. Schles. 3 (1), 375 (1887).
Kuhneola albida (Kühn) Magn., Bot. Zbl. 74, 169 (1898); Grove, Brit. Rust Fungi,
    p. 300; Gäumann, Rostpilze, p. 197.

Spermogonia epiphyllous, on reddish spots, subcuticular, pustular, large, prominent, 150–200μ diam., up to 100μ high. Aecidia uredinoid, epiphyllous, on pale yellow spots, surrounding the spermogonia, irregularly elongate and often confluent into rings, orange-yellow; aecidiospores globoid or obovoid, 19–23 × 18–20μ, wall colourless, 2–2·5μ thick, closely verrucose, pores obscure. Uredosori hypophyllous, sometimes on the petioles or caulicolous, scattered, pulverulent, on pale golden-yellow spots, without paraphyses, on the leaf minute, about 0·1 mm. diam., often covering the whole surface, soon naked, surrounded by the ruptured epidermis; on the stem subcortical, linear or narrowly oblong, up to 1 cm. long, 1–1·5 mm. wide, opening by a longitudinal, median rupture; on the petiole similar but smaller; uredospores obovoid 21–27 × 16–19μ, wall nearly colourless, 1·5–2μ thick, finely and closely verrucose-echinulate, pores indistinct, probably 3–4, equatorial. Teleutosori hypophyllous, scattered, isolated, or in roundish groups but never confluent, 0·1–0·5 mm. diam., soon naked, pulvinate, yellowish or whitish; teleutospores in clavate or elongate-cuneate chains which are straight or slightly curved, 40–120μ long, each of 2–12 (generally 5–7) superposed teleutospores, each teleutospore obtusely cuneate or ovoid-ellipsoid, often trapezoidal, 17–30μ long and 17–24μ wide (but some-

times wider than long), with colourless, smooth walls, the side walls becoming thicker from below upwards, the lower transverse wall thin and the upper thicker with irregular undulations or projections, the terminal spore of each chain

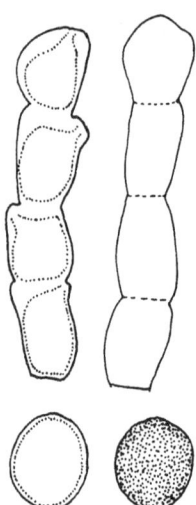

K. uredinis. Teleutospores and uredospores.

with a thick upper transverse wall undulate or with finger-like projections; in the lower spores the pore in lower cells superior on short lateral projection, pore of terminal spore apical; each chain with a short, thin-walled, basal cell. Brachy-form.

On Rubus fruticosus sensu lato. Great Britain and Ireland, frequent.

According to Klebahn (Z. Pfl.-Krankh., 22, 335, 1912) most Rubi can be easily infected with the uredospores of this rust, and Jørstad (1953 a, 12) has stated that in Norway it occurs on many of the sections of Rubus fruticosus.

Its life history is now generally known. The connection of *Uredo muelleri* with it was experimentally shown by Jacky (Zbl. Bakt. II, **18**, 91, 1907) who, by sowing teleutospores on the leaves on 24 July, produced spermogonia on 4 August and aecidia (primary uredosori) on 22 August; these are borne on a localized mycelium. According to Strelin (Myc. Zbl. **1**, 92 and 131, 1912) the aecidiospores germinate only after a resting period, usually in the following spring and infect leaves of the current year. Several generations of uredospores are borne on the under surface and these are followed by teleutospores from the middle of July until early autumn; these germinate at once and *in situ*. Elongated scattered uredosori are produced on the stems and occasionally on petioles of leaves of the previous year in April or May; presumably these are the result of infection by aecidiospores or uredospores during the previous summer or autumn, the mycelium remaining in a resting condition through the winter (see Klebahn, *loc. cit.*; Fischer & Johnson, Phytopath, **40**, 199, 1950).

Grove (*loc. cit.*) stated that this rust occurred on *R. caesius*. This was probably based on the statement by Hariot (Les Urédinées, 1908, 244) that it occurs on this species in France; it appears to be an error, although this host species is also included by Gaumann (*loc. cit.*). Jørstad (1953a) considers this rust to be rather wide ranging in the genus *Rubus*; there is insufficient information to draw any conclusions on the host range of the rust in Britain.

Xenodocus Schlecht.

Linnaea, **1**, 237 (1826).

Spermogonia subcuticular without paraphyses. **Aecidia** large, caeomoid, subepidermal, without paraphyses; aecidiospores catenulate, with colourless wall without pores. **Uredosori** unknown. **Teleutosori** subepidermal, black; teleutospores borne singly on fragile pedicels, of 3 to many superposed cells, pores 2 in a cell, opposite and superior, except in the terminal cell which has 1 apical pore, wall smooth, coloured.

A genus related to *Phragmidium* but differing in the arrangement of the germpores in the teleutospores. Only two species are known; both autoecious on the genus *Sanguisorba*.

Xenodocus carbonarius Schlecht.

Linnaea, **1**, 237 (1826); Grove, Brit. Rust Fungi, p. 302; Wilson & Bisby, Brit. Ured. no. 319; Gäumann, Rostpilze, p. 1207.

Phragmidium carbonarium Wint., Rabh. Krypt. Fl. Ed. 2, **1** (1), 227 (1882).

Spermogonia not known. **Aecidia** hypophyllous, on yellow or purple spots, on the petioles and veins of the leaves and stems, up to 1 cm. long, pulverulent, orange, without paraphyses; aecidiospores globoid or obovoid, orange, 17–

26×16–22μ; wall 2μ thick, densely verruculose, pores indistinct. **Uredosori** wanting. **Teleutosori** amphigenous, often confluent with the aecidia, soon naked, pulvinate, 1–3 mm. diam., black; teleutospores elongate-cylindrical, often curved, 4–22-celled, rounded at both ends, strongly constricted, smooth, dark-brown, 200–300×24–28μ; each cell with 2 opposite superior pores, the uppermost with one apical pore with a small hyaline papilla; wall brown, in basal cells often nearly colourless, 2μ thick, pedicel short, colourless, persistent. Opsis-form.

On *Sanguisorba officinalis*, June–October. England, Wales, Scotland, scarce.

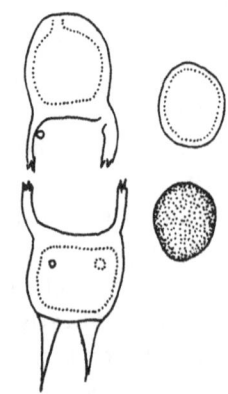

X. carbonarius. Apex and base of a multicelled teleutospore and aecidiospores.

This rust has been recorded quite frequently in southern England, but in Scotland only from Melrose. The whole chain of cells in the teleutospore is surrounded by a distinct subhyaline membrane which swells up considerably in lactic acid.

Frommea Arth.

Bull. Torr. Bot. Club, **44**, 503 (1917).

Spermogonia subcuticular with a flat hymenium without ostiolar filaments. **Aecidia** uredinoid, without peridium or paraphyses; aecidiospores borne singly on pedicels, the wall colourless, finely verrucose-echinulate, with obscure equatorial pores. **Uredosori** with short peripheral paraphyses or none; uredospores borne singly on pedicels, the wall finely echinulate with lateral pores. **Teleutosori** without paraphyses; teleutospores 2–7-celled, cinnamon-brown, smooth, with one apical pore in each cell; pedicel colourless except near the spore. Autoecious.

The single European species was placed by Grove in *Kuhneola* with *K. albida* but he stated that he was of the opinion that the two species were not congeneric. All the species are found on the *Rosaceae*.

Frommea obtusa (Str.) Arth.

Bull. Torr. Bot. Club, **44**, 503 (1917); Wilson & Bisby, Brit. Ured. no. 25; Gäumann, Rostpilze, p. 1174.

Uredo obtusa Str., Ann. Wetter. Ges. **2**, 107 (1810).
Phragmidium tormentillae Fuck., Jahrb. Nass. Ver. Nat. **23-4**, 46 (1870).
Kuehneola tormentillae (Fuck.) Arth., Res. Sci. Congr. Bot. Vienne, p. 342 (1906); Grove, Brit. Ured. no 301.

Spermogonia epiphyllous in small groups on reddish and rather hypertrophied spots. **Aecidia** uredinoid, epiphyllous, surrounding the spermogonia, orange-yellow when fresh; aecidiospores obovoid, 18–22 × 12–18 μ, wall verrucose-echinulate above, nearly smooth below, pores indistinct. **Uredosori** hypophyllous with a few peripheral, clavate paraphyses, small, punctiform, light yellow; uredospores globoid or obovoid, 18–25 × 14–20 μ, wall pale yellow, about 1·5 μ thick, finely echinulate, pores 3–4, indistinct. **Teleutosori** hypophyllous, similar to the uredosori but light brown; teleutospores cylindroid, fusoid or clavoid, 2–7- (mostly 5-) celled, often curved, thickened at the apex, slightly constricted, tapering below, 52–140 × 18–24 μ, wall thin, smooth, brown, with one pore in each cell, contents orange, pedicels varying in length, persistent, not much widened below. Auteu-form.

On *Potentilla erecta* and *P. reptans*,

July–October. Great Britain and Ireland, rare.

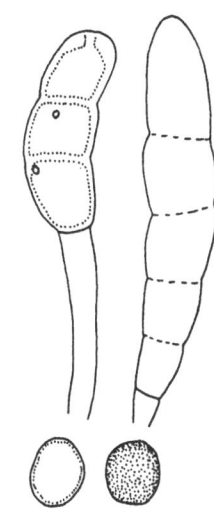

F. obtusa. Teleutospores and uredospores.

This rust has been reported from Norfolk on *Potentilla erecta* and *P. reptans* (Trans. Norf. Norw. Nat. Hist. Soc. **13**, 501, 1935) and from the Norwich area (TBMS. **20**, 10, 1936), Totnes, Devon (*ibid.* **21**, 10, 1938), in error as *Phragmidium potentillae*, Yorkshire (Mason & Grainger, 1937, 44), Scotland (Trail, 1890, 320; Boyd, *in litt.*) and in Ireland by O'Connor (Sci. Proc. Roy. Dub. Soc. **25**, 39, 1949).

This species resembles a *Puccinia* in some respects, especially in the thickening of the apex of the teleutospores, and the position of the solitary pore of each cell; the wall of each cell becomes darker upwards, the lower cell is nearly colourless, and the uppermost pale brown. They can germinate in September.

The development of the aecidium was described by Christman (Trans. Wis. Acad. Sci. **15**, 517, 1905). Olive (Ann. Bot. Lond. **22**, 331, 1908) described the cell fusion in the aecidium.

Triphragmium Link

Sp. Pl. 6 (2), 84 (1825).

Spermogonia subcuticular, flat, without paraphyses. **Aecidia** uredinoid, usually causing hypertrophy, without paraphyses; aecidiospores ellipsoid or obovoid, wall colourless, echinulate. **Uredosori** with peripheral paraphyses; uredospores globoid or obovoid, the wall pale yellow, echinulate, pores obscure. **Teleutosori** hypophyllous or caulicolous; teleutospores flattened laterally, triquetrously 3-celled, the walls coloured, more or less verrucose, with one apical pore in each cell. Autoecious on members of the *Rosaceae*. The genus differs from *Nyssopsora* only in the number of pores in the teleutospore.

Triphragmium filipendulae Pass.

N. Giorn. Bot. Ital. 7, 255 (1875); Wilson & Bisby, Brit. Ured. no. 259; Gäumann, Rostpilze, p. 1212.

[*Uredo* (*Uromyces*) *filipendulae* Lasch, Rabh. Herb. Myc. no. 580 (1844).]
As *Triphragmium ulmariae* (DC.) Link; Grove, Brit. Rust Fungi, p. 287 *p.p.*

[Spermogonia hypophyllous or on the petioles, flat, yellow.] Aecidia uredinoid amphigenous, usually on the petioles or nerves where they cause distortion, orange, 0·5–2 cm. long, pulverulent; aecidiospores variable in shape, ellipsoid, ovoid or piriform, contents orange, 22–28 × 16–22 μ, wall hyaline about 2 μ thick, echinulate, pores indistinct. Uredosori hypophyllous, scattered, up to 1 mm. diam., rounded, orange-yellow; uredospores resembling the aecidiospores. Teleutosori hypophyllous, scattered, minute, soon naked, pulverulent, brownish-black; teleutospores laterally flattened or elliptic, scarcely or not constricted at the septa, 3 (rarely 2-) celled with 1 basal and 2 apical cells, brown,

smooth, or with a few warts around the pores, 32–50 × 26–40 μ, cells equal in size, each with a single pore at the apex,

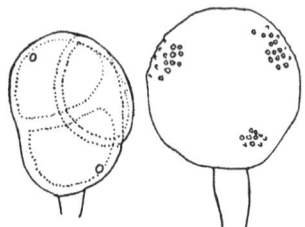

T. filipendulae. Teleutospores.

pedicel hyaline, equalling the spore in length, deciduous. Brachy-form.

On *Filipendula vulgaris*, England, rare.

An early published record of this rust in Britain records it from Lewes, Sussex (Cooke, Grevillea, **11**, 15, 1883) but it has also been found in Gloucestershire and Yorkshire. It was included in *Triphragmium ulmariae* by Grove and Arthur (1934, 98) but has been segregated by most subsequent authors. It differs from *T. ulmariae* in the variable form of the aecidiospores which have a rather thinner wall and in the poor development of the warts on the teleutospores. A form in which teleutospores replace the aecidiospores has been described in this species (Wilson, Proc. Roy. Soc. Edin. B, **63**, 177, 1948).

Triphragmium ulmariae (DC.) Link

Sp. Pl. Ed. 4, 6 (2), 84 (1825); Grove, Brit. Rust Fungi, p. 287 *p.p.*; Wilson & Bisby, Brit. Ured. no. 260; Gäumann, Rostpilze, p. 1210.

[*Uredo ulmariae* Schum., Enum. Pl. Saell. 2, 227 (1803).]
Puccinia ulmariae DC., Lam. Encycl. Meth. Bot. 8, 245 (1808).

Spermogonia epiphyllous, few, flat, reddish-yellow, inconspicuous; spermatia 6 μ long. Aecidia uredinoid, amphigenous, usually on the veins and petioles where they cause distortion, large, irregularly extended, without paraphyses, slightly pulverulent, orange-yellow; aecidiospores ellipsoid to obovoid, contents orange, $25-28 \times 18-21\,\mu$, wall colourless, echinulate grading to smooth at the base, $2-3\,\mu$ thick. **Uredosori** hypophyllous, small, round, scattered, lemon-yellow, very pulverulent, with peripheral paraphyses; uredospores globoid to obovoid, $20-30 \times 18-25\,\mu$, wall pale yellow, sharply echinulate above, less so below, $1-1\cdot5\,\mu$ thick, pores obscure. **Teleutosori** hypophyllous, small, round, soon naked, pulverulent, brownish-black; teleutospores usually 3-celled, subgloboid or more or less triangular, flattened, with 1 basal and 2 apical cells, chestnut-brown, scarcely constricted, with obtuse warts especially about the pores, chestnut-brown, $35-49\,\mu$, wall uniformly $1\cdot5-2\cdot5\,\mu$ thick with 1 apical

pore in each cell, pedicels colourless, one to one and a half times length of spore, deciduous. Brachy-form.

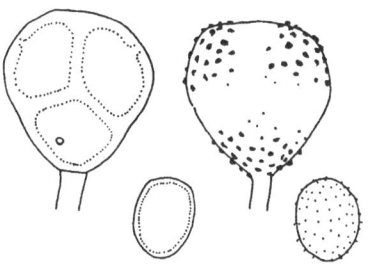

T. ulmariae. Teleutospores and uredospores.

On *Filipendula ulmaria,* aecidia, May–July; uredospores and teleutospores, August–November. Great Britain and Ireland, common.

A variety *alpinum* Lagerh., described by Lagerheim (Bot. Not. **1902,** 175) and recognised by the Sydows (Monog. Ured. **3,** 173), differs in possessing large teleutosori resembling the aecidia in form; these are found on the nerves and petioles and develop in the early summer; they are up to 3 cm. long. The occurrence of this variety was recorded in Germany by Dietel (Mitt. Thür. Bot. Ver. **8,** 10, 1895), in Russia by Kursanov (1922, 76) and in Norway by Lagerheim (*loc. cit.*) and Jørstad (1935, 43). Similar sori have been found in south-west Scotland (Wilson, Proc. Roy. Soc. Edin. B, **63,** 177, 1948) at a low elevation; Sydow (*loc. cit.*) also recorded them near Berlin at a low altitude. They are regarded as indicating the tendency for micro-forms to evolve under suitable conditions.

Klebahn (Z. Pfl.-Krankh. **17,** 142, 1907) proved that the aecidial stage is produced by infection by the basidiospores. The teleutospores can be found in the spring on last year's leaves, and germinate readily when placed in water. Some teleutospores are irregular with 2–5 cells; some have 3 superposed cells, and others 2 cells superposed or placed laterally.

The rust on *Filipendula vulgaris* is here regarded as a distinct species (see p. 112).

The cytology of spore development was described by Sappin-Trouffy (1897, 139) and Kursanov (1922, 76); Olive (Ann. Bot. Lond. **22,** 331, 1908) and Lindfors (1924, 19) investigated the development of the aecidium.

Nyssopsora Arth.

Res. Sci. Congr. Bot. Vienne, p. 342 (1906).

Spermogonia and **aecidia** not known with certainty. **Uredospores** formed singly on pedicels. **Teleutosori** black, naked; teleutospores flattened laterally, triquetrously 3-celled, opaque, with deeply coloured walls, spinose, with 2 or more pores in each cell, usually near the angles.

This genus was originally included in *Triphragmium*. The latter was revised by Tranzschel (J. Soc. Bot. Russie, **8**, 123, 1925) who separated it into two sections, the first *Triphragmium* Link having teleutospores with 1 pore in each cell and the second with 2 or more pores. The second section was again divided into *Triphragmiopsis* Naumov *emend.* and a group with dark, opaque teleutospores, *Nyssopsora* Arth. *emend.* The latter according to Tranzschel contains not only the microcyclic species *N. echinata* Arth. and *N. clavellosa* Berk. which occurs in North America but also two Japanese species with uredospores in addition to teleutospores; these are believed to be heteroecious with aecidia on the *Araliaceae*. *N. echinata* is the only European species.

Nyssopsora echinata (Lév.) Arth.

Res. Sci. Congr. Bot. Vienne, p. 342 (1906); Wilson & Bisby, Brit. Ured. no. 64.

Triphragmium echinatum Lév. Ann. Sci. Nat. Bot. Ser. 3, **9**, 247 (1848); Gäumann, Rostpilze, p. 1215.

Spermogonia unknown. **Aecidia** and **uredosori** wanting. **Teleutosori** amphigenous and on the petioles, at first small but soon becoming confluent, and then forming pustules on the petioles up to 2 cm. long, soon naked, surrounded by the conspicuous, ruptured epidermis, pulverulent, black; teleutospores flattened, trigonal, with 1 cell below and 2 above, scarcely constricted, with dark brown opaque contents, 24–35 × 24–30 μ, the cells equal in size, wall blackish-brown, 1·5–2·5 μ thick, each cell with slightly curved attenuate spines, some sharply pointed and others with minute dichotomous branches at the ends, brown but paler at the apex, 6–18 μ long, 2–3 μ thick; pores 2–4 (generally 2) usually situated near the inner angles; pedicel hyaline, more or less deciduous, up to 22 μ long. Micro-form.

On leaves, stems, petioles and fruits of *Meum athamanticum*, June–August. Scotland, rare.

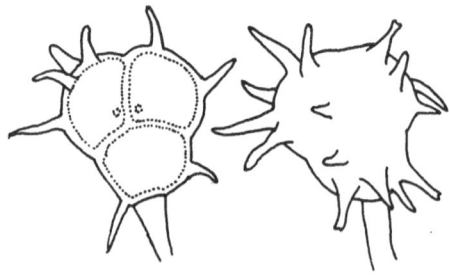

N. echinata. Teleutospores.

This species was first discovered by M. Y. Orr in Glen Lyon in 1939 (Trans. Bot. Soc. Edin. **33**, iv, 1940) and subsequently in other parts of

Perthshire (TBMS. **37**, 248, 1954). It is not very conspicuous on the leaf segments where the sori are small and although they are much larger on the petioles and stems they are often formed near the base of the host and may, in consequence, not be easily seen. The rust causes some swelling and distortion on the stems and petioles and frequently the portions of the leaf beyond the attacked parts become yellow and gradually die.

Abnormal teleutospores occur in which 2 or 3 superposed cells may be present. In a few cases spines divided near the end into short dichotomous branches have been observed; these resemble the spines of *N. clavellosa* Berk. all of which are branched. Hagen (Nytt Mag. Naturv. **82**, 126, 1941) noted similar branching.

Gymnosporangium Hedw. f.

DC. Fl. Fr. **2**, 216 (1805).

Spermogonia subepidermal, globose or flattened-globose, prominent and conspicuous, at first honey-coloured or golden-yellow, becoming blackish. **Aecidia** roestelioid, subepidermal in origin, more or less longly cylindrical, or rarely cup-shaped, apex at first closed, then dehiscent by apical or lateral rupture, peridial cells large, loosely joined; aecidiospores catenulate, globose to broadly ellipsoid, verrucose, usually deeply coloured, with distinct scattered pores. [**Uredosori** known in a few non-British species only.] **Teleutosori** subepidermal in origin, variously shaped, mostly flat, pulvinate, tongue-shaped or conical, pale yellow to dark brown, gelatinous, and expanding greatly when moistened; teleutospores ellipsoid, oblong or rarely fusoid, generally 2-celled, rarely 1- or several-celled, not or slightly constricted at septa, wall smooth, with 1 to several pores, usually 2, in each cell, pedicel usually long, colourless, the outer portion becoming gelatinous when moistened, germinating at once when mature; basidia 4-celled; basidiospores reniform or ovoid.

Chiefly heteroecious with aecidia usually on *Rosaceae* and teleutosori on *Cupressineae*; one species is autoecious on *Juniperus* in North America.

The four British species are heteroecious with teleutosori on *Juniperus*; they possess roestelioid aecidia and 2-celled teleutospores. Kern (Bot. Gaz. **44**, 445, 1910) described the peridial cells of the aecidia in detail and emphasised their taxonomic importance. Crowell considered the geographical distribution of the genus (Can. J. Res. **18**, 469, 1940). The development of the species present in the south of France with particular reference to phases of development of their respective hosts is considered in detail by Bernaux (Ann. Epiphyt. N.S. **7**. 1, 1956). The teleutospores vary considerably according to their position in the mass of spores. The outer spores are darker in colour with thicker walls and are broader than those in the centre. Kern (Mycol. **52**, 837, 1960) has reviewed the host relations and soral morphology of the genus.

Gymnosporangium clavariiforme (Pers.) DC.

Fl. Fr. 2, 217 (1805); Grove, Brit. Rust Fungi, p. 304; Wilson & Bisby, Brit. Ured.
no. 26; Gäumann, Rostpilze, p. 1153.
Tremella clavariaeformis Pers., Syn. Meth. Fung. p. 629 (1801).

Spermogonia chiefly epiphyllous and on fruits, in small clusters on red spots, yellow then dark brown, globoid-conoid, prominent, 130–140 μ high and 110 μ wide. **Aecidia** hypophyllous on the stems, and on fruits where they cover the whole or a great part of the surface, cylindrical, or elongate on the veins, 0·7–1·5 mm. high, 0·3–0·5 mm. wide, fimbriate above, soon becoming lacerate almost to the base, whitish-yellow, peridial cells in surface view from the inside, linear, 70–130 × 18–30 μ, in side view linear or oblong-linear, 15–25 μ in depth, in water often curved, with outer wall smooth, 1–2 μ thick, inner and side walls 5–7 μ thick, rather densely verrucose with rounded or irregular papillae; aecidiospores globoid or subgloboid, wall densely and minutely verruculose, pale brown, 22–30 × 18–26 μ with 6–10 pores. **Uredosori** wanting. **Teleutosori** ramicolous, on elongate, fusiform swellings of the branches, scattered or grouped, teleutospore mass cylindroid or slightly compressed with apex acute or slightly forked, yellowish-brown, 5–10 mm. long, 0·7–1·5 mm. wide; teleutospores lanceolate or fusiform, apex slightly attenuate but obtuse, slightly or not constricted at the septum, base attenuate, with walls about 1 μ thick, varying from brown, with spore size, 50–60 × 15–21 μ to pale yellow, with spore size, 100–120 × 10–12 μ, pores 2 in each cell near the septum, pedicels cylindrical, hyaline, very long; basidia 4-celled; basidiospores with orange contents, 16–18 × 9–11 μ. Heteropsis-form.
Spermogonia and aecidia on leaves

and fruits of Crataegus monogyna and C. oxyacanthoides, July–September; teleutosori on Juniperus communis, April–May. Great Britain and Ireland, frequent.

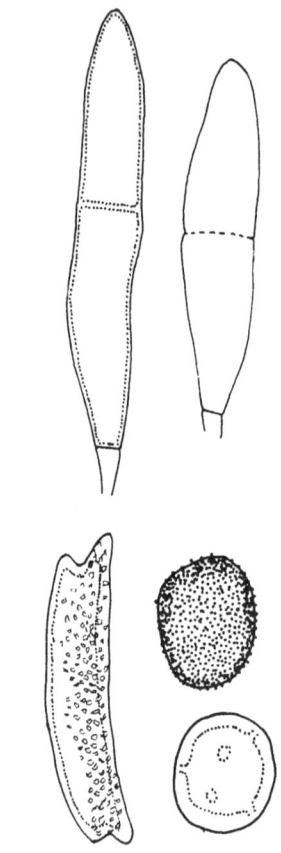

G. clavariiforme. Peridial cell, aecidospores and teleutospores (after Savulescu).

An aecidial stage on Pyrus communis has been recorded several times from the southern part of England as belonging to this species but these records should be transferred to Gymnosporangium fuscum.

The mycelium in the juniper branches is perennial and the teleutosori are formed in each succeeding spring so long as the branch survives. Under dry conditions they are small, rather hard and brittle and yellowish-brown but when wetted rapidly absorb water and swell, becoming much larger, softer and bright orange-yellow; this is largely brought about by the breaking-down and gelatinisation of the long pedicels of the teleutospores which, in consequence, almost entirely disappear. Under these conditions germination of the teleutospores with the production of the 4-celled basidium and basidiospores takes place very rapidly and usually within 48 hours. The basidiospores are shot off violently from the basidia. During rainy weather, owing to their increase in size and change in colour the teleutosori becomes very conspicuous.

Aecidia on hawthorn can be produced in 2–4 weeks after infection; as a result the twigs and leaves become much swollen, and owing to this and to the red and orange discoloration, the infections are easily seen. They were referred to by Malphigi in 1675 in his 'Anatome plantarum' which included, in one of the chapters headed 'Tumours and Excrescences', an illustrated account of a tumour bearing 'flowers' on twigs of English hawthorn.

The discovery of heteroecism in this species was first made by Oersted (Overs. K. Danske Vidensk. Selsk. Forh. **1866**, 185, 1867) who placed basidiospores on the leaves of *Crataegus oxyacanthoides* and produced spermogonia and aecidia. This culture has been frequently repeated both in Europe and America; it is one of the easiest to carry out successfully and has been frequently performed in Scotland. Infections of the juniper by aecidiospores were performed by Plowright (1889, 234) in June 1884 and teleutospores were produced in 1886. Tubeuf (Z. Pfl.-Krankh. **3**, 202, 1893) made similar cultures and obtained teleutospores in the year following infection. Bernaux (Ann. Epiphyt. N.S. **7**, 155, 1956) has described the life cycle and host relations in southern France.

Gymnosporangium confusum Plowr.

Monogr. Brit. Ured. Ustil. p. 232 (1889); Grove, Brit. Rust Fungi, p. 306; Wilson & Bisby, Brit. Ured. no. 27; Gäumann, Rostpilze, p. 1162.

[*Roestelia cydoniae* Thuem., Sacc. Syll. Fung. 7, 834 (1888).]

Spermogonia epiphyllous, orange, on thickened spots which are brown with a yellow, orange-streaked margin. **Aecidia** chiefly hypophyllous but also on the calyx and fruit, on thickened brownish spots with a yellow margin, cylindrical or obconical, curved, yellowish to pale brown, up to 4 mm. long and 0·1–0·3 mm. wide, opening at the summit and at length fimbriate, peridial cells in surface view lanceolate, in lateral view rhombic, 60–95 × 16–24μ, 17–22μ in depth, outer wall 1–1·5μ thick, smooth, inner and side walls 5–7μ thick, with rather large, elongate, obliquely arranged warts and ridges; aecidiospores globoid or ellipsoid, cinnamon-brown, 19–26 × 19–22μ,

walls, tapering above, slightly longer and narrower than (1); pores 2 in each cell near the septum, pedicels hyaline, cylindrical up to 120μ long; basidia 4-celled, cells often separating; basidiospores

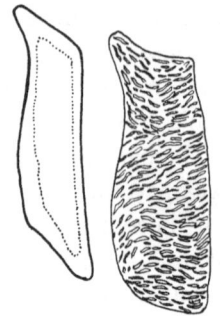 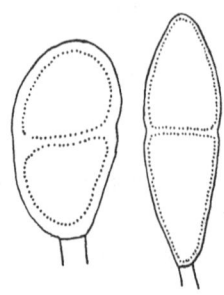

G. confusum. Peridial cells and teleutospores (after Savulescu).

wall 2·5–3·5μ thick, verruculose. **Uredosori** wanting. **Teleutosori** on small fusiform swellings on the branches, spore chains at first pulvinate and dark chocolate brown then swelling, irregularly conical, 5–8 mm. long; teleutospores ellipsoid to fusoid, smooth, scarcely or not constricted at septum, of two kinds, (1) with thick, dark brown walls and orange contents, rounded above about 30–50 × 20–25μ, (2) with thin hyaline

ellipsoidal, pointed towards apiculus, asymmetric. Heteropsis-form.

Spermogonia and aecidia on leaves, branches and fruits of *Crataegus monogyna, C. oxyacanthoides, Cydonia oblonga, Mespilus germanica* and occasionally on *Pyrus communis,* June–August; teleutospores on cultivated *Juniperus sabina* and its variety *prostrata,* April–May. England, scarce.

In 1887 Plowright (J. Linn. Soc. Bot. **24**, 97, 1888), as the result of cultures made with teleutospores from *Juniperus sabina* growing in the vicinity of King's Lynn, concluded that two species of *Gymnosporangium* exist on this host, one which forms its aecidia on *Pyrus communis* (*G. fuscum*) and another (*G. confusum*) which forms its aecidia on *Crataegus monogyna, C. oxyacanthoides* and the medlar (*Mespilus germanica*); later he also infected the quince, *Cydonia oblonga* and, on one occasion, the pear (*Pyrus communis*). There appears to be no evidence that Plowright found the aecidia in nature.

The rust was reported on the medlar (Misc. Publ. Min. Agric. **38**, 74 and 84, 1922) as frequent annually in chalk districts near Worthing, Sussex, and near Sevenoaks, Kent. It was collected at Ightham, Kent, on *Mespilus germanica* in 1921 and was reported by Wormald about 1928 (Rep. E. Malling Res. Sta. **1928–30**, 131, 1931) on *M. germanica* and *Cydonia oblonga* at East Malling, Kent. In 1938 Rilstone (J. Bot. Lond. **76**, 355, 1938) reported it in Cornwall on *J. sabina* (in herbaria labelled *J. sabina* var. *prostrata*) and on *C. monogyna*. Moore (TBMS. **28**, 13, 1945) has given an account of the rust on *J. sabina* and *M. germanica* at East Malling, where it occurred in 1943 and 1944 on the medlar on which it was found by Wormald in 1928.

Plowright's discoveries were confirmed by a long series of cultures made in Switzerland by Fischer (Z. Pfl.-Krankh. **1**, 193 and 260, 1891) and by Klebahn (Z. Pfl.-Krankh. **2**, 94 and 335, 1892) in Germany; these investigators occasionally infected *P. communis*.

The aecidia of *G. confusum* are quite different from those of *G. fuscum* (see below). They resemble those of *G. clavariiforme* but are usually shorter, less deeply torn and rather more inflated; the peridial cells have their side-walls marked with elongate, obliquely placed ridges whereas those of *G. clavariiforme* are coarsely warted; the aecidiospores of the latter species are larger than those of *G. confusum*. The teleutospores of *G. confusum* are shorter and broader than those of *G. clavariiforme*. *Gymnosporangium confusum* differs from *G. fuscum* in possessing smaller teleutosori and rather smaller teleutospores in which the apex of the upper cell is usually more rounded.

In studies of the reaction of pomaceous grafts to this rust, Sahli (Zbl. Bakt. II, **45**, 264, 1916) noticed no effects of the stocks upon the scions, and that in periclinal chimaeras of *Crataegus* and *Mespilus*, the reactions of the tissues involved are not changed.

The morphology and biology of this species in southern France has been fully described by Bernaux (Ann. Epiphyt. N.S. **7**, 105, 1956).

Gymnosporangium fuscum DC.

Fl. Fr. **2**, 217 (1805).

[*Tremella sabinae* Dicks., Pl. Crypt. Brit. **1**, 14 (1785).]
Puccinia juniperi Pers., Syn. Meth. Fungi, p. 228 (1801). (non *Gymnosporangium juniperi* Link, 1825).
[*Roestelia cancellata* Reb. Prod. Fl. Neom. p. 350 (1804).]
Gymnosporangium sabinae [Dicks.] Wint., Hedwigia, **19**, 55 (1880); Rabh. Krypt. Fl.
  Ed. 2, **1** (1), 232 (1882); Grove, Brit. Rust Fungi, p. 308; Wilson & Bisby, Brit.
  Ured. no. 28; Gäumann, Rostpilze, p. 1158.

Spermogonia epiphyllous on large, yellow, reddish or orange spots, very crowded, at length black. **Aecidia** hypophyllous, in groups on thickened and discoloured spots and on the petioles and fruits, rather large, 1–2·5 mm. high, 0·5–1·5 mm. wide, ovate conical, palebrown, cancellate or split to the base into laciniae which remain united at the apex, peridial cells in surface view rhomboid or linear-rhomboid, 65–100 × 20–28 μ, inner and lateral walls 14–20 μ, the inner wall papillate, papillae larger and more crowded at the upper end of the cell; aecidiospores globoid, subgloboid or ellipsoid, chestnut-brown

23–24 × 10–28 μ, wall finely verruculose 3–4·5 μ thick, with 6–10 pores. **Uredospores** wanting. **Teleutosori** at first pulvinate, dark brown, then becoming irregularly conical or laterally compressed, 8–10 mm. high, yellowish-brown, gelatinous; teleutospores ellipsoid to oblong, slightly constricted at the septum, slightly attenuate at each end, apex of upper cell bluntly conical, 38–48 × 20–30 μ, wall pale brown, 1·5–3 μ thick, each cell with 2 pores near the septum, pedicels hyaline, very long. Heteropsis-form.

Spermogonia and aecidia on *Pyrus communis*, July–September; teleutospores on cultivated *Juniperus sabina*, April–May. England and Scotland, scarce.

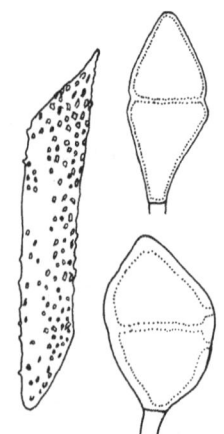

G. fuscum. Peridial cell and teleutospores (after Fischer).

This rust has been usually known as *Gymnosporangium sabinae* but this name is based on the pre-Persoonian description by Dickson. Most of the records are from the Midlands and south-west England; there is one from Scotland by Hopkirk in 1813 (Flora Glottiana, p. 149) as *Tremella sabinae* which he placed amongst the *Algae*.

Spermogonia were not mentioned by the Sydows (Monogr. Ured. 3, 51) or Fischer (1904, 394) although recorded and illustrated by Sowerby (Engl. Fungi, tab. no. 410, 1803) who described them as *Spheria*-like structures on the upper leaf surface.

The teleutospores of the species resemble those of *G. confusum* but may be distinguished by the bluntly conical apex of the upper cell.

There appears to be no doubt that this was the species upon which Micheli based his genus *Puccinia* in 1729, with a description and figure; the latter shows a heavily attacked swollen branch bearing teleutosori, but one twig which is not infected, bears adpressed leaves and a fruit and evidently represents *Juniperus sabina*.

The relation between the two stages was first shown Oersted (Bot. Zeit. **23**, 291, 1865). This has been repeatedly confirmed by de Bary and others, including Plowright (1889, 230) in this country.

The anatomical changes induced by this rust in *J. sabina* were investigated by Woernle (Forstl. Naturw. Z. 3, 156, 1894) and those in the leaves of *Pyrus communis* by Fentzling (Inaug. Diss., Freiburg, 1892, as quoted by Tubeuf & Smith, 1897, 398). Bernaux (Ann. Epiphyt. N.S. 7, 124, 1956) has summarised most of the data on this species and gives details of the susceptibility of various stages of host development.

Gymnosporangium cornutum Kern

Bull. N.Y. Bot. Gard. **7**, 444 (1911).
[*Aecidium cornutum* Pers., Syn. Meth. Fung. p. 205 (1801).]
Gymnosporangium juniperi Link, Sp. Pl. Ed. 4, **6** (2), 127 (1825), *nomen ambiguum*;
Grove, Brit. Rust Fungi, p. 307; Wilson & Bisby, Brit. Ured. no. 29.
Gymnosporangium aurantiacum Chev., Fl. Envir. Paris, **1**, 424 (1826), *nomen ambiguum*.
Gymnosporangium juniperinum Fr., Syst. Myc. **3**, 506 (1832), *nomen ambiguum*;
Gäumann, Rostpilze, p. 1170.
[*Gymnosporangium cornutum* [Pers.] Arth., Mycol. **1**, 240 (1909).]

Spermogonia epiphyllous, subepidermal, on large, reddish or orange spots, in roundish, crowded groups, yellowish, at length black. **Aecidia** hypophyllous on large yellow or orange, thickened spots in rounded groups, 3–10 mm. wide, broadly cylindric, attenuate towards the apex or horn-shaped, curved, 3–5 mm. high, 0·4–0·6 mm. wide, at length opening at the apex and slightly lacerate, often with rounded or irregular openings at the sides, yellowish-brown, peridial cells in face view widely lanceolate, 60–110 × 20–40 μ, in lateral view rhomboid, 30–40 μ in depth, outer wall 2 μ thick, smooth, inner wall 8–12 μ thick, covered with short somewhat elongate protuberances, the lateral wall 8–12 μ thick, coarsely rugose with short, ridge-like markings arranged obliquely to the long axis of the cell; aecidiospores globoid or subgloboid, chestnut-brown, 20–29 × 18–25 μ, wall finely verruculose, uniformly 2–2·5 μ thick, with 8–10 pores. **Uredosori** wanting. **Teleutosori** on the younger branches on fusiform swellings, occasionally on the leaves, scattered, solitary (principally on the leaves) or in groups and confluent, 1–3 mm. diam., low, spore mass applanate or hemispherical at first chocolate-brown then becoming orange, soft and gelatinous when moist; teleutospores ellipsoid, usually somewhat narrowed at both ends, septate in the middle, not or scarcely constricted, 32–52 × 18–28 μ, dark cinnamon-brown, pores 1–2 in each cell, in upper cell at or near the apex, in lower near the septum, each pore covered with a prominent hyaline papilla; pedicel moderately long. Heteropsis-form.

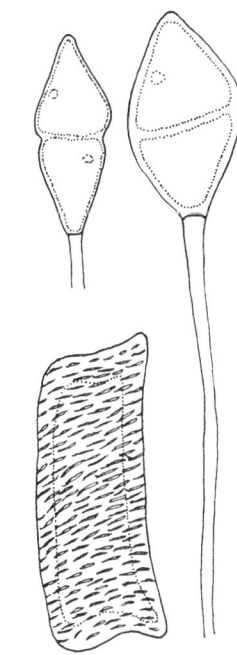

G. cornutum. Teleutospores and peridal cell (after Savulescu).

Spermogonia and aecidia on *Sorbus aucuparia*, July–October; teleutosori on *Juniperus communis*, April–June. England, Wales, scarce; Scotland, frequent.

Two kinds of teleutospores have often been described in this rust, (1) with thick brown walls (2) with thin, very pale yellowish walls; both kinds germinate at once when mature and produce 4-celled basidia. The relation of the two generations was shown first by Oersted (Overs. K. Danske Vidensk. Selsk. Forh. **1866**, 185, 1867) who, using basidiospores produced on *Juniperus communis*, infected *Sorbus aucuparia* and obtained spermogonia after 8 days and subsequently aecidia; also by placing aecidiospores on young twigs of the juniper he produced mycelium in the cortex. This was confirmed by other investigators, including Plowright (Grevillea, **11**, 54, 1882, and J. Linn. Soc. Bot. **24**, 93, 1888) who, by using teleutospores, produced aecidia on *S. aucuparia* and not on *Malus sylvestris*, *Sorbus aria* and *Cydonia vulgaris*; Brebner (J. Bot. Lond. **26**, 218, 1888) repeated the infection on 'Mountain Ash'. Recently in Scotland spermogonia and aecidia on *S. aucuparia* have been produced by infection with basidiospores of *G. cornutum*.

This rust is common in certain localities in central Scotland where the two hosts grow in proximity; it is often very conspicuous on *S. aucuparia* in the latter part of September, where on some trees, almost every leaf is discoloured and partially rolled up and shrivelled as the result of infection. A few scattered aecidia often occur on *Sorbus* far from the nearest Juniper.

Sappin-Trouffy (1896, 136) and Kursanov (1922, 30) described stages in the development of the aecidium of '*G. juniperinum*'.

Eriksson (K. Sv. Vetens. Handl. **59** (6), 1, 1919) gave a general description of the rust under the name *G. tremelloides* f. sp. *aucupariae*, with many illustrations and described numerous culture experiments.

Gymnosporangium juniperi-virginianae Schw.

Schr. Nat. Ges. Leipzig, **1**, 74 (1822).

This species, a native of North America, has been recorded occasionally on imported apples (Moore, 1959, 180) and there is one record of a *Gymnosporangium* on apple in Hampshire which may be this species but no material has been seen to substantiate the record.

Puccinia Pers.

Syn. Meth. Fung. p. 228 (1801).

Spermogonia subepidermal. **Aecidia** subepidermal in origin, aecidioid with a membranous peridium and catenulate spores or uredinoid with aecidiospores borne singly or endophylloid with aecidioid sori but the spores germinating to form basidia. **Uredosori** subepidermal, paraphysate or aparaphysate; uredospores borne singly, wall usually echinulate. **Teleutosori** subepidermal, persistently so in some species, teleutospores typically 2-celled, less commonly 1- or 3- or 4-celled, one pore in each cell, pedicels various, long or short, persistent or deciduous, wall coloured, basidia external, typically 4-celled.

The largest genus of the rust fungi with some three to four thousand species.

Puccinia calthae Link

Sp. Pl. Ed. 4, **6** (2), 79 (1825); Grove, Brit. Rust Fungi, p. 216; Wilson & Bisby, Brit. Ured. no. 112; Gäumann, Rostpilze, p. 790.

[*Aecidium calthae* Grev. Fl. Edin. p. 446 (1824).]

Spermogonia amphigenous, in groups, honey-coloured. **Aecidia** chiefly hypophyllous, in groups on yellowish spots 1–5 mm. diam. or on the stems on elongated swellings, up to about 6 mm. long, cupulate with a much incised, whitish peridium; aecidiospores globoid or ellipsoid, orange, 21–28 μ, wall colourless, finely verrucose, about 1 μ thick. **Uredosori** chiefly hypophyllous, minute, scattered, roundish, pulverulent, cinnamon-brown; uredospores globoid or ellipsoid, cinnamon-brown, 22–30 × 20–25μ; wall 1·5μ thick, echinulate, with 2, rarely 3 supra-equatorial pores. **Teleutosori** amphigenous, small, irregularly scattered or often circinate, pulverulent, chocolate-brown; teleutospores oblong-clavate or fusoid, hardly constricted, rounded or narrowed at both ends, chestnut-brown, 30–44 × 13–22μ, wall smooth 1·5–2μ thick at sides, slightly thickened above, with a hyaline papilla over the pores, upper pore apical, the lower at septum, pedicel hyaline, thick, persistent, up to 75μ long. Auteuform.

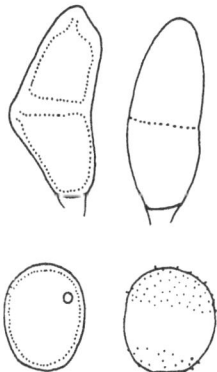

P. calthae. Teleutospores and uredospores.

On *Caltha palustris*, aecidia, May–June; teleutospores, July–October. Great Britain and Ireland, frequent.

For the differences between this species and *Puccinia calthicola* see the latter (p. 124).

Arthur (1934, 237) stated that the pedicels of the teleutospores are short and fragile in this species; other investigators (including Jørstad, 1934) all agree that they are long and persistent.

Winter (Hedwigia, **19**, 105, 1880) in Switzerland marked infected plants in the autumn and, in the following spring, examined aecidia upon them.

Puccinia calthicola Schroet.

Cohn, Beitr. Biol. Pfl. 3, 61 (1879); Wilson & Bisby, Brit. Ured. no. 113; Gäumann, Rostpilze, p. 904.

Puccinia zopfii Wint. Hedwigia, 19, 39 (1880); Grove, Brit. Rust Fungi, p. 217.

Spermogonia amphigenous or on the petioles, in little clusters of 6–10, brownish when old. Aecidia hypophyllous or on the petioles, usually surrounding the groups of spermogonia in scattered roundish clusters (elongated on the petioles, up to 15 mm. long), at first hemispherical, then shortly cupulate, yellow, with short, torn, scarcely reflexed peridia; aecidiospores globoid or ellipsoid, 16–24 × 20–28 μ, wall finely verruculose, about 1 μ thick. Uredosori generally hypophyllous, on small, pale yellow or brown spots, minute, scattered, surrounded by the erect epidermis, chestnut; uredospores ellipsoid, cinnamon-brown, 22–30 × 20–25 μ, wall echinulate 1·5–2 μ thick, with 2 or 3 approximately equatorial pores. Teleutosori generally hypophyllous, irregularly scattered, minute, soon naked, chocolate-brown; teleutospores ellipsoid or oblong, rounded or obtuse at both ends or narrowed below, slightly or not constricted at the septum, 35–60 × 24–35 μ, wall delicately verruculose, uniformly 2–2·5 μ thick with broad, flat, hyaline papilla over the apical pore, pore of lower cell more or less depressed, pedicels nearly hyaline, short, deciduous. Auteu-form.

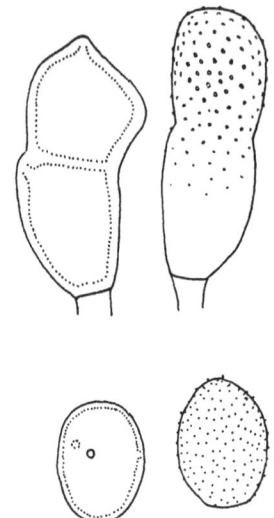

P. calthicola. Teleutospores and uredospores.

On Caltha palustris, aecidia in May; teleutospores, August–December. Great Britain and Ireland, frequent.

Puccinia calthicola can be distinguished from P. calthae by its broader teleutospores (less than 24 μ in P. calthae; more than 24 μ in P. calthicola) not narrowed towards the summit; they are darker in colour and have shorter pedicels. Furthermore, the teleutospore wall is smooth and the uredospores have 2 supra-equatorial pores in P. calthae whereas the teleutospores are slightly warted in P. calthicola and the uredospores have usually 3 equatorial pores. However, since the aecidia completely disappear before the formation of the teleutospores, there is some doubt whether the species can be distinguished in the aecidial stage and whether there has been correct association between aecidia and teleutosori in specimens and descriptions.

Fischer (1904, 91) described the structure of the aecidial peridium: in radial section the cells are almost right-angled with a small overlapping process on the outer side, the outer wall is thick (7–8 μ), the inner is thinner (2–4 μ) with small, close-set, well-developed warts; there is, however, some doubt whether the description applies to those of *P. calthae* or *P. calthicola*. Inoculation with aecidiospores in early June resulted in teleutospore production within a month (Krieg, Zbl. Bakt. ii, **15**, 259, 1906).

Puccinia eutremae Lindr.

Acta Soc. Fauna Fl. Fenn. **22** (3), 9 (1902); Wilson & Bisby, Brit. Ured. no. 150a.
Puccinia cochleariae Lindr., Acta Soc. Fauna Fl. Fenn. **22** (3), 10 (1902); Gäumann, Rostpilze, p. 909.

Spermogonia unknown, **aecidia** and **uredosori** absent. **Teleutosori** amphigenous and on the stems and petioles, scattered, 0·5–1·5 mm. diam. rounded, covered for a long time by the raised, silvery-grey epidermis, then naked, pulverulent, blackish-brown; teleutospores rounded at both ends, strongly constricted at the septum, the 2 cells readily falling apart, 33–50 × 13–17 μ, wall cinnamon-brown, smooth or minutely striate, 1·5 μ thick, with a small, hyaline papilla, up to 3–6 μ high over the pores, the pore of the upper cell apical, of the lower cell at the septum, pedicel hyaline, up to 50 μ long and 4 μ thick at attachment to spore, attenuated below, deciduous, a short portion breaking away with the liberated spore. Microform.

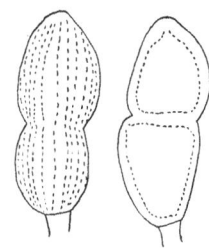

P. eutremae. Teleutospores.

On *Cochlearia alpina* and *C. danica*. England, Scotland, rare.

Jørstad described a specimen of this rust collected on Cross Fell, Cumberland on *Cochlearia alpina* by Samuelsson in June 1902. It has also been collected on *C. danica* just above sea level by E. A. Ellis (Kew Bull. **1957**, 408, 1958) in west Ross-shire in August 1957.

Lindroth (*loc. cit.*) and Sydow (Monogr. Ured. **1**, 894) distinguished *Puccinia eutremae* and *P. cochleariae* as distinct species. Arthur (1934, 291) and Jørstad (1932, 347) unite them as *P. eutremae*. Jørstad described specimens on *Cochlearia* sp. from Greenland with smooth and with striated teleutospores; in the specimens from Cross Fell and Scotland they are smooth.

Sydow described a few mesospores 26–29 × 11–13 μ on specimens on *Eutrema edwardsii*.

Micro-forms on *Cruciferae* are widely distributed in the northern circumpolar regions but are seldom abundant in any part of this range.

Puccinia fergussonii Berk. & Br.

Ann. Mag. Nat. Hist. IV, **15**, 35 (1875); Grove, Brit. Rust Fungi, p. 203; Wilson & Bisby, Brit. Ured. no. 153; Gäumann, Rostpilze, p. 916.

Spermogonia unknown, **aecidia** and **uredosori** absent. **Teleutosori** hypophyllous or on the petioles, on large roundish or irregular yellow spots, in suborbicular or (on the petioles) elongated clusters up to 1·5 cm. long, densely crowded and confluent, long-covered by the epidermis, then pulverulent, chocolate-brown; teleutospores irregular, generally oblong, attenuated or rarely rounded at both ends, slightly constricted 26–45 × 12–18 μ, wall smooth, pale brown, up to 6 μ thick at the apex, pedicels hyaline, thin, deciduous, up to 30 μ long; a few unicellular teleutospores occur occasionally. Micro-form.

On *Viola palustris*, May–August. Great Britain, scarce.

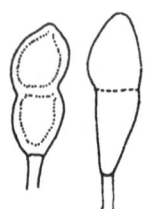

P. fergussonii. Teleutospores.

The large pulvinate groups of sori of this species make it prominent. The mycelium spreads considerably beyond the part occupied by the spores, and consequently causes large yellow patches, usually only one or at most two on each leaf, each the result of a separate infection by the basidiospores.

Kursanov (1922, 55) described the entire mycelium as binucleate.

Puccinia violae DC.

Fl. Fr. **6**, 62 (1815); Grove, Brit. Rust Fungi, p. 200; Wilson & Bisby, Brit. Ured. no. 243; Gäumann, Rostpilze, p. 912.

Uredo violae Schum., Enum. Pl. Saell. **2**, 233 (1803).
[*Aecidium depauperans* Vize, Gard. Chron. II, **6**, 175 and 361 (1876).]
Puccinia aegra Grove, J. Bot. Lond. **21**, 274 (1883); Grove Brit. Rust Fungi, p. 202; Gäumann, Rostpilze, p. 914; Wilson & Bisby, Brit. Ured. no. 80.
Puccinia depauperans [Vize] Syd. Monogr. Ured. **1**, 442 (1903).

Spermogonia amphigenous, crowded in small groups, yellowish. **Aecidia** on all parts of the plant, on the leaves often on swollen yellowish spots, chiefly hypophyllous, often on swollen, elongated or expanded parts of the petioles and stems which are sometimes bent, in rounded or irregular groups or scattered uniformly, especially on the stems, peridia cup-shaped with white torn revolute margin, aecidiospores globoid to ellipsoid, minutely verruculose, orange, 16–24 × 10–18 μ. **Uredosori** amphigenous, scattered or grouped, sometimes circinate, uredospores pulverulent, cinnamon-brown, globoid to ellipsoid, 20–28 × 17–23 μ, wall echinulate, with 2 pores. **Teleutosori** hypophyllous, often on yellowish spots, in groups or solitary, pulverulent, minute, rounded, chocolate-brown; teleutospores ellipsoid to oblong, rounded at each end or sometimes slightly attenuated at the base, slightly constricted at the septum,

20–40 × 15–23 μ, wall chestnut-brown, uniformly 1·5–2·5 μ thick, faintly verrucose with hyaline papilla over the pores, pore of upper cell apical, of the lower at septum or both slightly subapical. Auteu-form. On all green parts of *Viola canina*, *V. cornuta*, *V. hirta*, *V. lutea*, *V. odorata*, *V. reichenbachiana*, *V. riviniana*, *V. tricolor* and *V. tricolor* subsp. *curtisii*, and on *Viola* cultivars. Great Britain and Ireland, common on wild species but uncommon on cultivated plants.

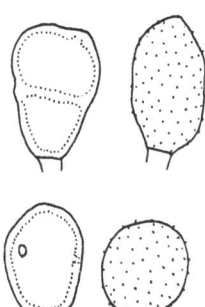

P. violae. Teleutospores and uredospores (after Savulescu).

The aecidial mycelium is more or less systemic and causes swellings on the stems and leaves while the mycelium producing uredospores and teleutospores is strictly limited and does not cause hypertrophy of the tissues.

A form which infects *Viola tricolor* and other species of the subgenus *Melanium* was distinguished as a distinct species, *Puccinia aegra* (*P. depauperans*) and was recognised by Sydow (Monogr. Ured. **1**, 439), Plowright (1889, 158), Grove and Gäumann; it was said to differ from *P. violae* in its smooth or very slightly punctate aecidiospores and teleutospores, but Dietel (Mitt. Thür. Bot. Ver. **6**, 46, 1894), Liro (1908, 579) and Jørstad have described the spores of *P. violae* and *P. aegra* as similar and considered the two rusts to be identical. The various symptoms seem to be due to differential host reactions. Thus on the *tricolor* group of species the mycelium producing aecidia is perennial in the rootstock, the affected shoots are much deformed and aecidia are produced upon them during the whole of the summer. Dietel (*loc. cit.*) pointed out that the effect of the rust on *V. riviniana* is somewhat similar for on this species aecidia are produced on all parts of infected parts, even on the sepals.

Jacky (Zbl. Bakt. II, **18**, 78, 1907) and Bock (Zbl. Bakt. II, **20**, 364, 1908) found no evidence of specialisation between rusts of the *tricolor* and *riviniana* groups.

Puccinia arenariae (Schum.) Wint.

Hedwigia, **19**, 35 (1880); Wilson & Bisby, Brit. Ured. no. 95; Gäumann, Rostpilze, p. 783.

Uredo arenariae Schum., Enum. Pl. Saell, **2**, 232 (1803).
Puccinia spergulae DC., Fl. Fr. **2**, 219 (1805).
Puccinia dianthi DC., Fl. Fr. **2**, 220 (1805).
Puccinia lychnidearum Link, Obs. **2**, 29 (1816); Grove, Brit. Rust Fungi, p. 218.
Puccinia stellariae Duby, Bot. Gall. **2**, 887 (1830).

Puccinia herniariae Unger, Einfl. Bodens, p. 218 (1836); Gäumann, Rostpilze, p. 788.
Puccinia saginae Fuck., Jahrb. Nass. Ver. Nat. 23–4, 51 (1870); Grove, Brit. Rust Fungi, p. 221.
Puccinia moehringiae Fuck., Jahrb. Nass. Ver. Nat. 23–4, 51 (1870).

Spermogonia unknown, **aecidia** and **uredosori** absent. **Teleutosori** hypophyllous, or on the stems, roundish or elongated, 0·25–1 mm., scattered or circinate, often on pale spots, sometimes confluent, pulvinate, pallid-brown, then darker, greyish-pulverulent from the numerous basidiospores; teleutospores oblong-fusoid or clavate-rounded or somewhat pointed above, slightly constricted, rounded or attenuate below, 30–50(63) × 14–21 μ, wall smooth, yellowish-brown, up to 10 μ thick at apex, pedicels hyaline, persistent, 60–85(140) μ long. Microform.

On *Arenaria serpyllifolia, Cerastium holosteoides, Cucubalus baccifer, Dianthus barbatus, D. deltoides, Gypsophila elegans, Herniaria ciliata, Moehringia trinervia, Sagina intermedia, S. maritima, S. nodosa, S. procumbens, S. saginoides, Silene alba, S. dioica, Spergula arvensis,*

Stellaria alsine, S. graminea, S. holostea, S. media, S. nemorum, S. palustris, May–November. Great Britain and Ireland,

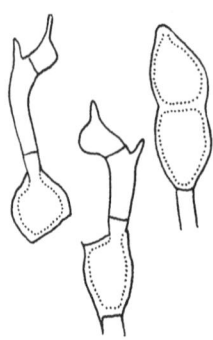

P. arenariae. Teleutospores with 2-celled basidia.

common on *Silene alba, S. dioica, Stellaria alsine,* and *S. media,* less so on other hosts.

This species is normally leptosporic although on some hosts the power of immediate germination appears to have been more or less lost; spores developed in the autumn may not germinate until the following spring. When mature the cells of the teleutospores separate with great ease. The numerous basidia and basidiospores produced give the teleutosori a greyish appearance. During germination the spores become slightly denticulate at the summit, apparently as the result of the throwing off of the greater part of the thickening. Grove (*loc. cit.*) figured the production of a normal 4-celled basidium on germination of teleutospores from *Silene dioica* and *Moehringia trinervia.* Lindfors (1924, 43) stated that on germination of teleutospores from *M. rubrum* (*S. dioica*) and *Arenaria serpyllifolia* a 2-celled basidium is produced, each cell having two nuclei; the two nuclei appeared to pass into the basidiospore. Jackson (1931, 22) stated that a 2-celled basidium occurs in American collections of this species on *Dianthus barbatus.*

De Bary stated (Ann. Sci. Nat. Bot. **20**, 89, 1863) that he had seen the germ-tubes of the basidiospores of '*P. dianthi*' enter the host-plant through the stomata; no similar case has been detected by any other observer.

The sori differ in appearance and arrangement on the various hosts and may or may not occur on discoloured spots but these differences probably depend on differences in the structure of hosts and, in consequence, should not be used to distinguish forms of the rust. The teleutospores are variable in form, size and colour but Sydow (Monogr. Ured. 1, 553), Jørstad (1932, 365) and others have shown that any apparent morphological distinctions break down completely when a long series of specimens is examined. Wille (Ber. D. Bot. Ges. 33, 91, 1915) carried out many infection experiments on different species and showed that spores from *Melandrium rubrum* (*Silene dioica*) would infect other species of *Melandrium* (*S. alba*) and species of *Stellaria*, *Dianthus* and *Sagina* while those from *Moehringia trinervia* would infect species of *Arenaria*, *Dianthus*, *Stellaria* and *Spergula*. He concluded that a sharp specialisation between the subfamilies *Silenoideae* and *Alsinoideae* does not exist. Wille emphasised the difficulty of obtaining positive infection results with *P. arenariae*; he pointed out that, as it is a leptosporic species, such results can only be obtained by applying whole sori to the youngest parts of the shoots. This may explain the failure of de Bary (*loc. cit.*) to obtain infection on *M. rubrum* (*S. dioica*) with basidiospores from *Dianthus barbatus*. Grove (*loc. cit.*) suggested that three forms of the rust may be distinguished, f. *lychnidearum* Link, f. *dianthi* DC. and f. *arenariae* Schum. but in view of the statements made by Sydow, Wille and Jørstad it appears that these cannot be maintained.

The record on *Cucubalus baccifer* is from Norfolk (Ellis, Trans. Norf. Norw. Nat. Hist. Soc., 17, 138, 1951) and that on *Dianthus deltoides* from Kent (Salmon and Ware, Rep. Wye Coll. Myc. Dept. 1934–5). The rust appears to be very rare on the latter host and this is the only record for Britain; it is not mentioned by Sydow (*loc. cit.*) but has been recorded by Tranzschel (1939) in the Ukraine.

The form on *Sagina procumbens* has been regarded as a distinct species, *P. saginae*, and Grove maintained the distinction as the spores which he examined did not appear to be thickened at the apex. Examination of a specimen collected by Vize shows, however, that the thickening is present in ungerminated spores and that the thickened cap usually falls off after germination. The rust on *Sagina* appears to be rare in this country but specimens are present in the Phillips Herbarium (Brit. Mus.) from near Wellington, in the Vize Herbarium, and in the Trail Herbarium in Aberdeen; it has also been recorded from Ayreshire (Scot. Crypt. Soc. Rep. 1921) and was collected in 1944 in Kircudbrightshire.

The form on *Spergula* has also been regarded as a distinct species and was recognized by Sydow (*loc. cit.*); this form does not appear to show any constant morphological distinctions and in view of the cultures by Wille cannot be maintained as a distinct species. It appears to be rare on this host; it was collected near Harlech in North Wales in 1935.

Puccinia arenariae was collected on *Herniaria ciliata* by Frost (TBMS. **40**, 159, 1957) on the Lizard Peninsula, Cornwall in June 1956 and has been found on herbarium specimens collected from the same locality in 1905. The rust on this host was regarded as a distinct species by Sydow (*loc. cit.*) and Liro (1908, 244) but has been placed by Hylander *et al.* (1953, 37) under *P. arenariae*.

A specimen of the rust on *Stellaria graminea* was found in the Trail Herbarium collected near Forres in 1880 (Wilson, 1934, 385). This rust was also collected in Ulster in 1924 and in both these specimens the sori are found on the leaves, sepals and stems, principally on the last; this form has also been found by E. A. Ellis in Norfolk (Trans. Norf. Norw. Nat. Hist. Soc. **15**, 371, 1942). The sori are grey or brownish-black, oblong or linear oblong, on the stems up to 6 mm. long and 1·5 mm. wide, and on the leaves up to 4 × 2 mm., at length surrounded by the ruptured epidermis and black. Although the specimens were collected in August the teleutospores showed no sign of germination. It appears to agree with *P. arenariae* var. *hysteriiformis* (Peck) Jørst. reported by Jørstad (1932, 367) on *S. graminea* in Norway and on *Arenaria capillaris* from Kamtchatka (Jørstad, 1934, 114) as a non-leptosporic variety. It was originally described from the mountains of western America by Peck (Bot. Gaz. **6**, 276, 1881) on *Arenaria* sp. as *P. hysteriiformis* and was regarded as a distinct species by Sydow (Monogr. Ured. **1**, 556), by Mayor (Bull. Soc. Bot. Suisse, **61**, 52, 1951) and also by Gäumann (1959, 787). The teleutospores are darker brown, usually rounded at both ends and are generally shorter and wider, 30–61 × 12–21 μ, usually 48–52 × 14–19 μ, wall at sides 2–2·5 μ thick, thickened at apex, 4–9(15) μ; pore of upper cell apical or slightly depressed, of lower close to the septum; pedicel persistent, up to 120 μ long, hyaline but brownish in upper third and 4–5 μ wide at attachment to the spore.

Treboux (Zbl. Myc. **5**, 120, 1915) stated that *P. arenariae* overwinters in the mycelial condition in *Moehringia trinervia* and Guyot (Rev. Path. Veg. **18**, 142, 1931) described the occurrence in France of witches' brooms on this host due to infection by this rust.

Puccinia behenis Otth

Mitt. Naturf. Ges. Bern, **1870**, 113 (1871); Grove, Brit. Rust Fungi, p. 222; Wilson & Bisby, Brit. Ured. no. 102; Gäumann, Rostpilze, p. 896.

[*Uredo behenis* DC., Fl. Fr. **6**, 63 (1815).]
[*Aecidium behenis* DC., Fl. Fr. **6**, 94 (1815).]
Puccinia silenes Schroet. apud Winter, Rabh. Krypt. Fl. Ed. 2, **1** (1), 215 (1882).

Uredosori amphigenous, scattered or circinate, sometimes confluent, on paler spots, minute, cinnamon-brown; uredospores subgloboid to ellipsoid, 20–26 × 17–22 μ, wall echinulate, pale brown, with 3 or 4 pores. Teleutosori similar, but black-brown; teleutospores oblong to ellipsoid, rounded at both ends, faintly constricted, with a small apical papilla, 25–40 × 16–26 μ, wall smooth, chestnut-brown, pore of upper cell apical, of lower half or more depressed, pedicel hyaline, short, deciduous. Auteu-, hetereu- or hemi-form?

On *Silene alba, S. maritima, S. dioica* and *S. vulgaris.* England and Scotland, scarce.

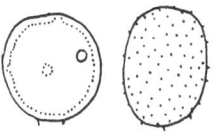

P. behenis. Uredospores.

There seems to be no reliable record in this country of the aecidial stage which has often been reputed to belong to this rust and teleutosori are very uncommon. The aecidium is said to differ from that of *Uromyces behenis* as follows: the early aecidia of *U. behenis* are situated on purplish spots and they are usually scattered singly and are often surrounded by the teleutosori, whilst those of *Puccinia behenis* are found usually in rather large, orbicular clusters, are rarely found singly and, if not clustered, are spread over the whole leaf; they are followed by the uredosori and are never accompanied by the teleutosori. It should be noted that the life cycle of *P. behenis* has never been confirmed and Tranzschel (Acta Inst. Bot. URSS. II, **4**, 336, 1940) suggests that it is a heteroecious species with aecidia on *Bupleurum* and that the aecidia which have been occasionally found on *Silene* belong to *U. behenis.* In Plowright's herbarium are some leaves of *S. dioica* covered by uredosori, which he mistakenly assigned to *P. arenariae* (see p. 127); there are no teleutospores of the latter, however, but a very few of *P. behenis* were found in the same sori. The circinate arrangement of the sori, on paler spots, is very similar in both species.

The rust has been reported on *S. maritima* by Hadden (J. Bot. Lond. **58**, 38, 1920) from Somerset and on *S. alba* by Mayfield (Trans. Suffolk Nat. Soc. 3, 6, 1935) in Suffolk where it was found growing with infected *S. vulgaris*; it has also been found on *S. alba* by Ellis in Norfolk (Trans. Norf. Norw. Nat. Hist. Soc. **15**, 208, 1940).

Puccinia malvacearum Mont.

Hist. Fis. Polit. Chili, **8**, 43 (1852); Grove, Brit. Rust Fungi, p. 206; Wilson & Bisby, Brit. Ured. no. 182; Gäumann, Rostpilze, p. 801.

Spermogonia unknown, **aecidia** and **uredosori** absent. **Teleutosori** hypophyllous and on petioles and stems, on conspicuous yellow or orange spots, closely scattered, hemispherical or on the stems elongated, pulvinate, compact, hard, small, at first pale reddish then reddishbrown; teleutospores oblong to subfusoid, attenuate at both ends or rarely rounded above, slightly or not constricted at the septum, 35–75 × 12–26 μ, wall smooth, yellowish-brown, 1·5–4 μ, thick at the sides, 5–10 μ at the apex, pore of upper cell subapical, superior in the lower cell, pedicel hyaline, persistent, short or up to 150 μ long and 10 μ thick; 1–3- or 4-celled teleutospores also occur. Micro-form.

On *Althaea rosea* cult. very common, less so on *Lavatera arborea, Malva moschata, M. neglecta, M. pusilla, M.*

sylvestris, *Sidalcea* sp. and *Brotex* sp., April–November. Great Britain and Ireland.

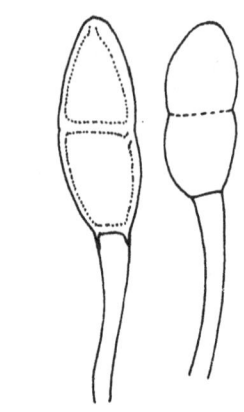

P. malvacearum. Teleutospores.

This species was first described in 1852 from Chile where it is presumably native. It was observed in Australia in 1857. In Europe it appeared in Spain in 1869, in 1872 in France and in 1873 in England (Gard. Chron. **33**, 946, 1873) and in Germany and in several other European countries between 1874 and 1890. It was found in South Africa in 1875 and in North America in 1886 and is now spread all over the world.

It is truly plurivorous attacking at least 40 host species of 10 different genera of the *Malvaceae*; on all these it appears to be identical. There appears to be no specialisation; infections have proved that it may be transferred from *Malva* to *Althaea* and *vice versa*.

The life history was investigated by Kellerman (Phys. Medic. Soc. Erlangen Sitzber. **6**, 157, 1874) and Plowright (Gard. Chron. **18**, 617, 1882); both found that the rust overwintered as pustules in sheltered places. Later investigators were Dandeno (Rep. Michigan Acad. Sci. **9**, 68, 1907) and Taubenhaus (Phytopath. **1**, 55, 1911).

It has been shown by many observers that the normal method of germination of the teleutospore is by the formation of a 4-celled basidium, each cell producing a basidiospore. A second method was, however, observed, especially by Eriksson (Overs. K. Vet. Akad. Förh. **47** (2),

1911; **62** (5), 1921) in which the growth from the cell of the teleutospore elongates considerably and cuts off terminal 'conidia'; this abnormal method takes place when water surrounds the germinating teleutospore. The two papers published by Eriksson (*loc. cit.*) were directed to the proof of his Mycoplasm Theory in connection with this rust. He stated that *Puccinia malvacearum* perennates in the form of 'mycoplasm' in the cells of the autumn buds at the base of the shoots and in the embryos of the seeds of the infected plants. In these it is said to develop mixed with the protoplasm of the host, and in the following year is transformed in the intercellular spaces into hyphae which produce the teleutosori. The 'mycoplasm' is formed from the 'conidia' produced on the germination of the teleutospores; they do not form a germ-tube but, without forming an opening, pour forth their protoplasm which passes through the outer wall of the epidermal cell and brings about infection. Klebahn (Z. Pfl.-Krankh. **24**, 20, 1914) carried out culture experiments to investigate this theory but failed to confirm it; it was also refuted by Bailey (Ann. Bot. Lond. **34**, 173, 1920). Both Klebahn (*loc. cit.*) and Hecke (Mitt. Landw. Lehrkanzeld. Hochsch. Bodenk. Wien **2**, 455, 1914) investigated spore germination and confirmed the earlier observations of Kellerman and Plowright that those formed late in autumn could germinate in the following spring. The latter investigator considered that the type of germination is determined entirely by external conditions.

Ashworth (TBMS. **16**, 177, 1931) showed by a series of observations that teleutospores may remain viable for months; she also noted that new infections occur during the winter and that all the pustules present in the spring are not necessarily the result of late autumn infections. Plowright (*loc. cit.*) suggested that the rust passed the winter as pustules fallen to the ground but Taubenhaus considered that there was no evidence for this and this was confirmed by Ashworth.

Whether mycelium can pass the winter in the living condition in young shoots is not quite clear; Grove (*loc. cit.* p. 49) considered it to be very likely that it does so in the young leaf rudiments at the base of last year's stems. In America Taubenhaus (*loc. cit.*) and Dandeno (*loc. cit.*) reported its occurrence in this position and Arthur (1929, 355) stated that the mycelium may overwinter from late autumn infections. Eriksson (1919) in Sweden found no evidence for this and Ashworth (*loc. cit.*) found no overwintering mycelium in *Althaea rosea* and *Malva pusilla* even in shoots with infected bud scales.

No mycelium is present in the embryo of the seed; the latter, however can carry infection by means of teleutospores formed on the bracts and carpels,

Robinson (Ann. Bot. Lond. **28**, 331, 1914) investigated the relations of the rust to its host, the tissues invaded and the effects of the fungus upon these tissues; he noted that the germ tubes of the basidiospores showed a negatively heliotropic response.

The cytology of teleutospore development has been investigated by Sappin-Trouffy (1897, 117), Blackman & Fraser (Ann. Bot. Lond. **20**, 42, 1906), Olive (Science, **33**, 194, 1911), Werth & Ludwigs (Ber. D. Bot. Ges. **30**, 525, 1912), Moreau (Le Botaniste, **13**, 190, 1914) and Lindfors (1924, 34); these are concerned principally with the origin of the dicaryophase.

Ashworth (*loc. cit.*) gave an account of the structure of the rust in monosporidial culture and showed that it is homothallic and this has been confirmed by Brown (Can. J. Res. **18**, 23, 1940); later Ashworth (Ann. Bot. Lond. **49**, 104, 1935) gave a description of the occurrence of receptive hyphae. Allen (Phytopath. **23**, 572, 1933) described the nuclear divisions in the development of the promycelium and basidiospores.

Puccinia argentata (C. F. Schultz) Wint.

Hedwigia, **19**, 38 (1880); Rabh. Krypt. Fl. Ed. 2, **1** (1), 194 (1882); Grove, Brit. Rust Fungi, p. 204; Wilson & Bisby, Brit. Ured. no. 96.

Aecidium argentatum C. F. Schultz, Prod. Fl. Starg. p. 454 (1806).
Puccinia noli-tangeris Corda, Icon. Fung. **4**, 16 (1840); Gäumann, Rostpilze, p. 938.

Spermogonia hypophyllous, scattered among the aecidia, honey-coloured. **Aecidia** hypophyllous, rather uniformly distributed on discoloured swollen spots, on the petioles and stems more scattered, white, with a deeply cut revolute peridium; aecidiospores $18–22 \times 13–20\mu$, contents golden-yellow. **Uredosori** hypophyllous, scattered or circinate, sometimes on minute yellowish spots, often confluent, covered by the silvery epidermis, then pulverulent, roundish, ochraceous; uredospores globoid to broadly ellipsoid, pale yellowish, $16–22 \times 14–20\mu$, wall delicately echinulate, with 4–7 (usually 6) pores. **Teleutosori** similar, but chestnut-brown; teleutospores ellipsoid to subclavoid, rounded or slightly attenuated at both ends, hardly constricted, pale brownish, $25–38 \times 12–22\mu$, wall smooth, pore of upper cell apical, of lower close to the septum, each with a colourless conical cap, pedicels hyaline, slender, short. Hetereuform.

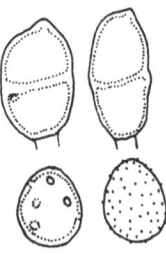

P. argentata. Teleutospores and uredospores (after Fischer).

Spermogonia and aecidia on *Adoxa moschatellina*, April–June; uredospores and teleutospores on *Impatiens capensis* May, August–October. England, very rare.

The aecidial stage has been recorded from Britain only recently (*in litt.*) probably because it has been confused with the aecidium of *Puccinia albescens* from which it is distinguished, according to Bubak (Zbl. Bakt. II, **12**, 412, 1904) by its gold-coloured spores (see under that species, p. 188). Uredospores and teleutospores were recorded from Surrey at Albury, Guildford, Shere and Kew Gardens before 1913 and by Sandwith on *Impatiens capensis* in Surrey. The records on '*I. noli-tangere*' almost certainly should refer to *I. capensis*. The teleutospores are at first produced in the same sori as the uredospores. The description given above of the spermogonia and aecidia is taken from Bubak (*loc. cit.*) who showed that the aecidiospores could produce the other stages on *Impatiens* in about 10 days and this has been repeated with British collections by Laundon (TBMS. **45**, 474, 1962). Bubak also showed (Zbl. Bakt. II, **16**, 150, 1906) that the aecidium could be produced on *Adoxa* by overwintered teleutospores; the incubation period was as long as one month, probably because the mycelium first permeated the whole plant, from the leaf to the stem, before producing spores. He also proved that the mycelium does not perennate in the rhizome but fresh infection must take place each spring. This has been confirmed by Moss (Mycol. **43**, 99, 1951). Klebahn (Z. Pfl.-Krankh. **22**, 324, 1912) confirmed Bubak's work and showed that infection takes place on the underground parts of the *Adoxa*, probably on the buds. Morgenthaler (Zbl. Bakt. II, **27**, 73, 1910) studied the conditions required for the germination of the teleutospores. Arthur (Mycol. **4**, 20, 1912) showed that the aecidium which occurs on *A. moschatellina* in the United States belongs to this species by infecting *I. aurea* (= *pallida*) with aecidiospores.

Puccinia buxi DC.

Fl. Fr. **6**, 60 (1815); Grove, Brit. Rust Fungi, p. 205; Wilson & Bisby, Brit. Ured. no. 111; Gäumann, Rostpilze, p. 769.

Spermogonia unknown, **aecidia** and **uredosori** absent. **Teleutosori** amphigenous, on indefinite spots, scattered or confluent, hemispherical, pulvinate, compact, soon naked, dark chestnut-brown or purplish-brown; teleutospores oblong to clavate, rounded above and not thickened, rounded or attenuate below, 55–90 × 20–35 μ, wall smooth, brown, pore of upper cell apical, of lower superior, pedicel hyaline, persistent, very long (reaching 160 μ). Micro-form.

On *Buxus sempervirens*, September, October, lasting through the winter and following spring. Great Britain and Ireland, frequent.

The spores of this species often fall apart into their constituent cells. Fischer (1904, 316) showed that it has only one spore form. In autumn or in the course of winter the teleutospores begin to develop and break

through the epidermis in spring. The rust is leptosporic and the teleutospores germinate in the
spring and early summer and infect the delicate
young leaves. The mycelium grows slowly and
during the summer and autumn the infected spot
becomes much thickened; this results from the
elongation of the palisade cells and the swelling
of the infected cells of the spongy parenchyma.
In this rust Dangeard & Sappin-Trouffy (1893,
123) discovered the fusion of nuclei in the teleutospore. Moreau (1914, 191) described the
development of the teleutospores.

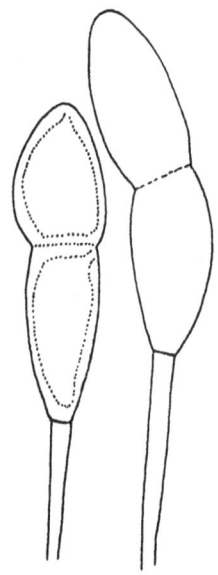

P. buxi. Teleutospores.

Puccinia umbilici Duby

Bot. Gall. **2**, 890 (1830); Grove, Brit. Rust Fungi, p. 211; Wilson & Bisby, Brit. Ured.
no. 237; Gäumann, Rostpilze, p. 918.

Puccinia rhodiolae B. & Br., Ann. Mag. Nat. Hist. Ser. 2, **5**, 462 (1850); Grove, Brit.
Rust Fungi, p. 210; Wilson & Bisby, Brit. Ured. no. 210; Gäumann, Rostpilze,
p. 917.

Spermogonia unknown, **aecidia** and **uredosori** absent. **Teleutosori** amphigenous
and on the petiole and stems, minute,
circular, at length confluent, at first compact, then pulverulent, dark reddish-
brown; teleutospores broadly ellipsoid,
rounded at both ends, not thickened
above, with a small, hyaline papilla over
the pores, not constricted, smooth, or
with a few, faint, verrucose striae, cinnamon-brown, $20–35 \times 17–26\mu$, cells
often broader than long and frequently
oblique, pore in upper cell apical, in
lower cell near pedicel, pedicel hyaline,
short, fragile. Micro-form.

On *Sedum roseum*, very rare, and
Umbilicus rupestris, locally common.
Great Britain and Ireland.

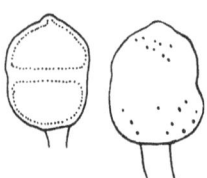

P. umbilici. Teleutospores.

Grove (*loc. cit.*) and Gäumann (*loc. cit.*) regard the rusts on *Sedum
roseum* and *Umbilicus rupestris* as distinct species but as they are
morphologically indistinguishable they have been united by Arthur
(1934, 292) and Jørstad (1932, 375). The latter has pointed out that the

pore of the lower cell of the teleutospore is near the pedicel or basal and not as figured by Grove. On *S. roseum* it appears to be very rare; it has been recorded in Scotland from a few localities in the north-east and there seems to be no recent record in this country. It has been found in Norway, Sweden, Switzerland, Greenland and in North America. Jørstad (*loc. cit.*) regards it as a relict species, formerly much more uniformly distributed than at the present time. It is frequently found on *U. rupestris* where the sori are often circinate and form large circular clusters up to 1 cm. diam. on yellowish spots.

Wilson (New Phytol. **36**, 185, 1937) described internal sori on this host. Kuhnholtz-Lordat *et al.* (Uredineana, **3**, 54, 1951) gave an account of the relationship of the mycelium to the tissues of *U. rupestris* and described the morphology of the teleutospore.

Puccinia chrysoplenii Grev.

Fl. Edin. p. 429 (1824); Grove, Brit. Rust Fungi, p. 214; Wilson & Bisby, Brit. Ured. no. 128; Gäumann, Rostpilze, p. 920.

Spermogonia unknown, **aecidia** and **uredosori** absent. **Teleutosori** of two kinds: (*a*) leptosporic, crustose type, amphigenous, chiefly hypophyllous, scattered or confluent, often circinate, small, circular 0·4–0·7 mm. diam., pulvinate, pale brown becoming greyish; teleutospores broadly fusoid and more or less conical at the apex, rounded or attenuate below, slightly constricted, contents very pale brown, 32–46 × 10–15 μ, wall smooth, colourless, 5–7 μ thick, pedicels hyaline, up to 50 μ long, persistent; (*b*) pulverulent type, amphigenous, mostly epiphyllous and on the petioles, scattered or in small groups, small, circular, 0·5 mm. diam., pulverulent, surrounded by the ruptured epidermis, brown; teleutospores oblong-ellipsoid and rounded at both ends, with a conical, colourless, apical papilla, dis-tinctly constricted, 35–42 × 10–14 μ, wall pale brown with very faint, rather irregular, longitudinal ridges, pedicels deciduous but leaving a very short portion

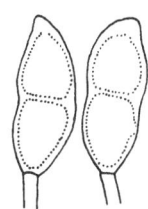

P. chrysosplenii. Teleutospores
(after Savulescu).

attached to the spore; basidium 4-celled, basidiospores colourless, about 7·5 × 5 μ. Micro-form.

On *Chrysosplenium alternifolium* and *C. oppositifolium*, March–September. Great Britain and Ireland, scarce.

In specimens collected near Edinburgh in May 1944 both types of spores germinated within 24 hours. The striations on the spores are very faint and in some appear to be absent. The two kinds of spores were figured by Dietel (Ber. D. Bot. Ges. **9**, 36, and t. 3, 1891). Savile (Can. J. Bot. **32**, 400, 1954) reviewed this and related rusts in North America, and lumped them all under *Puccinia heucherae*.

Puccinia saxifragae Schlecht.

Fl. Berol. **2**, 134 (1824); Grove, Brit. Rust Fungi, p. 212; Wilson & Bisby, Brit. Ured. no. 216; Gäumann, Rostpilze, p. 921.

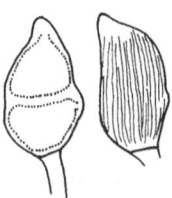

P. saxifragae. Teleutospores (after Savulescu).

Spermogonia unknown, **aecidia** and **uredosori** absent. **Teleutosori** generally hypophyllous on discoloured spots, scattered or in groups, often confluent, soon naked, pulverulent, dark brown; teleutospores ellipsoid or oblong, rounded at both ends or slightly attenuate below, slightly constricted, often with a large, pale, conical, apical papilla, 26—45 × 14–20 μ, wall pale brown, 1–2·5 μ thick, marked with faint, sometimes curved, longitudinal striae, pore of upper cell apical, of lower superior, pedicels hyaline, slender, deciduous, not as long as the spore. Micro-form. On *Saxifraga granulata, S. spathulata, S. stellaris* and *S. umbrosa*. Great Britain and Ireland, uncommon.

This rust is widely distributed on *Saxifraga stellaris* on the mountains of central and northern Scotland, rare in North Wales and Cumberland; it appears to be very rare on *S. granulata*, the only records being from Dovedale, Derbyshire (TBMS. **10**, 131, 1924), and Edinburgh district (Henderson, Notes R. B. G. Edinb. **23**, 503, 1961). It has been recorded on *S. umbrosa* from Clare Island, Galway, and on *S. spathularis* from Killarney by Dennis (*in litt.*).

Jørstad (1931, 384) made a special study of this rust and considered it best to maintain *P. saxifragae* as a separate species closely related to *P. chrysosplenii. P. saxifragae* occurs also on the British species, *S. cernua, S. rivularis* and *S. nivalis* in Norway, Greenland and northern Canada, but so far no infection of them has been discovered in Britain. Savile (Can. J. Bot. **32**, 400, 1954) advocated extensive lumping of the saxifragaceous rusts, reducing the species, *P. chrysoplenii* and *P. saxifragae*, to varieties of *P. heucherae* which he considered a wide ranging polymorphic species on many genera of the *Saxifragaceae*.

The striations on the spores are well marked on some host species and almost obsolete on others; the striations of the spores on *S. granulata* are very faint or the spores may be quite smooth. Although the sori are all pulverulent and the spores of one type, Fischer (Myk. Zbl. **1**, 4, 1912) and Jørstad (*loc. cit.*) have shown that some germinate directly and bring about infection during the summer while others overwinter. Fischer (1904, 151) infected *S. stellaris* with overwintered spores from this species but was unable to infect *S. nivalis* and other species; he concluded that a special form exists on *S. stellaris*. Poeverlein (Ann. Myc. **35**, 53, 1937) also suggested the existence of special forms in this rust. This species has no connection with *Melampsora vernalis* (see p. 67) which is also found on *S. granulata*.

Lindfors (1924) showed that the mycelium of this species is prevailingly dicaryotic.

Puccinia pazschkei Diet.

Hedwigia, **30**, 103 (1891); Grove, Brit. Rust Fungi, p. 213; Wilson & Bisby, Brit. Ured. no. 192; Gäumann, Rostpilze, p. 925.

Spermogonia unknown, **aecidia** and **uredosori** absent. Teleutosori epiphyllous, 0·5–1 mm., scattered or more often in orbicular groups, 2–3 mm. diam., a few occasionally hypophyllous, surrounded by the swollen and torn epidermis, pulverulent, dark reddish-brown; teleutospores ellipsoid or oblong, rounded at both ends, slightly thickened above or with a minute flat papilla, (2–3 μ thick), slightly constricted, 25–40 × 13–18 μ, wall 1·5–2·5 μ thick, irregularly verruculose or rugose, pale brown, with the pore of the upper cell usually apical, of the lower cell superior, pedicels hyaline, short, deci-duous, a few unicellular teleutospores occur. Micro-form.

On cultivated *Saxifraga aizoon*, *S.*

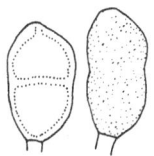

P. pazschkei var. *pazschkei*. Teleutospores (after Fischer).

cotyledon, *S. diapensioides*, *S. hostii*, *S. lingulata*, *S. longifolia*, *S. obristii* and *S. porophylla* and some of their hybrids and varieties. In gardens, uncommon.

The typical variety of this species occurs in Britain only on cultivated *Saxifraga* species of the sections *Engleria*, *Euaizoonia* and *Kabschia* and on these it has been found in gardens at Kew, Birmingham, Edinburgh, St Andrews and in Kent. There is good observational evidence that cross-infection between hosts in various sections of *Saxifraga* takes place under garden conditions in Edinburgh.

var. *jueliana* (Diet) Savile

Can. J. Bot. **32**, 411 (1954).

Puccinia jueliana Diet. Hedwigia, **36**, 298, (1897); Gäumann, Rostpilze, p. 923.

Teleutosori as in var. **pazschkei** but teleutospores with a very conspicuous cap on the pore of the upper cell, up to 6·5 μ thick, and a somewhat less conspicuous one on the lower pore.

On *Saxifraga aizoides* and *S. oppositifolia*. Scotland, uncommon, but certainly overlooked.

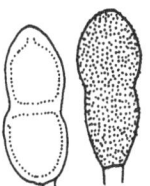

P. pazschkei var. *jueliana*. Teleutospores (after Fischer).

This variety was first collected in Britain by Ellis & Dennis in Sutherland on *Saxifraga aizoides* (Kew Bull. **1957**, 408, 1959). In the exceptionally dry summer of 1959 it was found quite abundantly on the same host in central and west Perthshire and in west Ross-shire and at the same time on *S. oppositifolia* at one locality in the Ben Lawers range in Perthshire. Infection appeared to be heaviest on the host plants most severely affected by drought. In the wet season of 1960 the rust could scarcely be found even on the same plants as were infected the previous year (Henderson, Notes R. B. G. Edin. **23**, 503, 1961).

The classification of the rusts of *Saxifraga* has been considered by Jørstad (1932, 378), Poeverlein (Ann. Myc. **35**, 53, 1937) and Savile (Can. J. Bot. **32**, 400, 1954). Jørstad considered *Puccinia pazschkei* a polymorphous species but did not recognise any subspecific taxa. Savile, working mainly with North American collections, reduced many previously recognised taxa to varietal rank and described one new one *P. pazschkei* var. *oppositifoliae* on *S. oppositifolia*. This variety according to his description is closer to var. *pazschkei* in lacking a conspicuously thickened apical cap than it is to var. *jueliana* on *S. aizoides*. The Scottish collections have a thick apical cap and agree exactly with the rust on *S. aizoides* and must be placed in var. *jueliana*.

There are no accounts of infection experiments with rusts of the *P. pazschkei* group although specialisation of the varieties is generally accepted.

Puccinia joerstadii Rytz described on *S. oppositifolia* from central Europe and *P. fischeri* which occurs on the same host are distinguished by the equatorial or subequatorial position of the lower pore.

Puccinia ribis DC.

Fl. Fr. **2**, 221 (1805); Grove, Brit. Rust Fungi, p. 212; Wilson & Bisby, Brit. Ured. no. 211; Gäumann, Rostpilze, p. 921.

Spermogonia unknown probably not formed, **aecidia** and **uredosori** absent. **Teleutosori** epiphyllous, orbicular, surrounded by a discoloured yellow zone, circinate and often confluent, pulverulent, rich chestnut-brown; teleutospores ovoid or oblong, rounded above and below, hardly constricted 20–30 × 15–20 μ, wall verrucose, chestnut-brown, uniformly 1·5–2·5 μ thick, with a hyaline papilla over the apical pore of the upper

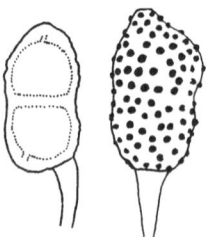

P. ribis. Teleutospores (after Viennot-Bourgin).

cell, pore of lower cell usually basal, without a papilla, pedicels hyaline, thin, deciduous, about as long as the spore; a few unicellular teleutospores present. Micro-form. On *Ribes spicatum.* Scotland, rare.

This rust was first found by Keith in Dallas manse garden Morayshire in July 1894 and was described by Plowright (TBMS. **1**, 57, 1898) as on 'red currant leaves'. The host was given by Grove (*loc. cit.*) as *Ribes rubrum* and this was the name accepted by Wilson (1934, 383), Noble & Gray (Gard. Chron. **122**, 92, 1947) and Wilson & Bisby (*loc. cit.*) Further investigation of the host has shown it to be a hybrid of *R. spicatum.* The host of this rust in Norway, according to Jørstad (1940, 78) is the red currant variety 'Viking', which belongs to the *R. spicatum* group and appears to be rather susceptible to *Puccinia ribis* especially in the coastal areas between 66–70° N. The rust is sometimes of economic importance by heavily attacking leaves and berries. The variety Viking was named by Hahn (U.S. D. A. Circ. **330**, 1935) and the Scottish host of *P. ribis* closely resembles this variety. *Puccinia ribis* has been found on very old bushes in gardens in Moray, Nairn (Noble & Gray, *loc. cit.*) and in Perthshire; up to the present it has not been found on the berries. As it appears possible that hosts and the rust have been imported into Scotland from Scandinavia, it is interesting that a red currant known as var. 'Norwegian' was introduced from Norway into Stirlingshire in 1880 (Thayer, Bull. Ohio Agr. Exp. Sta. 371, 1923).

Eriksson (Rev. Gén. Bot. **10**, 497, 1898) showed that the teleutospores do not germinate until they have overwintered. He considered the form on '*Ribes rubrum*' (i.e. *R. spicatum*) as biologically distinct from that on *R. nigrum* or *R. uva-crispa.*

Fischer (Mitt. Naturf. Ges. Bern. **1915**, 231, 1916) pointed out that the epiphyllous position of the teleutosori is unusual as there are no stomata on the upper epidermis of the leaf in *Ribes spicatum.*

Puccinia circaeae Pers.

Syn. Meth. Fung. p. 228 (1801); Grove, Brit. Rust Fungi, p. 198; Wilson & Bisby, Brit. Ured. no. 131; Gäumann, Rostpilze, p. 640.

Spermogonia unknown, **aecidia** and **uredosori** absent. **Teleutosori** hypophyllous, on sunken yellowish or purplish, round spots, minute, pulvinate, brown, then with a greyish bloom, scattered or circinate and at length confluent in a thick crust; teleutospores generally fusoid, rounded or conically attenuate at the apex, slightly constricted at the septum, $25–40 \times 9–13\mu$, wall smooth, yellowish-brown or brown, up to 12μ thick at the apex; pedicels hyaline, persistent, about as long as the spore. Micro-form.

On *Circaea* '*alpina*', *C. intermedia* and *C. lutetiana*, August–October. Great Britain and Ireland, frequent.

The sori of this species are of two different types: the first-formed are leptosporic, roundish, light brown, solitary or circinate and confluent, they become grey with the production of basidia and basidiospores; the later-formed, which appear round the others or on the stem and on the nerves of the leaves, are darker brown and never greyish. All the spores are of the same shape, but the paler leptosporic ones can germinate at once in the sorus, whereas the darker ones rest until the following spring.

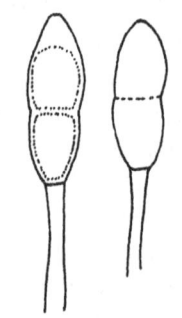

The scattered orange uredosori of *Pucciniastrum circaeae* often occur on the same leaves as the chocolate brown teleutosori of this species.

P. circaeae. Teleutospores.

Puccinia epilobii DC.

Fl. Fr. **6**, 61 (1815); Grove, Brit. Rust Fungi, p. 200; Wilson & Bisby, Brit. Ured. no. 150; Gäumann, Rostpilze, p. 865.

Spermogonia unknown, aecidia and uredosori absent. Teleutosori hypophyllous, scattered or rather crowded, often uniformly distributed over the whole surface of the leaf, rarely confluent, surrounded by the torn epidermis, pulverulent, reddish-brown; teleutospores ellipsoid, oblong or pyriform, rounded at both ends, much constricted, wall scarcely thickened at apex, minutely verruculose, 27–48 × 16–25 μ, pore of upper cell apical with a minute papilla, of the lower cell subequatorial, pedicel hyaline, 10–16 μ long. Micro-form.

On *Epilobium anagallidifolium, E. hirsutum, E. montanum, E. obscurum* and *E. palustre*, May–August. Great Britain, Ireland, scarce.

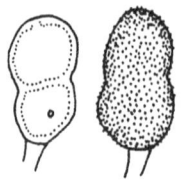

P. epilobii. Teleutospores.

This rust was recorded on *Epilobium obscurum* by Massee & Crossland (1905) from Yorkshire and by Wilson (TBMS. **9**, 137, 1924) from Kingussie, Inverness, on *E. anagallidifolium* by Gardiner (Fl. Forfar, 296) and near Edinburgh on *E. montanum* in 1927. It is a microcyclic form correlated with *Puccinia pulverulenta*. The mycelium is perennial and permeates the whole plant and deforms the shoots, making the leaves smaller and thicker. The warts of the wall of the teleutospore are sometimes hardly perceptible. The cytology of this rust was investigated by Lindfors (1924, 46). In collections from northern Europe on *E. palustre* and *E. davuricum* the spores are more finely verruculose according to

Urban (Preslia, **25**, 25, 1953) who described subsp. *palustre* for this type. The same distinctions seem to apply to British collections so far as they have been examined.

Puccinia pulverulenta Grev.

Fl. Edin. p. 432 (1824); Grove, Brit. Rust Fungi, 198; Wilson & Bisby, Brit. Ured. no. 207; Gäumann, Rostpilze, p. 929.

[*Uredo vagans* α *epilobii-tetragoni* DC., Fl. Fr. **2**, 228 (1805).]
Puccinia epilobii-tetragoni [DC.] Wint., Rabh. Krypt. Fl. Ed. 2, **1** (1), 214 (1882).

Spermogonia scattered among the aecidia, honey-coloured. **Aecidia** hypophyllous or, when very abundant, also epiphyllous, scattered rather closely over nearly the whole surface of the leaf, peridium white, cup-shaped, then revolute; aecidiospores very delicately verruculose, orange, 16–26 μ diam. **Uredosori** hypophyllous, scattered or circinate, sometimes confluent, pulverulent, chestnut-brown; uredospores globoid to ovoid, 20–28 × 15–25 μ, wall remotely echinulate, brown, with 2 pores. **Teleutosori** hypophyllous, often circinate, soon naked, pulverulent, dark-brown; teleutospores ellipsoid or ovoid, rounded at both ends, slightly constricted, 24–35 × 14–20 μ, wall smooth, brownish, up to 5 μ thick at apex, pore of upper cell apical, of the lower cell subequatorial, pedicels hyaline, fragile, deciduous. Auteu-form.

On *Epilobium adnatum*, *E. hirsutum*, *E. montanum* and *E. parviflorum*, aecidia, May–June, teleutospores, June–November. Great Britain and Ireland, common.

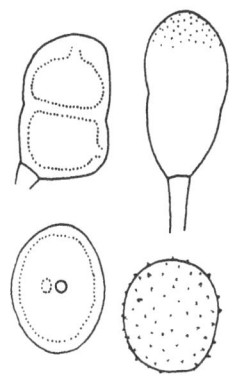

P. pulverulenta. Teleutospores and uredospores.

The aecidium-forming mycelium appears to be perennial, for the same plants are attacked year after year. The aecidia appear in May and cover leaf after leaf, as they are developed. The affected plants are easily recognisable by their much paler and yellowish colour. The sori of uredo- and teleutospores soon appear, at first on the same leaves as the aecidia but afterwards on the later-formed leaves.

In September and October the small last-formed leaves are thickly covered by the teleutospores. The mycelium of the uredo- and teleutosori is strictly localised. According to Nicolas & Aggery (C. R. Acad. Sci. Paris, **227**, 1068, 1948) the uredospores and teleutospores formed in late autumn are smaller and thinner-walled than those in September.

Plowright stated (1889, 151) that the aecidiospores sown on young seedlings of *E. hirsutum* give rise to aecidiospores in 17 days, but very possibly there is some oversight here, for Dietel (Flora, **81**, 401, 1895) obtained uredospores by sowing the aecidiospores from *E. adnatum* on the same host; he also suggested that the form on *E. tetragonum* is biologically distinct from that on *E. hirsutum*, since on the latter he obtained no results.

Gäumann (Ber. Schweiz. Bot. Ges. **51**, 341, 1941) as the result of numerous cultures has suggested the existence of two specialised forms, f. sp. *epilobii-tetragoni* Diet. which does not infect *E. hirsutum*, and f. sp. *epilobii-hirsuti* Gäum. on *E. hirsutum* and other species but not on *E. adnatum*.

Puccinia thesii Duby

Bot. Gall. **2**, 889 (1830); Grove, Brit. Rust Fungi, p. 229; Wilson & Bisby, Brit
    Ured. no. 231; Gäumann, Rostpilze, p. 874.
[*Aecidium thesii* Desv., J. Bot. Fr. **2**, 311 (1809), *p.p.*]

Spermogonia amphigenous, numerous amongst the aecidia. **Aecidia** amphigenous, scattered uniformly and rather thickly over the whole leaf surface, seldom in roundish or oblong groups, peridium cup-shaped to cylindrical, with a white recurved margin, peridial cells with a much thickened outer wall; aecidiospores angulato-globoid, orange, 16–24μ diam. **Uredosori** amphigenous or on the stems, distributed irregularly, minute, 0·2 mm. diam., roundish, long-covered by the epidermis, brown; uredospores globoid to broadly ellipsoid, 28 × 20–24μ, wall about 2μ thick, yellowish-brown, verruculose with 4–5 pores. **Teleutosori** similar but more compact, often elongate, at first covered by the epidermis; teleutospores, brown, clavoid, usually rounded above, very slightly constricted, rounded or attenuate below, both cells equal or the lower rather narrower and longer, 35–54 × 16–24μ, wall smooth, uniformly brown, up to 10μ thick at the apex and sometimes with a broad, clear cap, pore of upper cell apical, of lower superior, pedicel brownish, thick, persistent, up to 95μ long. Auteu-form.

On *Thesium humifusum*, spermogonia and aecidia, May–August, sometimes with the teleutospores in October, uredospores, June, teleutospores, August–October. England, rare.

This rust has been found in south and east England from Suffolk to Wiltshire and Dorset.

Puccinia aegopodii (Str.) Röhl.

Deutschl. Fl. Ed. 2, **3** (3), 131 (1813); Grove, Brit. Rust Fungi, p. 185; Wilson &
    Bisby, Brit. Ured. no. 79; Gäumann, Rostpilze, p. 965.
Uredo aegopodii Schum., Enum. Pl. Saell, **2**, 233 (1803).
Uredo aegopodii Str., Ann. Wetter. Ges. **2**, 101 (1810).

Spermogonia unknown, aecidia and uredosori absent. **Teleutosori** amphigenous, but chiefly on the petioles and nerves, on thickened, yellow spots, small, but arising in dense, irregular, often confluent clusters, at first black, covered by the shining epidermis which splits longitudinally, soon naked, pulverulent, blackish-brown; teleutospores oblong to ovoid, often obliquely asymmetrical, usually rounded above, hardly or not at all constricted (often broadest at the septum), more or less rounded below, 28–48 × 15–22 μ, wall smooth, chocolate-brown, pore of upper cell apical or subapical, of the lower cell basal, each with a cap 2–3 μ high, pedicels hyaline, occasionally up to 32 μ long, deciduous. Micro-form.

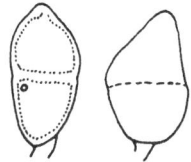

P. aegopodii. Teleutospores.

On *Aegopodium podagraria*, April–August. Great Britain and Ireland, frequent.

According to Tranzschel (see Lindroth, 1902, p. 113) a few isolated uredospores are to be found in the young sori; they are almost colourless, aculeate, 20–22 × 18 μ. This observation has not been confirmed. It appears that sometimes the whole pedicel remains attached to the teleutospore and is up to 32 μ long; usually it breaks off leaving only a short portion.

Semadeni (Zbl. Bakt. II, **13**, 531, 1904) proved that the spores of this fungus would infect *Aegopodium* but not *Astrantia*.

Kursanov (1922, 69) described the somatic origin of the dicaryons and the development of the teleutosorus.

Puccinia angelicae (Schum.) Fuck.

Jahrb. Nass. Ver. Nat. **23–4**, 52 (1869).

Uredo bullata Pers., Syn. Meth. Fung. 222 (1801).
Uredo angelicae Schum., Enum. Pl. Saell. **2**, 233 (1803).
Puccinia silai Fuck., Jahrb. Nass. Ver. Nat. **23–4**, 53 (1869); Grove, Brit. Rust Fungi, p. 191.
Puccinia bullata (Pers.) Schroet., Cohn, Beitr. Biol. Pfl. **3**, 74, 1879 (non *P. bullata* Link, 1816 = *P. ribis* DC.); Grove, Brit. Rust Fungi, p. 193; Wilson & Bisby, Brit. Ured. no. 109.

Spermogonia hypophyllous, subepidermal, mixed with the aecidia, yellow, 90–130 μ diam., with projecting paraphyses. **Aecidia** uredinoid, hypophyllous, chiefly on the swollen nerves and petioles, up to 3 cm. long, dark cinnamon-brown; aecidiospores resembling the uredospores. **Uredosori** hypophyllous, rarely epiphyllous, scattered, minute, punctiform, brown, uredospores globoid to obovoid, wall 2 μ thick at sides, 5–6 μ at apex, regularly and distantly echinulate, distance between spines 3–3·5 μ, brown, 25–40 × 18–28 μ, pores 3–4 equatorial with swollen caps. **Teleutosori** scattered, hypophyllous or on the stems, minute, often confluent on the stems,

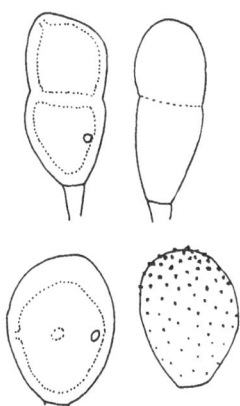

P. angelicae. Teleutospores and uredospores.

long-covered by the epidermis, blackish-brown; teleutospores oblong to obovoid, rounded above, slightly narrowed below, wall scarcely thickened at apex, hardly constricted, brown, 30–45 × 18–24 μ, pore of upper cell apical, and of lower inferior, pedicels hyaline, short, rather wide, deciduous. Auteu-form.

On *Angelica sylvestris*, *Peucedanum palustre*, *Selinum carvifolia* and *Silaum silaus*. Great Britain, scarce.

On *Peucedanum palustre* it was found at Shapwick Bog, Somerset, in 1883 and was collected by Sandwith from the same locality in 1940; it has also been recorded by Ellis from Norfolk and Mayfield (1935) from Suffolk on this host. On *Selinum carvifolia* it has been found in Cambridgeshire by Ellis in 1946 and by Brenan in 1950; it has been recorded on *Silaum silaus* from several localities.

Semadeni (Zbl. Bakt. II, **13**, 528, 1904) proved that the rust on *Angelica sylvestris* would infect *A. archangelica* amongst others but not *Aethusa cynapium* or *Peucedanum palustre*. The uredinoid aecidia are like those of *P. smyrnii*. *Puccinia apii* is also closely allied, but differs in the possession of a cup-shaped aecidium.

Puccinia apii Desm.

Cat. Pl. Omis. p. 25 (1823); Grove, Brit. Rust Fungi, p. 184; Wilson & Bisby, Brit. Ured. no. 94; Gäumann, Rostpilze, p. 943.

Spermogonia hypophyllous, mostly surrounded by the aecidia, often circinate, shining, reddish-brown. **Aecidia** hypophyllous or on the petioles, on minute, irregular, conspicuous yellowish spots, in roundish, or on the petioles, elongate clusters, shortly cylindrical, with white, torn peridium; aecidiospores delicately verruculose, orange, 17–24 μ diam. **Uredosori** hypophyllous, scattered, occasionally confluent, pulverulent, cinnamon-brown; uredospores ellipsoid to obovoid or even subclavoid, 24–35 × 20–26 μ, wall shortly echinulate, slightly thickened above (3–5 μ), brownish-yellow, with 3 equatorial pores. **Teleutosori** hypophyllous, rarely epiphyllous, if on the petioles sometimes very large, scattered or confluent, roundish, pulverulent, blackish-brown; teleutospores ellipsoid to oblong, rounded above, not thickened, hardly constricted, rounded or gently attenuate below, smooth, brown, 30–50 × 15–20 μ, pore of upper cell apical, of lower subequatorial,

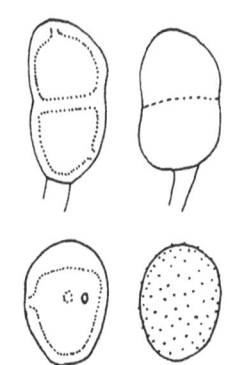

P. apii. Teleutospores and uredospores.

pedicels hyaline, thin, deciduous, about as long as the spore. Auteu-form.

On *Apium graveolens*, aecidia in May–June, teleutospores, September–November. England, Scotland, scarce.

Puccinia bulbocastani Fuck.

Jahrb. Nass. Ver. Nat. **23**–4, 52 (1869); Grove, Brit. Ured. no. 186; Wilson & Bisby, Brit. Ured. no. 108; Gäumann, Rostpilze, p. 989.
[*Aecidium bulbocastani* Cum., Atti Accad. Torino (1804–5).]
[*Aecidium bunii* DC., Syn. Pl. Gall. 51 (1806).]
Puccinia bunii Wint., Rabh. Krypt. Fl. Ed. 2, **1** (1), 197 (1882).

Spermogonia few, scattered amongst the aecidia, pale yellowish. Aecidia rarely on the leaves, hypophyllous, more often on the petioles and stems, densely crowded, causing considerable hypertrophy and curvature, cup-shaped to pustulate, whitish, with a white, irregularly torn peridium; aecidiospores yellowish, 15–22 μ diam., wall delicately verruculose. Uredosori absent. Teleutosori amphigenous, scattered, minute, roundish, sometimes on the petioles, confluent and elongate, long-covered by the epidermis, black; teleutospores ellipsoid to obovoid, generally rounded at both ends, not thickened above, hardly constricted, 25–42 × 14–24 μ, wall minutely reticulate, brown, pore of upper cell apical, of lower inferior, pedicels hyaline, thin, deciduous. Opsisform.

On *Bunium bulbocastanum*. England, rare.

This species has been found at Dunstable, Bedfordshire, and in Hertfordshire and Yorkshire. The record of *Aecidium bunii* on *Conopodium majus* by Plowright (1889, 270) probably refers to *Puccinia tumida* as regards teleutospores and an aecidial stage of *P. bistortae*. Gäumann (Ann. Myc. **32**, 300, 1934) obtained numerous teleutosori by infecting *Bunium* with aecidiospores. He also showed (Ann. Myc. **31**, 46, 1933) that this species differs from the closely allied *P. triniae* Gäumann on *Trinia glauca* in the size of the teleutospores, the average for *P. bulbocastani* being 31 × 17·7 μ while those of *P. triniae* are 39·4 × 22·2 μ.

Puccinia bupleuri Rud.

Linnaea, **4**, 514 (1829); Grove, Brit. Rust Fungi, p. 189. Wilson & Bisby, Brit. Ured. no. 110.
[*Aecidium falcariae* β *bupleuri-falcati* DC. Fl. Fr. **6**, 91 (1815).]
Puccinia bupleuri-falcati [DC.] Wint., Rabh. Krypt. Fl. Ed. 2, **1** (1), 212 (1882) *nomen nudum*; Gäumann, Rostpilze, p. 944.

Spermogonia amphigenous, numerous, generally scattered over the whole surface among the aecidia, subepidermal, yellowish-brown, 95–125 μ diam., paraphyses yellowish or hyaline, up to 65 μ long. Aecidia hypophyllous, or a few epiphyllous, uniformly scattered, cupshaped, with a torn, white, revolute peridium; aecidiospores globoid or ellipsoid, yellow, 16–24 μ diam. Uredosori amphigenous, scattered or occasionally circinate, on minute paler spots, small, circular, cinnamon; uredospores globoid to ellipsoid, yellow-brown, 19–24 × 17–22 μ, wall echinulate, with 3, 4 or even 5 pores. Teleutosori amphigenous, minute, scattered, roundish, on the stems often larger and oblong, occasionally confluent, subepidermal, at length naked, blackish-brown; teleutospores oblong to clavate, rounded at both ends, wall 2·5–4·5 μ, not thickened above, hardly constricted, smooth, brown, 25–44 × 16–30 μ, pore of upper cell apical, of lower near the pedicel, pedicel hyaline, thin, short, deciduous. Auteu-form.

On *Bupleurum tenuissimum*, uredospores and teleutospores, August. England, rare.

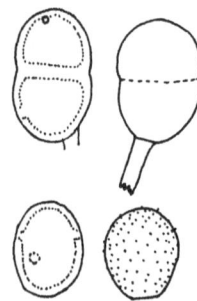

P. bupleuri. Teleutospores and uredospores.

This species was recorded from Essex, originally from Walton-on-the-Naze in 1887 and more recently by Ellis, Mayfield and others from several localities in Suffolk. The uredospores were found in small quantity among the teleutospores.

Fischer (1904, 123) said that the affected plants usually bear aecidia on every leaf, the leaves are narrower and paler, and the plants do not flower; the aecidial stage appears in May and June, often abundantly. This applies especially to the parasite on *Bupleurum falcatum*, which is probably identical with that on *B. tenuissimum*.

Puccinia chaerophylli Purton

Midland Flora, 3, 303 (1821); Grove, Brit. Ured. no. 195; Wilson & Bisby, Brit. Ured. no. 125; Gäumann, Rostpilze, p. 979.

Spermogonia pale yellow, roundish. **Aecidia** on the leaves and petioles, on the leaves scattered or circinate, on the petioles and nerves in dense elongate clusters causing a slight hypertrophy, yellowish, pustulate, with a poorly developed peridium; aecidiospores verruculose, orange, 18–35 × 16–26 μ. **Uredosori** hypophyllous, scattered, minute, roundish, pulverulent, cinnamon; spores globoid to obovoid, pale brownish-yellow, 20–30 × 18–25 μ, wall echinulate, with 3, usually equatorial pores. **Teleutosori** similar, but dark brown, on the petioles more elongate; teleutospores ovoid to oblong, rounded at both ends or gradually attenuate below, slightly constricted, yellowish-brown or brown, 24–36 × 16–25 μ, wall reticulate, not thickened apically, pore of upper cell apical, of lower subequatorial, pedicel hyaline, thin, as long as the spore. Auteu-form.

On leaves, petioles and stems of *Anthriscus sylvestris*, *Chaerophyllum aur-*

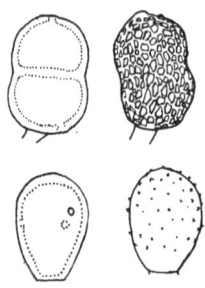

P. chaerophylli. Teleutospores and uredospores (after Fischer).

eum and *Myrrhis odorata*, aecidia, May–June; teleutospores, July–October. Great Britain and Ireland, scarce, except on the last host.

This species has been recorded on *Chaerophyllum temulum* but the records must be regarded as doubtful; no specimens can be found and umbelliferous hosts are notoriously misidentified. Macdonald (*in litt.*) has collected this species on *C. aureum* in Scotland but no longer has specimens for confirmation and the site of the collection has been destroyed.

Semadini (Zbl. Bakt. II, **13**, 215 and 338, 1904) obtained experimental evidence that this parasite is not identical with *Puccinia pimpinellae* nor with *P. heraclei*; he also confirmed the relation of the various spore stages. The feeble development of the peridium and the pustule-like, not cup-shaped, aecidia are paralleled by those of *P. heraclei*. The markings of the teleutospore are formed by a network of low ridges, forming small polygonal or rounded meshes. This ornamentation is similar to that of *P. pimpinellae* except that the meshes are a little smaller and not quite so easily seen. Semadeni (*loc. cit.*) showed that the spores from *Anthriscus silvestris* infected *Myrrhis odorata* readily but would not infect *C. temulum* nor *C. aureum*. *Puccinia svendseni*, a microcyclic species, on *A. sylvestris* with smooth teleutospores, should be sought for in Britain.

Puccinia cicutae Lasch

Rabh. Herb. Myc. no. 787 (1845); Grove, Brit. Rust Fungi, p. 183; Wilson & Bisby, Brit. Ured. no. 130; Gäumann, Rostpilze, p. 976.

Spermogonia amphigenous, between the aecidia, rounded, almost colourless, subepidermal, 100–125 μ diam., with projecting paraphyses. **Aecidia** amphigenous on the leaves, petioles and stems, on yellowish spots which become darker, in rounded, ellipsoidal or elongated groups, up to 1·5 cm. long, pustular, peridium cupulate or hemispherical, with entire margin, rather short, golden-yellow; aecidiospores globoid to ellipsoid, 17–26 × 10–20 μ, wall finely punctate, subhyaline. **Uredosori** generally hypophyllous, scattered, minute, pulverulent, cinnamon; uredospores subgloboid to ovoid, yellow-brown, 18–28 × 14–22 μ, wall echinulate, with 3 equatorial pores. **Teleutosori** similar, but blackish-brown; teleutospores ellipsoid, or oblong, rounded at both ends or rarely attenuated downwards, not thickened above, generally constricted, wall 28–46 × 18–30 μ, wall somewhat verru-culose or distinctly verrucose-reticulated, occasionally nearly smooth, brown, pore of upper cell apical, of lower cell sub-

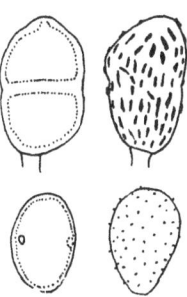

P. cicutae. Teleutospores and uredospores.

equatorial, pedicels hyaline, thin, short, deciduous. Auteu-form.

On *Cicuta virosa*, aecidia, June; uredospores and teleutospores, July–October. England, rare.

Plowright (TBMS. **2**, 26, 1902), who mentioned this species in his arrangement of British species, probably collected his material in Norfolk where the rust has been collected by Ellis (1934, 492); it has also been recorded in Suffolk by Mayfield (1935). The aecidia in Europe have been previously described as possessing a feebly developed peridium but in specimens collected at Wheatfen Broad near Norwich it is fairly well developed as described above.

Puccinia conii Lagh.

Tromsö Mus. Aarsh. **17**, 54 (1895).
[*Uredo conii* Str., Ann. Wetter. Ges. **2**, 96 (1810).]
[*Puccinia conii* [Str.] Fuck., Jahrb. Nass. Ver. Nat. **23–4**, 53 (1869).]; Grove, Brit. Rust Fungi, p. 196; Wilson & Bisby, Brit. Ured. no. 139; Gäumann, Rostpilze, p. 952.
Puccinia bullata [Pers.] Schroet., Cohn, Beitr. Biol. Pfl. **3**, 74 (1879) (non *P. bullata* Link, 1816 = *P. ribis* DC.)

Spermogonia and uredinoid **aecidia** known only in culture (Schroeter, *loc. cit.*). **Uredosori** hypophyllous, occasionally on the petioles, scattered, minute, rarely confluent, pulverulent, cinnamon; uredospores ellipsoid to obovoid, 24–36 × 17–26μ, wall echinulate in the upper part only, up to 7μ thick at apex, with 3 pores. **Teleutosori** similar, but blackish-brown, on the stems and petioles often larger and long-covered by the grey epidermis; teleutospores ovoid or ovoid-oblong or even clavoid, rounded at both ends or attenuate below, hardly constricted, 30–48 × 20–28μ, wall nearly or quite smooth, pale brown, not thickened at the apex, pore of the upper cell apical or slightly depressed, of the lower close below the septum, each with a small

hyaline cap, pedicels hyaline, short, deciduous. Brachy-form?

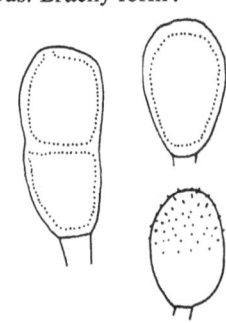

P. conii. Teleutospore and uredospores (after Savulescu).

On *Conium maculatum*, August–September. Great Britain and Ireland, scarce.

This species is readily recognised by its uredospores, which are echinulate only in the upper half; the spines gradually diminish in size downwards and the lower half is quite smooth. According to Grove the teleutospores are quite smooth when empty, even under the highest power, but the protoplasm is very granular and presents a misleading effect at first sight.

Puccinia heraclei Grev.

Scot. Crypt. Fl. 1, 42 (1823); Grove, Brit. Rust Fungi, p. 194; Wilson & Bisby, Brit. Ured. no. 163; Gäumann, Rostpilze, p. 986.

Spermogonia amphigenous, scattered amongst the aecidia. Aecidia hypophyllous frequently on the petioles and especially on the leaf veins, densely crowded in irregular clusters on thickened yellowish spots, often causing distortion, individual aecidia cup-shaped and pustulate, peridium feebly developed, opening by a pore, aecidiospores delicately verruculose, yellowish, 21–32 × 18–28 μ. Uredosori amphigenous, scattered, minute, chestnut-brown, uredospores globoid to ellipsoid, 25–32 × 19–27 μ, wall coarsely echinulate, pale brown, with 3 or 4 equatorial pores. Teleutosori small and scattered on the leaf surface or more or less confluent on the nerves, spore mass pulverulent, teleutospores blackish, ellipsoid, rounded at both ends, scarcely constricted, 26–37 ×

18–27 μ, wall finely reticulate, pore of upper cell superior, of lower subequatorial, pedicel hyaline, short, deciduous. Auteu-form.

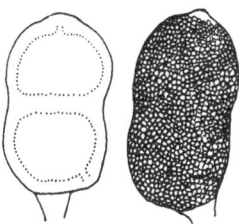

P. heraclei. Teleutospores.

On *Heracleum sphondylium,* aecidia, March–June, teleutospores, August. Not common.

This species is closely related to *Puccinia chaerophylli* and differs only in surface ornamentation of the teleutospores. In *P. heraclei* the areolae are round, 1 μ in diameter, whereas in *P. chaerophylli* they are more angular, often elongate in the longitudinal axis of the spore (up to 3 × 1 μ). The aecidia usually occur on the lower leaves, often of young plants.

Puccinia hydrocotyles Cooke

Grevillea 9, 14 (1880); Grove, Brit. Rust Fungi, p. 181; Wilson & Bisby, Brit. Ured. no. 169; Gäumann, Rostpilze, p. 992.

[*Caeoma hydrocotyles* Link, Sp. Pl. Ed. 4, 6 (2), 22 (1825).]
[*Trichobasis hydrocotyles* Cooke, J. Bot. Lond. 2, 343 (1864).]

[Spermogonia epiphyllous, arranged in small groups associated with the aecidia, 150–200 μ wide, honey-brown. Aecidia amphigenous and sparingly on the petioles in small groups, sulphur-yellow, peridia cupulate, yellowish, margins revolute, lacerate; aecidiospores globoid, 20–25 × 14–20 μ, wall hyaline, minutely and densely verruculose, 1 μ thick.] Uredosori amphigenous, scattered, sometimes confluent, often cir-

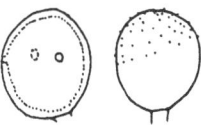

P. hydrocotyles. Uredospores.

cinate round a central larger one, very minute about 0·25 mm. diam., long-covered by the epidermis, at length naked, pulverulent, cinnamon-brown;

uredospores subgloboid or ellipsoid, 24–34 × 20–27μ, wall echinulate with 2 conspicuous equatorial pores. [Teleutosori chiefly epiphyllous, scattered, orbicular, 0·5–1 mm. diam., pulverulent, dark chestnut-brown, surrounded by the ruptured epidermis]; teleutospores ellipsoid, rounded at both ends, slightly thickened above and constricted, chestnut-brown, 30–44 × 18–28μ, wall smooth or with large, isolated, depressed, rounded warts, 2μ thick, pore of upper cell apical, of lower much depressed, each with a small hyaline papilla, pedicels hyaline, thin, deciduous. Auteu-form? On *Hydrocotyle vulgaris*, uredospores, July–September; teleutospores, October. Great Britain and Ireland, rare.

This species has been reported from a few localities in England and from one each in Scotland and Ireland. It is imperfectly known. Spermogonia and aecidia have been recorded only from South America and New Zealand. There is some disagreement in the description given of the spermogonia; they were described by Spegazzini from South America in 1881 (Anal. Soc. Ci. Argent. **12**, 80, 1881) under the name *Aecidiolum hydrocotyles*. Lindroth (1902, 76) suggested that this was *Darluca filum* attacking the uredosori but Fragoso (1924, 188) disagreed with this view. Later Spegazzini described *Aecidium hydrocotylinum* also from South America (Anal. Mus. Nac. Buenos Aires, **19**, 321, 1909); the Sydows (Monogr. Ured. **4**, 323) and Jackson (Mycol. **23**, 463, 1931) regarded this as probably the aecidial stage of *Puccinia hydrocotyles*. Aecidia found on *Hydrocotyle* in North America have been assigned to the heteroecious species, *Uromyces lineolatus*.

P. and H. Sydow (Monogr. Ured. **1**, 388) stated that the uredospores from all localities agree perfectly; Grove remarked that those from Hawaii agree exactly with our specimens, having the same peculiar colour, 'resembling a strong wash of raw sienna'.

In this country no teleutosori have been observed but a few teleutospores have been found in uredosori; a similar condition appears to exist throughout Europe and has been described by Cooke (Grevillea, *loc. cit.*) on *Hydrocotyle* sp. from Natal; definite teleutosori have been described in New Zealand.

The life history of this rust in South Africa has been discussed by Doidge (Bothalia, **2**, 88, 1926), in Tristan da Cunha by Jørstad (1947, 5), in New Zealand by Cunningham (1931) and northern America by Jackson (Mem. Torrey Bot. Club, **18**, 1, 1931).

Fromme (Bot. Gaz. **58**, 14, 1914) described the development of the aecidium.

Puccinia libanotidis Lindr.

Medd. Stockh. Hogsk. Bot. Inst. 4 (9), 2 (1901); Wilson & Bisby, Brit. Ured. no. 177; Gäumann, Rostpilze, p. 963.

Spermogonia scattered around the uredinoid aecidia, subepidermal, pale yellow, 90–110μ diam. with projecting paraphyses. Aecidia uredinoid, chiefly along the nerves, brick-brown, large, up to 3 cm. long; aecidiospores similar to the uredospores. Uredosori amphigenous, mostly hypophyllous, scattered, minute, punctiform, pulverulent, cinnamon; uredospores ovoid or ellipsoid, 26–34 × 21–28μ, wall evenly echinulate, apex strongly thickened (4–8μ), pale brown, with 3 or 4 thickened equatorial pores. Teleutosori amphigenous chiefly hypophyllous, scattered, minute, punctiform, pulverulent, dark brown; teleutospores ellipsoid, ovoid-ellipsoid, rounded at apex, attenuated below, slightly constricted, 32–50 × 15–24μ, wall smooth, brown, scarcely thickened at the apex, pores slightly papillate, pore of upper cell usually apical or subapical, of the

lower cell usually superior; pedicels hyaline, short, deciduous. Auteu-form. On *Seseli libanotis*. England, very rare.

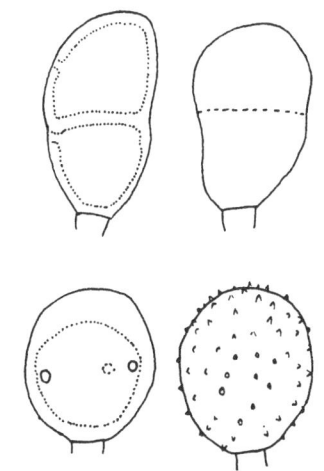

P. libanotidis. Teleutospores and uredospores.

This rust was found by Brenan near Cambridge in 1946 and later on a specimen of *Seseli libanotis* in the Edinburgh Herbarium collected by C. E. Salmon in September 1910 in Sussex. Secondary uredosori and teleutosori were present on the leaves. The species resembles *Puccinia conii* in the position of the pores of the teleutospore but can be distinguished from it by the smoother and thinner wall of the teleutospore.

Puccinia nitida (Str.) Röhl.

Deutschl. Fl. Ed. 2, 3 (3), 130 (1813).
[*Uredo petroselini* DC., Fl. Fr. 2, 597 (1805).]
Uredo nitida Str., Ann. Wetter. Ges. 2, 100 (1810).
Puccinia aethusae Mart., Prod. Fl. Mosq. Ed. 2, 225 (1817); Grove, Brit. Rust Fungi, 190; Wilson & Bisby, Brit. Ured. no. 81; Gäumann, Rostpilze, p. 953.
Puccinia petroselini [DC.] Lindr., Acta Soc. Fauna Fl. Fenn. 22, 84 (1902).

Spermogonia hypophyllous, in small groups, surrounded by uredinoid aecidia, subepidermal, yellowish-brown or almost hyaline, 80–95μ diam. Aecidia uredinoid, chiefly hypophyllous, in

rather small clusters, very small, occasionally confluent, circinate and larger, pulverulent, cinnamon-brown; aecidiospores resembling the uredospores. Uredosori like the aecidia, hypophyllous, scattered; uredospores globoid to ellipsoid, 22–29 × 21–25 μ, wall distantly echinulate all over or only in the upper part, or in upper part and base, leaving the central part smooth, 2·5–3 μ, thickened above (5–6 μ), yellowish or brownish-yellow, with 3 (rarely 2) equatorial pores with conspicuous caps. Teleutosori similar to uredosori but dark brown, on the petioles and stems, often larger, confluent and elongate; teleutospores ellipsoid or ovoid, rounded at both ends or slightly attenuate below, hardly constricted, 28–48 × 18–25 μ, wall not thickened above, smooth or nearly so, brown, pore of upper cell apical or slightly depressed, of lower cell much depressed, pedicel hyaline, thin, short, deciduous. Brachy-form.

On *Aethusa cynapium* and *Petroselinum crispum* cult., June–October. Great Britain and Ireland, scarce.

It is possible that the forms of these two hosts are distinct species, or at least biological races. Semadeni showed (Zbl. Bakt. ii, **13**, 443, 1904) that, while he could infect several (non-British) species of *Umbelliferae* with uredospores from *Aethusa cynapium*, he could not infect *Petroselinum crispum* although there is no morphological difference between the two forms. *Conium maculatum* became only very weakly infected. It was recorded on parsley in Yorkshire in 1932 but is rare on this host (Moore, 1959, 296).

Puccinia nitida is closely allied to *P. conii*, but differs from it in its relatively broader uredospores and in the lower position of the pore in the lower teleutospore cell.

Puccinia physospermi Pass.

Rabh. Fungi Europ., no. 1969 (1875); Gäumann, Rostpilze, p. 974.

Spermogonia, hypophyllous, honey-coloured, subepidermal, spherical, scattered among and between the teleutotosori. **Teleutosori** hypophyllous, chiefly along the under surface of the veins, 1–2 mm. long, 0·5 mm. broad, erumpent, surrounded by the ruptured epidermis; teleutospores in mass chocolate-brown, pulverulent, broadly ellipsoid, 37–46 × 24–27 μ, pore of upper cell apical with a conspicuous hyaline papilla, pore of lower cell median, spore wall irregularly undulate, pedicel hyaline, slender, rarely more than 4 μ broad, up to 40 μ long. Micro-form.

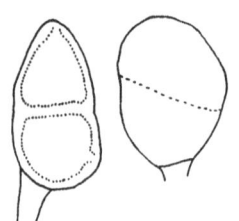

P. physospermi. Teleutospores.

Spermogonia and teleutosori on *Physospermum cornubiense.* Southern England, rare.

The only British material of this species is a specimen in the Grove collection in the British Museum from Buckinghamshire collected in 1932 which Grove labelled *Puccinia bullata.* However, it is easily distinguished from that group by the undulate spore wall. The pedicels of the teleutospores in the British collection are rather shorter than is usual for the species but the sori are not fully developed. The host occurs in the warmer parts of Europe and the rust has been recorded from the Mediterranean regions and the Balkans.

Puccinia pimpinellae (Str.) Röhl.

Deutschl. Fl. Ed. 2, **3** (3), 131 (1813).

Uredo pimpinellae Str., Ann. Wetter. Ges. **2**, 102 (1810).
Puccinia pimpinellae Mart., Prod. Fl. Mosq. Ed. 2, 226 (1817); Grove, Brit. Rust Fungi, p. 188; Wilson & Bisby, Brit. Ured. no. 197; Gäumann, Rostpilze, p. 977.

Spermogonia amphigenous, in groups or scattered amongst the aecidia, sub-epidermal, yellowish, 125–140 μ diam., paraphyses projecting up to 50 μ. Aecidia hypophyllous, in smaller or larger groups, often along the nerves and causing slight hypertrophy, between cup-shaped and pustulate, yellowish, peridia sunken, the projecting part seldom up to 0·5 mm. wide, shining white with irregular, laciniate margin, peridial cells very unequal, and irregularly arranged, most-ly 4-angled or rectangular, with outer wall distinctly and irregularly warted and slightly thicker than the inner; aecidiospores subgloboid or ellipsoid, hyaline with verruculose wall, 22–29 × 20–26 μ. Uredosori hypophyllous, scattered, minute, pulverulent, cinnamon; uredospores globoid to ellipsoid, brown, 22–32 × 20–26 μ, wall echinulate with 2 (rarely 3) equatorial pores. Teleutosori similar, but blackish-brown; teleutospores ellipsoid, hardly constricted,

28–37 × 19–25 μ, wall reticulate, pore of upper cell apical, of lower basal, pedicels hyaline, deciduous, rather short. Auteu-form.

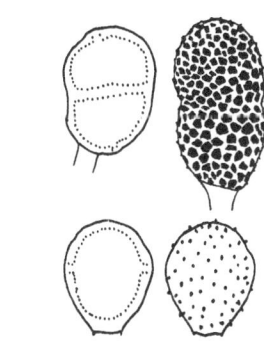

P. pimpinellae. Teleutospores and uredospores (after Fischer).

On *Pimpinella major* and *P. saxifraga*; aecidia, May–June; teleutosori, July–October. Great Britain, frequent.

Puccinia pimpinellae is very similar to *P. chaerophylli* but is distinguised from it by the uredospores, which have for the most part a thicker and darker membrane with only 2 pores. The peridium of the aecidia is better developed, and the teleutospores are plainly but not so densely reticulate. Klebahn (Jahrb. Wiss. Bot. **34**, 404, 1900) proved by cultures that *P. pimpinellae* is distinct from *P. chaerophylli* and Semadeni (Zbl. Bakt. II, **13**, 215, 1904) similarly proved its difference from that species and from *P. heraclei*; the latter showed it could be transferred to other species of the genus *Pimpinella*, but not to other genera of the *Umbelliferae*.

'*Aecidium bunii* var. *poterii* on *Poterium sanguisorba*' (Cooke, Handb. Brit. Fungi, p. 540) upon examination of the specimen in Kew Herbarium, has proved to be the aecidial stage of *P. pimpinellae* on *Pimpinella saxifraga*.

Puccinia rugulosa Tranz.

S.B. St Petersburg Naturf. Ges. **1** (1892); Wilson & Bisby, Brit. Ured. no. 213; Gäumann, Rostpilze, p. 996.

Puccinia umbelliferarum var. *peucedani-parisiensis* DC., Fl. Fr. **6**, 58 (1815).
Puccinia auloderma Lindr., Medd. Stockh. Hogsk. Bot. Inst. **4** (9), 2 (1901).
Puccinia peucedani-parisiensis (DC.) Lindr., Acta Soc. Fauna Fl. Fenn. **22**, 79 (1902).

Spermogonia usually epiphyllous, subglobose, pale yellow, with projecting paraphyses. **Aecidia** uredinoid, hypophyllous, mostly on the veins and leafstalks, up to 1 cm. long, cinnamon; aecidiospores resembling uredospores. **Uredosori** hypophyllous, scattered, minute, punctiform, for a long time covered by the epidermis, brown; uredospores subgloboid to ellipsoid, $26–33 \times 18–26\,\mu$, wall minutely echinulate, more or less thickened (up to $6\,\mu$) at the apex, base also thickened (up to $4\,\mu$) with 3 or rarely 4 pores. **Teleutosori** hypophyllous, minute or confluent in long patches on the leaf-stalks, at first covered by the epidermis then naked, pulverulent, blackish-brown; teleutospores oblong-ellipsoid or oblong-clavoid, apex rounded, not or scarcely thickened, slightly constricted, slightly narrowed below, $35–52 \times 19–28\,\mu$, wall with numerous, narrow, more or less parallel lines or ridges which occasionally anastomose, brown,

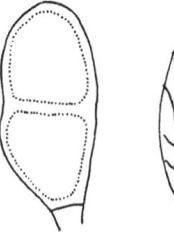

P. rugulosa. Teleutospores (after Savulescu).

pore of upper cell apical, of lower cell superior, pedicels hyaline, thin, deciduous. Brachy-form.

On *Peucedanum officinale*. England, very rare.

This rust was first recorded in Britain by Lindroth (*loc. cit.*) in 1902 from Faversham (in error 'Tewersham'), Kent. It was included as a British species by Massee (Mildews, Rusts and Smuts, 1913, 122), and then described by Grove (J. Bot. Lond. **59**, 16, 1921) from near Whitstable, Kent. It has been found on a specimen of *Peucedanum officinale* in Edinburgh Herbarium collected by J. B. French near Faversham in 1848.

It differs from *P. conii* chiefly in its longitudinally ridged teleutospores.

Puccinia saniculae Grev.

Fl. Edin. p. 431 (1824); Grove, Brit. Rust Fungi, p. 182; Wilson & Bisby, Brit. Ured.
no. 214; Gäumann, Rostpilze, p. 947.

[*Aecidium saniculae* Cooke, J. Bot. Lond. 2, 39 (1864).]

Spermogonia amphigenous, in little
groups, 124–145 µ diam., with projecting
paraphyses, brownish-yellow. Aecidia
hypophyllous or on the petioles on
brown or purple spots, in small clusters
2–4 mm. diam., cup-shaped or somewhat elongate on the nerves and petioles,
with a whitish, lobed, revolute peridium;
aecidiospores ellipsoid, 18–26 × 15–22 µ,
hyaline, delicately verruculose. Uredosori hypophyllous, on pale, minute spots
2–3 mm. diam., scattered, rarely aggregated, minute, punctiform, pale-cinna-
mon; uredospores globoid to ellipsoid,
brown, 25–38 × 18–27 µ, wall up to 3·5 µ
thick, echinulate, with 2 (rarely 3)
equatorial pores, with slightly swollen
caps. Teleutosori similar, but darker;
teleutospores ellipsoid, sometimes slightly thickened above with a small papilla,
hardly constricted, brown, 26–45 ×
18–26 µ, wall smooth with pore of upper
cell apical, of lower basal; pedicels hyaline, thin, deciduous. Auteu-form.

On *Sanicula europaea*, aecidia, April–
June; teleutospores, August–October.
Great Britain and Ireland, common.

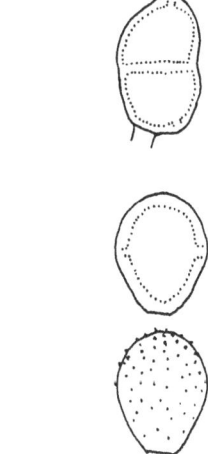

P. saniculae. Teleutospore and uredospores
(after Fischer).

Sometimes all three kinds of spores may be found together in sheltered
places as early as April.

Puccinia smyrnii Biv.-Bernh.

Stirp. Rar. Sicil. 4, 30 (1816); Wilson & Bisby, Brit. Ured. no. 225.

[*Aecidium bunii* var *smyrnii-olusatri* DC., Fl. Fr. 6, 96 (1815).]
Puccinia smyrnii Corda, Icon. Fung. 4, 18 (1840); Grove, Brit. Rust Fungi, p. 197.
Puccinia smyrnii-olusatri Lindr., Acta Soc. Fauna Fl. Fenn. 22, 9 (1902).

Spermogonia epiphyllous, on sunken
spots, subepidermal, about 110–190 µ
diam., with projecting paraphyses 30–
40 µ long. Aecidia amphigenous, in
rather irregular clusters or on the
petioles and stems in elongated groups,
on yellow spots, hemispherical, yellow,
opening by an irregular pore in the
nearly entire peridium; aecidiospores

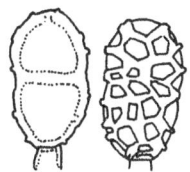

P. smyrnii. Teleutospores.

globoid or ovoid to fusiform or pyriform, delicately verruculose, yellowish, 16–40 × 16–20 μ. **Uredosori** wanting. **Teleutosori** hypophyllous on small yellow spots, scattered or a few together, minute, pulverulent, dark brown; teleutospores ellipsoid, hardly constricted, brown, 30–48 × 17–26 μ, wall rather thick, not thicker above, coarsely and remotely reticulate and tuberculate, pore of upper cell apical, of lower cell subequatorial, pedicels hyaline, thin, deciduous, up to 60 μ long. Opsis-form.

On *Smyrnium olusatrum*, aecidia, August–June; teleutospores, June–August. England, Wales, Ireland, frequent.

Grove (*loc. cit.*) stated that this was rather common near the coast but this is but an effect of host distribution. Ramsbottom (J. Bot. Lond. **52**, 185, 1914) found that at Torquay aecidia can be collected during every month from August to April; the teleutospores germinate overnight and the basidiospores often germinate *in situ*.

The aecidia and teleutosori may occur on the same or on separate plants. The markings on the teleutospore form a wide-meshed network, which bears wart-like tubercles at the angles of the meshes. The aecidiospores are rather like uredospores but they are produced in chains with intercalary cells in the usual way. The peridial cells are grossly verrucose on the inner surface, and are not arranged in regular rows. This irregular arrangement prevents the peridium splitting into laciniae; dehiscence is by an apical pore.

Puccinia tumida Grev.

Fl. Edin. p. 430 (1824); Grove, Brit. Rust Fungi, p. 187; Wilson & Bisby, Brit. Ured. no. 236; Gäumann, Rostpilze, p. 971.

Spermogonia and **aecidia** wanting. **Uredospores** very few, oval, pale yellow, sparsely verruculose, 20–25 × 15–18 μ, mingled with the teleutospores. **Teleutosori** on the leaves, more often on the petioles and nerves, minute, but many crowded together and confluent in thickened elongated masses (up to 1 cm. long), long-covered by the ash-coloured epidermis, black-brown; teleutospores ellipsoid to ovoid, hardly constricted, brownish, 26–36 × 14–26 μ, wall smooth or with a few groups of verrucae, not thickened at apex, pedicels hyaline, short, deciduous; occasional mesospores are found, 17–32 μ diam. Micro-form.

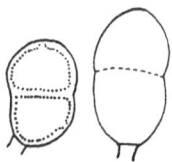

P. tumida. Teleutospores.

On *Conopodium majus*, April, May. Great Britain and Ireland, frequent.

This species was at first confused with *Puccinia bulbocastani* in which, however, the teleutosori are usually isolated on the leaves and cause no swelling of the affected part. Moreover, *P. bulbocastani* is an opsis-form with aecidia and uniformly verruculose teleutospores.

Plowright (1889, 206) stated that the mycelium is perennial but this is doubtful. Grove thought that the sori were confined to the radical leaves and that attacked plants did not flower; sori certainly appear first on the radical leaves but afterwards develop on the flowering stems where the groups of sori are much elongated. A few uredospores may be found in the sori on the radical leaves but not in the stem sori.

The rather uncommon aecidium of *Puccinia bistortae* (see p. 160) is sometimes to be found on the same plant as the teleutosori of *P. tumida*.

Puccinia acetosae Körnicke

Hedwigia, **15**, 184 (1876); Grove, Brit. Rust Fungi, p. 223; Wilson & Bisby, Brit. Ured. no. 77; Gäumann, Rostpilze, p. 894.

[*Uredo acetosae* Schum., Enum. Pl. Saell. **2**, 231 (1803).]

Spermogonia and **aecidia** unknown. **Uredosori** amphigenous, scattered, minute, roundish on the leaves, elongated on the stems and petioles, soon naked, cinnamon-brown; uredospores globoid to obovoid, $24-30 \times 20-23\,\mu$, wall sparsely echinulate with colourless spines $2-3\cdot5\,\mu$ apart, yellowish-brown, $1\cdot5-2\cdot5\,\mu$ thick, with 2 supra-equatorial pores. **Teleutosori** amphigenous, minute, rounded, elongated on the stems, chocolate-brown; teleutospores ellipsoid, oblong or subclavoid, rounded at both ends or slightly attenuated below, slightly constricted at the septum, chestnut-brown, $30-46 \times 19-26\,\mu$, wall delicately verruculose, not thickened at the apex, pore apical in the upper cell with a broad pore cap, pore of lower cell superior, pedicel hyaline, slender, deciduous, up to $35\,\mu$ long. Hemi-form? Uredospores and teleutospores on leaves and stems of *Rumex acetosella*; uredospores only on *R. acetosa*, July–

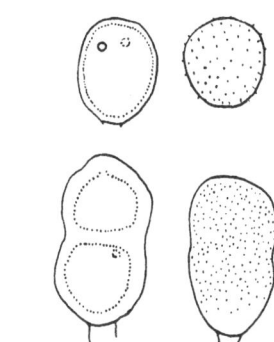

P. acetosae. Teleutospores (after Savulescu) and uredospores.

October. Great Britain and Ireland, frequent on *R. acetosa*, rare on *R. acetosella*.

This has been often confused with *Uromyces acetosae* but the latter is distinguished by its densely verruculose uredospores with 3 equatorial pores.

Teleutospores have been rarely found but were described by Plowright (TBMS. **1**, 57, 1889) from Yorkshire.

The rust has been found in Sutherland on Ben Loyal at an altitude of about 1000 ft. and has been recorded from the Shetland Isles by Dennis & Gray (Trans. Bot. Soc. Edin. **36**, 220, 1954). It is rarely found on *Rumex acetosella* but there is a specimen on this host from near

Edinburgh collected in 1821 and it has been recorded recently from several localities in Scotland.

The aecidial host is unknown and it is evident that the rust must persist through the winter by means of uredospores or mycelium.

Puccinia bistortae DC.

Fl. Fr. **6**, 61 (1815); Wilson & Bisby, Brit. Ured. no. 104.
Uredo polygoni var. *bistortae* Str., Ann. Wetter. Ges. **2**, 103 (1810).
Puccinia polygoni var. *bistortae* (Str.) Röhl., Deutschl. Fl. Ed. 2, 3, (3), 132 (1813).
Puccinia polygoni-vivipari Karst., Not. Sällsk. Fauna Fl. Fenn. Förh. **8**, 221 (1866); Grove, Brit. Rust Fungi, p. 226; Gäumann, Rostpilze, p. 886.
Puccinia conopodii-bistortae Kleb., Z. Pfl.-Krankh. **6**, 331 (1896); Grove, Brit. Rust Fungi, p. 225; Gäumann, Rostpilze, p. 884.
Puccinia cari-bistortae Kleb., Z. Pfl.-Krankh. **9**, 157 (1899); Gäumann, Rostpilze, p. 881.
Puccinia angelicae-bistortae, Kleb., Z. Pfl.-Krankh. **12**, 142 (1902); Gäumann Rostpilze, p. 883.

Spermogonia amphigenous, honey-coloured. **Aecidia** hypophyllous, in rounded groups, and on the stems and petioles where the groups are elongate, on large, bright yellow or orange, thickened areas, crowded but rarely confluent, rather large, pustular; aecidiospores globoid or polyhedroid, yellowish-orange, 18–26 × 15–23 μ, wall verruculose. **Uredosori** hypophyllous, on slightly yellowish spots, small, 0·3–1 mm. diam., roundish, yellowish-red, soon naked; uredospores globoid or ellipsoid, pale yellowish-brown, 20–25 × 18–20 μ, wall 1·5–2·5 μ thick, finely echinulate, with about 6, indistinct, scattered pores. **Teleutosori** hypophyllous, on pale spots, scattered or united in roundish groups, 0·3–0·5 mm. diam., soon naked and pulverulent, dark brown; teleutospores broadly ellipsoid or subclavoid, rounded at both ends or slightly oblique at the apex, slightly constricted, yellowish-brown, 24–42 × 16–25 μ, wall 1–2 μ thick, smooth

or finely verrucose in a few longitudinal or oblique lines, pores without prominent papillae, the upper apical, the lower variously placed but usually equatorial

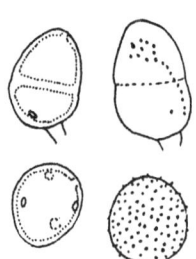

P. bistortae. Teleutospores and uredospores.

or subequatorial, pedicel hyaline, short, deciduous. Hetereu-form.

Aecidia on *Angelica sylvestris* and *Conopodium majus*, May–June; uredospores and teleutospores on *Polygonum bistorta* and *P. viviparum*, July–August. England, Scotland, scarce.

This rust was regarded by Jørstad (1932, 340) as a polymorphous species consisting of several physiologic races which also differ slightly morphologically. They produce uredospores and teleutospores on species of *Polygonum* belonging to the subgenus *Bistorta* and aecidia on species belonging to certain genera of the *Umbelliferae*.

The first discovery in this country of *Puccinia bistortae* was probably on *P. viviparum* by Greville at Mar Lodge, Aberdeenshire, in 1822; the specimen is in the Edinburgh Herbarium and is labelled *P. vivipari* in Greville's handwriting, but this name was never validly published by Greville. Stevenson (Myc. Scot. 1879, 234) mentioned specimens on this host collected by Keith in Moray and by Buchanan White in Glenshee, Perthshire.

The first record on *Polygonum bistorta* appears to be that by Cooke (Grevillea, **2**, 161, 1874) from near Liverpool.

The following races of *Puccinia bistortae* have been found in Britain:

(1) On *Polygonum viviparum* in Scotland north of latitude 56°; there is no record of its aecidial stage in this country. Trail (1890, 315) collected it in 1879 and 1882 at Braemar and later he recorded it in the Tay, Dee and Moray areas. Since that time it has been found on several mountains in Perthshire, Argyllshire, Glen Affric, Cairngorm, in Inverness-shire (TBMS. **30**, 2, 1939) and in Morayshire.

No culture experiments have been carried out with this rust in Scotland; in Sweden, Juel (Overs. K. Vet. Akad. Förh. **1899**, no. 1, 5, 1899) made a series of cultures with overwintered teleutospores from *Polygonum viviparum* and produced on *Angelica sylvatica* aecidia identical with *Aec. angelicae* Rostr. (Bot. Tidsskr. **15**, 230, 1886); no spermogonia were formed. He also made reciprocal cultures with the aecidiospores thus produced, on *Polygonum viviparum* with positive results; similar inoculations on *P. bistorta* gave no result. These culture results were confirmed by Klebahn (1904, 320).

Both Juel and Klebahn agreed that this race has unusually small teleutospores, the measurements being 23–29×18–$22\,\mu$; this is also true of the rust in Scotland and Scandinavia.

This circumpolar arctic-alpine race is widely spread in the northern hemisphere. It has often been recorded from Norway by Jørstad (1932, 340); he has also described it from Kamtchatka (1934, 112) and Iceland (1951, 25). An account of its synonomy and distribution in Scandinavia is given by Hylander *et al.* (1953, 39). Jørstad (1962) has reported it north of latitude 70° in Finmark. Its aecidial stage on *Angelica sylvatica* is of rare occurrence in Norway, Sweden, Finland and parts of northern Russia; spermogonia are not present; it appears to occur more commonly on the mountains of Central and Southern Europe but in these regions its host relationships are often not clear.

As no aecidial stage of this rust has been found in Britain it is probable that the mycelium is present in the green leaves of *Polygonum viviparum* which persist through the winter and develop uredospores in the early spring as described by Jørstad (1932, 342) in Norway. Vleugel (Sv. Bot. Tidskr. 5, 341, 1911) in Sweden described the occurrence of uredospores and teleutospores on the bulbils and considered that the rust hibernates in this way.

The structure of the aecidium is unusual and has been described and figured by Juel.

(2) The race on *Polygonum bistorta* with its aecidium on *Conopodium majus* was discovered at Hebden Bridge and near Leeds by Soppitt (Grevillea, 22, 45, 1893; Gard. Chron. 18, 773, 1895). Probably it was this race that was found near Keswick, Cumberland in 1935 on *P. bistorta*; both stages occurred at Strachur, Argyll in 1943 and in Westmorland in 1951.

The aecidia are produced on the petioles and segments of the radical leaves, on the stems and sometimes on the cauline leaves; the portions of the leaf above the point of infection frequently die off; no spermogonia are present. In Scotland uredospores found in May on green and dying leaves of the previous year which had persisted through the winter germinated readily; this supports the Sydows' suggestion (Monogr. Ured. 1, 571) that the aecidium on the Umbellifer is merely facultative. The teleutospores soon follow the uredospores and may be found on the same spot 8 or 10 days after the latter are mature. The yellowish spots on which both the uredospores and teleutospores develop are found on the upper surface immediately above the sori as well as on the lower surface; sometimes 'green islands' are seen on the dying leaf.

Soppitt (*loc. cit.*) infected *Conopodium denudatum* with teleutospores and from the resulting aecidiospores produced uredospores and later teleutospores on *P. bistorta* while *P. persicaria, P. aviculare* and *P. viviparum* remained uninfected. These cultures were confirmed by M. R. Gilson (*in litt.*) in Westmorland in 1951 on *C. majus* and on *P. bistorta*.

The race of *P. bistortae* with its aecidium on *Angelica sylvestris* has been found only once in Britain, near Sheffield by J. Webster (TBMS. 42, 328, 1959) who produced infections with aecidiospores in the spring of 1958 obtaining uredospores soon followed by teleutospores on *P. bistorta*. No spermogonia were present with the aecidia.

Puccinia oxyriae Fuck.

Jahrb. Nass. Ver. Nat. 29–30, 14 (1876); Fungi Rhen. no. 2635 (1874) *nomen nudum*; Grove, Brit. Rust Fungi, p. 224; Wilson & Bisby, Brit. Ured. no. 190; Gäumann, Rostpilze, p. 895.

Spermogonia and aecidia unknown. Uredosori amphigenous, on minute purple spots, scattered or aggregated, sometimes confluent, rounded or irregular, surrounded by cleft epidermis, cinnamon-brown; uredospores ellipsoid or obovoid, yellowish-brown, 23–30 × 20–26 μ, wall echinulate, 2–2·5 μ thick with 4–6 indistinct, scattered pores. Teleutosori chiefly hypophyllous and on the petioles and peduncles on similar spots, scattered or in minute groups, rounded, elongate on petioles and peduncles, pulverulent, chestnut-brown; teleutospores broadly ellipsoid, medianly constricted, light chestnut-brown, 30–46 × 15–25 μ, wall uniformly 2–3 μ thick with irregular flat warts which are larger and more distinct around the pores, pore of upper cells apical, of lower cell close to septum, pedicels almost hyaline, rather short and fragile, deciduous. Hemi-form.

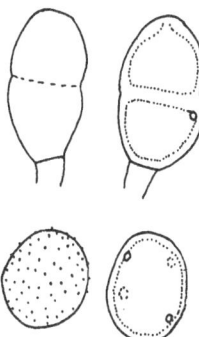

P. oxyriae. Teleutospores and uredospores.

On *Oxyria digyna*. Scotland, Wales, rare.

This is a circumpolar, arctic-alpine rust which occurs on the mountains of central and northern Scotland and has been recorded by Ellis (*in litt.*) in the Snowdon area.

The life cycle of *Puccinia oxyriae* is not completely known; it is possibly heteroecious but is certainly independent of host alternation in this country. Single, heavily infected plants often occur in the midst of uninfected populations of the host, a distribution strongly suggesting some form of uredo-perennation. The teleutospores are usually abundant and are said to germinate only after overwintering.

Jørstad (1951, 41) stated that in Iceland the spore stages often occur on the basal and subterranean parts of the petioles; he has suggested that this may be connected with the method of overwintering.

In Switzerland Gäumann & Müller (Phytopath. Zeitschr. 30, 327, 1957) failed, with teleutospores, to infect *Ranunculus glacialis* which they considered might be the alternate host of this rust. They investigated the method of overwintering and have excluded the possibility of the existence of latent mycelium in the host and shown that uredospores are unable to survive the winter. They were able, in carefully controlled cultures with overwintered teleutospores, to infect the host in the spring with the production of uredospores on the petioles and leaf blades, 12–14 days afterwards. No production of spermogonia was observed. They conclude that *P. oxyriae* is a hemi-form, but leave the question open if this shortened development holds good for the whole of the species or whether they worked with a specialised, perhaps apomictic, Swiss race.

Puccinia polygoni-amphibii Pers.

Syn. Meth. Fung. 227 (1801); Grove, Brit. Rust Fungi, p. 227; Wilson & Bisby, Brit. Ured. no. 202; Gäumann, Rostpilze, p. 771.

[*Aecidium sanguinolentum* Lindr., Bot. Not. **1900**, 241 (1900).]

Spermogonia amphigenous, few, roundish, 90–150 μ high, 90 μ wide. **Aecidia** hypophyllous, mostly in concentric groups on deep-red, blood-red or purple spots, often surrounded by a conspicuous greenish-yellow zone sometimes occupying the greater part of the leaf without hypertrophy of the host tissues, peridium shortly cylindrical or cup-shaped with finely laciniate margin, peridial cells strongly developed and firmly connected, polygonal or roughly quadrate, 17–30 × 15–25 μ, arranged in regular rows with finely warted outer wall 6–7 μ thick, inner wall thinner, aecidiospores globoid to broadly ellipsoid or polyhedroid, 18–23 μ diam., with uniformly and very finely verruculose hyaline wall, contents yellowish. **Uredosori** amphigenous, mostly hypophyllous, occasionally on the sheathing stipules, scattered or sometimes concentrically arranged, rounded, soon naked, pulverulent, brownish; uredospores globoid, ellipsoid or obovoid, 17–30 × 15–22 μ, wall faintly and rather distantly echinulate, 1–1·5 μ thick, cinnamon-brown, with 2 or sometimes 3 supra-equatorial pores, contents colourless, pedicel long, colourless. **Teleutosori** amphigenous and on the sheathing stipules, slightly convex, brown or brownish-black; of two kinds: (1) in sori which have already borne uredospores; amphigenous, naked, with numerous colourless uredospore pedicels beyond which the teleutospores project, the latter regularly rounded at the apex;

(2) sori chiefly hypophyllous, concentrically arranged around the uredosori or often scattered over almost the whole of the under surface and on the sheathing stipules, at first embedded in the leaf tissue, then remaining long-covered by the epidermis, protuberant and warty, developing teleutospores only which are often irregularly truncate or obliquely

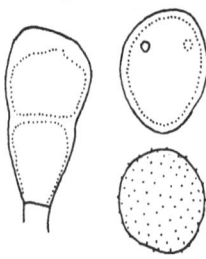

P. polygoni-amphibii. Teleutospore and uredospores.

conical at the apex; all teleutospores oblong to clavoid, gradually tapering below, slightly constricted at the septum, the lower cell usually longer and narrower than the upper, 35–52 × 16–22 μ, wall yellowish-brown, smooth, strongly thickened (5–22 μ) at the apex, pore of upper cell apical, of lower close to the septum, pedicels almost hyaline to yellowish-brown, persistent, almost as long as the spore. Auteu-form.

[Spermogonia and aecidia on *Geranium sylvaticum*]; uredospores and teleutospores on *Polygonum amphibium*, '*P. lapathifolium*' and '*P. persicaria*', July–October. Great Britain and Ireland, frequent.

Puccinia polygoni-amphibii has been found only on the terrestrial form of *Polygonum amphibium*. The records on *P. lapathifolium* and *P. persicaria* are doubtful and are probably due to misidentification of the hosts. Both Tranzschel (Trav. Mus. Bot. Acad. Imp. Sci. St Petersburg, 2, 28, 1905) and Klebahn (Z. Pfl.-Krankh. 15, 70, 1905) failed to infect *P. lapathifolium* and they considered it very doubtful whether this species was a host for the rust in Europe; they suggested that the host was misidentified. On *P. persicaria* it was recorded by E. A. Ellis from Norfolk and by O'Connor from Eire. In Scandinavia the rust is facultatively alternating with *Geranium sylvaticum* (Jørstad, 1960, 130). This rust has been united with *P. polygoni-convolvuli* by Plowright, Sydow, Tranzschel (1939, 170) and Wilson & Bisby; Arthur (1934) and Hylander *et al.* (1953, 64) have regarded *P. polygoni-convolvuli* as a variety of *P. polygoni-amphibii* and this procedure is adopted here.

var. convolvuli Arth.

Rusts U.S. Can. 233 (1934).

[*Uredo betae β convolvuli* Alb. & Schw., Consp. Fung. Nisk. 127 (1805).]
Puccinia polygoni Alb. & Schw., Consp. Fung. Nisk. 132 (1805); Gäumann, Rost-
 pilze, p. 775.
Puccinia polygoni-convolvuli DC., Lam. Encycl. Meth. Bot. 8, 251 (1808); Grove,
 Brit. Rust Fungi, p. 228.

Spermogonia and aecidia on *Geranium dissectum*, June; uredospores and teleu-tospores on *Polygonum convolvulus*, August–September. England, scarce.

The aecidial stage was found by W. G. Bramley at Bolton Tracy, Yorkshire in June 1946. Aecidiospores from these specimens were placed on leaves of *Polygonum convolvulus* at Edinburgh and produced uredospores 10 days afterwards and subsequently teleutospores; the uredospores failed to infect *Polygonum amphibium*.

The aecidia on *Geranium dissectum* are borne on pale greenish spots which later become reddish. The teleutosori are distinguished from those on *Polygonum amphibium* by their compact, pulvinate form and by being soon uncovered by the epidermis; the teleutospores are darker at the summit and, if conical, are less oblique.

Many cultures of this rust and its variety have been made. Tranzschel (Trav. Mus. Bot. Acad. Imp. Sci. St Petersburg, 2, 28, 1905) showed that *Polygonum convolvulus* was not infected by *Puccinia polygoni-amphibii* var. *polygoni-amphibii* and this was confirmed by Klebahn (Z. Pfl.-Krankh. 22, 327, 1912) who considered that the two rusts on *Polygonum* spp. were distinct. Tranzschel (*loc. cit.*, p. 76) proved the heteroecism by sowing teleutospores from *P. amphibium* on *G. palustre* and *G. pratense*, obtaining aecidia. The same investigator made successful cultures with var. *convolvuli* by sowing teleutospores from *Polygonum convolvulus* on *G. pusillum*.

These results have been confirmed and extended by Bubak (Ann. Myc. **2**, 16, 1904), Klebahn (Z. Pfl.-Krankh. **15**, 70, 1905, and **22**, 327, 1912), Treboux (Ann. Myc. **10**, 305 and 557, 1912), Jacob (Zbl. Bakt. II, **44**, 617, 1916), Mayor (Bull. Soc. Neuchâtel Sci. Nat. **48**, 382, 1923), Viennot-Bourgin (see Guyot, Ann. Ecole Nat. Agric. Grignon, Ser. 2, **1**, 45, 1937) and Dupias (Bull. Soc. Hist. Nat. Toul. **85**, 37, 1950).

The aecidial stage of *Uromyces geranii* which also occurs on the leaves of several *Geranium* species produces pale greenish spots which later become reddish and are much hypertrophied; it may also be distinguished from that of *Puccinia polygoni-amphibii* by the inner and outer walls of the peridial cells being of about equal thickness.

Sappin-Trouffy (1896, 115) gave an account of nuclear division, and of the cytology of uredospores and teleutospore development.

Puccinia septentrionalis Juel

Overs K. Vet. Akad. Förh. **1895** (6), 383 (1895); Wilson & Bisby, Brit. Ured. no. 220; Gäumann, Rostpilze, p. 877.

Spermogonia unknown. **Aecidia** amphigenous or on the petioles on swollen, rather large, dark violet spots, numerous, cupulate, opening widely with revolute peridium, peridial cells isodiametric with a wide lumen; aecidiospores subgloboid, orange, 18–20μ diam. **Uredosori** hypophyllous, producing pale spots on the upper surface of the leaf, scattered, minute, pulverulent, yellowish; uredospores globoid or ellipsoid, with orange-coloured contents, 20–22μ diam., wall echinulate, brown, with 3–4 pores. **Teleutosori** hypophyllous, occasionally epiphyllous, scattered, minute, pulverulent, dark brown; teleutospores at first developed in the uredosori, later in separate sori, ovoid-ellipsoid or pyriform, not constricted at the septum, with colourless contents, 28–48 × 13–23μ; wall smooth, thickened at the apex with a rather large, hyaline, papilla, pore of lower cell near the septum, covered by a small papilla; pedicel hyaline, very short. Hetereu-form.

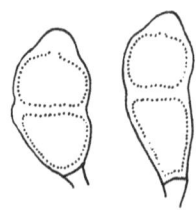

P. septentrionalis. Teleutospores
(after Fischer).

Aecidia on *Thalictrum alpinum*, May–June, uredospores and teleutospores on *Polygonum viviparum*, July–October. Scotland, frequent.

The aecidial stage was found on Ben Lui, Perthshire, in June 1913 and the teleutospores in the same locality in October 1914 (Wilson, J. Bot. Lond. **53**, 45, 1915). The rust is a characteristic feature of the higher mountains in Scotland at altitudes from 600 to 1000 m. where the two hosts are found in proximity; it has also been found in the Shetlands (TBMS. **7**, 82, 1921, and **9**, 135, 1924).

The aecidial stage is conspicuous on account of the violet swellings it produces on the leaves and petioles and occasionally on the flowering stems of its host. The uredospores are generally found only in small quantities and later in the season may be entirely lacking. It is obviously always heteroecious.

Juel (*loc. cit.*) proved the connection of both stages by cultures and also infected *Polygonum bistorta* with aecidiospores. These culture experiments were confirmed in Scotland (Wilson, TBMS. **9**, 135, 1924). The teleutospores of *P. septentrionalis* are quite smooth whereas those of *P. bistortae* are sparsely verrucose. The aecidial stage of grass rusts of the *P. recondita* type which occurs on *Thalictrum alpinum* does not produce swellings or discoloration.

Juel (*loc. cit.*) described and figured the structure of the aecidial peridium and also the germination of the teleutospores.

Puccinia primulae Duby

Bot. Gall. **2**, 891 (1830); Grove, Brit. Rust Fungi, p. 179; Wilson & Bisby, Brit. Ured. no. 205; Gäumann, Rostpilze, p. 998.

[*Uredo primulae* DC., Fl. Fr. **6**, 68 (1815).]

Spermogonia unknown. **Aecidia** hypophyllous on yellowish spots, densely but irregularly clustered in roundish groups, shortly cylindrical, with a broad, much cut revolute, white peridium; aecidiospores verruculose, orange, 17–23 × 12–18 μ. **Uredosori** hypophyllous, minute, scattered or circinate, roundish, soon naked, brown; uredospores subgloboid to ovoid, 20–23 × 16–19 μ, wall echinulate, pale brown, with 3–4 pores. **Teleutosori** similar, but long-covered by the grey epidermis, often confluent, or in circles round the aecidia or uredosori, blackish-brown; teleutospores ovoid or oblong, rounded at both ends, 22–30 × 15–18 μ, wall hardly thickened above, pore of upper cell apical, of lower much

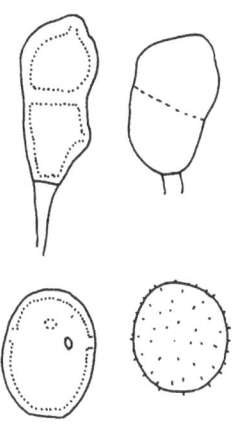

P. primulae. Teleutospores and uredospores.

depressed, with a colourless papilla on each, smooth, pale brown, pedicels hyaline, short, deciduous; mesospores occasionally present. Auteu-form.

On *Primula vulgaris*, aecidia, May; teleutospores, June–October. Great Britain and Ireland, frequent.

All three spore-forms may be found on the same leaf. The teleutospores are rather irregular in shape; 1-celled teleutospores are not infrequent, and Fischer (1904) described and figured 3-celled spores. There are very doubtful records on the continent on *Primula elatior* and *P. veris*.

Puccinia soldanellae Fuck.

Symb. Myc. Nachtr. 3, 14 (1875); Grove, Brit. Rust Fungi, p. 180; Wilson & Bisby, Brit. Ured. no. 226; Gäumann, Rostpilze, p. 999.

Spermogonia amphigenous, numerous, punctiform, spherical. **Aecidia** hypophyllous, scattered uniformly over nearly the whole leaf-surface, shortly cylindrical or urceolate, with a white, denticulate, revolute margin; spores delicately verruculose, yellowish, 18–26 μ diam. [**Uredosori** generally epiphyllous, without spots, scattered or circinate, minute, surrounded by the torn epidermis, brown; spores globoid to ellipsoid, pale brown, 20–30 × 18–28 μ, wall 2–3 μ thick, echinulate, with 3 pores. **Teleutosori** similar but black-brown; teleuto-spores ellipsoid or ovoid-oblong, gently constricted, usually rounded below, 35–55 × 20–34 μ, wall smooth, chestnut-brown, thickened at apex up to 5–8 μ, pore of upper cell slightly depressed, of lower subequatorial or basal, both with broad, well-developed, hyaline papillae, pedicel hyaline, deciduous, up to 50 μ long.] Auteu-form.

On *Soldanella alpina*, all spore stages are said to occur together in July and August. England, Scotland, rare and introduced.

There is a record of this rust from the Botanic Garden, Glasgow, before 1836 and another from the Cambridge Botanic Garden in April 1926, both no doubt on recently imported plants; of the latter a specimen is in the Kew Herbarium. Only the aecidial stage appears to have occurred in Britain. This stage is the more common; its mycelium is perennial and systemic, causing a conspicuous change in the leaves; they become smaller, paler, and longer stalked. The description of the uredosori and teleutosori is after Sydow (Monogr. Ured. 1, 349).

Puccinia vincae Berk.

Smith, Engl. Fl. 5 (2), 364 (1836); Grove, Brit. Rust Fungi, p. 176; Wilson & Bisby, Brit. Ured. no. 242; Gäumann, Rostpilze, p. 1020.
[*Uredo vincae* DC., Fl. Fr. 6, 70 (1815).]

Spermogonia hypophyllous or rarely epiphyllous, scattered, arising from systemic mycelium, flask-shaped, honey-coloured. **Aecidia** uredinoid, hypophyllous, thickly scattered over the leaf, rather irregular, often elongate and slightly curved, up to 2–3 × 1 mm., first-formed spores in these sori, ovoid or ellipsoidal, 38–40 × 22–24 μ, wall hyaline, echinulate, 4–4·5 μ thick, without differentiated pores, later-formed spores ellipsoidal, 34–40 × 23–25 μ, wall echinulate, brown, 2–3 μ thick, with 3 equatorial or slightly supra-equatorial pores. **Uredosori** on older host leaves, scattered, without spermogonia and lacking the hyaline type of spore. Primary **teleutosori** usually formed by the replacement of aecidiospores; teleutospores ellipsoidal to clavate, scarcely constricted at the septum, 37–41 × 21–23 μ, wall brown, with verrucae arranged in irregular longitudinal lines, pore of upper cell apical of lower basal, pedicel hyaline, up to 10 μ long. Secondary teleutosori similar to the primary but scattered on

local infections on the older leaves of the host. Brachy-form.

On *Vinca major*, infrequent.

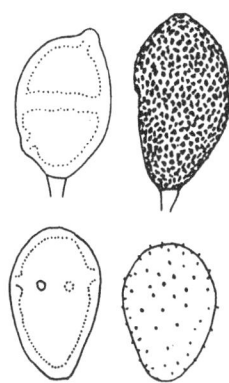

P. vincae. Teleutospores and uredospores (after Fischer).

This species was first recorded in Britain in 1887 (Gard. Chron. **2**, 227, 1887). The life history is not yet completely understood. Its elucidation has been obscured by the fact that the sori are very frequently attacked by the hyperparasite *Tuberculina sbrozzii* whose effuse orange sporodochia often completely replace the rust sori. These sporodochia were taken by Plowright (1889, 161) for aecidia and Grove (*loc. cit.*) only hinted at their real nature. Biraghi (Boll. Staz. Pat. Veg. **20**, 71, 1940) has examined the association between rust and *Tuberculina* and regarded it as one of mutualistic symbiosis.

Infected stems are deformed, tend to a more erect posture and are sterile. The mycelium is obviously systemic and the spermogonia and aecidia and teleutosori appear early, covering most of the leaves. The uredosori and secondary teleutosori develop later on the older leaves, presumably by infection by aecidiospores, but these secondary stages have been less frequently collected. The hyaline spores which are present at first in aecidia appear non-functional but only experiment could clarify this point. They are often the only spores present in aecidia attacked by *Tuberculina* where attack has stopped spore formation at an early stage in the life of the sorus.

The inclusion of *Vinca minor* as a host in Britain by Grove (*loc. cit.*) remains unconfirmed and is best excluded.

Puccinia gentianae Röhl.

Deutschl. Fl. Ed. 2, 3(3), 131 (1813); Grove, Brit. Rust Fungi, p. 178 *p.p.*; Wilson & Bisby, Brit. Ured. no. 155; Gäumann, Rostpilze, p. 1017.

Uredo gentianae Str., Ann. Wetter. Ges. 2, 102 (1811).

[**Spermogonia** honey-coloured, in small clusters. **Aecidia** hypophyllous or on the stems, on circular brown spots, in irregular clusters, cup-shaped, with torn, white margin; aecidiospores delicately verruculose, orange, 16–23 × 14–17 μ, wall, hyaline, verruculose.] **Uredosori** scattered or circinate, epiphyllous, less commonly amphigenous, roundish, at first covered by the epidermis, pale chestnut; uredospores globoid to ovoid, 20–30 × 18–24 μ, wall aculeate, brownish-yellow, with 2 equatorial pores. **Teleutosori** similar and also on the stems, but pulverulent and dark brown; teleutospores ellipsoid to ovoid, rounded at both ends, not constricted, sometimes with a low, broad, apical papilla, 28–38 × 24–30 μ; wall smooth, dark chestnut, pore of upper cell apical, of lower near the septum or slightly depressed, pedicels hyaline, thin, rather long, deciduous; mesospores sparingly present. Auteu-form.

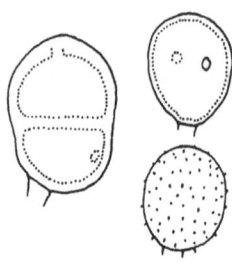

P. gentianae. Teleutospore and uredospores.

On cultivated *Gentiana acaulis* and *G. verna*, August–October. England, Wales, rare.

This species has been recorded from Kew Gardens and Horsham (Grove, *loc. cit.*), Cardiff (TBMS. **7**, 79, 1921) and Cheshire (in Herb. Kew). The aecidia are said to appear in April and June and more frequently on the stems and peduncles than on the leaves. The specimens from Cardiff were found on *Gentiana acaulis* and *G. verna*; *G. asclepiadea* growing near them was not infected and this agrees with the observation of Bock (Zbl. Bakt. II, **20**, 564, 1908) that this species, when found in the vicinity of other infected gentians in Switzerland, was not attacked.

This species is almost exclusively restricted to *Gentiana; Uromyces eugentianae* (q.v. p. 347) with which it was previously confused, is found only on *Gentianella* (Jørstad, Nytt Mag. Bot. **3**, 103, 1954).

Puccinia polemonii Diet. & Holw.

Bot. Gaz. **18**, 255 (1893); Gäumann, Rostpilze, p. 806.

Spermogonia unknown, **aecidia** and **uredosori** wanting. **Teleutosori** mostly confined to the petioles but spreading along the midribs of the leaflets, up to 2 mm. long, often confluent into soral lines, 1 cm. long, spore mass dark rusty-brown, pulverulent; teleutospores 34–48 × 13–15 μ, pale brown, oblong-cylin-

dric or clavoid not or slightly constricted at the septum, apex pointed,
hyaline, up to 6 μ thick, wall smooth,
pore in upper cell apical, in lower
superior, pedicels 25 μ long, deciduous.
Micro-form.
On *Polemonium caeruleum*. England,
rare.

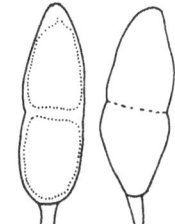

P. polemonii. Teleutospores.

This species was recorded by Piggott (J. Ecol. **46**, 507, 1958) from three localities in Derbyshire. It appears to be rare throughout its range in the Old World. Webster (TBMS. **42**, 328, 1959) described the British collections and pointed out that Dietel & Holway (*loc. cit.*) described two types of spore, one which is the type present in the British material with coloured spores and deciduous pedicels and the other, leptosporic, with hyaline walls and persistent pedicels which germinate immediately. This latter type is not known in Europe (Jørstad, 1932, 361).

Puccinia convolvuli Cast.

Obs. **1**, 16 (1843); Grove, Brit. Rust Fungi, p. 175; Wilson & Bisby, Brit. Ured. no. 140; Gäumann, Rostpilze, p. 805.

[Spermogonia epiphyllous, in groups, minute, subepidermal, globoid or ovoid-globoid, yellowish, 18–100 μ broad, 90–130 μ high, ostiolar filaments up to 100 μ long.] Aecidia hypophyllous, on brownish or purplish spots, more or less circinate, often on the petioles and then in elongated patches, peridia cup-shaped, minute, white and recurved and torn at the margin; aecidiospores delicately verruculose, pallid-yellow, 17–28 μ. Uredosori scattered or circinate, minute, often confluent, soon naked, brown; uredospores more or less ellipsoid, rarely ovoid, pale brown, 22–30 × 18–26 μ, wall echinulate, with 2 or 3 supra-equatorial pores. Teleutosori similar, but long-covered by the grey epidermis, black-brown; teleutospores ellipsoid to oblong, obtuse or rounded above, slightly constricted, rounded below, chestnut-brown, 38–66 × 18–30 μ, wall smooth, up

to 9 μ thick at the apex; with them are intermixed (according to Sydow) ovoid mesospores, much thickened at the apex, brown, 25–35 × 20–26 μ, pedicels brown-

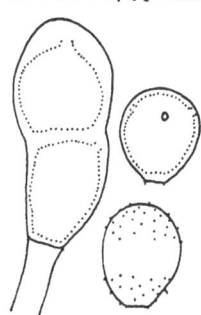

P. convolvuli. Teleutospore and uredospores (after Fischer).

ish, thick, persistent, up to 35 μ long. Auteu-form.

On *Calystegia sepia*. England, very rare.

The only justification for including this species is Plowright's account (1889, 146) of an early collection by Miss Jelly, the locality unspecified. No specimens can be found to substantiate the record.

Puccinia albulensis Magn.

Ber. D. Bot. Ges. 8, 169 (1890); Wilson & Bisby, Brit. Ured. no. 87; Gäumann, Rostpilze, p. 815.

As *Puccinia porteri* auct. (non *P. porteri* Peck apud Porter & Coulter, Syn. Fl. Colo. p. 164 (1874)); Wilson, Ured. Scot. p. 377.

As *Puccinia veronicarum* DC. Grove, Brit. Rust Fungi, p. 169.

Spermogonia unknown, **aecidia** and **uredosori** wanting. **Teleutosori** hypophyllous or on the stems, sometimes epiphyllous, generally densely gregarious and often covering the whole surface of the leaf, rarely solitary, minute or medium-sized, confluent, rounded, oblong or irregular, pulverulent in the resting form, cinnamon-brown, pulvinate in the leptosporic form, becoming cinereous by germination; teleutospores rounded at both ends, slightly constricted, ellipsoid or oblong, 26–46 × 11–18 µ, wall cinnamon-brown, smooth or very faintly punctate-rugose, 1–1·5 µ thick, at the apex 3–6 µ (rarely up to 9 µ), pore of upper cell apical, of the lower near the septum; pedicel hyaline, up to 40 µ long. Micro-form.

On *Veronica alpina*. Scotland, rare.

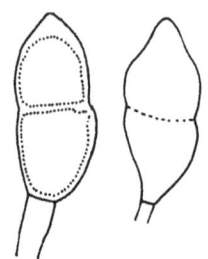

P. albulensis. Teleutospores.

This is probably the only rust attacking *Veronica alpina*. According to Jørstad (1932, 357) the teleutospores of the two kinds of sori, are not quite similar; in the pulvinate sori they are hyaline or nearly hyaline, usually with smooth walls and generally do not exceed 37 µ in length, possess persistent pedicels and germinate at once; in the pulverulent the wall is light chestnut-brown and often, in the upper cell, faintly punctate-rugose, the pedicels are deciduous and the spores germinate only after the winter's rest.

Magnus (*loc. cit.*) suggested that this rust develops a systemic mycelium; Jørstad confirmed this and stated that both kinds of sori may be developed on it; a localised mycelium on which only pulverulent sori develop, may also be present.

Both kinds of sori have been found in Scotland. The specimens in the Kew Herbarium collected by Berkeley on Ben Alder, Inverness-shire (and also from Perthshire), possess both kinds of sori; these collections were placed in *Puccinia veronicarum* by Grove (*loc. cit.*) on account of their smooth walls. Another specimen in the Kew Herbarium collected by J. Ferguson has no locality given. Specimens in the Trail Herbarium collected by Buchanan White in the Corrie of Loch Kander, Braemar, in August 1886 possess pulvinate sori only, borne on stems and upper and lower leaf surfaces; they appear to arise from a systemic mycelium; these are probably referred to in the Scot. Nat. N.S. 3, 41 (1887–8). Trail (Scot. Nat. N.S. 4, 319, 1890) included the rust on *V. alpina* under *P. veronicae*. This rust was again collected on the southern slopes of Ben Alder in September 1940; both kinds of sori are present and the teleutospores from the pulverulent sori are minutely verrucose; those from pulvinate sori germinated readily. Systemically infected sterile shoots of *V. alpina* with pulvinate sori were found in 1953 in Caenlochan Glen, Angus (TBMS. 37, 250, 1954).

In 1932 Jørstad (*loc. cit.*) noted that, in the mountains of central Europe, apparently only pulverulent sori are developed whereas pulvinate sori are found not only in Norway, but elsewhere in northern Europe and in North America and this is confirmed by Gäumann (1959, 817). The pulvinate sori have been described by Johanson (Bot. Not., **1886**, 164, 1886) from Sweden, by Liro (1908, 328) from the Kola Peninsula, by Rostrup (Medd. Grønl. 3, 534, 1888) from Greenland, by Arthur (1934, 334) from North America and by Jørstad (1951, 24) from Iceland.

Considering the distribution of the two kinds of teleutosori the British populations are closer to those of northern Europe than to those of central Europe.

Puccinia antirrhini Diet. & Holw.

Hedwigia, **36**, 298 (1897); Wilson & Bisby, Brit. Ured. no. 93; Gäumann, Rostpilze, p. 808.

Spermogonia and **aecidia** unknown. **Uredosori** chiefly hypophyllous, scattered, circular, surrounded by the ruptured epidermis, pulverulent, chestnut-brown; uredospores globoid or ellipsoid, 16–24 × 21–30μ, wall light chestnut-brown, 1·5–2·5μ thick, shortly echinulate, with 2, rarely 3 equatorial pores. **Teleutosori** amphigenous, generally hypophyllous, on pale spots which become dry, scattered or confluent, of medium size or minute, circular, surrounded by the ruptured epidermis, pulvinate, blackish-brown; teleutospores oblong or ellipsoid, obtuse, rounded or truncate above, slightly constricted, narrowed below, chestnut-brown, 36–54 × 17–26μ, wall 1·5–2·5μ at sides, up to 12μ thick at

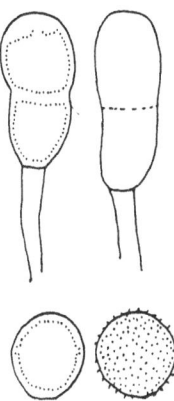

P. antirrhini. Teleutospores and uredospores (after Arthur).

the apex, smooth, pedicel subhyaline, equal in length to the spore. Hemiform.

On *Antirrhinum majus* and *A. gluti-* *nosum*, uredospores common, teleutospores rare. Great Britain and Ireland, frequent.

This rust, first discovered in California in 1896, is now known throughout the United States, Canada and Bermuda. It was first definitely observed in England (Kent) in 1933 (Green, Gard. Chron. **94**, 131, 1933) although there had been previous unconfirmed reports of the disease as early as 1921 (Pethybridge, J. Min. Agric. **41**, 336, 1934; Cuthbertson, Gard. Chron. **84**, 136, 1928). Its spread was remarkably rapid; it reached south, east and south-west England in 1933, Northumberland, East Lothian and Ireland in 1935 and it now appears to be generally distributed in England, Wales and southern Scotland. It was first found on the continent in 1931 in France (Viennot-Bourgin, Rev. Path. Vég. **20**, 280, 1933) and rapidly spread throughout Europe reaching Sweden in 1935, Egypt in 1936 and Palestine in 1937. It was recorded from Rhodesia in 1941 and South Africa in 1939; in both the African localities it was believed to have been introduced on seed. It spread and distribution is traced in Comm. Myc. Inst. Map, no. 20. Repeated unsuccessful attempts have been made by Mains (Phytopath. **14**, 281, 1924) and others to infect *Antirrhinum* plants by using teleutospores; the species is considered to be heteroecious, but without suggestion as to the alternate host. Both in this country and in North America the rust can overwinter by mycelium and by uredospores on plants kept from one season to the next.

Dietel (quoted by Andres, Ber. D. Bot. Ges. **52**, 614, 1934) stated that teleutospores borne on the stems differ from those on the leaves in their larger size (up to 75μ long) and in possessing conical apices and longer pedicels (up to 130μ long).

The rust causes a serious disease of *Antirrhinum* in North America and Europe which brings about decrease in seed production and often results in death of the host; plants kept from one season to the next frequently become heavily infected. It has been reported on the calyx and on the capsules, destroying the ovaries and seeds (Pape, Nachr.-Bl. D. Pfl.-Sch. Dienst, **14**, 113, 1934).

It has been suggested that in Sweden the rust may have been introduced on seed (Palm, Sv. Bot. Tidskr. **31**, 288, 1937) although Peltier (Bull. Illinois Agric. Exp. Sta. **221**, 535, 1919), Pethybridge (*loc. cit.*) and others could not confirm that the disease is seed-borne. It is listed by Noble *et al.* (Ann. List of Seed-Borne Diseases, CMI, 1958).

The optimum temperature for uredospore germination and infection is 10° C. (Doran, Bull. Mass. Agric. Exp. Sta. **202**, 39, 1921) and it was suggested that this unusually low temperature relation may partially account for the rapid spread of the rust in Europe, but conversely Moore (TBMS. **24**, 264, 1940) associated heavy attacks with a succession of hot summers and Hawker (Physiology of Reproduction in Fungi, 1957) associates unusually hot summers with teleutospore production. In Germany Andres (Ann. Myc. **33**, 353, 1935) described infection taking place during January and February. Teleutospores germinate after a period, quite freely in winter at 50°–60° F. but less readily in spring.

The rust was reported on *A. orontium* from France by Viennot-Bourgin (Ann. Ecole Nat. Agric. Grignon, III, **1**, 1938–9) but this host is said to be highly resistant in Germany (Laubert, Gartenwelt, **39**, 574, 1935). J. A. Macdonald (*in litt.*) found it on *Antirrhinum glutinosum* in the Botanic Garden, St Andrews.

Individual plants resistant to the rust have been found and breeding work has produced resistant horticultural varieties. Yarwood (Phytopath. **27**, 113, 1937) described two specialised forms of the rust in California.

Puccinia clintonii Peck

Rep. N.Y. State Mus. **28**, 61 (1876); Wilson & Bisby, Brit. Ured. no. 134.

Spermogonia unknown, **aecidia** and **uredosori** wanting. **Teleutosori** amphigenous and on stems, petioles and calyx, usually scattered and solitary on discoloured, often sunken spots, round, large, 0·5–1·5 mm. diam. pulverulent, chestnut-brown, surrounded by a conspicuous epidermis, the leptosporic form compact, pulvinate, at first light cinnamon-brown, becoming cinereous on germination; teleutospores ellipsoid, oblong or broadly clavoid, rounded or obtuse above, rounded or narrowed below, usually strongly constricted at the septum, 15–18 × 26–36 μ, wall golden-brown or nearly colourless in the germinating form, cinnamon-brown in the resting form, uniformly 1·5–2 μ thick, pore of the upper cell apical, with a hyaline papilla, of the lower at septum, finely striate in resting form, smooth in the germinating form, pedicel colourless, short, fragile in the resting spores, long

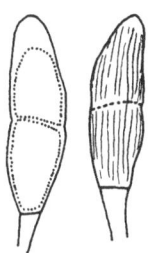

P. clintonii. Teleutospores.

and stout in the germinating form. Micro-form.

On *Pedicularis palustris* and *P. sylvatica*. Scotland and Ireland, rare.

The rust was discovered in Northern Ireland by Dovaston & Batts (TBMS. **35**, 129, 1952) on *Pedicularis palustris* in September 1948 and since then on several occasions in West Ross, West Perthshire and Argyllshire on *P. sylvatica*.

The early collections bore only leptospores but resting pulverulent sori were found in Argyll in 1959 on *P. sylvatica*.

Puccinia veronicae Schroet.

Cohn, Beitr. Biol. Pfl. **3**, 89 (1878); Grove, Brit. Rust Fungi, p. 169; Wilson & Bisby, Brit. Ured. no. 240; Gäumann, Rostpilze, p. 812.

Spermogonia unknown, **aecidia** and **uredosori** wanting. **Teleutosori** hypophyllous, on circular, brown spots, minute, scattered or circinate in groups, 1–2 mm. diam., rounded or slightly elongate, pulvinate, at first yellowish-brown then brown; teleutospores fusoid, generally rounded above, hardly constricted, tapering below, yellowish or very pale brown, 28–52 × 10–16 μ, lower cell generally longer and narrower than the upper, wall almost hyaline, thickened at apex up to 7 μ and 1 μ or less at the sides, pore of upper cell apical, of the lower close to the septum, pedicels hyaline, persistent, as long as the spore. Micro-form. On *Veronica montana* and *V. spicata*

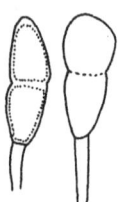

P. veronicae. Teleutospores.

subsp. *hybrida*, June–October. Great Britain and Ireland, frequent.

This species is leptosporic, producing only one kind of teleutosorus, on a limited mycelium. Grove's statement that it occurs on *Veronica alpina* appears to be an error (see under *Puccinia albulensis*, p. 172). It is known to occur in Europe on *V. officinalis* but no British specimen has been seen on this plant and Grove's record is doubtful. Johnston's record of *P. veronicarum* on *V. chamaedrys* (Fl. Berwick, **2**, 194) appears to be based on a specimen now preserved in Kew Herbarium; on examination it proved to be *P. veronicae* on *V. montana*. The Berkeley & Broome specimen assigned by them to *P. veronicarum* (Ann. Mag. Nat. Hist. **4**, 6, no. 1310, 1870) has also proved to be *P. veronicae* on *V. montana*.

This rust appeared on plants of *V. spicata* subsp. *hybrida* collected at Humphrey Head, north Lancashire and grown near York for several years; it was also collected in 1950 on this host in Denbigh, Wales, by N. Robertson. These specimens were previously regarded as *P. veronicarum* but Gäumann (Ann. Myc. **39**, 38, 1941) transferred the rust on this host to *P. veronicae*, as only leptosporic teleutosori are produced and its teleutospores, although longer than those on *V. montana*, do not differ significantly from the latter.

Gäumann has shown that two *formae speciales* occur; f. sp. *montanae* on *V. montana* with teleutospores 31–38 × 13–17μ and f. sp. *spicatae* on *V. spicata* subsp. *hybrida* with teleutospores 41–45 × 13–13·5μ. Each of these will infect only its own host. The teleutospores of the Humphrey Head specimens measure 31–52 × 13–16μ.

Puccinia annularis (Str.) Röhl.

Deutschl. Fl. Ed. 2, 3(3), 134 (1813); Grove, Brit. Rust Fungi, p. 175; Wilson & Bisby, Brit. Ured. no. 91; Gäumann, Rostpilze, p. 823.

Uredo annularis Str., Ann. Wetter. Ges. 2, 106 (1819).

Spermogonia, aecidia and uredosori wanting. Teleutosori hypophyllous, on indefinite yellowish or brownish concave spots, at first minute, roundish, covered by the epidermis, in orbicular clusters, then naked, confluent, and forming a thick pulvinate mass, rusty-brown; teleutospores oblong, rounded or attenuate at the apex, slightly constricted, rounded or attenuate at the base, 30–54 × 14–21μ, wall smooth, up to 8μ thick at apex, very pale yellowish-brown, pore of upper cell apical, of lower close to septum, pedicels hyaline, persistent, up to 80μ long. Micro-form.

On *Teucrium scorodonia*, July–October. Great Britain, Ireland, frequent.

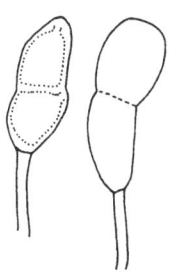

P. annularis. Teleutospores.

The teleutospores in British collections which have been examined are all leptospores, which germinate without a period of rest. The statement in Grove that there are also resting spores seem inaccurate. The teleutospores are borne in fascicles according to Kuhnholtz-Lordat (Uredineana, 3, 27, 1951).

Puccinia betonicae DC.

Fl. Fr. 6, 57 (1815); Grove, Brit. Rust Fungi, p. 174; Wilson & Bisby, Brit. Ured. no. 103; Gäumann, Rostpilze, p. 1012.

Puccinia anemones var. *betonicae* Alb. & Schw., Consp. Fung. Nisk. 131 (1801).

Uredo betonicae DC., Lam. Encycl. Méth. Bot. 8, 247 (1808).

Spermogonia, aecidia and uredosori wanting. Teleutosori hypophyllous, occasionally epiphyllous, on pallid irregular spots, numerous, aggregated in patches, or more generally spreading over nearly the whole of a leaf, more or less crowded on the nerves, minute, perfectly round, surrounded by the torn

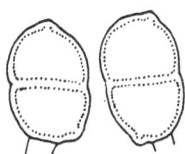

P. betonicae. Teleutospores (after Fischer).

erect epidermis, pulverulent, reddish-brown; teleutospores ellipsoid to ovoid, rounded at both ends, slightly constricted, 27–45 × 15–24 μ, wall yellowish-brown, smooth, uniformly thick, pore of upper cell apical with a hyaline papilla, of the lower cell equatorial, mesospores present; pedicel thin, hyaline, about half as long as the spore. Microform.

On *Stachys officinalis*, May–September. England frequent, Scotland rare.

The rust appears early in the summer on the young leaves and continues until late autumn. The affected leaves are paler on the upper side and stand more erect than the healthy ones. The mycelium is perennial. Teleutospores with 3 or more cells occur occasionally (Grove, Gard. Chron. **24**, 180, 1885).

Puccinia glechomatis DC.

Lam. Encycl. Méth. Bot. **8**, 245 (1808); Grove, Brit. Rust Fungi, p. 173; Wilson & Bisby, Brit. Ured. no. 158; Gäumann, Rostpilze, p. 822.

Puccinia glechomae DC., Fl. Fr. **6**, 56 (1815).

Spermogonia absent. **Teleutosori** hypophyllous or on the petioles, with or without accompanying brownish spots, 0·5–1 mm. diam., rounded, solitary and scattered, or more often subconfluent into rounded clusters as much as 4 mm. diam., on the stem and petioles, often elongate, pulvinate, at first yellowish, then chestnut, and at last blackish; teleutospores ellipsoid or oblong, with an acute apex, faintly constricted, rounded below, 30–48 × 15–24 μ, wall smooth, pale brown, up to 8–12 μ thick at apex, pore of upper cell apical, of lower superior, pedicels hyaline, persistent, up to 75 μ long; mesospores occasionally present. Micro-form.

On leaves petioles and stems of *Nepeta hederacea*, June–October. Great Britain and Ireland, frequent.

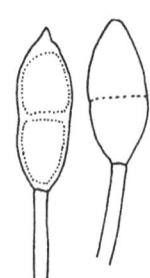

P. glechomatis. Teleutospores.

The sori are especially large, round and compact late in the season, when they produce spores which are darker and will not germinate immediately (as the others do) but only after the winter's rest. Grove's material reputed to be this species on *Prunella vulgaris* is *Uromyces valerianae* on young plants of *Valeriana officinalis*. Keith & Trail's record (Scot. Nat. **2**, 331, 1885–6), also reputed to be on *Prunella*, is almost certainly similarly erroneous.

Puccinia menthae Pers.

Syn. Meth. Fung. p. 227 (1801); Grove, Brit. Rust Fungi, p. 170; Wilson & Bisby, Brit. Ured. no. 183; Gäumann, Rostpilze, p. 1003.

Spermogonia scattered or arranged in little groups among the aecidia, honey-coloured, with flexuous hyphae and periphyses; spermatia 2·5–3·5 × 3–4·5 μ. Aecidia hypophyllous and often on the stems, arranged on the leaves in clusters on orange or purplish spots, or forming elongate patches on the thickened and deformed stems and petioles, opening irregularly, margin scarcely torn, erect or slightly incurved, peridial cells elongate and compressed, not firmly connected, walls rather thin, in section striate, outer about 2 μ thick, finely punctate, inner about 3 μ thick, verrucose; aecidiospores pallid-yellow, ellipsoid, 24–40 × 17–28 μ, wall colourless, verruculose, 1·5–2·5 μ thick. Uredosori hypophyllous, on yellowish or brownish spots (or without spots), 1·5 mm. diam., roundish, scattered or aggregated, soon naked, surrounded by the ruptured epidermis, sometimes confluent, cinnamon; uredospores ellipsoid rarely subgloboid, 17–28 × 14–19 μ, wall 1·5–3 μ thick, echinulate, pallid-brown, with 3 equatorial pores. Teleutosori similar but dark brown; teleutospores subgloboid to obovoid, rounded at both ends, not or scarcely constricted, dark brown, 26–35 × 19–23 μ, wall indistinctly verruculose, sometimes smooth, pore of upper cell apical, of lower cell close to the septum, each with a broad hyaline papilla;

pedicels slender, longer than the spore. Auteu-form.

On *Calamintha ascendens*, *Clinopodium vulgare*, *Mentha aquatica*,

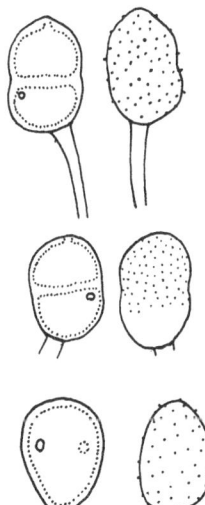

P. menthae. Teleutospores from *Mentha aquatica* and *Origanum vulgare* and uredospores.

M. arvensis, *M.* × *cordifolia*, *M. longifolia*, *M.* × *niliaca*, *M.* × *piperita*, *M. rotundifolia*, *M. spicata*, *Origanum vulgare* and *Satureja hortensis*, May–October, teleutospores from August. Great Britain and Ireland, very common on some *Mentha* hosts.

There can be little doubt that this is a collective species and Baxter (Lloydia, **22**, 242, 1959) has proposed segregation of three American, but non-British variants, as varieties. Points of difference are found in the finer or coarser warts of the teleutospores and in the length of the pedicel. The form on *O. vulgare* shows less difference from that on *Mentha aquatica* than those on species of *Mentha* do from one another. Cruchet (Zbl. Bakt. II, **17**, 212, 1906) was unable to infect any one of the four: *Mentha arvensis*, *M. aquatica*, *M. longifolia*, *Origanum vulgare*.

except by spores from the same species. As the result of his experiments, he divided *Puccinia menthae* on *Mentha* and *Calamintha* into eight biological races; and the form on *Origanum* is also biologically distinct. Niederhauser (Bull. Cornell Univ. Agric. Exp. Sta. **263**, 1945), using uredospores from cultivated *M. piperita*, failed to infect several wild species of *Mentha*, *Nepeta hederacea* and *Prunella vulgaris*. Baxter & Cummins (Phytopath. **43**, 178, 1953) showed that in North America this species is composed of at least 15 physiologic races. Susceptibility to one or more races infecting cultivated mints was shown by *Satureja hortensis* and *M. arvensis*. Underground aecidial lesions were found at depths ranging from 1·3 to 16·5 cm. below the surface of the soil.

The rust of this type on *Ajuga reptans* mentioned by Johnston (Fl. Berwick, **2**, 203, 1829) and referred to by Plowright (1889, 157) is an error; the specimens are of *Uromyces valerianae* on *Valeriana dioica*.

In garden mint (*M. spicata*) the mycelium of the aecidial stage is spread throughout the whole plant, even the rhizome; Klebahn (1904, 57) was able to trace the hyphae in some plants nearly up to the growing point. There is, however, no evidence that the mycelium is perennial in the rhizome according to Ross (Z. Pfl.-Krankh. **34**, 101, 1924), on *M. piperita* by Vergovsky (Bull. Med. Techn. Pl. Simferopol, **3**, 5, 1935; RAM. **15**, 527, 1936), and on *M. spicata* by Niederhauser (*loc. cit.*). The mycelium of the other two stages is purely local.

Sappin-Trouffy (1897, 112) briefly described the development of the uredospores and cytology of this species and suggested that it was heterothallic.

Puccinia thymi (Fuck.) Henderson

Notes R. B. G. **27**, 359 (1966).

Aecidium thymi Fuck. Fungi Rhen. no. 2113 (1868).
Puccinia caulincola Schneid., Jahresb. Schles. Ges. Vaterl. Kult. **48**, 120 (1871) (non *P. caulincola* Spreng. 1827); Grove, Brit. Rust Fungi, p. 172; Gäumann, Rostpilze, p. 1015.
Puccinia schneideri Schroet., Schneid. Herb. Schles. Pilze, no. 448 (1879) (n.v.); Wilson & Bisby, Brit. Ured. no. 217.
Puccinia ruebsaamenii Magn., Ber. D. Bot. Ges. **22**, 344 (1904); Wilson & Bisby, Brit. Ured. no. 212; Gäumann, Rostpilze, p. 1011.

Spermogonia, aecidia and **uredosori** wanting. **Teleutosori** on the stems and petioles, rarely on the leaves and then usually on the midribs, scattered, occasionally confluent, roundish or elongate, long-covered by the epidermis and then appearing black, pulverulent, at length cinnamon-brown, teleutospores ellipsoid, rounded at both ends, rather constricted, minutely verrucose, wall uni-

formly 1·5–2μ thick, pores depressed,
covered with low, broad, hyaline caps,
24–33 × 15–24μ, pedicels hyaline, thin,
more or less persistent, as long as the
spores or longer; a few mesospores
occasionally present. Micro-form.
  On *Thymus drucei*, *T. pulegioides* and
Origanum vulgare, June–October. England, Scotland, rare.

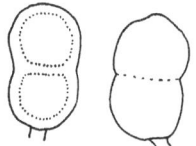

P. thymi. Teleutospores.

The mycelium appears to be perennial and causes an annual witches'
broom. On *Thymus* the internodes are lengthened and the leaves fewer
and the stems sterile so that affected plants can be readily distinguished.
The effects on *Origanum* are slightly different: the internodes are slightly
shorter than those of the healthy shoots and at almost every node two
axillary shoots are produced, giving the broom a very characteristic
appearance. On both genera the sori are confined almost entirely to the
stems where they cause a slight thickening and are more frequent at the
nodes than elsewhere.

The rust was first recorded in Britain on '*T. serpyllum*' (= *T. drucei*)
from Aberdeen and Ballater by Trail (Scot. Nat. N.S. **2**, 331, 1885–6). It
has been found occasionally since then on that host; Ayrshire (Glasgow
Nat. **1**, 111, 1909), Callater, Aberdeenshire (by Ogilvie who infected
plants of *Thymus serpyllum* agg. at Cambridge in 1922 with teleutospores), Beinn Eighe, West Ross-shire and in Buckinghamshire by C. D.
Piggott who also collected it on *T. pulegioides* in Norfolk.

On *Origanum vulgare* this rust was found by Sandwith in two localities
in Surrey in September 1941. *Thymus serpyllum* growing abundantly in
the neighbourhood of the infected plants was not attacked. It has been
recorded on *O. vulgare* from Gloucestershire, Somerset and Kent.

The rusts on *Thymus* and *Origanum* are frequently treated as distinct
('*P. schneideri*' and *P. ruebsaamenii* respectively) but the differences
appear slender although there is circumstantial evidence of host
specialised races. Guyot, Massenot & Bulit (Uredineana, **4**, 257, 1953)
give the mean size of the teleutospores of *P. ruebsaamenii* as 31–33 ×
21–23μ, those of *P. caulincola* (*P. thymi*) 27–30 × 19–21μ.

The distribution of the races on *Thymus* and *Origanum* is of some
interest. It is probable that the *Origanum* rust has been present in
Britain for only a short time, for the witches' brooms which it produces
are conspicuous and could hardly have escaped notice for a long period.
It appears significant that its discovery in Surrey in 1941 was quickly
followed by records from Kent, Gloucestershire and Somerset. Its

distribution in northern France was described by Guyot in 1930 (Rev. Path. Vég. 17, 360, 1930) who recorded it in the Départments of the Somme and Seine-Inférieure and noted its recent appearance further south in Seine and Oise; he considered it to be spreading rapidly. Later (*ibid.* 20, 274, 1933) he recorded it as very common in these areas. Viennot-Bourgin (*ibid.* 20, 286, 1933) confirmed these reports and added earlier records in Seine-Inférieure in 1910 and 1923 and in Belgium (Tournai) in 1918; he referred to the rust which for twenty years had been regarded as very rare, as widespread.

It appears from the recent records in the south of England that the western extension of this rust already described in northern France is still continuing, probably by the transfer of airborne spores across the Channel.

Magnus (*loc. cit.*) and Guyot (Bull. Soc. Linn. Nord France, **1928**, 246, 1929) described the 'witches' broom' and the distribution of the mycelium in the infected shoots of *Origanum*. Rostrup (Bot. Tidsskr. **14**, 242, 1885, and **17**, 230, 1890) had already described the 'witches brooms'. Stämpfli (Hedwigia, **49**, 230, 1910) described the changes in the morphology and anatomy of the stem of *T. serpyllum* which result from infection.

Puccinia campanulae Berk.

Smith, Engl. Fl. **5**, 365 (1836); Grove, Brit. Rust Fungi, p. 159; Wilson & Bisby, Brit. Ured. no. 114; Gäumann, Rostpilze, p. 1030.
Puccinia campanulae-rotundifoliae Gäumann & Jaag, Hedwigia, **75**, 121 (1935); Gäumann, Rostpilze, p. 1032.

Spermogonia unknown, **aecidia** and **uredosori** wanting. **Teleutosori** hypophyllous, rarely epiphyllous, often on the petioles and stems, scattered or circinate, sometimes (especially on the stems) confluent and larger, long covered by the epidermis, then surrounded by it, roundish or irregular, ferruginous-brown; teleutospores ellipsoid or oblong rounded or obtuse at both ends, slightly constricted at the septum, 26–45 × 12–22μ, wall pale brown, 1·5–2·5μ thick at the sides, finely verruculose, pore of upper cell apical, of lower close to the septum, each with a hyaline papilla, pedicels hyaline. Micro-form. On *Campanula rotundifolia*, *C. rapun-*

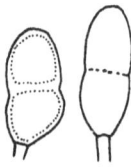

P. campanulae. Teleutospores.

culus, *Jasione montana* and '*C. persicifolia*', June–August. Great Britain, scarce.

Carmichael's type specimen in the Kew Herbarium is on *C. rotundifolia*. Gäumann & Jaag described several micro-species of this rust upon different species of *Campanula* which differ slightly in the measurements of their teleutospores. They assumed incorrectly that the type specimen is upon *C. rapunculus* and called the form on this host, *P. campanulae*, and named the form on *C. rotundifolia*, *P. campanulae-rotundifoliae* and Gäumann (1959, 1030) continued this interpretation.

The early record of this rust on *Jasione montana* which originated from Berkeley (Ann. Mag. Nat. Hist. **7**, 24, 1850 'On *Jasione*, Lampeter, J. Ralfs') appears to be correct although doubted by the Sydows (Monogr. Ured. **1**, 196). The specimen is in the Kew Herbarium; the host is undoubtedly *J. montana* and a rust is present on the leaves. Only teleutosori are present, these are hypophyllous, circinate, confluent, long covered by the epidermis, greyish-brown, up to 1·3 mm. diam.; the teleutospores agree with those of *P. campanulae* but are more strongly verruculose, 31–33 × 14–19 μ with wall about 2·5 μ thick. This rust must be regarded as a form of *P. campanulae*. The record on *Campanula persicifolia* is from Moore (1959, 299).

Puccinia galii-cruciatae Duby

Bot. Gall. **2**, 888 (1830).

Puccinia celakovskyana Bub., S.B. Kon. Böhm. Ges. Wiss. **28**, 11 (1898); Grove, Brit. Rust Fungi, p. 166; Wilson & Bisby, Brit. Ured. no. 122.
Puccinia punctata f. sp. *celakovskyana* (Bub.) Gäumann, Ann. Myc. **35**, 215 (1937); Gäumann, Rostpilze, p. 844.

Spermogonia amphigenous, in small groups, dark honey-coloured. **Aecidia** uredinoid, hypophyllous, on yellow spots, rather large, circinate around the spermogonia, long-covered by the epidermis, later naked, dark brown; aecidiospores similar to the uredospores. **Uredosori** amphigenous, minute, scattered, soon naked, rounded or elongate, pulvinate, pale chestnut-brown or pale cinnamon; uredospores globoid to obovoid 21–33 × 18–25 μ, wall distinctly echinulate, brown, with 2 nearly equatorial pores. **Teleutosori** hypophyllous or on the stems, on brownish spots, thickly scattered, minute, roundish, on the stems linear and up to 1 mm. long, soon naked, pulvinate, brownish-black; teleutospores oblong to clavoid, rounded above, slightly constricted, 24–66 × 18–26 μ, wall up to 11 μ thick at apex and darker, smooth, brown, pore of upper cell subapical, of lower superior; pedi-

cels hyaline, thick, rigid, persistent, up to 50 μ long. Brachy-form.

On *Galium cruciata*, uredosori, May–July; teleutosori August, September. England, Scotland, scarce.

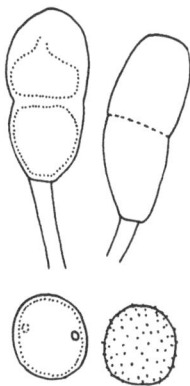

P. galii-cruciatae. Teleutospores and uredospores.

According to Bubak almost the only distinction of this species from *Puccinia punctata* is the presence of the uredinoid aecidium. Wurth (Zbl. Bakt. II, **14**, 212 and 310, 1905) and Gäumann (1937, 44, and 1959, 847) showed that this rust cannot be transferred to other British species of *Galium* and the latter regarded it as only a specialised brachy-form of *P. punctata*; Grove considered that the teleutosori are larger, more numerous and more compact than those of *P. punctata* on *Galium palustre* but this is probably only a reflection of the leaf type of the host species.

Puccinia difformis Kunze

apud Kunze & Schm. Myk. Hefte, **1**, 71 (1817); Grove, Brit. Rust Fungi, p. 168; Wilson & Bisby, Brit. Ured. no. 144; Gäumann, Rostpilze, p. 849.

[*Aecidium galii* var. *ambiguum* Alb. & Schw., Consp. Fung. Nisk. p. 116 (1805).]
Puccinia galii Wint., Rabh. Krypt. Fl. Ed. 2, **1** (1), 210 (1882), *p.p.*
[*Puccinia ambigua* [Alb. & Schw.] Bub., S.B. Kon. Böhm. Ges. Wiss. 1898, 14 extr. (1898).]

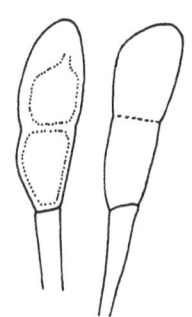

P. difformis. Teleutospores.

Spermogonia hypophyllous, scattered or grouped between the aecidia, yellow. **Aecidia** hypophyllous, on yellow spots, solitary or irregularly disposed over the whole leaf, whitish-yellow, with torn reflexed margin; aecidiospores verruculose, orange, 13–25 μ diam. **Uredosori** wanting. **Teleutosori** hypophyllous or on the stems, small elliptic, solitary or clustered, on the stems often elongate and confluent, long-covered by the ash-coloured epidermis, then naked, firm, black; teleutospores ellipsoid to clavoid, much thickened above, hardly constricted, tapering below, 35–55 × 15–25 μ, wall smooth, brown above, paler downwards, pedicels brownish, persistent, as long as the spore or longer. Opsis-form.

On *Galium aparine*, July–August. England, Scotland, scarce.

Bubak (*loc. cit.*) and Treboux (Ann. Myc. **10**, 305, 1912) found that the aecidia repeat without spermogonia; they may be found together with the teleutospores right up to September, often on the same spot. This species is very different from the others on *Galium*; the teleutosori, as Cooke says, are 'firm and compact like little spots of pitch', and may be accompanied by swellings and distortion.

Puccinia punctata Link

Mag. Ges. Naturf. Fr. Berlin, **7**, 30 (1815); Grove, Brit. Rust Fungi, p. 164; Wilson & Bisby, Brit. Ured. no. 208; Gäumann, Rostpilze, p. 844.

Puccinia asperulae-odoratae Wurth., Zbl. Bakt. II, **14**, 314 (1905); Grove, Brit. Rust Fungi, p. 163; Wilson & Bisby, Brit. Ured. no. 99.

Spermogonia epiphyllous in groups, honey-coloured. **Aecidia** hypophyllous in groups, on rounded or oblong spots, cupulate, with a short, white, recurved peridium, peridial cells rhombic, inner wall 3–4 μ thick, outer wall 8–10 μ thick, with striated structure, inner with small warts; aecidiospores globoid or ellipsoid, orange-yellow, 16–23 μ, wall 1 μ thick, finely verrucose. **Uredosori** amphigenous, minute, cinnamon-brown, roundish, on the stems linear, often confluent; uredospores globoid to ovoid, pale brown, 22–30 × 17–23 μ, wall aculeolate, 1·5–2 μ thick, with 2 equatorial pores. **Teleutosori** amphigenous, oblong or suborbicular, compact, chocolate-brown to black; teleutospores ellipsoid to clavoid, truncate, rounded or conically attenuate above and often darker, slightly constricted, tapering below, 30–56 × 14–24 μ, wall brown, smooth, pore of upper cell apical, of lower superior, pedicel about as long as the spore; a few meso-spores occasionally present. Auteu-form. On *Galium cruciata*, *G. mollugo*, *G. odoratum*, *G. palustre*, *G. saxatile*, *G. uliginosum*, *G. verum* and sparsely on *G. aparine*. Great Britain and Ireland, frequent.

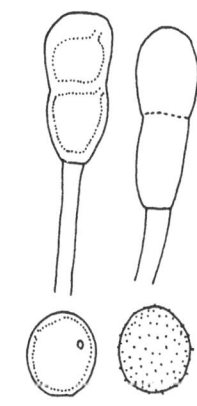

P. punctata. Teleutospores and uredospores.

The rust on *Galium odoratum* has been regarded as a distinct species, *Puccinia asperulae-odoratae*, by Wurth (Zbl. Bakt. II, **14**, 209, 1905), Grove and others. It is uncommon on this host in this country. Grove was doubtful of its occurrence on *G. cruciata* but aecidia have been found twice on this host in Scotland (Wilson, 1934, 375) which could not belong to *P. galii-cruciatae*. Gäumann (Ann. Myc. **35**, 194, 1937) showed that the aecidium on *G. cruciata* could not belong to a heteroecious species and is that of *P. punctata* and that the rust is a special form confined to this host. According to Grove the teleutospores are larger in the rust on *G. saxatile*.

The aecidium is more uncommon than the other spore forms; it has been seen on *G. cruciata*, *G. mollugo*, *G. palustre*, *G. uliginosum* and *G. verum*. Mesospores have been found on *G. saxatile*, *G. palustre* and *G. verum*.

Gäumann (*loc. cit.*) described several specialised forms of this rust.

Puccinia galii-verni Ces.

Klotsch, Herb. Viv. 1092 (1846) and Bot. Zeit. **4**, 879 (1846).
As *Puccinia valantiae* Pers., Syn. Meth. Fung. 227 (1801) *nomen confusum* =
P. *celakovskyana* Bub.; Grove, Brit. Rust Fungi, p. 167; Wilson & Bisby, Brit.
Ured. no. 238; Gäumann, Rostpilze, p. 852.

Spermogonia, aecidia and uredosori wanting. **Teleutosori** hypophyllous, rather thick, pulvinate, compact, orbicular, scattered, circinate or confluent, up to 2 mm. diam. at first yellowish then chestnut-brown, at length greyish-brown, often elongate and causing distortion of the stems; teleutospores fusoid or slightly oblong, attenuate at both ends, slightly constricted, 35–60×12–$17\,\mu$; wall smooth, yellowish, thin, apex up to $9\,\mu$ thick, pore of upper cell apical, of lower superior, pedicel hyaline, persistent, up to $80\,\mu$ long. Micro-form.

On *Galium cruciata*, *G. saxatile*, *G. uliginosum* and *G. verum*, May–September. Great Britain and Ireland, frequent.

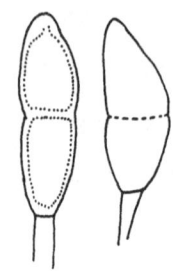

P. galii-verni. Teleutospores.

This leptoform is distinguished by the fusiform shape, thin wall and pale colour of the teleutospores; on germination the teleutospores often become devoid of the thickening at the apex, by the dropping off of the pale cap.

It appears to be rare on *Galium verum* on which it was collected by Alex. Smith near Aberdeen, but is frequent on *G. saxatile*, especially in early spring.

Fischer (1904, 366) noted that infection can take place only on the young growing parts of the host where the mycelium can still cause deformation.

The misapplication of the name *Puccinia valantiae* is fully explained by Jørstad (Nytt Mag. Bot. **6**, 135, 1958).

Puccinia adoxae DC.

Fl. Fr. **2**, 220 (1805); Grove, Brit. Rust Fungi, p. 160; Wilson & Bisby, Brit. Ured. no. 78; Gäumann, Rostpilze, p. 1027.

Spermogonia unknown, **aecidia** and **uredosori** wanting. **Teleutosori** amphigenous, roundish, united in large clusters, often confluent, on discoloured spots on the leaves, up to 12 mm. diam., and on the petioles forming elongate swollen patches, long-covered by the silver-grey epidermis, at length naked,

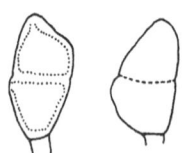

P. adoxae. Teleutospores.

pulverulent, dark brown; teleutospores ellipsoid to broadly fusoid, rounded or attenuate above, mostly rounded below, scarcely constricted, chestnut-brown, $25–35 \times 15–20\,\mu$; wall smooth, uniformly $1\cdot5–2\,\mu$ thick, pore of the upper cell apical with a hyaline papilla, pore of the lower cell superior or equatorial, pedicel hyaline, short, fragile, deciduous. Microform.

On *Adoxa moschatellina*, March–May. Great Britain, frequent.

The rust affects all parts of the plant, rhizomes, petioles, leaves, peduncles and flowers.

Soppitt (see Plowright, 1889, 207) first proved that overwintered teleutospores could infect healthy plants on germination and that teleutospores were produced in about 10 days. The rust is a microcyclic species and Dietel (Zbl. Bakt. II, **48**, 493, 1918) pointed out its correlation with *Puccinia argentata*. Whether the widespreading mycelium is perennial or not is uncertain. W. G. Smith (Gard. Chron. **24**, 21, 1885) raised seedlings of *Adoxa* from fruits of an infected plant; the seedlings exhibited the *Puccinia* from the earliest stages of growth, but he did not state what precautions were taken to prevent infection from outside. He also found teleutosori on fusiform swellings of the underground parts of the plant (peduncles and petioles), and also on rhizomes and scales, in March; the spores were irregular, 1-, 2-, or 3-celled. In April the leaves and in May the flowers and young fruits were infected. No mycelium could be found in the rhizome. If Plowright's ascription of a perennial mycelium is incorrect, the infection must have taken place, in this instance, underground on the young growth, and the mycelium gradually spread upwards. That this is probably the case is shown by Fischer's (1904, 146) experiment; he kept plants which had borne teleutospores, in pots; if he removed all the leaves, they produced healthy growth next spring, whereas, if the leaves were left on and allowed to fall upon the soil, one plant at least (out of four) showed teleutospores on the new shoots.

Moss (Mycol. **43**, 99, 1951) by infections in Canada in June 1942 produced teleutospores in the summer of the following year; he considered that *P. adoxae* is a microcyclic species without spermogonia. He found hypertrophy on the flowers as well as on the leaves and supported the view that it is systemic but not perennial.

The cytology was investigated by Blackman & Fraser (Ann. Bot. Lond. **20**, 35, 1906).

Puccinia albescens Plowr.

Monogr. Brit. Ured. Ustil. p. 153 (1889); Grove, Brit. Rust Fungi, p. 162; Wilson & Bisby, Brit. Ured. no. 86; Gäumann, Rostpilze, p. 1026.

[*Aecidium albescens* Grev., Fl. Edin. p. 444 (1824).]

Spermogonia scattered amongst the aecidia, yellowish. **Aecidia** scattered uniformly over the whole surface of the plant, the leaves and stems becoming stunted and whitish, peridium shortly cylindrical, margin deeply cut, revolute, whitish-yellow; aecidiospores subgloboid finely warted, pale yellowish, 15–22μ diam. **Uredosori** hypophyllous, minute, scattered singly or in small clusters, long-covered by the epidermis, brown; uredospores globoid to ellipsoid, pale brown, 22–28 × 18–25μ, wall echinulate with 2 pores. **Teleutosori** hypophyllous, small, round, scattered or long-covered by the epidermis; teleutospores at first in the uredosori, teleutospores ellipsoid to subfusoid, rounded or attenuate above, hardly constricted, chestnut-brown, 32–45 × 14–25μ, wall smooth, pore of upper cell apical with a conspicuous hyaline papilla, of lower cell, superior, pedicels hyaline, delicate, short, deciduous. Auteu-form.

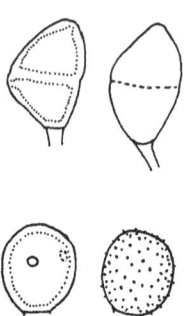

P. albescens. Teleutospores.

On *Adoxa moschatellina*, aecidia, March–April; uredospores and teleutospores, May–June. England and Scotland, scarce.

The uredosori are seldom produced and teleutosori are also rather uncommon. Schroeter (Beitr. Biol. Pfl. **3**, 75, 1879), Nielsen (Bot. Tidsskr. **2**, 41, 1877–9) and Soppitt (see Plowright, *loc. cit.*) all produced them in cultures using aecidiospores; Soppitt's inoculations on 26 April resulted in both uredo- and teleutospores on 16 May. Their mycelium is localised, but that of the aecidium permeates the whole plant; it is a disputed point whether it perennates in the rhizome or not; Plowright (1889, 153) affirmed, Fischer (1904, 144) denied this and Bubak (Zbl. Bakt. II, **16**, 150, 1906) thought it probably not perennial. Fischer (1904, 144) recorded 3- and 4-celled teleutospores.

The distinction between this species and *Puccinia adoxae* lies not only in the presence of the aecidium and uredo, but also in the appearance and character of the teleutosori. In *P. albescens* these are widely scattered and mostly single, and follow the aecidium towards the end of May—in *P. adoxae* they are crowded in larger groups, on more or less deformed parts of the plant, and can be found as early as April or even March; moreover there are no uredospores in them.

There is no evidence of the occurrence of this species in America; the aecidium on *Adoxa moschatellina* which is found there and which also occurs in Britain belongs to *P. argentata* (see. p. 134).

Puccinia commutata P. & H. Syd.

Monogr. Ured. **1**, 201 (1904); Wilson & Bisby, Brit. Ured. no. 137.

Spermogonia chiefly epiphyllous, in groups surrounded by the aecidia. **Aecidia** amphigenous, mostly hypophyllous and on the petioles and stems, on the leaves in rounded or irregular groups which are often elongated along the veins, on the petioles and stems forming elongated groups, 1–10 cm. long, seated on pale spots, crowded, cup-shaped with laciniate, white, recurved margin; aecidiospores polygonal or globoid, minutely verrucose, orange, 14–19 μ diam., wall 1–1·5 μ thick. **Uredosori** wanting. **Teleutosori** arranged similarly to the aecidia and often among them, small, confluent into long striae, pulverulent, dark brown; teleutospores oblong, subfusoid or subclavoid, apex rounded or obtuse, slightly constricted, tapering downwards or rounded, smooth, chestnut-brown, wall 1·5–2 μ thick at the side, slightly thicker at the apex, pore of the upper cell apical, of lower close to the septum, with a pale brown papilla over each, 45–66 × 20–30 μ, average 55 × 25 μ, pedicels up to 42 μ

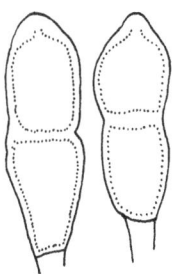

P. commutata. Teleutospores (after Fischer).

long, average 16 μ, hyaline, deciduous; occasional mesospores are present. Opsis-form.

On *Valeriana officinalis.* Scotland, rare.

This rust is known from two localities only, both in west and central Scotland (Mull and Lake of Menteith). The host at both sites bears rather large and few pinnae corresponding to the plant known as *Valeriana officinalis* var. *sambucifolia* by continental authors but not recognised as taxonomically distinct in Britain. Dietel reported repeating aecidia. The overwintered teleutospores germinate in spring and from infection by their basidiospores spermogonia and aecidia develop. Spores from these aecidia produce a second generation of aecidia and these are followed by teleutospores on the same mycelium. Jackson (1931, 52) and Arthur discussed the relations of this species. With the knowledge now available this species appears to be correlated with the auteu-form, *Uromyces valerianae*, the micro-form, *Puccinia valerianae*, and the endoforms known as *Endophyllum valerianae-tuberosae* and *E. centranthi-rubri*.

KEY TO UREDOSPORE AND TELEUTOSPORE COLLECTIONS OF
PUCCINIAS ON COMPOSITAE IN BRITAIN

1. Uredosori and aecidia never or rarely present; teleutosori either black and
   pulvinate or immersed with columnar paraphyses   2
   Uredosori or aecidia usually present   5

2. Teleutospore pedicels deciduous   *P. glomerata*, p. 202
   Teleutospore pedicels persistent   3

3. Teleutosori exposed, erumpent, without paraphyses   4
   Teleutosori with brown columnar paraphyses   *P. virgae-aureae*, p. 216

4. Teleutospores smooth   *P. cnici-oleracei*, p. 197
   Teleutospores finely verruculose; uredospores present on some hosts
                                                          P. tanaceti, p. 212

5. Aecidia abundant; on *Senecio vulgaris*   *P. erechthites*, p. 213
   Aecidia sparser; on other hosts   6

6. Uredospores with 2, very rarely 3, pores   7
   Uredospores with 3 or 4 pores   13

7. Uredospore pores supra-equatorial   8
   Uredospore pores equatorial   9

8. Uredospore pores strongly supra-equatorial   *P. hieracii* var. *hieracii*, p. 203
   Uredospore pores only slightly supra-equatorial (on *Hieracium* section
   Pilosella)   *P. hieracii* var. *piloselloidarum*, p. 206

9. Aecidioid aecidia or their remains usually present or uredospores less than
   22 μ in greatest diameter   10
   Aecidioid aecidia absent; uredospores always over 22 μ in greatest diameter   12

10. Uredospores less than 22 μ in greatest diameter, on *Lapsana* *P. lapsanae*, p. 208
    Uredospores greater than 22 μ in greatest diameter   11

11. Teleutospores usually more than 38 μ long   *P. major*, p. 210
    Teleutospores usually less than 38 μ long   *P. variabilis*, p. 214

12. Uredinoid aecidia present; on *Hypochoeris*   *P. hieracii* var *hypochoeridis*,
                                                                        p. 206
    Aecidial stage unknown in Britain; on *Crepis*   *P. crepidicola*, p. 200

13. Teleutospores ovoid to subgloboid, not medianly constricted   14
    Teleutospores oblong ellipsoid, medianly constricted   15

14. Uredospores pores with conspicuous hyaline caps   *P. maculosa*, p. 209
    Uredospores pores lacking such caps   *P. cyani*, p. 201

15. Systemic mycelium present, bearing spermogonia and uredinoid aecidia
    on deformed host organs   *P. punctiformis*, p. 210
    Systemic mycelium absent   16

16. Uredospore pores with conspicuous hyaline swollen caps; species with
    aecidioid aecidia or aecidia absent   17
    Uredospore pores without conspicuous caps; aecidia either aecidioid or
    uredinoid   18

17. Teleutospores verrucose   *P. hysterium*, p. 207
    Teleutospores finely verruculose or echinulate   *P. cnici*, p. 196

18. Uredospores ovoid or ellipsoid; teleutospores and other spore forms usually
absent in Britain; on *Chrysanthemum*   *P. chrysanthemi*, p. 194
Uredospores subgloboid to broadly ellipsoid   19
19. Teleutospore wall thicker at apex; on *Tanacetum* and *Artemisia*
                                                          P. tanaceti, p. 212
Teleutospore wall uniform, not thicker at apex; on various genera of
Compositae   *P. calcitrapae*, p. 194

Puccinia calcitrapae DC.

Fl. Fr. 2, 221 (1805).

Puccinia centaureae DC., Fl. Fr. 6, 59 (1815); Grove, Brit. Rust Fungi, p. 139;
Wilson & Bisby, Brit. Ured. no. 123; Gäumann, Rostpilze, p. 1058.
Puccinia bardanae Corda, Icon. Fung. 4, 17 (1840); Grove, Brit. Rust Fungi, p. 138;
Wilson & Bisby, Brit. Ured. no. 101; Gäumann, Rostpilze, p. 1047.
Puccinia cirsii Lasch, Rabh. Fungi Europ. no. 89 (1859); Grove, Brit. Rust Fungi,
p. 142; Wilson & Bisby, Brit. Ured. no. 132; Gäumann, Rostpilze, p. 1072.
Puccinia carlinae Jacky, Z. Pfl.-Krankh. 9, 289 (1899); Grove, Brit. Rust Fungi,
p. 137; Wilson & Bisby, Brit. Ured. no. 121; Gäumann, Rostpilze, p. 1050.
Puccinia carduorum Jacky, Z. Pfl.-Krankh. 9, 288 (1899); Grove, Brit. Rust Fungi,
p. 141; Wilson & Bisby, Brit. Ured. no. 117; Gäumann, Rostpilze, p. 1048.
Puccinia cardui-pycnocephali P. & H. Syd., Monogr. Ured. 1, 34 and 852 (1904);
Grove, Brit. Rust Fungi, p. 142; Wilson & Bisby, Brit. Ured. no. 116; Gäumann,
Rostpilze, p. 1049.

Spermogonia amphigenous, but generally hypophyllous and often on petioles
and midribs of the leaves, often on
yellowish spots, scattered in small
groups, generally amongst the aecidia,
pale at first, becoming orange-red or
honey-coloured. **Aecidia** uredinoid, amphigenous but usually epiphyllous and
often on the petioles or midribs of
the leaves, sometimes on yellowish spots
up to 0·5 cm. diam., variable in size,
scattered, often confluent in roundish,
concentric or ring-like groups around
the spermogonia, cinnamon-brown, pulverulent; aecidiospores globoid, subgloboid or ellipsoid, brown, 22–30 × 16–
28 μ, wall echinulate, 2–2·5 μ thick, with
3 equatorial pores. **Uredosori** amphigenous, but generally hypophyllous,
scattered, minute, pulverulent, brown;
the uredospores resemble the aecidiospores. **Teleutosori** amphigenous but
generally hypophyllous and sometimes
on the petioles and midribs of the leaves,
scattered, small, sometimes confluent in
large, irregular groups, dark brown;
teleutospores oblong to ovoid. rounded
above and below or sometimes attenuate

downwards, hardly constricted, chestnut-brown, 24–50 × 16–27 μ, wall faintly
verruculose, 2·5–3 μ thick, pore in upper

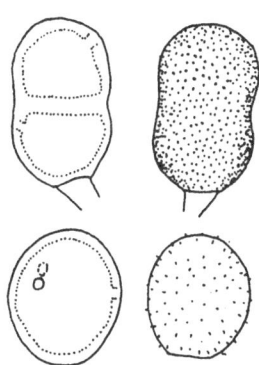

P. calcitrapae. Teleutospores and uredospores.

cell apical or subapical, in lower superior,
pedicels thin, hyaline, up to 40 μ long,
deciduous. Brachy-form.

On *Arctium lappa, A. minus, A. pubens,
Carduus acanthoides, C. nutans, Carlina
vulgaris, Centaurea nigra, C. scabiosa,
Cirsium acaule, C. dissectum, C. heterophyllum* and *C. palustre.*

Puccinia calcitrapae is here regarded as containing a number of races or *formae speciales*; they are auteocious and occur on various species of the tribe *Cynareae* of the family *Compositae*. They are characterised by uredinoid aecidia accompanied by spermogonia, uredospores with thin, echinulate walls with 3 equatorial pores and thin-walled, rounded, more or less verruculose teleutospores without apical thickening. *Puccinia calcitrapae* in the restricted sense of many continental authors, a parasite of *Centaurea calcitrapa*, does not occur in Britain, but continental collections on that host certainly cannot be distinguished from the majority of collections on the many hosts here treated as one species.

Arctium

The rust on *Arctium minus*, *A. lappa* and *A. pubens* is rare in England. It was recorded on *A. lappa* by Plowright (1889, 185) under *P. hieracii* and was included by Grove (*loc. cit.*) as a doubtful native. It was found in 1940 by Ellis in Norfolk (Trans. Norf. Norw. Nat. Hist. Soc. **15**, 371, 1943) and in Suffolk in 1950 and collected in 1945 by Wiltshire in Kent. Recently the aecidial stage has been found by Ellis in June on *A. lappa* in Norfolk (*in litt.*). The aecidia on *Arctium* are epiphyllous, on yellowish-brown rounded spots, up to 0·5 cm. diam., often confluent in ring-like groups which surround the brownish spermogonia, soon naked and pulverulent, cinnamon-brown.

The races on *Cirsium spinosissimum* and *C. oleraceum* will not infect *Arctium lappa* (Jacky, Z. Pfl.-Krankh. **9**, 193, 1899) nor will the *Arctium* race infect species of *Cirsium* or *Taraxacum* (Jacky, Zbl. Bakt. II, **9**, 796, 1899).

Carduus

The rust on *Carduus crispus* and on *C. nutans*, previously known as *P. carduorum*, appears to be scarce but has been recorded from Yorkshire, Norfolk, Gloucestershire, Derbyshire, near Edinburgh and from Clare Island, Ireland and other localities. The aecidial stage is mentioned by Grove (*loc. cit.*) but his description was taken from Sydow; it appears to have been found only in Derbyshire in 1935 (TBMS. **21**, 2, 1938) and near Edinburgh (Wilson, Proc. Roy. Soc. Edin. **53**, 177, 1948). The spermogonia are amphigenous, the aecidia usually hypophyllous, and both kinds of sori are mostly developed on the midrib and veins of the leaf in May and June. The aecidia are elongated on the veins, about 5 × 1 mm. while those on the lamina are rounded, up to 3 mm. diam. In both localities numerous 'primary teleutospores' were found mixed with the aecidiospores, sometimes completely replacing them. Sori consisting entirely of aecidiospores were also found.

Jacky (Z. Pfl.-Krankh. **9**, 288, 1889) showed that forms on *Carduus* could not infect *Cirsium* and Probst (Ann. Myc. **6**, 97, 1908) proved experimentally that the form on *Carduus crispus* could not be transferred to *C. nutans*. Gäumann (1959, 1049) came to a similar conclusion in experiments with the race on *C. crispus*.

A few of the specimens of *P. syngenesiarum* issued by Cooke & Vize and also a specimen of the same species in the Johnston Herbarium at Berwick (Johnston, Fl. Berwick, **2**, 197) have been shown to be *P. calcitrapae* on *Carduus crispus* and *C. nutans* (Wilson, TBMS. **24**, 244, 1940).

The rust on *Carduus tenuiflorus*, previously known as *P. carduipycnocephali* is scarce in Great Britain; it is closely related to those known as *P. carduorum* and *P. cirsii* and, like these, it probably possesses uredinoid aecidia. It is distinguished from the former by its larger and slightly punctate teleutospores. In Scotland in 1946, at the end of July, uredospores were scanty while teleutospores were present in abundance. The Sydows (Monogr. Ured. **1**, 34) at first described two species on *Carduus pycnocephalus* Jacq., *P. cardui-pycnocephali* from Italy and *P. galatica* from Asia Minor but later (*loc. cit.* p. 853) considered them to be identical. In this country *C. pycnocephalus*, which is closely allied to *C. tenuiflorus* has been recorded only from a few localities and at Plymouth and it appears that no rust has been found upon it. Sydow (*loc. cit.* p. 852) also recorded *P. carduorum* on *C. tenuiflorus*.

Centaurea

On *Centaurea* as *P. centaureae*, common on *C. nigra*, scarce on *C. scabiosa*, May–November. Great Britain and Ireland.

The aecidial stage appears in May resulting from infection by overwintered teleutospores. It was mentioned by Grove and the Sydows and described more fully by Plowright (1889, 186); the spermogonia are hypophyllous on yellow spots with a purple margin; these are rounded on the leaves and elongated on the petioles; they surround the rather large aecidia which are often confluent. The aecidiospores are distributed by the wind and Plowright has shown that these, by infection, produce the uredospores in about 14 days.

This species has been recorded in Norfolk on *C. scabiosa* by E. A. Ellis (1934, 494); it has also been found in Kent and Angus on this host. Guyot (Uredineana, **3**, 67, 1951) who raised Hasler's f. sp. *scabiosae* to specific rank stated that the spores were slightly larger than those on *C. nigra*. Hasler (Zbl. Bakt. II, **48**, 221, 1918) showed that *C. nigra* and *C. scabiosa* each possess a specialised form of the rust which is not transferable to other host species. It should be noted that *P. hieracii* also occurs on *Centaurea* in Britain and can be recognised by the 2 supra-equatorial pores in the uredospores.

Cirsium

This rust has been recorded frequently as *P. cirsii* on *C. palustre*, less so on *C. heterophyllum*, *C. acaule* and *C. dissectum*. It occurs in Great Britain and Ireland. The aecidial stage does not appear to have been found in this country but has been described by Bubak (1908, 137). Gäumann's tabulation (1959, 1704) of the experimental results of Jacky and Mayor show the specialised races existing within this group.

Carlina

The rust on *Carlina vulgaris* is scarce in England and there appears to have been one record from Scotland. The spermogonia are epiphyllous and are surrounded by the aecidia. The latter are on yellowish spots up to 0·5 cm. in diameter. These stages do not appear to have been found in this country.

The uredospores are similar to the aecidiospores and possess hardly perceptible spines with 3 irregularly placed pores with well-developed papillae. The teleutospores are often narrowed below. No cultures appear to have been carried out with this form, except for Jacky's experiments (Z. Pfl.-Krankh. **9**, 222, 1899) which showed that the rust on *Carlina acaulis* would not infect species of *Cirsium*.

Puccinia chrysanthemi Roze

Bull. Soc. Myc. Fr. **16**, 92 (1900); Grove, Brit. Rust Fungi, p. 131; Wilson & Bisby, Brit. Ured. no. 127; Gäumann, Rostpilze, p. 1140.

[*Uredo chrysanthemi* Roze, Bull. Soc. Myc. Fr. **16**, 78 (1900).]

Uredosori generally hypophyllous, on irregular, pallid-yellow or brownish spots, scattered or in clusters, about 1–1·5 mm. diam. often circinate, pulverulent, snuff-brown; uredospores globoid to ellipsoid, 24–52 × 17–27μ, wall delicately echinulate, brown, with 3 equatorial pores. Teleutospores mixed with the uredospores, oblong or ellipsoid, rounded and slightly thickened above, usually rounded or somewhat tapering at base, scarcely constricted, chestnut-brown, 35–57 × 20–25μ, wall delicately verruculose, pedicels thick, hyaline persistent, 35–60μ long; mesospores subgloboid or pyriform, slightly thickened at the summit, 32–37 × 20–21μ. Hemiform.

On leaves of cultivated Chrysanthemums in greenhouses all the year round, occasionally on outdoor plants. Great Britain and Ireland.

This species is said to be native in Japan (Hiratsuka, Sydowia, Beih. **1**, 34, 1956). It was first observed in England in 1895 and has been found in other European countries and in North America. In 1904 it reached Australia and New Zealand and South Africa in 1905; The optimum temperature for spore germination (15–21° C.) has been used to explain this rust's distribution in North America (Campbell & Dimock, Phytopath. **45**, 644, 1955).

In Japan it produces teleutospores in separate sori which are

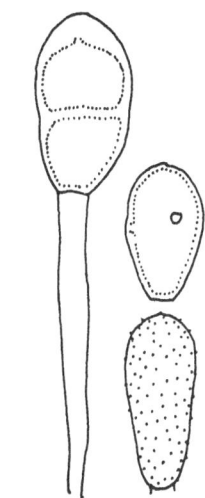

P. chrysanthemi. Teleutospore and uredospores (after Fischer).

hypophyllous, roundish dark brown and naked, but in Europe the teleutospores have been rarely seen, although mesospores occasionally occur; teleutospores have not been found in North America. Abnormal 1- and 2-celled uredospores (as well as 3- and 4-celled teleutospores) have been described and figured by Roze (*loc. cit.*), Jacky (Z. Pfl.-Krankh. **10**, 132, 1900; Zbl. Bakt. II, **10**, 369, 1903, and **18**, 88, 1907) and Fischer. Hammarlund (Bot. Not. 211, 1928) found teleutospores in Sweden and described their rapid germination in a decoction of chrysanthemum leaves; the basidiospores are hyaline, 5–6 μ in diameter and rapidly produce germ-tubes. He also described bicellular uredospores and uredo-teleuto 'twins', the upper cell consisting of a typical uredospore while the lower one was a teleutospore. Abnormal spores have not been described in British specimens.

Since, under the conditions in which the plants are grown here, the young shoots appear above ground before the old ones die away, it is probable that the parasite maintains itself by the uredospores alone: the alternative would be the possession of a perennial mycelium, which has not been found (Gibson, New Phytol. **3**, 188, 1904). It has been shown by Jacky (*loc. cit.*) that uredospores which have wintered in the open can germinate in the spring and it is believed by Pape (Gartenwelt, **32**, 623, 1928) that the uredospores can remain viable through the winter on woodwork, glass, or fallen leaves or in the soil. Affected leaves fall prematurely and diseased plants are stunted and produce few flowers.

The race on cultivated chrysanthemums in Japan will not infect many of the wild species of south-eastern Asia but *C. pacificum* var. *radiatum*, *C. makinoi* and *C. indicum* are slightly susceptible and cultivars of *C. morifolium* var. *sinense* extremely so (Hiratsaka, Sydowia, Beih. 1, 34, 1956). Bailey (J. Roy. Hort. Soc. 76, 322, 1951) obtained rather similar results in England and classified 44 cultivars according to susceptibility.

Puccinia cnici Mart.

Prod. Fl. Mosq. Ed. 2, 226 (1817); Wilson & Bisby, Brit. Ured. no. 135.

Puccinia cirsii-lanceolati Schroet., Cohn, Krypt. Fl. Schles. III, 1, 317 (1887); Grove, Brit. Ured. no. 143; Gäumann, Rostpilze, p. 1069.

Puccinia cirsii-eriophori Jacky, Z. Pfl.-Krankh. 9, 275 (1899); Gäumann, Rostpilze, p. 1068.

Spermogonia epiphyllous, flask-shaped, brownish, in groups on the same spots as the aecidia. Aecidia hypophyllous sometimes epiphyllous, seated on minute, yellow spots, mostly solitary, at first closed and then opening by a circular pore, at length widely cup-like, peridium poorly developed; aecidiospores globoid or ellipsoid, orange-yellow, 22–35 × 20–28 μ, wall verruculose, 2–3 μ thick. Uredosori amphigenous, scattered or often grouped and confluent, surrounded by the ruptured epidermis, pulverulent, ferruginous; uredospores globoid or ovoid, brown, 24–36 × 20–26 μ, wall echinulate, 1·5–3 μ thick with 3 equatorial pores with conspicuous thickened caps. Teleutosori amphigenous, scattered or grouped, minute, surrounded by the ruptured epidermis, pulverulent, dark brown; teleutospores ellipsoid, not or scarcely constricted, base rounded or slightly attenuate, pale brown, 30–40 × 22–25 μ, wall finely verrucose, 1·5–3 μ thick with small hyaline papillae over the pores, apex unthickened, pore of upper cell apical, of lower superior, pedicels short, hyaline. Auteu-form.

On *Cirsium eriophorum* and *C. vulgare*. England, Scotland, Ireland, frequent.

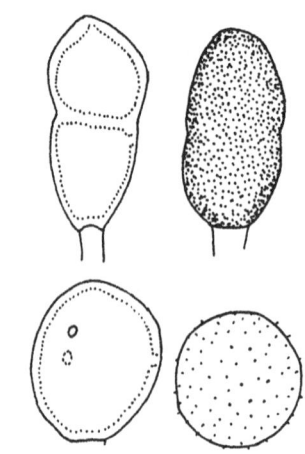

P. cnici. Teleutospores and uredospores (after Fischer).

The rather uncommon aecidial stage has been discovered in England and Scotland (Grove & Chesters, TBMS. 18, 271, 1934); it has also been recorded, from Ireland.

Inoculation by teleutospores from *Cirsium vulgare* to plants of the same species produced aecidia (Kellerman, J. Mycol. 9, 229, 1903).

The rust was discovered on *C. eriophorum* by Ellis & Rhodes at Woodstock, Oxon. The form on this host was regarded by Jacky (*loc. cit.*), Bubak and Gäumann (*loc. cit.*) as a distinct species (*Puccinia cirsii-eriophori* Jacky) but as there are no constant morphological differences between the forms on the two hosts and as they can be transferred from one host to the other, they are here treated as one. In this rust the cells making up the aecidial wall are only loosely connected and closely resemble aecidiospores. Olive (Ann. Bot. Lond. 22, 331, 1908) described nuclear division and the development of the aecidium in this species.

Puccinia cnici-oleracei Pers. ex Desm.

Cat. Pl. Omis. p. 24 (1823); Grove, Brit. Rust Fungi, p. 144; Wilson & Bisby, Brit. Ured. no. 136; Gäumann, Rostpilze, p. 682.

Puccinia cirsiorum var. *cirsii-palustris* Desm., Cat. Pl. Omis. p. 25 (1823).

Puccinia asteris Duby, Bot. Gall. 2, 888 (1830); Wilson & Bisby, Brit. Ured. no. 100; Gäumann, Rostpilze, p. 660.

Puccinia tripolii Wallr., Fl. Krypt. Germ. 2, 223 (1833); Grove, Brit. Rust Fungi, p. 129.

Puccinia millefolii Fuck., Jahrb. Nass. Nat. Ver. 23–24, 55 (1869); Grove, Brit. Rust Fungi, p. 131; Wilson & Bisby, Brit. Ured. no. 184; Gäumann, Rostpilze, p. 649.

Puccinia leucanthemi Pass., Hedwigia, 13, 47 (1874); Grove, Brit. Rust Fungi, p. 133; Wilson & Bisby, Brit. Ured. no. 176; Gäumann, Rostpilze, p. 675.

P. andersoni B. & Br., Ann. Mag. Nat. Hist. ser. 4, 15, 35 (1875); Grove, Brit. Rust Fungi, p. 146.

Puccinia cardui Plowr., Monogr. Brit. Ured. Ustil. p. 216 (1889).

Puccinia lemonnieriana Maire, Bull. Soc. Myc. Fr. 16, 65 (1900); Wilson & Bisby, Brit. Ured. no. 174; Gäumann, Rostpilze, p. 684.

Puccinia cirsii-palustris (Desm.) Wilson, Trans. Brit. Myc. Soc. 24, 244 (1940).

Spermogonia, aecidia and uredosori wanting. Teleutosori amphigenous, usually hypophyllous on pale yellow spots, minute, circinate, often confluent, in large groups, pulvinate, long-covered by the epidermis, dark brown; teleutospores clavoid, apex rounded, sometimes acutely conical or truncate, constricted at septum, base rounded or often attenuate, yellowish-brown, 38–56 × 14–21 μ, lower cell equal to or longer than the upper, wall smooth, 8–14 μ thick at apex, pore of upper cell apical, often slightly oblique, of lower close below the septum, pedicel persistent, thick, yellowish-brown or almost hyaline, about as long as the spore. Micro-form.

On *Achillea millefolium, Aster tripolium, Chrysanthemum leucanthemum, C. segetum, Cirsium heterophyllum* and

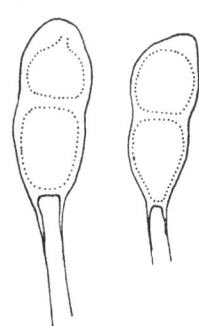

P. cnici-oleracei. Teleutospores (after Fischer).

C. palustre, July–November. Great Britain, frequent.

This species consists of a number of races of microcyclic lepto-forms in which the teleutosori become greyish as the result of direct germination. They are certainly most closely related as a group to the *Carex–Compositae Puccinias* from which they have no doubt been derived.

The rust on *Achillea millefolium* is scarce in England and southern Scotland and extends as far north as the Dee area in Aberdeenshire (Trail, 1890, 319); it forms minute scattered sori on indistinct spots on both surfaces of the leaf. Plowright (1889, 216) showed that it would not infect *Aster tripolium* and this has been confirmed by Magnus and others. It has been recorded from central and western Europe, Japan and western North America on *Achillea millefolium*.

On *Aster tripolium* (as *Puccinia tripolii*) records have been made on the east coast of England and Scotland from Norfolk to the Moray Firth and from Gloucestershire by Carleton Rea (see Sprague, 1954, 8). Cooke (Micro-Fungi, 207, 1878) and Grove assigned the British rust on *A. tripolium* to *P. tripolii* Wallr. Grove stated that the latter differs from the American specimens of *P. asteris* in the absence of spots, in the occurrence of the sori all over the leaf surface and in minor differences in the teleutospores. Wilson & Henderson (TBMS. 1954, 250) noted that the stomata are generally distributed over the leaf surface; as the sori arise beneath the stomatal openings they have a similar distribution. The absence of spots is merely a host reaction. The differences in teleutospore structure mentioned by Grove do not appear to be taxonomically significant.

The cytology of this form was investigated by Olive (Ann. Bot. Lond. **22**, 331, 1908) and Walker (Trans. Wis. Acad. Sci. **23**, 567, 1927). The latter stated that the binucleate condition arises at the base of the sorus by nuclear division without subsequent cell division.

A rust on *Chrysanthemum leucanthemum* was collected in England by F. J. Chittenden at Lamorna Cove, Cornwall, in September 1906; it had unusually long teleutospores, $40–70 \times 14–24\mu$, with apical thickening of the upper cell up to 14μ and with pedicels $40–60\mu$ long. It appears to be correlated with the heteroecious *P. aecidii-leucanthemi* Fisch. with teleutospores on *Carex montana* and to date not recorded in Britain. Viennot-Bourgin (1956, 41) illustrates this rust.

An almost similar rust has been recorded recently on *Chrysanthemum segetum* by Ellis from Norfolk (Trans. Norf. Norw. Nat. Hist. Soc. **17**, 138, 1950). Sydow (Ann. Myc. **28**, 430, 1930) recorded a rust on this host from north-western Germany where it was found growing in the vicinity of plants of *Anthemis arvensis* infected with *P. anthemidis* Syd. and he suggested that spores from the latter had probably passed on to *Chrysanthemum segetum* and brought about infection. He noted that the teleutospores on *C. leucanthemum* are slightly larger both in length and breadth than those on *C. segetum*.

The rust on *Cirsium heterophyllum*, previously known as *P. andersoni* has been recorded from Yorkshire and several places in Scotland. The sori occur on the leaves on rounded yellow spots with a brown border, 1–1·5 cm. diam. and are almost concealed by the pubescence of the leaf. In Finland Liro (1908, 396) recorded this form on *Cirsium oleraceum* and *C. heterophyllum* and on the hybrid between the two. Stein (Zbl. Bakt. II, **80**, 411, 1930) working in Switzerland transferred the rust from *C. oleraceum* to *C. heterophyllum*, *C. acaule* and *C. acaule × oleraceum* but could not infect *Achillea millefolium*. Fischer (Tschirch-Festschr, Leipzig-Tauchnitz, 415, 1926) showed that although this rust is a lepto-form it can overwinter by means of its teleutospores. Stein (*loc. cit.*) confirmed this and cultivated it through six generations during the summer. Kuhnholtz-Lordat (Ann. Epiphyt. **13**, 43, 1947) described the structure of the teleutosori.

The *Cirsium* race is a rust of scarce occurrence on *C. palustre* in England and central and southern Scotland; it does not appear to have been found on this host in Scandinavia. It has been recorded as *P. cardui* Plowr. and as *P. lemonnieriana* (Wilson, 1934, 371). The small teleutosori usually form confluent groups on the under surfaces of the radical leaves of the host and are borne on yellow spots which are seen both on the under and upper surfaces. The wall of the teleutospores is strongly thickened, up to 14 μ, at the apex.

Grove (1913, 144) stated that most of the specimens of *P. syngenesiarum* issued by Cooke belong to *P. cnici-oleracei* and in this he was correct. Unfortunately Cooke wrongly identified the host of these specimens as *Cirsium lanceolatum* (= *C. vulgare*) whereas they were *Cirsium palustre* (see Wilson, TBMS. **24**, 244, 1940) and this incorrect identification appears to have been accepted by Plowright (under *P. cardui*), the Sydows and Grove and possibly by others. There is no evidence that *P. cnici-oleracei* occurs on *C. lanceolatum* or *Carduus crispus* (for the rust on the latter host, see *P. calcitrapae*, p. 191).

The statement that *P. cnici-oleracei* has been found on *Cirsium dissectum* in Ireland (Wilson & Bisby, 1954, 70) requires confirmation.

Puccinia crepidicola Syd.

Oest. Bot. Zeit. **51**, 17 (1901); Wilson & Bisby, Brit. Ured. no. 142; Gäumann, Rostpilze, p. 1084 *p.p.*

As *Puccinia crepidis* Schroet. Grove, Brit. Rust Fungi, p. 156.

Spermogonia and **aecidia** unknown. **Uredosori** amphigenous or caulicolous, scattered, small, punctiform, pulverulent, pale brown; uredospores globoid or subgloboid, pale brown, 21–26 × 18–21 μ, wall echinulate, 1·5–2·5 μ thick, with 2 equatorial pores. **Teleutosori** amphigenous or caulicolous, minute, long-covered by the epidermis, blackish-brown; teleutospores ellipsoid to ovoid, rounded at both ends, not thickened above, scarcely or not at all constricted, dark brown, 30–36 × 22–26 μ, wall 3–4 μ thick, finely verrucose, pore of upper cell sub-apical, of lower equatorial, pedicels very short, hyaline. Hemi-form.

On *Crepis biennis*, *C. capillaris* and

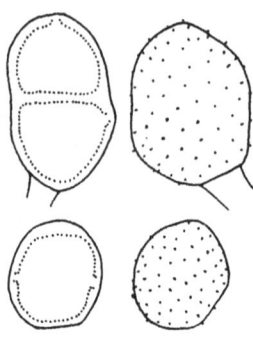

P. crepidicola. Teleutospores and uredospores.

C. vesicaria subsp. *taraxacifolia*, July onwards. Great Britain and Ireland, frequent.

The species of *Puccinia* on *Crepis* in Europe can be divided into three main types: (1) with numerous aecidia borne on a systemic mycelium all over the pale, slender leaves, *P. crepidis*; (2) those with aecidia in discrete groups on thickened spots on the leaves, *P. major* and *P. praecox*; and (3) a number of forms in which aecidia are unknown, *P. crepidicola* and allied species.

The rusts on *Crepis capillaris* and *C. vesicaria* subsp. *taraxacifolia* were recorded as *P. crepidis* by Grove (*loc. cit.*) but there is no evidence of the occurrence of aecidia of this type in Britain; apart from the discrete aecidial groups of *P. major* on *Crepis paludosa* only uredospore- and teleutospore-bearing collections are known on *Crepis*. It should be noted, however, that in Scandinavia the diplont stages of *P. crepidis* usually occur on plants other than those bearing the systemic aecidial stage (Jørstad, 1940, 59). Nevertheless, the aecidial stage is so conspicuous that it should have been discovered were it present. The distinction frequently made between *P. crepidis* and other *Crepis* rusts in teleutospore sizes is not valid. Schroeter's original measurements (20–30 × 17–22 μ) are obviously too low; recent continental material of *P. crepidis* give measurements of 34–39 × 18–24 μ which are comparable with the other *Crepis* rusts.

The record of *P. praecox* on *Crepis biennis* (Wilson & Henderson,

TBMS. **37**, 248, 1954) is here placed under *P. crepidicola.* Although aecidia have been described for the rust on this host none has been found in Britain. The uredospore and teleutospore stages in British collections cannot be separated from the other British *Crepis* rusts, and its only claim to specific recognition lies in presumed host specialisation. In a long series of experiments Hasler (Zbl. Bakt. II, **48**, 221, 1918) proved close host specialisation for most of the common European *Crepis* rusts. Many of these are recognised by Gäumann at specific rank. Experimental studies would certainly reveal some of them in Britain.

Puccinia cyani Pass.

Rabh. Fung. Europ. no. 1767 (1874); Grove, Brit. Ured. p. 140; Wilson & Bisby, Brit. Ured. no. 143. Gäumann, Rostpilze, p. 1063.

[*Uredo cyani* DC., Lam. & DC., Syn. Pl. Gall. p. 47 (1806).]

Spermogonia scattered on leaves and stems, honey-coloured. **Aecidia** uredinoid, amphigenous, and on the stems, systemic, covered at first by shining white epidermis, rounded or elongate then naked, pulverulent, dark cinnamon-brown; aecidiospores globoid to ellipsoid, 22–30 × 19–24 μ, wall yellow-brown, finely echinulate, 1·5–2 μ thick, with 2 equatorial pores. **Uredosori** scattered over the leaves, resembling the aecidia; uredospores resembling the aecidiospores. **Teleutosori** amphigenous, often occurring with the aecidia, apparently systemic, rounded, scattered, soon naked, pulverulent, black; teleutospores broadly ellipsoid, rounded at both ends, not thickened above, not constricted, 30–35 × 22–27 μ, wall chestnut-brown, very finely verruculose, 1·5–3 μ thick, pore of upper cell subapical, of lower equatorial, pedicels colourless, short, fragile. Brachy-form.

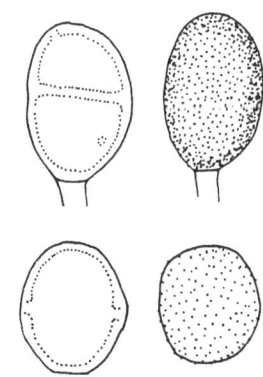

P. cyani. Teleutospores and uredospores (after Savulescu).

On *Centaurea cyanus.* England and Scotland, in gardens, uncommon.

The aecidial mycelium penetrates the whole plant and forms spermogonia, aecidia and teleutosori on the leaves and stems. There is also a secondary localised mycelium formed probably by infection by aecidiospores on which scattered uredosori and teleutosori develop.

Puccinia glomerata Grev.

Fl. Edin. p. 433 (1824); Berkeley in Smith, Engl. Fl. **5** (2), 365 (1836).
Puccinia expansa Link, Sp. Pl. Ed. 4, **6** (2), 75 (1825); Grove, Brit. Rust Fungi, p. 136; Wilson & Bisby, Brit. Ured. no. 151; Gäumann, Rostpilze, p. 1039.

Spermogonia, aecidia and **uredosori** wanting. **Teleutosori** amphigenous mostly hypophyllous on round, yellowish or brownish spots, densely crowded, confluent in roundish groups, up to 1 cm. diam., at first covered by the epidermis then dehiscing with a pore-like opening; teleutospores ellipsoid, rounded at the base and apex, not or scarcely contracted at the septum, $25-48 \times 19-30\mu$, both cells about equal in length and breadth but sometimes the upper broader, yellowish-brown, wall not thickened at apex, smooth, pore of upper cell apical, of lower cell usually superior, pores covered by small, colourless papillae, pedicels short colourless; mesospores occasionally present. Microform.

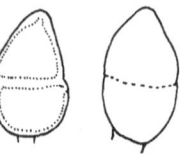

P. glomerata. Teleutospores.

On *Senecio aquaticus* and *S. jacobaea*, April–September. Great Britain and Ireland, frequent.

Puccinia glomerata and *P. expansa* were regarded as distinct species by the Sydows (Monogr. Ured. **1**, 146 and 148), Grove (*loc. cit.*) and Wilson (1934, 369) but the distinction can no longer be upheld.

Gäumann (*loc. cit.*) summarised his previous work on this species in which he divided it into the following four biological races which show only slight morphological differences:

1. *adenostyles* Gäum. which occurs only on *Adenostyles* spp., and is leptosporic with systemic mycelium in the rhizome.
2. *senecionis alpini* Gäum. on *S. aquatica* and *S. jacobaea* and other *Senecio* spp. is leptosporic with perennating mycelium.
3. *senecionis doronici* Gäum. found only on *S. doronicum.*
4. *petasites* Gäum. on *Petasites* spp.

Only race 2 seems to be represented in Britain. Gäumann described rudimentary peridia and aecidiospores in race 3 but these have not been observed in the British material of race 2. Gäumann's observations clearly show interesting examples of transitions between micro- and opsis-forms.

Puccinia helianthi Schw.

Syn. Fung. Carol. Super. 47 (1822); Massee, Mildews, Rusts and Smuts, p. 102 (1913); Wilson & Bisby, Brit. Ured. no. 162; Gäumann, Rostpilze, p. 1034.

The occurrence of this species in Britain is extremely doubtful, Massee's record is the only one and no material is known.

Puccinia hieracii Mart.

Prod. Fl. Mosq. Ed. 2, 226 (1817); Grove, Brit. Rust Fungi, p. 158; Wilson & Bisby, Brit. Ured. no. 164; Gäumann, Rostpilze, p. 1094.

Uredo hieracii Schum., Enum. Pl. Saell. 2, 232 (1803).

[Uredo cichorii DC. Fl. Fr. 6, 74 (1815).]

Puccinia jaceae Otth, Mitt. Naturf. Ges. Bern, 1865, 173 (1866).

Puccinia cichorii Kickx, Fl. Crypt. Flandres, 2, 65 (1867); Grove, Brit. Rust Fungi, p. 148; Wilson & Bisby, Brit. Ured. no. 129; Gäumann, Rostpilze, p. 1067.

Puccinia endiviae Pass., Hedwigia, 12, 114 (1873); Wilson & Bisby, Brit. Ured. no. 149.

Puccinia picridis Hazl., Math. Term. Köslem. p. 152 (1877); Wilson & Bisby, Brit. Ured. no. 196; Gäumann, Rostpilze, p. 1113.

Puccinia taraxaci Plowr., Monogr. Brit. Ured. Ustil. p. 186 (1889); Grove, Brit. Rust Fungi, p. 154; Wilson & Bisby, Brit. Ured. no. 230; Gäumann, Rostpilze, p. 1125.

Puccinia leontodontis Jacky, Z. Pfl.-Krankh. 9, 339 (1899); Wilson & Bisby, Brit. Ured. no. 175; Gäumann, Rostpilze, p. 1106.

Puccinia tinctoriae Magn., Abh. Nat. Ges. Nürnb. 13, 37 (1900) (non P. tinctoriae Speg. 1886); Grove, Brit. Rust Fungi, p. 139.

Puccinia tinctoriicola Magn., Oest Bot. Zeit. 52, 491 (1902); Wilson & Bisby, Brit. Ured. no. 232; Gäumann, Rostpilze, p. 1119.

Spermogonia amphigenous, usually on the midribs and veins, numerous and crowded into groups among the aecidia, on pale yellow, oval spots up to 5 mm. long, honey-coloured; spermatia globoid or ovoid, 1–2 μ diam. Aecidia uredinoid, amphigenous on reddish-yellow, rounded or oval spots which are thickened and deformed, usually on the midribs and veins, rather large and sometimes confluent, covered by the greyish epidermis but soon naked, pulverulent, cinnamon-brown; aecidiospores resembling the uredospores. Uredosori amphigenous and on the stems, on small yellowish spots, minute, sometimes confluent, scattered, pulverulent, cinnamon-brown; uredospores globoid to ellipsoid, 21–30 × 16–25 μ, wall echinulate, 1–2·5 μ thick, with 2 supra-equatorial pores. Teleutosori similar to the uredosori but more often on the stems where they are elongate, blackish-brown; teleutospores ovoid or ellipsoid, rounded at both ends, scarcely constricted at the septum, 26–42 × 18–29 μ, wall chestnut-brown, uniformly 1–3 μ thick, finely verrucose, pore of upper cell subapical of lower equatorial, pedicel colourless,

short in most races, fragile. Brachy-form.

On Centaurea nigra, Cichorium endivia (cult.) C. intybus, Hieracium spp. (subg. Euhieracium), Hypochoeris macu-

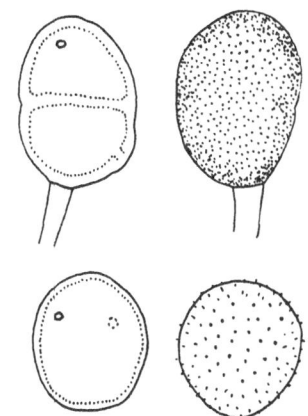

P. hieracii. Teleutospores and uredospores.

lata, Leontodon autumnalis, L. hispidus, L. taraxacoides, Picris hieracioides, Serratula tinctoria and Taraxacum spp., spermogonia and aecidia, April–May; uredospores and teleutospores, June–November. Great Britain and Ireland, frequent.

This is a collective species made up of 3 varieties and a number of specialised forms, with hosts chiefly in the tribe *Cichorieae* of the *Compositae*. These rusts are autoecious brachy-forms with uredinoid aecidia borne on a limited mycelium. Spermogonia have been described as infrequent but this is probably due to the infrequent discovery of the aecidial stage or to its incomplete description. The 2 pores in the uredospores and the uniformly thin-walled teleutospores with a depressed pore in each cell, are important characters for identification.

The 3 varieties and several of the specialised forms have been previously regarded as distinct species. Early investigations by Jacky (Zbl. Bakt. II, **18**, 82, 1901) and by Probst (*ibid.* II, **22**, 676, 1909) showed that a high degree of specialisation exists amongst them; it was generally found that a rust on one host genus would not infect another genus and that sometimes a form could be specialised to a single host species. Their results as regards the rusts on *Euhieracia* are well tabulated by Gäumann (1959, 1036).

Puccinia hieracii closely resembles *P. calcitrapae*, the latter is distinguished by the 3 equatorial uredospore pores. Species of *Centaurea* in Britain may be infected by both species, but *P. hieracii* is less frequent; *Puccinia calcitrapae* is the common rust on this host. Jørstad (1936, 25) has recorded *P. hieracii* on *Centaurea nigra* in Norway, this race has been known as *P. jaceae*.

The rust is scarce in England on *Cichorium intybus*; it has been found in Cornwall, Devon, Norfolk and Surrey and, in 1950, in Suffolk (Moore, 1959, 300). It has been recorded on the cultivated endive (*C. endivia*) in Cornwall and Devon. On the latter host species the pedicels of the teleutospores may be very long, up to 76μ, but on *C. intybus* long pedicels may also occur. No aecidial stage has been found in this country but Mayor (Bull. Soc. Neuchâtel. Sci. Nat. **46**, 26, 1920–1) produced spermogonia and aecidia on both *C. intybus* and *C. endivia* by infecting these plants with teleutospores from the latter.

The rust is commonly found on a large number of species and forms of *Hieracium* subg. *Euhieracium* and on these spermogonia and rather large aecidia have been described. The host range on *Hieracium* requires investigation in Britain. The form on *Hypochoeris maculata* which appears to be a new host record in this country for the rust, is included here and not under the var. *hypochoeridis* as the pores in the uredospores are distinctly supra-equatorial. The specimen is in the Edinburgh Herbarium and was collected on the Lizard, Cornwall, in

1891. Jacky (1901, *loc. cit.*) failed to infect *H. maculata* with teleuto-spores from *H. radicata*.

The rust has been recorded frequently on *Leontodon autumnalis* and *L. hispidus* but is scarce on *L. taraxacoides*. Spermogonia and aecidia have been described on the hosts but there appears to be no record of them in this country. The form on *L. taraxacoides* which was described as f. *thrinciae* by Klebahn (Krypt. Fl. Mark Branb. 424, 1914) was recorded by Ellis in Norfolk (Trans. Norf. Norw. Nat. Hist. Soc. **13**, 494, 1933–4). There appears to be no definite distinction between this form and the type (*P. leontodontis*) in which it was included by the Sydows, who noted that the rust on *Leontodon* spp. may be very variable in the size and shape of its teleutospores. Those on different host species and even those on the same host plant may differ considerably in length and width and the relative sizes of the two cells of the spore may also vary. This variation was also noted and figured by Grove (1913, 150). Jørstad (1951, 39) noticed similar variability in Iceland where he dis-tinguished two types of teleutospores on *Leontodon*, one with much larger teleutospores than the other and Rytz (Veröff. Geobot. Inst. Rubel Zürich, **4**, 1, 1927) in Switzerland has also described similar varia-tions. A considerable variation in the size of the teleutospores on *L. hispidus* has been found in this country.

The rust on *Picris hieracioides* is scarce in England and Wales. It was found near Swansea in 1915 and has been recorded from Gloucestershire (TBMS. **6**, 327, 1920) by Sprague (1954, 95), from Suffolk by Mayfield (1935, 4) and from Kent. The warts on the teleutospores are described as slightly longer than those of the form on *Hieracium*. Jacky (Z. Pfl.-Krankh. **9**, 339, 1899) showed that this rust will not infect species of *Hieracium*, *Hypochoeris*, *Crepis* or *Lapsana*. Mayor (Bull. Soc. Neuchâtel. Sci. Nat. **54**, 53, 1929) recorded spermogonia and obtained spermogonia and aecidia by infection with teleutospores but gave no description of the aecidia.

The rust on *Serratula tinctoria* has been rarely found in England; it was found near London by Cooke about 1865 and was recorded from Worcestershire in 1915 (TBMS. **5**, 254, 1916) and by Hadden from near Lynton, Devon, in 1914. There appears to be no reference to spermo-gonia or aecidia in this form.

The rust on species of *Taraxacum* (more precise host records are required) is common throughout the British Isles and spermogonia and the large aecidia can be found usually in April and May, chiefly on the upper surface of the leaves. Even in lowland areas amongst the hills in

Peeblesshire a considerable number of teleutospores, called 'primary teleutospores' by Jørstad (1940, 70), may be found among the aecidiospores. No aecidia containing only teleutospores have been found but up to 50 % are frequently present (Wilson, 1948, 180). Plowright (1889, 150) showed that *T. officinale* is not infected with teleutospores of *P. lapsanae* from *Lapsana communis* nor by teleutospores of *P. hieracii* on *Leontodon autumnalis* or uredospores on *Centaurea nigra*. Jacky (*loc. cit.*) infected *T. officinale* with uredospores on the same host but failed to infect *Cichorium intybus* and *C. endivia*. Schiller (Sydowia, **3**, 201, 1949) isolated ten biotypes from *P. taraxaci* which attack various forms of *Taraxacum officinale* s. lat.

var. *piloselloidarum* (Probst) Jørst.

K. Norske Vidensk. Selsk. Skr. **1935**, 38, 27 (1936); Wilson & Bisby, Brit. Ured. no. 165.

Puccinia piloselloidarum Probst, Zbl. Bakt. II, **22**, 712 (1909); Gäumann, Rostpilze, p. 1098.

On *Hieracium pilosella*, spermogonia and aecidia, May; uredospores and teleutospores, June onwards. Great Britain and Ireland, frequent.

This variety was previously regarded as a distinct species; it differs from Var. *hieracii* in the only slightly supra-equatorial position of the pores of the uredospores and thus holds an intermediate position between var. *hieracii* and var. *hypochoeridis*. It is restricted to species of *Hieracium* subg. *Pilosella* Tausch. and in this country, where it is very common, occurs only on *H. pilosella*.

Probst (*loc. cit.*) distinguished seven *formae speciales* but most of these are on species not found in this country. One of his isolates from *H. pilosella* was restricted to that host whereas another from the same host species had a wider host range.

var. *hypochoeridis* (Oud.) Jørst.

K. Norske Vidensk. Selsk. Skr. **1935**, 38, 27 (1936).

[*Uredo hyoseridis* Schum., Enum. Pl. Saell. **2**, 233 (1803).]

Puccinia hypochoeridis Oud., Nederl. Kruidk. Arch. II, **1**, 175 (1874); Grove, Brit. Rust Fungi, p. 148; Wilson & Bisby, Brit. Ured. no. 170; Gäumann, Rostpilze, p. 1099.

Puccinia hyoseridis [Schum.] Liro, Bidr. Känned. Finl. Nat. Folk, **65**, 369 (1908).

On *Hypochoeris glabra* and *H. radicata*, spermogonia and aecidia, April–May; uredospores and teleutospores, June–October. Great Britain and Ireland, frequent.

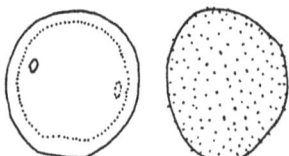

P. hieracii var. *hypochoeridis*. Uredospores.

This variety was previously regarded as a distinct species. The pores of the uredospores are depressed more than those of the var. *pilosel-loidarum* and are nearly or quite equatorial The aecidial stage is well defined and spermogonia are commonly present; it produces considerable deformation on the midrib and veins of the leaf and is also present on the lamina. It has been found rather frequently in Scotland in May, at low altitudes in mountainous areas in Stirlingshire and Argyllshire. In several specimens where aecidia are present a variable number of teleutospores are found among the aecidiospores; in some the percentage of teleutospores is low, in others high, and in a few only teleutospores have been found. Sori developing on the midrib usually contain more teleutospores than those on the lamina. The aecidiospores germinate at once, producing long, septate hyphae; the teleutospores failed to germinate and evidently require a resting period before they can do so. Teleutospores produced in early summer in the aecidia have been called primary by Jørstad (1940, 27). Later, minute secondary uredosori and teleutosori less than 1 mm. diam., appear on both sides of the leaf. The latter are rarely produced but have been found in several localities in the south of England.

The rust is rare on *Hypochoeris glabra* but has been found by Salisbury in Surrey in 1944 and by Ellis in Norfolk in 1951. Jacky (*loc. cit.*) showed that in Switzerland *H. glabra* can be infected with teleutospores from *H. radicata*; the infection is severe, spermogonia and aecidia being produced on both sides of the leaf on the midrib where the aecidia are up to 2 mm. long.

Puccinia hysterium (Str.) Röhl.

Deutschl. Fl. Ed. 2, **3** (3), 131 (1813); Wilson & Bisby, Brit. Ured. no. 171.
[*Aecidium tragopogi* Pers., Syn. Meth. Fung., p. 211 (1801).]
Uredo hysterium Str., Ann. Wetter. Ges. **11**, 102 (1810).
Puccinia tragopogonis Corda, Icon. Fung. **5**, 50 (1842); Gäumann, Rostpilze, p. 1127.
Puccinia tragopogi [Pers.] Wint., Hedwigia, **19**, 44 (1880); Grove, Brit. Rust Fungi, p. 150.

Spermogonia epiphyllous and on the stems, honey-coloured. **Aecidia** hypophyllous, without spots, scattered uniformly over the whole surface and on other green parts, cup-shaped, with a white, laciniate, revolute peridium, peridial cells with strongly thickened inner wall and thinner outer wall, inner wall with small warts; aecidiospores globoid or ellipsoid with orange contents, 20–30 × 18–24 μ, wall pale orange, densely verrucose, with 3 equatorial pores. **Uredosori** absent. **Teleutosori** amphigenous and on the stems, without

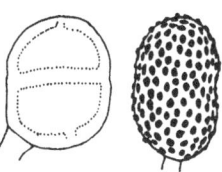

P. hysterium. Teleutospores.

spots, scattered or in small rounded or elongate groups long-covered by the epidermis, pustulate, becoming pulverulent, dark brown; teleutospores broadly ellipsoid, rounded at both ends, slightly constricted at the septum, chestnut-brown, of very variable size, 26–45 × 18–32μ, mostly 30–44 × 22–30μ, wall not thickened above, tuberculate, pore of upper cell apical or slightly depressed, of lower cell equatorial, pedicel short, colourless; mixed with a few globoid or ellipsoid uredospores which are faintly echinulate, brownish, 24–30μ diam. and with a few mesospores. Opsis-form.

On *Tragopogon pratensis* and its subsp. *minor*, April–September. England, Scotland, frequent, but absent north of the Forth–Clyde valley.

The life history of this species was first worked out by de Bary (1863, 80). The mycelium arising from the infection of young plants by the basidiospores permeates the whole of the host, so that aecidia are produced on every part—stems, leaves, bracts, and receptacles—and the infected plants are noticeable for their paler colour and distorted form. The teleutospores are born on a localised mycelium.

Plowright stated that seeds from infected plants give rise to healthy seedlings but he also observed that seedlings in the autumn may bear aecidia. Dietel (Flora, **81**, 400, 1895) in consequence, suggested that in this species the aecidiospores may sometimes reproduce aecidia. The observation made by Dietel that numerous spermogonia occur with the first-formed aecidia in the spring and that they are very few or absent from later produced aecidia is additional evidence for this suggestion.

The cytology of the aecidial development was investigated by Lindfors (1924, 11).

Puccinia lapsanae Fuck.

Jahrb. Nass. Ver. Nat. **15**, 13 (1860); Grove, Brit. Rust Fungi, p. 147; Wilson & Bisby, Brit. Ured. no. 173; Gäumann, Rostpilze, p. 1105.

[*Aecidium lapsanae* Schultz, Prod. Fl. Starg. p. 454 (1806).]

Spermogonia crowded in little clusters, epiphyllous, honey-coloured. **Aecidia** amphigenous, somewhat crowded on large, roundish, purple spots, flattish, with torn, white, reflexed peridium; aecidiospores subgloboid or ovoid, nearly smooth, orange, 16–21 × 13–17μ. **Uredosori** amphigenous, very minute, round, often confluent, pulverulent, chestnut-brown; uredospores globoid to ovoid, pale brown, 17–22 × 15–18μ, wall delicately echinulate, with 2 equatorial pores. **Teleutosori** amphigenous, minute, scattered, pulverulent, blackish-brown; teleutospores ellipsoid or ovoid,

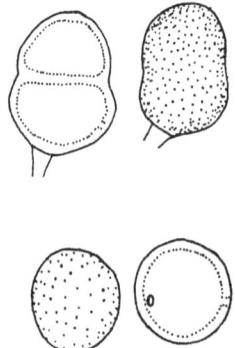

P. lapsanae. Teleutospores and uredospores.

scarcely constricted, chestnut-brown, 22–33 × 17–26 μ, wall very delicately echinulate, pore of upper cell almost apical, of lower equatorial, pedicels hyaline, slender, short, often oblique. Auteu-form.

On leaves and stems of *Lapsana communis*, aecidia, March–May; uredospores, April–June; teleutospores, June–September. Great Britain and Ireland, very common.

The aecidia of this species occur very early in the year especially on the leaves of young seedlings. Plowright (1889, 149) demonstrated that all three spore-forms belong to the same life cycle. The mycelium of the aecidia, when occurring on the petioles, causes them to become pale and swollen; on the leaves it often produces conspicuous purple spots which bear the spermogonia on the upper surface.

Puccinia maculosa (Str.) Röhl.

Deutschl. Fl. Ed. 2, 3 (3), 131 (1813).
[*Aecidium prenanthis* Pers., Syn. Meth. Fung. p. 208 (1801).]
Uredo prenanthis Schum., Enum. Pl. Saell. 2, 232 (1803).
Uredo maculosa Str., Ann. Wetter. Ges. 2, 101 (1810).
Puccinia prenanthis Kunze, Fl. Dres. 2, 250 (1823).
Puccinia chondrillae Corda, Icon. Fung. 4, 15 (1840); Grove, Brit. Rust Fungi, p. 151; Wilson & Bisby, Brit. Ured. no. 126; Gäumann, Rostpilze, p. 1066.
Puccinia prenanthis Lindr., Acta Soc. Fauna Fl. Fenn. 20 (9), 6 (1901).

[**Spermogonia** amphigenous, rather large, between the aecidia.] **Aecidia** hypophyllous or on the petioles, rarely a few on the upper surface, seated on large, yellow and purple spots, in clusters as much as 1 cm. broad, at first hemispherical, opening by a pore, then flattened, whitish or yellow, sometimes with a purplish tinge, peridium quite rudimentary, consisting of separate, rounded or elongate cells with coarsely warted wall; aecidiospores globoid to ellipsoid, finely verruculose, pale orange, 13–24 μ diam. **Uredosori** hypophyllous, on pallid irregular spots, scattered, minute, punctiform, pulverulent, pallid-brown; uredospores more or less globoid, yellow-brown, 16–24 μ, wall echinulate with 3 (rarely 4) pores each with a colourless, convex cap. **Teleutosori** similar, blackish-brown; teleutospores ellipsoid, not con-

stricted, brown, 26–36 × 16–24 μ, wall very delicately verruculose, pore of upper cell apical, of lower approximately

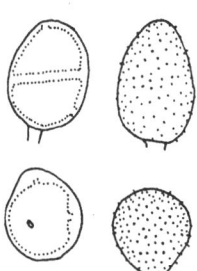

P. maculosa. Teleutospores and uredospores.

equatorial, pedicels hyaline, very short. Auteu-form.

On *Mycelis muralis*, aecidia, May, uredospores May–June; teleutospores from July. Great Britain, scarce.

Spermogonia do not appear to have been seen in this country but were described by Bubak (1908, 78) and are mentioned by Jørstad (1962) in Norway.

The connection of the three spore forms has been proved experimentally by Jacky (Zbl. Bakt. II, **9**, 842, 1902).
For the aecidium on *Lactuca sativa* see p. 244.

Puccinia major Diet.

Mitt. Thür. Bot. Ver. **6**, 46 (1894); Grove, Brit. Rust Fungi, p. 157; Wilson & Bisby, Brit. Ured. no. 181; Gäumann, Rostpilze, p. 1088.

Puccinia lampsanae var. *major* Diet., Hedwigia, **27**, 303, (1888).

Spermogonia generally hypophyllous, on reddish or yellowish spots. **Aecidia** hypophyllous, often surrounding the spermogonia, in roundish clusters or more often forming oblong patches on the nerves and petioles, shortly cylindrical, with white, torn, erect peridium; aecidiospores ovoid or rarely subgloboid, delicately verruculose, orange, $20–30 \times 16–24\,\mu$. **Uredosori** amphigenous, solitary, minute, cinnamon, spores subgloboid to ovoid, brownish, $24–30 \times 21–26\,\mu$, wall distinctly echinulate, with 2 equatorial pores. **Teleutosori** chiefly hypophyllous, standing singly, scattered over nearly the whole leaf-surface, blackish-brown; teleutospores ellipsoid to ovoid, hardly constricted, chestnut-brown, $33–48 \times 22–30\,\mu$, wall thin, delicately verruculose, pore subapical in upper cell, in lower cell subequatorial, pedicels short, deciduous. Auteu-form.

On *Crepis paludosa*, aecidia in June, uredospores mixed with teleutospores in

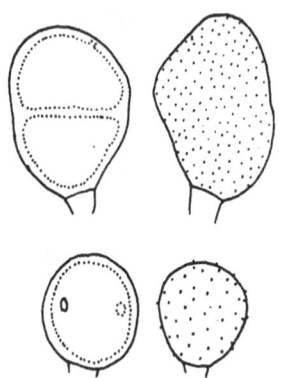

P. major. Teleutospores and uredospores (after Savulescu).

August–September. Northern England scarce, Scotland frequent; also in Wales and Ireland.

This species was formerly considered a large-spored variety of *P. lapsanae*. Dietel (*loc. cit.*) proved by cultures that the uredo- and teleutospores are connected with the aecidial generation. Hasler (Zbl. Bakt. II, **48**, 252, 1918) showed by infection experiments that this species is strictly confined to *Crepis paludosa*. It is distinguished from *P. crepidicola* by the size of the spores and by the limitation of the aecidial mycelium to certain parts of the leaf.

Puccinia punctiformis (Str.) Röhl.

Deutschl. Fl. Ed. 2, **3** (3), 132 (1813).

[*Uredo suaveolens* Pers., Syn. Meth. Fung. p. 221 (1801).]
Uredo punctiformis Str., Ann. Wetter. Ges. **2**, 103 (1810).
[*Hypodermium* (*Uredo*) *obtegens* Link, Mag. Ges. Naturf. Fr. Berlin, **7**, 27 (1816).]
[*Puccinia obtegens* Tul., Ann. Sci. Nat Bot. Ser. 4, **2**, 87 (1854)] [III not described];
     Grove, Brit. Rust Fungi, p. 145.

Puccinia obtegens Fuck., Jahrb. Nass. Ver. Nat. **23–24**, 54 (1869); Wilson & Bisby, Brit. Ured. no. 188.
Puccinia suaveolens Rostr., Forh. Skand. Naturf. Møde, Kjobenh. **11**, 339 (1874); Gäumann, Rostpilze, p. 1511.

Spermogonia chiefly hypophyllous, a few epiphyllous, crowded, covering the whole surface of the leaf, bright honey-yellow colour and with a pleasant smell. **Aecidia**, uredinoid, hypophyllous, scattered thickly over the whole surface of the leaf, minute, often confluent, reddish-brown, then darker. **Uredosori** more scattered, spores globoid to broadly ellipsoid, pale brown, 21–28 μ, wall echinulate with 3 irregularly placed pores. **Teleutosori** similar, always dark brown; teleutospores ovoid to ellipsoid, somewhat tapered below, hardly constricted, brown, 26–42 × 17–25 μ, wall thin, delicately verruculose, pedicels hyaline, thin, short. Brachy-form. On *Cirsium arvense*, April–Novem-

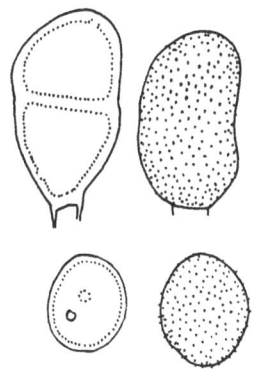

P. punctiformis. Teleutospores and uredospores.

ber. Very common throughout Great Britain and Ireland.

The life history of this common species is interesting. In spring the mycelium permeates the host in every part. The affected plants can be recognised immediately by their pale-green colour and spindly appearance; they never flower. The spermogonia are first seen towards the end of April, and are easily detected by their bright colour, and their strong sweet smell; the uredinoid aecidia occasionally mixed with a few teleutospores, follow on the same leaves during May. As a result of aecidiospore infection of other plants, uredospores and teleutospores develop in July in sori which are more scattered, never confluent, and darker brown: teleutospores are abundant from September to November. This generation is not accompanied by spermogonia. The mycelium of these sori is confined to distinct spots and the host does not suffer so severely.

The pores are very easy to see in the uredospores of this species and its allies. Each pore is made conspicuous by a thickened border.

Olive (Ann. Myc. **11**, 297, 1913) showed that the mycelium hibernates in the rhizome, as previously stated by Plowright. In the spring Olive found that the infected plants contain the haploid and the dicaryotic mycelia intermingled, both being systemic; the former produces the spermogonia, the latter the aecidia and rarely teleutospores; occasionally only the systemic dicaryotic mycelium is present and then spermogonia are not produced. The second generation of infected plants contains only the localised dicaryotic mycelium.

Kursanov (1922) found that a number of aecidiospores are produced on the uninucleate mycelium and that during their development a nuclear migration takes place similar to that described by Blackman (1904) in other species.

Bryzgalova (Morb. Plant. Leningrad, **17**, 101, 1928, Abs. in RAM. **8**, 790, 1929) found that in Siberia infected shoots died after the formation of teleutospores and that the latter hibernate in the dead tissue and germinate only in spring. Satisfactory control of the thistle by the rust cannot be expected owing to the production of new stolons which become only slowly infected.

Buller & Brown (Phytopath. **31**, 4, 1941) sowed aecidiospores on seedlings and produced uredospores; buds formed on roots of those infected grew up and the shoots produced spermogonia and aecidia, sometimes spermogonia only. In the latter cases when nectar from other infected plants was applied aecidiospores were formed at points of mixing, indicating that the rust is heterothallic.

Puccinia tanaceti DC.

Fl. Fr. **2**, 222 (1805); Grove, Brit. Rust Fungi, p. 135; Wilson & Bisby, Brit. Ured. no. 229; Gäumann, Rostpilze, p. 1144.

Uredo absinthii DC., Lam. Encycl. Méth. Bot. **8**, 245 (1808).
Puccinia absinthii (Hedw. f.) DC., Fl. Fr. **5**, 56 (1815); Grove, Brit. Rust Fungi, p. 134; Wilson & Bisby, Brit. Ured. no. 76; Gäumann, Rostpilze, p. 1131.
Puccinia artemisiella P. & H. Sydow, Monogr. Ured. **1**, 14 (1902); Gäumann, Rostpilze, p. 1133.

[**Spermogonia** epiphyllous, in groups. **Aecidia** chiefly hypophyllous, uredinoid, cinnamon-brown; aecidiospores ellipsoid, 27–31 × 17–24 μ, wall pale cinnamon-brown, 1–2 μ thick, sparsely echinulate, with 3 equatorial pores, each covered with a large, flat, smooth, hyaline cap.] **Uredosori** usually hypophyllous, cinnamon-brown; uredospores ellipsoid, 24–28 × 20–22 μ, wall 1·5–2 μ thick, finely echinulate with 3 equatorial pores. **Teleutosori** hypophyllous but sometimes amphigenous or on the stems where they are up to 2 mm. long, 0·5 mm. wide, similar to the uredosori but occasionally confluent, soon naked,

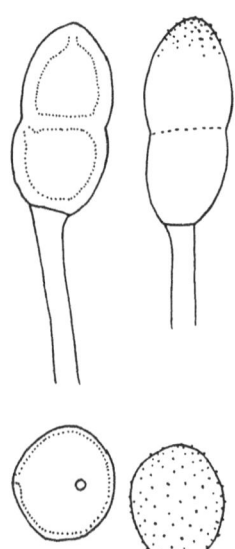

P. tanaceti. Teleutospores and uredospores.

dark brown or blackish; teleutospores ellipsoid to clavoid, constricted, slightly attenuate below, wall of the upper cell finely verruculose, 3–7μ thick at apex, of the lower frequently smooth, especially at the base, 38–62 × 16–27μ, pedicels hyaline, persistent, up to 120μ long and 10μ thick. Brachy-form.

On *Artemisia absinthium*, *A. maritima*, *A. vulgaris*, *Chrysanthemum coccineum* and *C. vulgare*, July–September. England, Wales, southern Scotland, scarce.

The rusts on *Artemisia* and *Chrysanthemum* which were generally united by early writers, including Plowright, were separated by Fischer, the Sydows, Grove, Wilson & Bisby and others but are here united as there are not clear morphological differences between them.

Spermogonia and aecidia have not been discovered in this country but have been found in the United States. Arthur (Mycol. **1**, 243, 1909, and **4**, 21, 1912) showed that at least in America this species is a brachy-form by sowing teleutospores from *Artemisia dracunculoides* Pursh. on the same host with the production of spermogonia followed directly by uredinoid aecidia.

The rust on *A. maritima* has been regarded by Fahrendorff (Ann. Myc. **39**, 158, 1941) as a distinct species, *Puccinia artemisiae-maritimae*, on account of differences in the teleutospore, the wall being thicker (2–3μ as against 1·5–2μ), the warts longer and the apical pore very distinct but we have not observed this distinction in collections on this host in Britain. Nor have we detected British material which might be segregated as *P. artemisiella* Syd. described on *A. vulgaris* and said to differ from *P. absinthii* in its shorter and narrower teleutospores (39–43 × 18–20μ as against 46–53 × 23–24μ in the latter). The teleutospores of the few British collections on *C. vulgare* are, on the average, slightly shorter and narrower with longer pedicels.

The rust on *Artemisia maritima* frequently occurs in England especially in the south but is scarce in Scotland where uredospores are rarely produced. It appears to be rare on *A. vulgaris*. It is scarce on *Chrysanthemum* (*Tanacetum*) *vulgare* in England and appears to have been only once recorded for Scotland (Scot. Nat. ser. 3, p. 31, 1891) on this species.

Treboux (Ann. Myc. **10**, 306, 1912) has found that in Russia, this rust can overwinter by means of uredospores on both *A. absinthium* and *A. maritima*.

Puccinia lagenophorae Cooke

Grev. Nea. **13**, 6 (1884).
Puccinia terrieriana Mayor, Ber. Schweiz. Bot. Ges. **72**, 266 (1962).

Spermogonia absent. **Aecidia** amphigenous in large orange groups on the stems and leaves, later surrounding

teleutosori on the stems, cupulate, with poorly developed peridia; aecidiospores subgloboid, 10–16μ diam., wall hyaline, finely verruculose, less than 1μ thick, contents orange. **Teleutosori** sparse on the stems, usually associated with aecidial groups, 1–2 mm., pulvinate, dark brown long-covered by the epidermis; teleutospores bluntly clavate to broadly ellipsoid, 32–41 × 12–18μ, wall smooth but for 1 or 2 longitudinal, raised ridges and occasional transverse ridges on some spores, dark brown, 1·5–2μ thick and 5–8μ thick at the apex, pore of upper cell apical, of lower cell superior, pedicel persistent, hyaline, 20–40 × 4–5μ. Some sori have up to 20% 1-celled teleutospores, 20–30 × 12–15μ. Opsis-form.

Aecidiospores and teleutospores on *Senecio squalidus* and *S. vulgaris* and by inoculation on *Bellis perennis, Calendula officinalis, S. cruentus* and *S. viscosus.*

Western Britain north to Ayrshire, Eire, Scilly Isles, known since 1961 and still spreading.

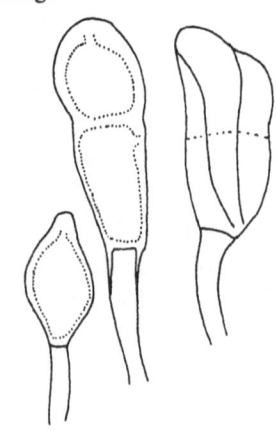

P. lagenophorae. Teleutospores.

This species, a native of Australia, was unknown in Europe until its discovery in 1961. It appeared in that year in central France and was collected at Dungeness by Dennis. In 1962 it was found in south-western England from Southampton to the Lizard, Wales and in the Scilly Isles (Wilson & Manners, *in litt.*)

Attack by the aecidial stage is heavy and conspicuous and it seems very unlikely that the rust had been overlooked previously. Teleutospores occur in the centre of aecidial groups on the stems and are rather inconspicuous. Mayor (*loc. cit.*) described the early French collections as a new species but there is little doubt from the investigation of Wilson (Nature, **494**, 383, 1963) and her collaborators that it is the same as the rusts which in Australian occurs on introduced *Senecio vulgaris* and several other species of *Senecio, Lagenophora* and *Erechtites.*

Puccinia variabilis Grev.

Scot. Crypt. Fl. **2**, pl. 75 (1824); Grove, Brit. Rust Fungi, p. 152; Wilson & Bisby Brit. Ured. no. 239; Gäumann, Rostpilze, p. 1124.

[*Aecidium taraxaci* Grev., Fl. Edin. p. 444 (1824).]
[*Aecidium grevillei* Grove, Jour. Bot. Lond. **23**, 129 (1885).]

Spermogonia absent? **Aecidia** amphigenous, on indeterminate, yellow or purplish spots, solitary or a few loosely aggregated, cup-shaped with white, laciniate peridium, peridial cells not in distinct rows, on the outer side project-ing over the cells below, apparently soon falling apart, outer walls thin, the inner walls up to 7μ thick, with small, rather close warts; aecidiospores subgloboid or ovoid with orange contents, verruculose, 20–25 × 15–20μ. **Uredosori** amphigenous, on very minute, yellow or purplish spots, scattered, minute, punctiform, soon naked, brown; uredospores not abundant, globoid or ovoid, or of irregular form, wall echinulate, brown, 22–32 × 19–26 μ with 2 equatorial pores. **Teleutosori** similar but darker, teleutospores ellipsoid, ovoid or almost spherical, often distorted in various ways, not constricted, brown, 28–40 × 18–25 μ; wall finely verruculose, pores of upper and lower cells subapical, pedicels hyaline, about as long as the spore, deciduous. Auteu-form.

On *Taraxacum officinale* agg. and *T. palustre* agg., July–October. Great Britain, frequent.

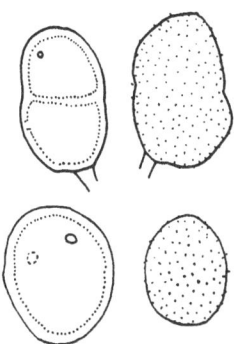

P. variabilis. Teleutospores and uredospores.

The presence of spermogonia, although mentioned by Fischer (1904) and Grove (*loc. cit.*) seems very doubtful; they have not been found in Britain and Jørstad (1960) has not found them in Norway.

Plowright and Soppitt both proved, by laying leaves affected with the aecidium of this species on healthy plants of *Taraxacum*, that the uredo- and teleutospores were produced in about 14 days. In July the three spore-forms may be found on the same leaf. The Sydows described uredosori but uredospores are usually scanty and are often produced in the teleutosori; Jørstad (1951, 52) stated that in Iceland uredospores occur only in teleutosori.

Two aecidia are to be found on *Taraxacum* in Britain; those of *Puccinia variabilis* (*Aecidium grevillei* Grove = *A. taraxaci* Grev. non K. & S.) spreads uniformly over the whole leaf in 'numerous little clusters with single ones scattered between them', as Greville describes it (*loc. cit.* p. 444)—the other of *P. dioicae* (*Aecidium taraxaci* K. & S.) form large round clusters. These differences are not sufficient to separate the two, however. Fischer pointed out many years ago that the aecidia differ in the form of their peridial cells, those of *P. variabilis* having the inner wall thickened, whereas those of *P. dioicae* have the outer wall more strongly thickened. In addition spermogonia accompany aecidia on *P. dioicae* but are lacking in *P. variabilis*.

Greville (Scot. Crypt. Fl. p. 75) figured a number of abnormal teleutospores in which either one or both of the cells is divided by a vertical septum.

Puccinia virgae-aureae (DC.) Lib.

Pl. Crypt. Arduenn. 4, 393 (1837); Grove, Brit. Rust Fungi, p. 130; Wilson & Bisby, Brit. Ured. no. 244; Gäumann, Rostpilze, p. 542.

Xyloma virgae-aureae DC., Lam. & DC., Syn. Pl. Gall. p. 63 (1806).
Puccinia clandestina Carm., Berk. Eng. Fl. 5 (2), 365 (1836); Grove, Brit. Rust Fungi, p. 390.

Spermogonia, aecidia and **uredosori** wanting. **Teleutosori** hypophyllous, minute, crowded in dendritic clusters, on round, yellowish, purple or black-centred spots, compact, shining, black; teleutospores ellipsoid, clavoid or fusoid, hardly constricted, wall smooth, up to 12μ thick at the apex which is paler than the rest of the wall, 30–56 × 12–20μ, pedicels somewhat hyaline, half as long as the spore. Micro-form.

On *Solidago virgaurea*, August–September. England and Scotland, especially in the west, but scarce.

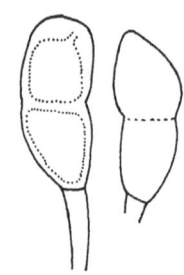

P. virgae-aureae. Teleutospores.

Although this species is most frequent on the west coast particularly of Scotland it has been recorded from the east by Trail (1890, 317) and at Shere, Surrey in 1865 (in Herb. Kew). In western England it has been found in Devon and Somerset by Hadden (J. Bot. Lond. **54**, 52, 1916, and *idem*, **58**, 37, 1920) and in Cornwall by Rilstone (J. Bot. Lond. **76**, 354, 1938, see below). The groups of sori and their discoloured spots occasionally extend to the midrib of the leaf suggesting an origin from a systemic mycelium. However, the petioles of such leaves do not show any mycelium and infected plants overwintered in pots in Edinburgh gave no evidence of a perennial mycelium.

The development of the soral groups proceeds centrifugally and even after the spores in the central sori are dark considerable growth takes place at the margin so that sections through the groups show sori and teleutospores in all stages of development. Each sorus is surrounded by a palisade of dark brown hyphae, which according to Arthur (Bot. Gaz. **39**, 219, 1905) are not true paraphyses but modified hyphae of the general mycelium. The sori remain long-covered by the epidermis and have a superficial resemblance to a *Dothidea* or *Asteroma*; *A. atratum* Chev. and *D. solidaginis* var. *virgaureae* Fr. are additional synonyms for this rust.

An examination of the type specimen of *Puccinia clandestina* Carm. (see Grove, *loc. cit.*) in the Kew Herbarium showed that this rust is identical with *P. virgae-aureae*. Carmichael considered that the host of *P. clandestina* was *Succisa pratensis* but in this he was mistaken; the latter and *Solidago virgaurea* closely resemble one another before flowering and grow together in the locality (Port Appin, Argyllshire) where Carmichael collected his specimen of *P. clandestina*. *P. virgae-aureae* has been found in this locality.

Rilstone (*loc. cit.*) recorded *P. clandestina* from Cornwall on *Succisa pratensis*; examination of the specimen collected in June 1929, and now in the Plowright Herbarium at the British Museum has shown it to be an early stage of development of *P. virgae-aureae* on *Solidago virgaurea*.

Dietel (Zbl. Bakt. II, **48**, 492, 1918) and Arthur (1934, 202) considered that this species is correlated with *Puccinis extensicola solidaginis* but, as pointed out by Gäumann (Ber. Schweiz. Bot. Ges. **49**, 168, 1939) this is an error, the teleutospores of the two species being quite dissimilar. Jørstad (Nytt Mag. Naturv. **86**, 1, 1948) discussed differences between this and several allied species and the existence of a possible correlated hetereu-form.

Corda (Icon. Fung. 1840, 14) gave an exceptionally good early account of the general structure of this rust.

Puccinia allii Rud.

Linnaea, **4**, 392 (1829); Gäumann, Rostpilze, p. 435.
[*Uredo porri* Sow., Engl. Fungi, pl. 411 (1810).]
[*Uredo alliorum* DC., Fl. Fr. **6**, 82 (1815).]
Puccinia mixta Fuck. Fungi Rhen. no. 377 (1863).
Puccinia porri [Sow.] Wint., Rabh. Krypt. Fl. Ed. 2, **1** (1), 200 (1882); Gäumann, Rostpilze, p. 432.

[**Spermogonia** amphigenous, among the aecidia. **Aecidia** amphigenous, on pale spots, circinate or in rounded or elongate groups, peridium shortly cylindrical, white, with torn revolute margin; aecidiospores globoid, 19–28 μ diam., wall yellow, verruculose, 1–2 μ thick.] **Uredosori** amphigenous, on indeterminate yellow spots scattered or more or less in rows, small, at first covered by the swollen epidermis, yellowish or reddish-yellow; uredospores globoid to ellipsoid, yellowish, 23–29 × 20–24 μ, wall 1–2 μ thick, very delicately echinulate, with 5–10 pores. **Teleutosori** amphigenous or caulicolous, scattered, small, on the stems confluent, larger,

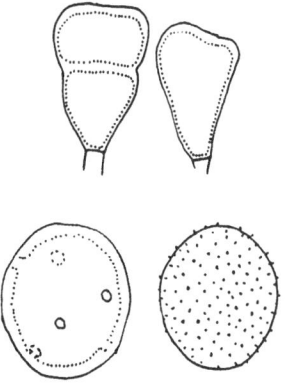

P. allii. Teleutospore, mesospore (after Fischer) and uredospores.

long-covered by the lead-coloured epidermis, black-brown, with few or many paraphyses; teleutospores ellipsoid to obovoid, slightly constricted, 28–45 × 20–26 μ, wall smooth, brown, 1–2 μ thick, slightly or not thicker above, pedicels nearly colourless, fragile, short; mesospores often very numerous, subgloboid, to ovoid, 22–36 × 15–23 μ, wall as in teleutospores, pedicels up to 30 μ long. Auteu-form.

Uredospores and teleutospores on *Allium cepa*, *A. porrum* (II only) *A. cepa × porrum* (II), *A. schoenoprasum*, *A. vineale*, *A. vineale* var. *compactum* and imported *A. sativum*. Great Britain, scarce.

The three autoecious rusts often recognised on *Allium* in western Europe, *Puccinia alli*, *P. mixta* (the valid name for *P. porri* [Sow.] Wint.) and *Uromyces ambiguus* form a series from rather strongly stromatic types with brown teleutosoral paraphyses and few unicellular teleutospores (*P. allii*) to pulverulent superficial teleutosori containing all unicellular teleutospores (*U. ambiguus*), with *P. mixta* holding a somewhat intermediate position. *Puccinia mixta* certainly merges with *P. allii* and is here treated under the latter name.

In Britain the rust on *Allium vineale* is quite frequent and is invariably strongly stromatic. On leek only uredosori have been found. In southern England it is common on chives and rather infrequent on onions (Moore, 1959). The records of '*P. porri*' on *A. schoenoprasum* from Orkney (Wilson, 1934, 391) are in error, the specimens bear *U. ambiguus*. Corsican collections of the rust on *A. sativum* have been treated as *P. blasdalei* by Mayor and Viennot-Bourgin (Rev. Myc. **15**, 80, 1950).

Aecidia have not been found in this country; they were obtained with the race *P. allii* in the restricted sense by infection on *A. schoenoprasum* by Schneider (Zbl. Bakt. II, **32**, 452, 1912) and by Sydow (Ann. Myc. **36**, 320, 1938). Von Tavel (Ber. Schweiz. Bot. Ges. **41**, 123, 1932) also obtained them with rudimentary spermogonia on *A. schoenoprasum*, *A. flavum* and *A. fistulosum*; Lind (1913) found them on the last species in Denmark; spermogonia and aecidia are also recorded in North America (Arthur 1934, 221). Tranzschel (Ann. Myc. **8**, 415, 1910) claimed to have sown basidiospores from *A. schoenoprasum* upon the same host and obtained uredosori without the appearance of spermogonia or aecidia. It is known, however, that uredospores are often found in the teleutosori and the possibility of the presence of uredospores in the material used for infection was not excluded. Jørstad (1932, 336) has suggested that some of the forms of *P. mixta* (*sens. lat.*) are in a labile condition and that the spore forms of the gametophytic stage may be represented by either spermogonia and aecidia (or aecidia alone) or by uredinoid aecidia.

Puccinia asparagi DC.

Fl. Fr. **2**, 595 (1805); Grove, Brit. Rust Fungi, p. 233; Wilson & Bisby, Brit. Ured. no. 98; Gäumann, Rostpilze, p. 594.

Spermogonia caulicolous, honey-coloured, in small groups. **Aecidia** caulicolous, scattered, often in elongate groups, long closed, then shortly cupulate with a whitish, erect laciniate peridium, peridial cells oblong, in section outer wall striate, 8μ thick with punctate surface, inner wall 3μ thick, verrucose; aecidiospores globoid, finely verruculose, 15–18μ diam., wall colourless, 1μ thick, contents orange-red. **Uredosori** caulicolous, oblong, narrow, long-covered by the epidermis, cinnamon-brown; uredospores globoid or ellipsoid, 20–30 × 17–25μ, wall golden-brown, 1·5–2μ thick, echinulate with usually 4 equatorial pores. **Teleutosori** caulicolous, rarely on the phylloclades, oblong or linear, often in groups which are more or less confluent and up to 0·5 mm. or more long, blackish-brown; teleutospores oblong, ellipsoid or clavoid, slightly constricted, yellowish-brown, lower cell longer than upper with slightly thinner and paler wall, or both cells similar in length and width, 35–42 × 17–26μ; mesospores often numerous, 29–35 × 19–23μ, wall smooth, up to 8μ thick at the apex, pores rather indistinct, of upper cell apical, of lower sub-

apical, pedicel one-half to twice the length of the spore. Auteu-form.

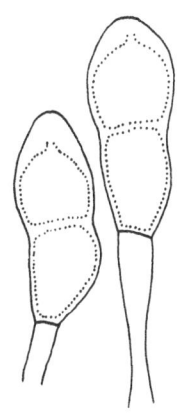

P. asparagi. Teleutospores (after Fischer).

On native and cultivated *Asparagus officinalis*, spermogonia and aecidia, May; uredospores and teleutospores, September–December. England, scarce, Scotland, very rare.

The first British record of this rust was by Greville near Edinburgh in 1824 (Fl. Edin. 1824, 429) and this has remained the only record from Scotland. It was first found in England at Swanscombe, Kent, in 1865 and since that time it has appeared sporadically, particularly in East Anglia. A map showing its distribution up to December 1946 was published by Moore (Bull. Min. Agric. London, **139**, 42, 1948).

Ogilvie (*in litt.*) in Somerset produced spermogonia and aecidia on forced plants of asparagus by infection with overwintered teleutospores in March 1939.

Ogilvie & Croxall (*in litt.*) also artificially infected the Egyptian tree onion, *Allium cepa* var. *aggregatum*, with teleutospores of the asparagus rust producing spermogonia and aecidia upon it. Walker (Phytopath. **11**, 87, 1921) had previously carried out a similar infection in California.

Ogilvie, Croxall & Hickmann (Rep. Agric. Hort. Res. Sta. Long Ashton, Bristol, 1938, 91) investigated the viability of the spores; uredospores remain viable on cut shoots for 2 weeks, the teleutospores on the stems become capable of germination from December onwards;

the time of germination appears to be connected with the time at which they are formed and also on the weather conditions during the winter. The rust was introduced into the United States in 1896 and into Norway in 1937.

Puccinia liliacearum Duby

Bot. Gall. **2**, 891 (1830); Grove, Brit. Rust Fungi, p. 234; Wilson & Bisby, Brit. Ured. no. 178; Gäumann, Rostpilze, p. 856.

Spermogonia numerous, especially at the apex of infected leaves, yellowish, conical. **Aecidia** and **uredosori** wanting. Teleutosori amphigenous, embedded in yellowish parts of the leaf, densely crowded, often confluent and forming rings round the groups of spermogonia, long-covered by the ash-coloured epidermis, at length naked, pulverulent, reddish-brown; teleutospores ellipsoid, apex not thickened, slightly or not constricted at the septum, attenuate below, 40–75 × 22–35 μ, wall equally 2 μ thick, smooth, pale brown, pores not obvious, pedicel thick, hyaline, rather long. Micro-form.

On *Ornithogalum pyrenaicum* and *O. umbellatum*, March–May. England and Scotland, scarce.

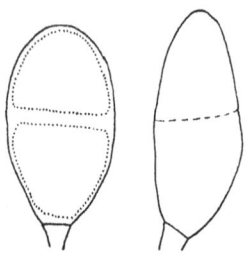

P. liliacearum. Teleutospores.

Plowright, the Sydows and Grove described aecidia in this rust but these (*A. ornithogalum* Bubak) are now known to belong to *P. hordei* (p. 264).

The infected parts of the leaves are swollen, compact and harder than the healthy parts; infected plants do not usually flower.

Fischer (Zbl. Bakt. II, **15**, 230 and **17**, 206, 1906) studied the specialisation of this rust; he also found 3- and 4-celled teleutospores. He considered that reinfection takes place each year by basidiospores produced from teleutospores on dead leaves lying on the ground.

Its cytology was investigated by Sappin-Trouffy (1897, 33 and 111), and by Maire (Bull. Soc. Myc. Fr. **18**, 33, 1902).

Puccinia prostii Duby

Bot. Gall. **2**, 891 (1830); Wilson & Bisby, Brit. Ured. no. 206; Gäumann, Rostpilze, p. 860.

Spermogonia amphigenous, numerous, amongst the teleutosori, yellowish-brown, 120–140 μ diam., spermatia 10

× 5 μ. **Teleutosori** amphigenous, at first forming yellow, oval spots, which later become grey and finally open by a longitudinal slit, surrounded by the ruptured epidermis, dark brown or black; teleutospores ellipsoid, scarcely constricted, brown, 56–62 × 17–19 μ, wall 2–3 μ thick, covered with acute, hyaline spines about 6 μ long, pedicels variable, up to 80 μ long, hyaline, deciduous. Micro-form. On *Tulipa sylvestris* Scotland, and cultivated *Tulipa* sp. England, very rare.

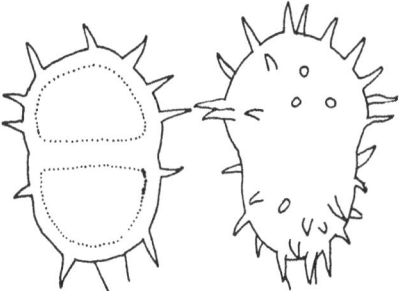

P. prostii. Teleutospores (after Savulescu).

This species was recorded in England by Massee (Mildews, Rusts and Smuts, 1913, 139) on cultivated tulips and was recorded in the Royal Botanic Garden, Edinburgh, in 1914 on *Tulipa sylvestris* (Notes R. B. G. Edin. 1914, 219; and J. Bot. Lond. **53**, 43, 1915) on which it had probably been introduced from Italy. Although it has been present in the Botanic Garden since 1914 it has not spread to cultivated tulips in its vicinity; there is probably a systemic mycelium and, as all attempts to germinate the teleutospores have failed, it appears that the spread of the fungus has taken place only by the vegetative propagation of the host plant. The species is striking on account of its very large, spiny teleutospores.

The mycelium is uninucleate and the dicaryon is formed by hyphal fusion according to Lamb (Trans. Roy. Soc. Edin. **58**, 143, 1934) who described the cytological development in detail.

Puccinia cancellata Sacc.

Rev. Myc. **9**, 26 (1881); Wilson & Bisby, Brit. Ured. no. 115; Gäumann, Rostpilze, p. 866.

[*Uredo cancellata* Dur. & Mont., Fl. Alger. **1**, 314 (1846).]

Spermogonia and aecidia unknown. Uredosori rather large, elongate, often surrounding the stem, dehiscing with a longitudinal slit, often confluent, pulverulent, chestnut-brown, containing both uredospores and teleutospores; uredospores globoid or ellipsoid, pale brown, 22–34 μ, wall 3–5 μ thick, verruculose-echinulate; teleutospores ellipsoid or oblong, truncate at base and apex, constricted at the septum, pale brown, 35–46 × 19–28 μ, wall smooth, 3–4 μ thick, slightly thicker at the apex, pedicel flexuous, slender, hyaline, often inserted obliquely, deciduous, up to 50 μ long. Hetereu-form?

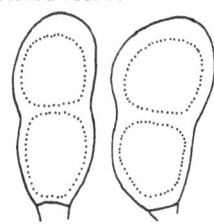

P. cancellata. Teleutospores (after Savulescu).

On *Juncus acutus*. England, Wales, Channel Islands, rare.

There is a specimen, probably collected by Rhodes in 1931, at Harlech, as *Uromyces junci* in the Grove Herbarium. This species was recorded by E. A. Ellis, from Vazon Bay, Guernsey, and from Herm in the Channel Islands, in 1939; it has also been found at Braunton, Devon.

Puccinia luzulae Lib.

Pl. Crypt. Arduenn. p. 94 (1830). Gäumann, Rostpilze, p. 610.

[*Caeoma oblongata* Link, Mag. Ges. Naturf. Fr. Berlin, p. 27 (1816).]
Puccinia oblongata Wint., Rabh. Krypt. Fl. Ed. 2, **1** (1), 183 (1882); Grove, Brit. Rust Fungi, p. 238; Wilson & Bisby, Brit. Ured. no. 186.

Spermogonia and **aecidia** unknown. **Uredosori** amphigenous, on irregular and confluent reddish-brown or blackish-brown spots, scattered, oblong, long-covered by the epidermis, ferruginous; uredospores ellipsoid-ovoid to pyriform or clavoid, contents orange, 30–44 × 12–15 μ, wall 3–5 μ thick, hyaline, smooth, rarely aculeate at the summit, with 4 indistinct equatorial pores. **Teleutosori** similar, soon naked, compact, blackish-brown; teleutospores clavoid, slightly constricted tapering downwards, 44–80 × 16–24 μ wall smooth, brown, 10–25 μ thick at apex, pedicels hyaline, persistent, equalling the spore or shorter. Hetereu-form?

On *Luzula pilosa*, uredospores May–July; teleutospores September–Novem-ber. Great Britain and Ireland, uncommon.

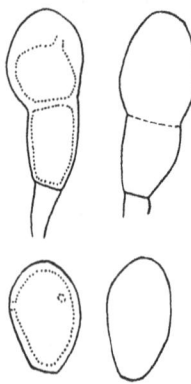

P. luzulae. Teleutospores and uredospores.

As pointed out by Wilson & Bisby this species has been often confused with *Puccinia obscura*; there is no evidence of its occurrence on *Luzula campestris*. It appears to have a more northern distribution than *P. obscura* from which it is distinguished by its longer, narrower, light-coloured, nearly smooth, rather thick-walled uredospores with 4 indistinct, equatorial pores and by its slightly larger teleutospores with the great thickening of the apical wall of the upper cell; there is no evidence for Grove's opinion (*loc. cit.*) that *P. luzulae* is merely an abnormal development of *P. obscura*.

No aecidial stage is known and there appears to be no evidence for the suggestion made by Gäumann (*loc. cit.*) that the aecidial stage may be on a member of the *Compositae*; Tranzschel (Trav. Mus. Bot. Acad. Imp. Sci. St Petersburg, **2**, 79, 1905, and **7**, 19, 1910) and Ito (Bot. Mag. Tokyo, **48**, 536, 1934) have failed to infect *Bellis* and several other genera.

In this species in northern Europe, teleutospores are commonly developed on the basal leaves of its host; but as the uredo-stage is able to pass the winter on living basal leaves it is independent of host alternation.

Puccinia obscura Schroet.

Just's Bot. Jahresber. **5**, 162 (1879) (in N. Giorn. Bot. Ital. **9**, 256 (1877) *nomen nudum*); Grove, Brit. Rust Fungi, p. 236; Wilson & Bisby, Brit. Ured. no. 187; Gäumann, Rostpilze, p. 605.
Puccinia luzulae-maximae Diet., Ann. Myc. **17**, 57 (1919); Gäumann, Rostpilze p. 609.

Spermogonia amphigenous, minute, in dense roundish clusters, honey-coloured. **Aecidia** amphigenous and on the petioles, on roundish or irregular yellow spots in loose clusters or scattered, peridium cup-shaped or cylindrical with torn margin; aecidiospores globoid to angular, finely verruculose, yellowish, 16–22 μ diam. **Uredosori** generally hypophyllous, on irregular confluent purplish-brown spots, scattered, elliptical or linear, long-covered by the epidermis, pulverulent rusty-yellow; uredospores ellipsoid to ovoid, 18–26 × 15–22 μ, wall rather thick, echinulate, pale to cinnamon-brown with 2 supra-equatorial pores often smooth below the pores, contents almost colourless. **Teleutosori** similar, but compact, pulvinate, covered or surrounded by the cleft epidermis, blackish-brown; teleutospores broadly ellipsoid, rounded rarely truncate or conical above, slightly constricted, attenuate below, 30–48 × 14–20 μ, wall smooth, brown, 4–5 μ thick at apex, pedicels subhyaline, persistent, up to 30 μ long; mesospores frequently present. Hetereuform.

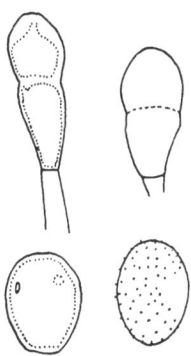

P. obscura. Teleutospores and uredospores.

Spermogonia and aecidia on *Bellis perennis*, September–December; uredospores and teleutospores on *Luzula campestris*, *L. forsteri*, *L. multiflora*, *L. pilosa* and *L. sylvatica*, June–November. Great Britain and Ireland, frequent.

On *Luzula forsteri* it has occurred in Cornwall (Rilstone, 1938) and in Sussex (P. H. Davis, *in litt.*).

The rust on *L. sylvatica* was previously regarded as a distinct species, *Puccinia luzulae-maximae*, by Dietel (Ann. Myc. **17**, 57, 1919) on account of its darker coloured uredospores. It appears to be a special form which does not attack any other species of *Luzula*.

The aecidial stage has not been recorded frequently, possibly because it is rather inconspicuous and develops in the autumn and winter.

It has been noted by several observers (Bubak, 1908; Jørstad, 1940;

and others) that the rust can overwinter by means of the uredo mycelium. This, perhaps, partly accounts for the observations made by Vestergren, and the Sydows (Monogr. Ured. **1**, 898) that they have been unable to find the aecidia on *Bellis perennis* in localities where the rust is abundant on *Luzula*.

Teleutospores are rarely produced but have been found on dead leaves usually in sites where alternation with daisies takes place. They germinate after a short resting period, so the spermogonia usually develop in September or October appearing on yellow spots and are then slowly followed by the aecidia.

The heteroecism of this rust was first demonstrated by Plowright in 1883–4 and further cultures were made in 1887. Maire (Bull. Soc. Hist. Nat. Afrique Nord. **8**, 150, 1917) showed that in North Africa aecidiospores from *B. silvestris* gave infection on *L. graeca*.

Gäumann (Angew. Bot. **12**, 290, 1937) made many measurements and cultures of the uredospores; his figures agree closely with those of Dietel (*loc. cit.*) but he could not correlate uredospore size with host range. He showed that five specialised forms of this species are found in Switzerland; of these his f. sp. *campestris* on *L. campestris, L. forsteri, L. spicata* and occasionally *L. pilosa*, f. sp. *multiflorae* chiefly on *L. multiflora* and f. sp. *pilosae* on *L. pilosa* and occasionally *L. campestris* may be expected to occur in Britain.

Puccinia schroeteri Pass.

N. Giorn. Bot. Ital. **7**, 255 (1875); Grove, Brit. Rust Fungi, p. 232; Wilson & Bisby, Brit. Ured. no. 218; Gäumann, Rostpilze, p. 862.

Teleutosori epiphyllous, large, oblong or elliptic, surrounded by a brownish-violet discoloration, 1–3 mm. long, solitary or in small clusters, long-covered or half-uncovered and surrounded by the lead-coloured epidermis, blackish-brown; teleutospores ellipsoid or oblong, rounded at both ends, hardly constricted, 40–60 × 25–29 μ, wall reticulate, golden-brown, then chestnut, pedicels hyaline, short, thick, deciduous; mesospores occasional. Micro-form.

On cultivated *Narcissus jonquilla*, *N. majalis* and *N. pseudonarcissus*. England very rare.

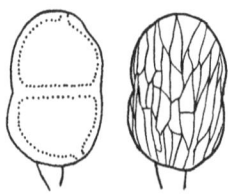

P. schroeteri. Teleutospores.

This species was found at Malpas, Cheshire, in May 1889 by Wolley-Dod (J. Roy. Hort. Soc. **12**, liii, 1890) and W. G. Smith (Gard. Chron. III, **5**, 725, 1889) on *N. majalis* and *N. jonquilla* and on *N. pseudonarcissus* at Kew, 1947–9. Specimens 'on daffodil leaves' in the British Museum Herbarium, labelled *P. liliaceanum* by Cooke in 1894, proved to be this species.

Plowright (TBMS. **1**, 57, 1898) observed that the spores would not germinate at once, but, by securing the infected leaves during the winter near some plants of *N. majalis* he found infection the next year and for 8 or 9 years afterwards, though only on the tips of the leaves. Mesospores and other abnormal spores were recorded by Fischer (1904, 78). Schneider (Zbl. Bakt. II, **72**, 261, 1927) in Switzerland showed that the rust on '*N. angustifolius*' (= *N. poeticus*?) will infect *N. pseudonarcissus*.

Puccinia gladioli (Duby) Cast.

Obs. Pl. Acotyl. **2**, 17 (1843); Wilson & Bisby, Brit. Ured. no. 157; Gäumann, Rostpilze, p. 438.

Uredo gladioli Duby, Bot. Gall. **2**, 901 (1830).

[**Spermogonia** generally hypophyllous, yellow at length blackish, globose, paraphysate. **Aecidia** from a systemic mycelium, generally hypophyllous and spread uniformly over the leaf surface, occasionally epiphyllous and isolated, peridium cylindric or somewhat cupulate, margin laciniate, 250–300μ diam., peridial cells rhomboid, 20–34μ long, 17–24μ wide with external wall transversely striate, 3–6μ thick and internal wall verrucose, 2–3·5μ thick; aecidiospores subgloboid, angular-globoid or ellipsoid, 17–24 × 13–18μ, wall minutely verrucose, 1·5–3μ thick.] **Uredosori** wanting. **Teleutosori** amphigenous on oblong, reddish-brown spots, limited by the veins, minute, rounded densely-crowded or confluent and forming a crust up to 1 cm. long and generally covering the whole leaf, compact, black; teleutospores ellipsoid to clavoid, apex rather acute, slightly constricted at the septum, narrowed below, 36–56 × 16–22μ, wall smooth, pale brown, up to 8μ thick at the apex, pedicels hyaline, persistent, 10–60μ long; paraphyses linear, or

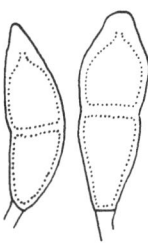

P. gladioli. Teleutospores (after Viennot-Bourgin).

slightly clavate, brown, up to 80μ long; mesospores sometimes present. Opsisform.

[Spermogonia and aecidia on *Valeriana* sp.], teleutospores on cultivated *Gladiolus*. England, very rare.

This rust has been found only once in this country, in November 1924 on cultivated *Gladiolus* in Cornwall as recorded by Beaumont (Rep. Dep. Pl. Path. Seale Hayne Agric. Coll. **1933–34**, 55, 1935). The aecidial stage has not been found in Britain but occurs in southern Europe, south-east Asia, North Africa and North America.

The connection of the spore stages was proved by d'Oliviera (Nature, **164**, 239, 1949) in Portugal in 1946; innoculation with aecidiospores from *Aecidium valerianellae* produced infection on *Gladiolus* after 10 days; he also proved that a number of species of *Valerianella* were susceptible to attack by the rust. The aecidial mycelium is systemic during the winter and early spring but localised during the summer. He also showed that the rust is heterothallic.

The heteroecism of the rust was confirmed by Dupias (Bull. Soc. Myc. Fr. **69**, 224, 1953) in southern France by infecting *G. segetum* with aecidiospores from *V. rimosa*; he noted that infected plants of *Valerianella* usually do not flower. Several native British species of *Valerianella* are listed by Gäumann (1959) as aecidial hosts.

The black crust-like growth on *Gladiolus* has been figured by Dietel (Jahresb. Ver. Naturk. Zwick. 29, 1928) and by Dupias (*loc. cit.*); it was described by Arthur (1929, 6 and 150) and others as a compound teleutosorus, consisting of groups of teleutospores, each group developing under a stomatal opening and each surrounded by much modified, subepidermal paraphyses.

The name '*Uredo gladioli* Requien in herb. DC.' validly published by Duby, was probably based on two specimens of infected *Gladiolus* leaves in the De Candolle Herbarium now in the Delessert Herbarium in Geneva, the original label attached to the 'specimens bearing the words "*Xyloma gladioli nob*"' with the addition by De Candolle 'Mr. Requien, 1819'. These specimens have been examined by Ainsworth (TBMS. **32**, 255, 1949) who found that they bear teleutospores of *P. gladioli*.

Puccinia iridis Rabh.

Deutschl. Krypt.-Fl. **1**, 23 (1844); Grove, Brit. Rust Fungi, p. 230; Wilson & Bisby, Brit. Ured. no. 172; Gäumann, Rostpilze, p. 599.

[*Uredo iridis* DC., Lam. Encycl. Méth. Bot. **8**, 224 (1808).]

[**Spermogonia** chiefly epiphyllous, in small groups, spherical, subepidermal, sunk in tissues of leaf, 80–140 μ diam. with projecting ostiolar paraphyses. **Aecidia** hypophyllous in small groups on non-thickened leaf tissue, peridium cup-shaped, white with laciniate margin, peridial cells comparatively loose but in distinct rows, quadratic or angular, 16–25 μ long, 16–21 μ broad, outer walls 1–2 μ thick, with tesselate structure, inner walls 2–3 μ thick with coarse, very regu-lar, tesselate structure; aecidiospores in distinct chains, subgloboid or poly-hedroid, 15–20 × 12–16 μ, wall colourless, scarcely 1 μ thick, finely verruculose.] **Uredosori** amphigenous, solitary or in groups, rounded or elongate, minute, long-covered by the epidermis, not pul-verulent, reddish-brown; uredospores globoid to ovoid, ochraceous-brown, 24–31 × 19–24 μ, wall echinulate, 2–3·5 μ thick with 2 or 3 equatorial pores. **Tele-utosori** hypophyllous, irregularly scattered, sometimes confluent, linear or oblong, compact, soon naked, black; teleutospores clavoid or broadly ellipsoid, apex rounded, truncate, or acuminate, slightly constricted at the septum, base generally attenuate 30–52×14–22μ, wall smooth, pale brown becoming darker above, up to 14μ thick at apex, pore of upper cell apical, of lower, superior, pedicels brownish, thin-walled, persistent. Auteu-form. [Spermogonia and aecidia on *Urtica dioica* and *U. urens*]; uredospores and teleutospores on *Iris foetidissima*, '*I. pseudacorus*' and *I. germanica*, *I. tolmeiana* and other cultivated species. Great Britain and Ireland, scarce.

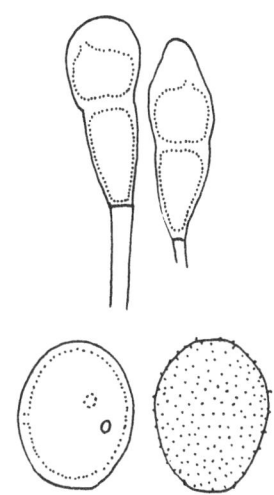

P. iridis. Teleutospores (after Fischer) and uredospores.

No aecidia belonging to *Puccidia iridis* are known in this country. Plowright (1889, 190) reported that the uredospores occurring on certain cultivated irises (*Iris sibirica*, *I. tolmieana* and others) repeatedly failed to infect *I. foetidissima* and *I. pseudacorus*; also that he was unable to find teleutospores on cultivated species. In consequence he considered that the form which occurs on the latter is different from that on our native species. Grove (*loc. cit.*) pointed out that the teleutospores on many species are difficult to find as they appear only on dying leaves, especially towards the base, at the end of the season; they can be easily recognised by being naked while the uredosori remain long-covered. There appears to be no doubt that the uredospores which are unusually thick-walled, can survive the winter.

Teleutospores were recorded (Ellis, 1934, 495) on cultivated species and doubtfully on *I. pseudacorus* in Norfolk. Mayfield (1935, 7) reported uredospores and teleutospores from Suffolk on *I. foetidissima*. It is possible that the aecidial stage of *P. iridis* may occur in this country, especially where teleutospores have been found on the native hosts. As pointed out by Jørstad & Roll-Hansen in Norway, if *Urtica* is the aecidial host, the aecidia in question might easily be mistaken for those on *Urtica* belonging to *P. caricina* growing in the vicinity; the two aecidia are similar but can be distinguished by the microscopic structure of their peridial cells.

Tranzschel (Notulae Syst. Crypt. Petrop. 2, 83, 1923) showed by culture that the aecidial host of a race on *I. sibirica* in eastern Siberia is

Valeriana officinalis, whereas Jørstad & Roll-Hansen (Nytt Mag. Naturv. **87**, 61, 1949) have proved by culture that the aecidial host of a form on *I. sibirica* grown in a garden in Norway is *Urtica dioica* and reciprocal cultures made later with aecidiospores obtained on this host, showed that *I. sibirica* became infected but not *I. graminea*. In the same garden various other irises, including *I. pseudacorus*, were free from the rust; teleutospores from this form on *I. sibirica* failed to infect *Valeriana officinalis* var. *latifolia*.

Jørstad & Roll-Hansen (*loc. cit.*) described considerable variation in the size of the uredospores and have distinguished three types; the largest averaging 33·8 × 27·6μ, found on two host species in southern Russia and Roumania; a medium-sized type occurring on 19 host species, averaged 25·6–30·1 × 20·7–25·4μ, while small-spored types occurring on *I. graminea*, *I. ruthenica* and *I. sibirica* averaged 23·4 × 20·9μ. They considered that their data would hardly justify a dividing of *P. iridis* into definite uredospore-size groups. The form on *Iris sibirica* from Norway is a small-spored type. The uredospores of British collections on wild *I. foetidissima* and cultivated *I. germanica* all lie close together within the range 24–31 × 19–24μ, and seem to correspond most closely to the medium-sized type.

Gäumann (Phytopath. Zeitschr. **25**, 99, 1955) investigated the host alternation of this rust on *I. graminea* in Switzerland. It was morphologically identical with the form on *I. sibirica* in Norway, but strongly specialised to its host and failed to attack *I. sibirica*, *I. pseudacorus*, *I. foetidissima*, *I. germanica* and other species; its teleutospores readily infected *U. dioica* and *U. urens* but not *V. officinalis*.

Gäumann also failed to infect *U. dioica*, *U. urens* and *V. officinalis* with a small number of teleutospores from cultivated species of *Iris* which he obtained from Italy and Switzerland; he pointed out that this confirmed the investigations of Mains (Amer. J. Bot. **21**, 23, 1934; Phytopath. **28**, 67, 1938) who could not infect *Valeriana officinalis* with teleutospores from some cultivated irises.

As the result of these investigations, Gäumann suggested that two types of *P. iridis* may be recognised: (1) forma typica, on *I. pallida* and *I. pumila* and generally on cultivated species, with large uredospores 26–30 × 21–25μ, which may survive the winter, forming few or no teleutospores, with aecidial stage unknown; (2) on wild hosts, with small uredospores, 21 × 20–22μ, generally forming teleutospores, with aecidial stage on *U. dioica*, *U. urens* and *V. officinalis*.

Dupias (Bull. Soc. Hist. Nat. Toul. **93**, 228, 1958) confirmed alteration of a rust he calls *P. urticae-xiphioides*, between *U. dioica* and *I. xiphioides* in the French Pyrenees. Its uredospores are rather larger (25–30 × 21–27 μ) than the small-spored *Urtica* alternating race of the other authors. Kursanov (1922, 74) described the early stages in the development of the uredospores and Treboux (Ann. Myc. **10**, 306, 1912) the overwintering by the latter in Russia.

Puccinia satyrii Sydow

Monogr. Ured. **1**, 594 (1903); Wilson & Bisby, Brit. Ured. no. 215.

Spermogonia and **aecidia** wanting. **Teleutosori** hypophyllous, without spots, generally scattered over the whole surface of the leaf, minute, rounded, compact, yellowish-brown; teleutospores oblong or fusiform, narrowed at both ends, strongly thickened at the apex (up to 11 μ), not or slightly constricted at the septum, smooth, pale yellow, 32–52 × 13–19 μ, pedicel hyaline, persistent, equalling the spore in length; [uredospores present with the teleutospores, globoid or subgloboid, 16–24 μ diam., wall echinulate, golden-yellow.] Hemiform?

On *Satyrium aureum*. Rare, introduced.

The rust occurred in the Orchid House, Kew Gardens, in 1929 on dying leaves of plants imported from Africa; only teleutospores were present.

Puccinia cladii Ellis & Tr.

Bull. Torr. Bot. Club, **22**, 61 (1895); Wilson & Bisby, Brit. Ured. no. 133; Gäumann, Rostpilze, p. 616.

Uredosori amphigenous, scattered or grouped, confluent, up to 8 mm. long, surrounded by the ruptured epidermis, dark cinnamon-brown; spores broadly ellipsoid or obovoid, cinnamon-brown, 27–39 × 20–29 μ, wall very finely echinulate 2–2·5 μ thick, with 4 (rarely 3) equatorial pores. [**Teleutosori** similar, blackish-brown; spores ellipsoid or clavoid, rounded or obtuse above, usually narrowed below, somewhat constricted, dark chestnut-brown, paler below, 42–51 × 18–26 μ, wall 1–2 μ thick at sides, much thickened (up to 9 μ) above, smooth, pedicel brownish, up to one-third the length of the spore, persistent.] Hetereu-form?

On *Cladium mariscus*, July–September. England, rare.

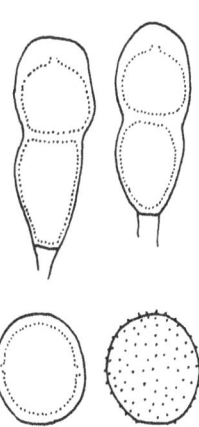

P. cladii. Teleutospores and uredospores (after Arthur).

This species was first found in Britain at Wheatfen Broad, near Norwich by E. A. Ellis in 1939 and later in the same year at Wicken Fen, Cambridge, by M. R. Brown. Only the uredospores have been discovered. The fine echinulation of the uredospores is difficult to see unless the material is fully mature. A comparison with American material was made by Dr C. L. Shear who agreed on the identity of the English and American specimens.

Guyot (*Uredineana*, **1**, 74, 1938) discovered in 1937 a *Puccinia* on *Cladium mariscus* in northern France which he considered to be a new species and named *P. cladiana*. This, he stated, differed from *P. cladii* in its rather larger teleutospores more or less verrucose at the apex and in small numbers in the uredosori rather than in separate sori. He suggested, however, that this may be only a European form of *P. cladii*. It appears doubtful whether it should be regarded as a distinct species.

Puccinia eriophori Thuem.

Bull. Soc. Imp. Nat. Moscou, **55**, 208 (1880).

Puccinia confinis Syd., Ann. Myc. **18**, 154 (1920); Wilson & Bisby, Brit. Ured. no. 138; Gäumann, Rostpilze, p. 615.

Spermogonia epiphyllous, pale yellow. **Aecidia** hypophyllous, in small groups on pale spots, peridia white with revolute laciniate margin; aecidiospores 16–20 × 15–18 μ. **Uredosori** few, scattered, cinnamon brown, long-covered by the epidermis, uredospores obovoid or subgloboid, 20–24 × 18–22 μ, wall distantly echinulate with 2 supra-equatorial pores. **Teleutosori** scattered, compact, surrounded by the ruptured epidermis, black, 0·5–1 mm. long; teleutospores clavate, at the apex rounded or acuteconical rarely truncate, more or less constricted at the septum, attenuated towards the base, smooth, yellowish-brown, paler above and strongly thickened at the apex (7–14 μ), 40–62 × 17–24 μ, pedicel up to 50 μ long, thick, coloured. Hetereu-form.

Spermogonia and aecidia on *Solidago virgaurea*, June–September; uredospores and teleutospores on *Scirpus cespitosus* subsp. *germanicus*. Scotland, very rare.

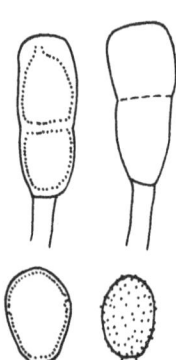

P. eriophori. Teleutospores and uredospores.

This rust was first described on *Scirpus cespitosus* by Sydow from specimens collected at Libau in Latvia in 1917. The aecidium on *S. virgaurea* was found in western Norway by Jørstad (Nytt Mag. Naturv. **83**, 100, 1942) who in September 1939 collected the teleutospores on

S. cespitosus and discovered old aecidial spots on *S. virgaurea*. In 1940 he infected *S. virgaurea* with overwintered teleutospores and obtained spermogonia and aecidia. In 1936 Heslop-Harrison (*in litt.*) discovered aecidia on *S. virgaurea* on the Island of Raasay and there was an unidentified collection of aecidia in Grove's herbarium collected in Inverness-shire in 1931. In September 1942 they were found on Ben Vaar, Argyll, at an altitude of about 2800 ft.

Teleutospores and sparse uredospores associated with aecidia on *Solidago* were discovered on the Applecross peninsula Wester Ross, in 1959. The rarity of this rust in Britain is probably due to the infrequency with which *Solidago virgaurea* grows close to *Scirpus cespitosus*. The rust is clearly obligatorily heteroecious.

KEY TO PUCCINIA ON CAREX

The British material of rusts on *Carex* has been revised (Henderson, Notes R. B. G. Edin. **23**, 223, 1961) and four species recognised which may be keyed out as follows:

1. Amphispores abundant   *P. microsora,* p. 245
   Amphispores absent or scarce   2

2. Uredospores with 3 or more pores   *P. caricina,* p. 232
   Uredospores with 2, rarely 3, pores (with three pores; always less than 20μ
   long)   3

3. Uredospores pores supra-equatorial, uredospores more than 20μ long
                                           P. dioicae, p. 241
   Uredospores pores equatorial, uredospores small, less than 20μ long
                                           P. opizii, p. 244

Determination of collections at infra-specific rank depends partly upon the identity of the host but as has been shown by many workers (Gäumann, Hasler, and Mayor) the races on any one host may be readily distinguished morphologically, usually by uredospore characters.

The following list summarises the varieties under their hosts and indicates methods of distinguishing them when more than one variety is present on any sedge in Britain. These varieties are dealt with in alphabetical order under the appropriate species.

Carex host	Rust varieties
acuta	*caricina* var. *pringsheimiana*
acutiformis	*caricina* var. *urticae-acutiformis*. Uredospores mostly more than 34 × 26μ
	(*caricina* var. *magnusii*). Uredospores mostly less than 30 × 22μ, not yet recorded in Britain

Carex host	Rust varieties
appropinquata	*opizii*
arenaria	*dioicae* var. *arenariicola*. Aecidia on *Centaurea*
	dioicae var. *schoeleriana*. Aecidia on *Senecio*
bigelowii	*caricina* var. *paludosa*
binervis	*caricina* sens. *lat.*
capillaris	*dioicae* var. *silvatica*
demissa	*dioicae* sens. *lat.*
dioica	*dioicae* var. *dioicae*
elata	*caricina* var. *paludosa*. Uredospores less than $30 \times 22\,\mu$, subgloboid, with 3 pores
	caricina var. *urticae-acutae*. Uredospores more than $34 \times 26\,\mu$, with 4 pores
extensa	*dioicae* var. *extensicola*
flacca	*caricina* var. *urticae-flaccae*
hirta	*caricina* var. *urticae-hirtae*
juncella	*caricina* sens. *lat.*
laevigata	*caricina* sens. *lat.*
lasiocarpa	*caricina* var. *ribis-nigri-lasiocarpae*
lepidocarpa	*dioicae* sens. *lat.*
maritima	*dioicae* var. *schoeleriana*
nigra	*caricina* var. *urticae-acutae*
	caricina var. *paludosa*
	caricina var. *pringsheimiana*
	caricina var. *uliginosa*
ovalis	*dioicae* sens. *lat.*
pairaei	*opizii*
pallescens	*caricina* sens. *lat.*
paniculata	*caricina* var. *ribis-nigri-paniculatae*. Uredospores with 3 equatorial pores
	opizii. Uredospores with 2 equatorial pores
pendula	*caricina* var. *ribesii-pendulae*
pseudocyperus	*caricina* var. *caricina*. Uredospores less than $30\,\mu$ diam., with 3 equatorial pores
	caricina var. *urticae-acutiformis*. Uredospores often more than $30\,\mu$, with 3–4 pores
riparia	*caricina* var. *magnusii*. Teleutospores usually less than $60\,\mu$ long
	caricina var. *urticae-ripariae*. Teleutospores usually more than $60\,\mu$ long
rostrata	*caricina* var. *urticae-inflatae*
vesicaria	*caricina* var. *urticae-vesicariae*. Amphispores absent
	microsora. Amphispores abundant

Puccinia caricina DC.

Fl. Fr. **6**, 60 (1815); Wilson & Bisby, Brit. Ured. no. 118.

Uredo caricis Schum., Enum. Pl. Saell. **2**, 231 (1803) (non *Uredo caricis* Pers., Syn. Meth. Fungi, p. 225 (1801) = *Cintractia caricis*.)

[*Uredo caricina* DC., Fl. Fr. **6**, 83 (1815).]

[*Aecidium grossulariae* DC., Fl. Fr. **6**, 92 (1815), *p.p.*]

Puccinia caricis (Schum.) Schroet., Jahresb. Schles. Ges. Vaterl. Cult. p. 103 (1873) and Krypt. Fl. Schles. **3**, 327 (1887) (non *Puccinia caricis* Rebent Prod. Fl. Neom. 356 (1804), = *P. dioicae*); Grove, Brit. Rust Fungi, p. 241.
Puccinia grossulariae Lagh., Tromsö Mus. Aarsch. **17**, 60 (1895).

Spermogonia epiphyllous, in small clusters, honey-coloured. Aecidia hypophyllous and occasionally epiphyllous, often on stems and petioles, on calyces of *Parnassia* and on fruits of *Ribes*, on reddish, yellow or purplish spots, in dense clusters of various sizes often very large and causing great swellings and distortion on stems, cup-shaped with torn, white, recurved peridium, peridial cells in longitudinal rows, almost quadratic, outer wall about 7μ thick, inner wall $3-5\mu$ thick, with distinct tesselate structure; aecidiospores polygonal with orange contents, $16-26 \times 12-20\mu$, wall colourless, verruculose, 1μ thick. Uredosori amphigenous, generally hypophyllous, scattered, oblong, about 0.5 mm. long, pale brown; uredospores subgloboid to ovoid, yellow-brown, wall $2-3\mu$ thick echinulate, with 3 or 4 equatorial pores. Teleutosori generally hypophyllous, scattered or arranged in lines, oblong or linear and confluent, up to 1 mm. long, pulvinate, compact, blackish-brown; teleutospores clavoid, usually rounded above, the upper cell much broader and shorter than the lower one, tapering downwards, constricted, $35-66 \times 14-23\mu$, wall smooth, brown,

darker and thickened $(5-10\mu)$ at the apex, $1.5-2\mu$ at the sides; pedicels yellowish, persistent, about half as long as the spore or less. Hetereu-form.

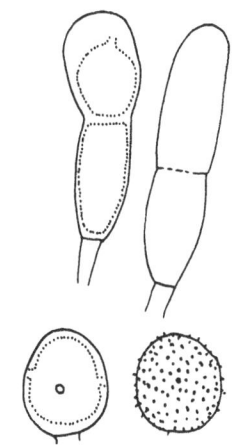

P. caricina var. *urticae-hirtae*. Teleutospores and uredospores.

Spermogonia and aecidia on *Parnassia palustris* (aecidia only), *Pedicularis palustris*, *Ribes nigrum*, *R. sanguineum*, *R. sylvestre*, *R. uva-crispa*, *Urtica dioica* and *U. urens*; uredospores and teleutospores on many species of *Carex*. Many races common throughout Britain.

Aecidia of the rust previously known as *Puccinia uliginosa* have been found on *Parnassia palustris* in several localities in Scotland, in the Tweed, Forth, Tay, Dee, Clyde and Argyle areas and also in Orkney; they have also been recorded from Ireland. In 1894 Juel in Sweden successfully infected *Carex nigra* with aecidiospores from *Parnassia palustris*. Up to the present in Scotland, the uredospores and teleutospore stages have been found only on *C. nigra* but a rust recently collected by R. W. G. Dennis on *C. bigelowii* near Glen Shee at about 800 m. probably also forms its aecidia on *Parnassia*. The rust on this host has been recorded in Norway by Jørstad (1940, 54) who has stated that it produces its aecidia on *Parnassia*.

Aecidia of the rust previously known as *P. paludosa* have been found on *Pedicularis palustris* in various localities in Scotland and in Norfolk.

Plowright infected *P. palustris* with aecidiospores from *C. nigra* and conversely aecidiospores applied to the latter species produced uredospores and teleutospores. The same infection experiments have been carried out in Scotland (Wilson, TBMS. **9**, 136, 1924) and in Switzerland Hasler (Ann. Myc. **28**, 356, 1930) has infected only *C. nigra* and *C. elata* with aecidiospores from *P. palustris*.

The aecidium on *Ribes* has been found frequently in Great Britain and Ireland; it appears to have been first recorded in this country by Greville (Scot. Crypt. Fl. **2**, 62) on the leaves and fruits of *Ribes uva-crispa* and was described in 1890 by Plowright (J. Roy. Hort. Soc. **12**, cix, 1890) on the leaves of *R. nigrum*. It was recorded on the leaves and fruits of *R. sylvestre* in Sussex, Kent and Oxfordshire in 1933 (Bull. Min. Agric. **126**, 66, 1943) and in Norfolk by Ellis (1934, 496) in the same year; it has since been found frequently on these hosts. On *R. sanguineum* it appears to have been recorded only in Skye and Mull (Wilson, 1934, 391) where leaves and fruits were infected. Moore (1959, 310) suggests that aecidial infection is heaviest following a dry March which delays teleutospore germination.

In this country Soppitt in 1897 (Gard. Chron. **24**, 145, 1898) with teleutospores from *C. acuta* collected in Cumberland, produced aecidia on *R. uva-crispa*; in 1898 he infected *C. nigra* with aecidia from the gooseberry. In 1931 Wilson (1934, 392) using teleutospores of *C. nigra* collected near Stirling, infected *R. uva-crispa*. In 1935 in Northern Ireland, when Saunderson & Cairns (Ann. Appl. Biol. **24**, 17, 1937) placed gooseberry bushes in pots in a meadow in which *C. nigra*, *C. panicea*, '*C. flava*' and *C. rostrata* had been heavily infected with rust in the previous year, heavy infection of the gooseberries resulted; of these species only rusts on *C. nigra* (see Klebahn, 1904) and on *C. rostrata* (Eriksson, Arkiv. Bot. **16** (11) 1921) are known to produce aecidia on *R. uva-crispa*.

The first cultures with sedge rusts were carried out in 1872 by Magnus (Verh. Bot. Ver. Brand. **14**, xi, 1872) who infected *C. hirta* with aecidiospores from *U. dioica*. In 1892 Klebahn (Z. Pfl.-Krankh. **3**, 199, 1893) first proved the connection between the rust on *C. nigra* and the aecidium on *Ribes*. Series of cultures were subsequently carried out by Klebahn (see 1904, 295–301) in Germany who gave names to three species connected with *Urtica* and five with *Ribes*, and by Eriksson (*loc. cit.*) in Sweden who placed these rusts under three species and three subspecies.

The aecidium on *Urtica dioica* is found commonly in Great Britain and Ireland especially where *Carex acutiformis* and *C. hirta* are infected.

Plowright (1889, 170) in many cultures produced the aecidiospores by placing teleutospores from *C. hirta* on *U. dioica*. Klebahn (Z. Pfl.-Krankh. **5**, 267, 1895) first produced aecidia on *U. urens* from teleuto-spores on *C. nigra*; Ellis (Trans. Norf. Norw. Nat. Hist. Soc. **15**, 427, 1943) found aecidia on *U. urens* in Norfolk and with teleutospores from *C. hirta* produced them in culture. Grove (J. Bot. Lond. **51**, 42, 1913), showed that the rust could be easily introduced into a new locality by placing in January or February a bundle of leaves of *C. acutiformis* infected by the rust on the ground where a patch of nettles was known to occur; from his experiments he concluded that the efficiency of wind distribution of the spores was much reduced by increased distance and by tall vegetation.

The following fifteen varieties of *P. caricina* occur in Britain.

var. *caricina* DC.

Puccinia ribesii-pseudocyperi Kleb., Jahrb. Wiss. Bot. **34**, 391 (1900); Gäumann, Rostpilze, p. 637.

Uredospores $26-28 \times 18-20\mu$ with 3 equatorial pores. Teleutospores $45-52 \times 18-20\mu$, apex $6-8\mu$.

Known from only a few localities in England on the type host, *Carex pseudocyperus*.

There is no record of the aecidial stage in Britain, which in Europe may develop on *Ribes alpinum*, *R. aureum*, *R. nigrum*, *R. sylvestre* and *R. sanguineum* and occasionally *R. uva-crispa* (Gäumann, 1959).

var. *magnusii* (Kleb.) Henderson

Notes R. B. G. Edin. **23**, 235 (1961).

Puccinia magnusii Kleb., Z. Pfl.-Krankh. **5**, 79 (1895); Gäumann, Rostpilze, p. 631.

Uredospores $24-30 \times 20-22\mu$ with 3 equatorial pores. Teleutospores $50-60 \times 18-21\mu$, apex $6-10\mu$ thick.

On *Carex riparia* in England and Ireland, uncommon; on *Carex acutiformis* in continental Europe.

The aecidial stage, not recorded in Britain, forms on *Ribes alpinum*, *R. aureum*, *R. nigrum* and *R. sanguineum* according to Klebahn's experimental results (Z. Pfl.-Krankh. **9**, 137, 1899).

var. *paludosa* (Plowr.) Henderson

Notes R. B. G. Edin. **23**, 236 (1961).

Puccinia paludosa Plowr., Monogr. Brit. Ured. Ustil. p. 174 (1889); Grove, Brit. Rust Fungi, p. 248, Wilson & Bisby, Brit. Ured. no. 191; Gäumann, Rostpilze, p. 645.

Uredospores broadly ellipsoid to sub-globoid, $22-26 \times 20-24\mu$, wall $2-3\mu$ thick, with 3 equatorial pores. Teleutospores $44-62 \times 17-22\mu$, apex $8-10\mu$ thick.

On *Carex nigra* and *C. panicea*, frequent, less so on *C. elata* and *C. bigelowii*.

When *Carex nigra* is infected teleutospores predominate and the aecidial host usually occurs nearby—clear indications of obligatory heteroecism. On *C. panicea* teleutospores are infrequent and the infected plants are not usually closely associated with *Pedicularis palustris*. Plowright carried out successful reciprocal inoculations with this race using *P. palustris* and *C. nigra* and Wilson (TBMS. 9, 136, 1924) and Hasler (1930) repeated these.

var. *pringsheimiana* (Kleb.) Henderson

Notes R. B. G. Edin. 23, 237, (1961).

Puccinia pringsheimiana Kleb., Z. Pfl.-Krankh. 5, 79 (1895), sub *P. caricina*; Grove, Brit. Rust Fungi, p. 242; Wilson & Bisby, Brit. Ured. no. 119; Gäumann, Rostpilze, p. 633.

Uredospores subgloboid 20–25 μ diam., with 3 equatorial pores, wall 2 μ thick. Teleutospores 50–75 × 15–20 μ, apex 5–8 μ thick. On *Carex nigra* widespread and on *C. acuta* from Windermere by Soppitt (Gard. Chron. 24, 145, 1898) who demmonstrated the alternation with cultivated gooseberry. Common.

This is probably the rust which Saunderson & Cairns (Ann. Appl. Biol. 24, 17, 1937) obtained when they placed gooseberry plants in a field in Ireland where *Carex nigra, C. panicea, C. flava* and *C. rostrata* had been heavily infected the previous year, and obtained heavy aecidial infection.

var. *ribis-nigri-lasiocarpae* (Hasler) Henderson

Notes R. B. G. Edin. 23, 237 (1961).

Puccinia ribis-nigri-lasiocarpae Hasler, Ann. Myc. 28, 350 (1930); Gäumann, Rostpilze, p. 635.

Uredospores 28–30 × 20–22 μ with 3 equatorial pores. Teleutospores 40–50 × 20–22 μ, apex 4–5 μ. On *Carex lasiocarpa*, Scotland, rare.

This variety is known from only one collection of *Carex lasiocarpa* from the Scottish Highlands (Henderson, *loc. cit.*) The aecidium is unknown in Britain, Hasler (*loc. cit*) showed that it occurs on *Ribes nigrum* and *R. petraeum* and less abundantly on *R. aureum* and *R. rubrum*. Gäumann (1959, 636) has records of this variety only from Switzerland.

var. *ribis-nigri-paniculatae* (Kleb.) Henderson

Notes R. B. G. Edin. 23, 237 (1961).

Puccinia ribis-nigri-paniculatae Kleb., Jahrb. Wiss. Bot. 34, 393, (1900); Gäumann, Rostpilze, 636.

Uredospores 26–30 × 17–20 μ. Teleutospores 45–60 μ, apex 8–10 μ. On *Carex paniculata*, frequent.

This race is known from scattered localities in Scotland and England on *Carex paniculata* and is probably present wherever the host occurs in quantity. Both uredospores and teleutospores are present, but in Britain there is no information on aecidial hosts, which are *Ribes alpinum*, *R. aureum*, *R. nigrum*, *R. rubrum* and *R. sanguineum* and less abundantly *R. uva-crispa* according to Gäumann (1959). Teleutospores are often scanty and the race undoubtedly perennates by uredospores. Collections on *C. paniculata* should be carefully examined for this host also houses *P. opizii* which has much smaller uredospores with two equatorial pores and aecidial hosts belonging to the *Compositae*.

var. *ribesii-pendulae* (Hasler) Henderson

Notes R. B. G. Edin. **23**, 237 (1961).

Puccinia ribesii-pendulae Hasler, Ber. Schweiz. Bot. Ges. **55**, 15 (1945); Gäumann, Rostpilze, p. 637.

Uredospores 26–32 × 20–26μ, with 3 equatorial pores. **Teleutospores** 42–54 × 18–21μ, apex 6–10μ.

On *Carex pendula*, Great Britain, common.

Teleutospores are rather infrequently produced and the variety is certainly usually independent of alternation. It forms aecidia on *Ribes alpinum*, *R. aureum*, *R. rubrum*, and *R. uva-crispa* in Europe according to Hasler (*loc. cit.*) but they are unknown in Britain.

var. *uliginosa* (Juel) Jørst.

Skr. Norske Vidensk.-Akad. Oslo, **1951**, 30 (1952); Wilson & Bisby, Brit. Ured. no. 120.

Puccinia uliginosa Juel., Overs. K. Vet. Akad. Förh. **51**, 410 (1894); Grove, Brit. Rust Fungi, p. 249; Gäumann, Rostpilze, p. 627.

Uredospores 20–24 × 21–23μ, wall 2μ thick, with 3 equatorial pores. **Teleutospores** 34–50 × 15–22μ, apex 4–8μ thick.

On *Carex nigra*, Great Britain, frequent.

In Britain the only diplont host for this variety is *Carex nigra*, although it has been shown to be capable of infecting *Carex bigelowii*, *C. juncella* and *C. rariflora*. It usually forms abundant teleutospores. The alternation with *Parnassia palustris* which usually grows in the vicinity can be confirmed by the presence of aecidia on the leaves of the latter.

This variety differs somewhat from others of the *P. caricina* group; spermogonia are lacking and the teleutospores are rather shorter than in the other varieties.

var. *urticae-acutae* (Kleb.) Henderson

Notes R. B. G. Edin. **23**, 238, 1961.

Puccinia urticae-acutae Kleb., Z. Pfl.-Krankh. **9**, 152 (1899); Gäumann, Rostpilze, p. 620.

Uredospores 25–34 × 25–30 μ with 3 equa-   On *Carex nigra* frequent, on *C. elata*
torial pores, wall 1–2 μ thick. Teleuto-   rare.
spores 48–60 × 20–22 μ, apex 8–10 μ thick.

On *Carex nigra* this variety occurs throughout Britain, teleutospores usually occur and alternation probably takes place regularly with the aecidial host *Urtica dioica*. On *Carex elata* it is less common and known only from England. It occurs on *C. acuta* on the continent and has there also been shown to develop aecidia on *U. urens* in addition to the common nettle. Gäumann (1959) summarises the results of numerous infection experiments.

var. *urticae-acutiformis* (Kleb.) Henderson

Notes R. B. G. Edin. **23**, 239 (1961).

Puccinia (urticae-) caricis f. *urticae-acutiformis* Kleb., Z. Pfl.-Krankh. **15**, 70, 1905; Gäumann, Rostpilze, p. 621.

Uredospores 34–38 × 26–30 μ with 4   Uredospores and teleutospores on
or occasionally 3 equatorial pores.   *Carex acutiformis*, common; uredo-
Teleutospores 50–60 × 18–22 μ, apex 8–   spores only on *C. pseudocyperus*.
10 μ thick.

This variety is frequent in England but less so in Scotland where the usual host, *Carex acutiformis*, is less common. On *C. acutiformis* half the collections bear teleutospores in addition to uredospores and the variety probably often alternates. The only other host in Britain is *C. pseudocyperus* on which only uredospores are formed. This variety is readily differentiated from others by the large uredospores with usually more than three equatorial pores. The connection with aecidia on *Urtica dioica* has been frequently demonstrated in Europe (Gäumann, 1959).

var. *urticae-flaccae* (Hasler) Henderson

Notes R. B. G. Edin. **23**, 239 (1961).

Puccinia urticae-flaccae Hasler, Ber. Schweiz. Bot. Ges. **55**, 6 (1945); Gäumann, Rostpilze, p. 622.

Uredospores 24–30 × 20–24 μ, wall 2 μ   On *Carex flacca*, Great Britain, com-
thick with 3 equatorial pores. Teleuto-   mon.
spores 40–50 × 17–21 μ, apex 5–8 μ thick.

This variety whose alternation with *Urtica dioica* was proved by Hasler (*loc. cit.*) with Swiss material occurs throughout Britain. Teleutospores usually occur but there is no information on its alternation in this country.

var. *urticae-hirtae* (Kleb.) Henderson

Notes R. B. G. Edin. **23**, 240, (1961).

Puccinia urticae-hirtae Kleb., Z. Pfl.-Krankh. **9**, 152 (1899); Gäumann, Rostpilze, p. 623.

Uredospores 26–28 × 18–20 μ with 3 equatorial pores. **Teleutospores** 45–67 × 18–24 μ, apex 8–11 μ thick.
On *Carex hirta*, common.

This variety, known throughout Britain on *Carex hirta*, is probably obligatorily heteroecious as in all collections teleutospores are abundant. Plowright (1889) produced aecidia on *Urtica dioica* using teleutospore material on this sedge and Ellis (Trans. Norf. Norw. Nat. Hist. Soc. **15**, 427, 1943) successfully inoculated *Urtica urens* with similar material.

var. *urticae-inflatae* (Hasler) Henderson

Notes R. B. G. Edin. **23**, 240 (1961).

Puccinia urticae-inflatae Hasler, Mitt. Aarg. Naturf. Ges. **17**, 64 (1925); Gäumann, Rostpilze, p. 623.

Uredospores subspherical, 26–28 × 23–26 μ, wall 2 μ thick with 3 equatorial pores. **Teleutospores** 50–65 × 18–21 μ, apex 7–9 μ thick.
Frequent on *Carex rostrata* in Scotland and England.

Teleutospores are usually present, and alternation is probably with *Urtica dioica* but the only proof of this relation is due to Hasler (*loc. cit.*) working in Switzerland. The uredosori and teleutosori of this variety develop on the upper surface of the host leaves contrary to almost all other sedge rusts, but as Gäumann (1959) noted the stomata are restricted to this surface in *Carex rostrata*.

var. *urticae-ripariae* (Hasler) Henderson

Notes R. B. G. Edin. **23**, 240 (1961).

Puccinia urticae-ripariae Hasler, Ber. Schweiz. Bot. Ges. **15**, 8 (1945); Gäumann, Rostpilze, p. 626.

Uredospores not collected in Britain. **Teleutospores** 58–70 × 16–21 μ, apex 8–12 μ.
On *Carex riparia*, frequent.

This variety is known from only two localities in Britain which is somewhat surprising in view of the frequency of the sedge host. The prevalence of teleutospores suggests an obligatory heteroecious rust whose proved relation with *Urtica dioica* and *U. pilulifera* has been summarised by Gäumann (1959).

var. *urticae-vesicariae* (Kleb.) Henderson

Notes R. B. G. Edin. **23**, 241 (1961).

Puccinia (*urticae-*) *caricis* f. *urticae-vesicariae* Kleb., Pfl.-Krankh. **15**, 70 (1905); Gäumann, Rostpilze, p. 627.

Uredospores 28–30 × 18–20 μ with 3 equa- 20 μ, apex 4–6 μ.
torial pores. **Teleutospores** 40–46 × 18– On *Carex vesicaria*, uncommon.

Hasler (Ber. Schweiz. Bot. Ges. **55**, 5, 1945) has confirmed the connection with *Urtica dioica* demonstrated by many previous workers. *Puccinia microsora* occurs on the same host but is readily distinguished by its abundant amphispores and lack of typical brown teleutospores.

Puccinia caricina sensu lato

In addition to these varieties there are a few sedge rusts in Britain about which there is insufficient information either here or in foreign literature to place them accurately in any described taxa. These are referred to here under their hosts.

On *Carex binervis*. An obviously uredo-perennating rust, never producing teleutosori, is common on this sedge from the lowlands to over 1000 m. on the mountains. A similar rust occurs on this host throughout Europe and both Gäumann (1959) and Jørstad (1954) have noted the lack of teleutosori. The spores of all the British collections are very similar, 26–34 × 20–23 μ, wall 2·5 μ thick with 3 equatorial pores and there is little doubt that only one race is present.

On *Carex juncella*, one collection is known, made by Marshall in Surrey. Only teleutospores are present.

On *Carex laevigata*, several collections bearing uredospores are known on herbarium specimens of the host. This may represent a type similar to that on *C. binervis* which has become completely independent of alternation but we have no knowledge of it in the field.

On *Carex pallescens*, a rust with uredospores only occurs. The uredospores are 26–30 × 22–24 μ, wall 2 μ thick, with 2, rarely 3, equatorial pores. This rust is somewhat intermediate between *P. opizii* and *P. caricina* in uredospore pore characteristics but the spores are much larger than those of *P. opizii*. It is probably a specialised non-alternating race similar to the one on *C. binervis*.

Puccinia dioicae Magn.

Amtl. Ber. 50 Versamml. Deut. Naturf. Arzte, München, 199 (1877), *n.v.*
Puccinia caricis Reb., Prod. Fl. Neom. p. 35 (1804) (*nomen ambiguum*).

Spermogonia amphigenous in small clusters, honey-coloured. **Aecidia** hypophyllous, on roundish, yellow or brown spots, in clusters 2–5 mm. diam. cupulate with torn white margin, peridial cells in distinct rows, outer walls punctate up to 10μ thick, inner thinner, aecidiospores angular-globoid, finely verruculose, orange, 18–25μ diam. **Uredosori** chiefly hypophyllous, scattered, minute, punctiform, brown; uredospores pale brown, subgloboid to broadly ellipsoid, 22–28 × 16–24μ, wall echinulate, 2μ thick with 2 supra-equatorial pores. **Teleutosori** hypophyllous, scattered, roundish or oblong, 1 mm. long, soon naked, surrounded by the cleft epidermis, pulvinate, black; teleutospores clavoid, slightly constricted, darker at the apex, 36–60 × 16–24μ, wall smooth, 4–12μ thick at apex, pore of upper cell apical, of lower superior; pedicels brownish, persistent, up to 50μ long.

Spermogonia and aecidia on *Aster*

tripolium, *Centaurea nigra*, *Cirsium palustre*, *C. dissectum*, *Senecio jacobaea* and *Taraxacum officinale* agg., uredo-

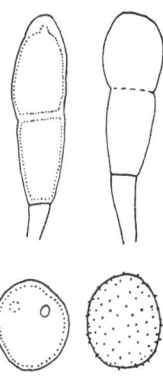

P. dioicae. Teleutospores and uredospores.

spores and teleutospores on *Carex arenaria*, *C. capillaris*, *C. demissa*, *C. dioica*, *C. extensa*, *C. lepidocarpa*, *C. maritima* and *C. ovalis* (uredospores only).

var. *dioicae* Magn.

Grove, Brit. Rust Fungi, p. 244, Wilson & Bisby, Brit. Ured. no. 145. Gäumann, Rostpilze, p. 677.

Uredospores 24–27 × 20–22μ, wall 2μ thick with 2 supra-equatorial pores. **Teleutospores** 36–48 × 16–19μ, apex 6–8μ thick.

Aecidia on *Cirsium palustre* and *C. dissectum*, uredospores and teleutospores on *Carex dioica*, Scottish Highlands, Norfolk and Ireland.

In the presence of the thistle teleutospores are abundantly present on the sedge and the rust is habitually heteroecious but where the thistle is not associated uredospores may be the only stage present. Aecidia of this rust were collected in Ireland on *Cirsium palustre* and *C. dissectum* but the diplont stages which are rather inconspicuous have not been discovered there. The rust on *Carex rostrata* recorded by Ellis (1934, 496) as *Puccinia dioicae* is probably a variety of *P. caricina*; no uredospores are present in the collections to confirm this but no rust of the *P. dioicae* type has ever been confirmed on this host.

var. *arenariicola* (Plowr.) Henderson

Notes R. B. G. Edin. 23, 243, (1961).

Puccinia arenariicola Plowr., J. Linn. Soc. Bot. 24, 90 (1888), Grove, Brit. Rust Fungi, p. 247; Wilson & Bisby, Brit. Ured. no. 146; Gäumann, Rostpilze, p. 667.

Uredospores 22–26 × 20–22μ, wall faintly echinulate with 2 supra-equatorial pores.	**Teleutospores** 40–50 × 20μ (*fide* Plowright).

This variety has not been re-collected in Britain since Plowright's first investigation of it in 1888 using material he found on *Carex arenaria* and *Centaurea nigra* at Hemsby in Norfolk. In its almost smooth uredospores it closely approaches var. *schoeleriana*. The rust no longer appears to be present at Hemsby (Ellis, *in litt.*).

var. *extensicola* (Plowr.) Henderson

Notes R. B. G. Edin. 23, 343 (1961).

Puccinia extensicola Plowr., Monogr. Brit. Ured. Ustil. p. 181 (1889); Wilson & Bisby, Brit. Ured. no. 151; Gäumann, Rostpilze, p. 655.

Uredospores broadly ellipsoid, 22–28 × 19–20μ, wall pale, faintly echinulate, with 2 supra-equatorial pores. **Teleutospores** 42–60 × 16–22μ, apex 6–10μ thick.	Spermogonia and aecidia on *Aster tripolium*; uredospores and teleutospores on *Carex extensa*. England and Wales, scarce.

This variety was originally described from Norfolk by Plowright who proved the alternation between *Carex extensa* and *Aster tripolium*. The variety appears to be confined to south-east and south-west England. It has not been collected in Scotland where the diplont host is somewhat scarcer. It is probably obligatorily heteroecious.

var. *schoeleriana* (Plowr. & Magn.) Henderson

Notes R. B. G. Edinb. 23, 244 (1961).

Puccinia schoeleriana Plowr. & Magn., Quart. Journ. Micro. Sci. 25, 170 (1885); Grove, Brit. Rust Fungi, p. 246; Wilson & Bisby, Brit. Ured. no. 146; Gäumann Rostpilze, p. 697.

Uredospores 27–30 × 18–24μ, wall almost smooth with 2 supra-equatorial pores. **Teleutospores** 50–60 × 18–24μ, apex 8–12μ thick.	Spermogonia and aecidia on *Senecio jacobaea*; uredospores and teleutospores on *Carex arenaria*, common.

The alternation of this variety between *Carex arenaria* and *Senecio jacobaea* was the subject of a long series of experiments by Plowright (*loc. cit.*). It is often very abundant on sand dunes where the two hosts are abundant. The aecidia on *Senecio* are often very conspicuous. The same race is probably represented by old collections of uredospores and teleutospores found in the Edinburgh Herbarium on *Carex maritima* from the coast of Morayshire.

var. *silvatica* (Schroet.) Henderson

Notes R. B. G. Edin. **23**, 245 (1961).
Puccinia silvatica Schroet., Cohn, Beitr. Biol. Pfl. **3**, 68 (1878); Grove, Brit. Rust Fungi, p. 111; Wilson & Bisby, Brit. Ured. no. 147 (records doubtful in both); Gäumann, Rostpilze, p. 702.

Uredospores sparse, about $24 \times 16\mu$, wall 2μ thick, sparsely echinulate with 2 supra-equatorial pores. Teleutospores $38–44 \times 16–18\mu$, apex $4–5\mu$. On *Carex capillaris*, Scotland, rare.

The evidence for this variety in Britain is slender. On Ben Lawers in Perthshire an aecidium of this type has been collected on *Taraxacum officinale* but no diplont stages have been found there although *Carex capillaris*, a host in Scandinavia, grows in the vicinity. The only diplont collection was made by B. Flannigan in Sutherland in 1955 on *Carex capillaris*.

This rust has been much confused with autoecious species on *Taraxacum*. There is no material in Grove's herbarium which undoubtedly belongs here and it is extremely doubtful if he ever saw a British collection. Soppitt's collection which he illustrated is *P. variabilis*. Spermogonia accompany the aecidia of the sedge rust on *Taraxacum* but are absent in *P. variabilis*. Moreover, as Fischer (1904) pointed out many years ago, the outer wall of the peridial cells is thicker than the inner in *P. dioicae* whereas the converse is true for *P. variabilis*.

Puccinia dioicae sensu lato

A number of collections cannot be assigned to particular varieties due to insufficient information. These are arranged according to their sedge hosts.

Carex lepidocarpa

Uredospores $22–25 \times 18–20\mu$, wall 2μ thick, with 2 supra-equatorial pores. Teleutospores $45–60 \times 20–23\mu$, apex $7–8\mu$ thick.

Rust is quite frequent on this host in Scotland and teleutospores occur frequently but there is no information on aecidial hosts.

Carex demissa

One teleutospore-bearing collection is known on this host from Wicklow.

Puccinia opizii Bubak

Zbl. Bakt. II, **9**, 925 (1902); Wilson & Bisby, Brit. Ured. no. 189; Gäumann, Rostpilze, p. 686.

Spermogonia epiphyllous, on discoloured spots. **Aecidia** in groups up to 5 mm. diam., cupulate, peridial cells with outer wall thicker than the inner, aecidiospores globoid or slightly angular, mostly 16–20μ diam., wall finely verrucose. **Uredosori** hypophyllous on leaves and culms, usually long-covered by the silvery epidermis, spore mass dark brown; uredospores ellipsoidal, 18–22 × 16–17μ, wall finely echinulate with 2, very rarely 3, equatorial pores. **Teleutosori** hypophyllous on the leaves or on the culms, similar to the uredosori but the spore mass black; teleutospores clavoid, constricted at the septum, rounded at the apex, 44–48 × 14–16μ, wall smooth, up to 8–10μ thick at apex. Hetereu-form.

Spermogonia and aecidia on *Lactuca*

sativa and *L. virosa*, Norfolk; uredospores and teleutospores on *Carex paniculata* and *C. appropinquata*, England, on *C. muricata*, England and Scotland.

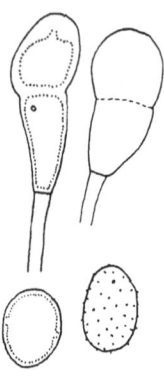

P. opizii. Teleutospores and uredospores.

This species was first recorded in England (Pethybridge *et al.*, Bull. Min. Agric. **79**, 58, 1934) on lettuces imported from Holland. Later Ellis recorded it abundantly (Trans. Norf. Norw. Nat. Hist. Soc. **17**, 137, 1951) on lettuce in markets and gardens in Norfolk and produced aecidia by inoculating from *Carex appropinquata*. Since then the species appears to have died out in Norfolk and has not been found since in spite of persistent searching (Ellis, *in litt.*). Bubak (*loc. cit.*) showed experimentally that the aecidia of this species on *Lactuca serriola* and *L. muralis* belonged to a rust on *C. muricata*. Later investigations by Tranzschel & Mayor (Bull. Soc. Myc. Fr. **36**, 97, 1920) showed that the rust on *C. muricata* can form aecidia on several species of *Lactuca* (including *Mycelis muralis*), *Crepis*, *Sonchus* and *Lapsana communis*.

Puccinia microsora Körnicke ex Fuckel

Symb. Myc. Nacht. 3, 14 (1876); Gäumann, Rostpilze, p. 709.

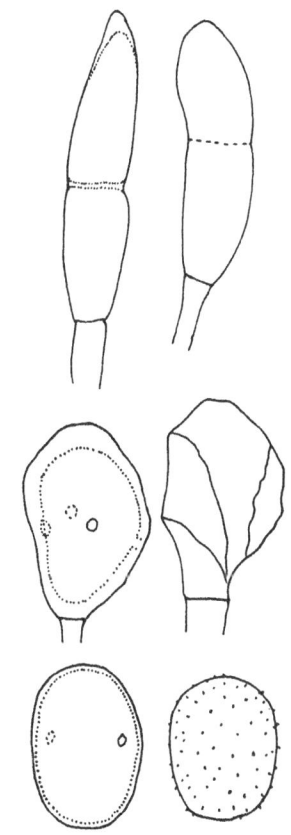

Sori hypophyllous minute, up to 1 mm. long, but numerous, often arranged in lines, all the spore forms occur in the same sori, the uredospores first and then the amphispores and teleutospores; uredospores 28–32 × 20–22 μ, thin-walled, echinulate with 2 equatorial pores; amphispores 34–43 × 23–28 μ, unicellular, angular, brown, with an irregularly thickened brown wall, 1·5–3 μ thick, with 2 equatorial pores; Teleutospores 55–60 × 16–18 μ, hyaline, thin-walled.

Uredospores, amphispores and teleutospores, on *Carex vesicaria*, Scotland, rare.

This species is known only from the Aviemore district in the Scottish Highlands where it was found in 1959. It has been recorded sparingly in northern Europe and North America, but everywhere is uncommon. There is no information on its alternation but the teleutospores do not appear functional and the rust is probably a good example of complete loss of alternation coupled with the production of resistant amphispores which effectuate overwintering.

P. microsora. Teleutospores, amphispores and uredospores.

Puccinia scirpi DC.

Fl. Fr. 2, 223 (1805); Grove, Brit. Rust Fungi, p. 239; Wilson & Bisby, Brit. Ured. no. 219; Gäumann, Rostpilze, p. 612.

Spermogonia epiphyllous, in roundish clusters. Aecidia epiphyllous, on roundish, yellow spots, in orbicular clusters, up to 1 cm. diam., surrounding a group of spermogonia, scutelliform, yellow, with a slightly irregular, laciniate, narrow, peridium, peridial cells in regular rows, the inner walls up to 4 μ thick, the outer up to 7 μ thick; aecidiospores angular-globoid, verruculose, orange, 12–20 μ diam. Uredosori scattered or in rows, often confluent, oblong, ellipsoid or linear, long-covered by the swollen epidermis which is at length longitudinally split, ferruginous, uredospores subgloboid to ovoid, often flattened on one side, 19–32 × 12–24 μ, wall echinulate, pale brown, with 2 pores. Teleutosori

similar, generally numerous and confluent, brownish-black; teleutospores
ellipsoid or subclavoid, apex rounded,
truncate or subconical, hardly constricted, attenuate downwards, 30–60 ×
12–24 μ, wall smooth, brown, at apex
5–9 μ thick, pores indistinct, that of
upper cell subapical, of the lower cell
superior; pedicel yellowish, persistent,
24–45 μ long; mesospores often numerous, 24–40 μ long. Hetereu-form.

Spermogonia and aecidia on *Nymphoides peltata*, July; uredospores and
teleutospores on *Scirpus lacustris*, July–
November. England, rare.

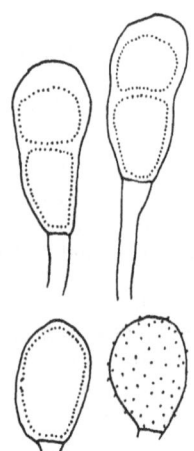

P. scirpi. Teleutospores and uredospores
(after Savulescu).

The teleutospores were first found in this country on *Scirpus lacustris*
floating down the River Ouse at King's Lynn, in November 1877, by
Plowright (TBMS. **1**, 58, 1898) but it was not until 1894 that he found
infected plants *in situ* near Earith, Huntingdonshire, and aecidia were
discovered there in July 1895 in the Old Bedford Level (Gard. Chron.
18, 96 and 135, 1895). Since then it has been found at Symonds Yat,
Hereford, on the River Wye (TBMS. **5**, 15, 1914), several times at
Woodstock, Oxfordshire, on the River Glyme and recently at Flatford
Mill, Suffolk (*ibid*. **33**, 180, 1950), Oxford (*ibid*. **33**, 375, 1950) and at
Llangorse Lake, Brecon.

Plowright (1889, 191) suggested that this rust was probably a heteroecious species and Chodat (Arch. Sci. Phys. Nat. **22**, 387, 1889) proved
its connection with the aecidium on *Nymphoides peltata*; he observed
the occurrence of the two stages in the same pond in the Botanic Gardens
in Geneva and then carried out infection of *S. lacustris* with aecidiospores from *N. peltata*. Klebahn (Abh. Nat. Ver. Bremen, **12**, 365, 1892)
observed a similar association of the two infected hosts at Bremen.
Bubak (Oest. Bot. Zeit. **48**, 14, 1898) independently confirmed the connection of the two stages.

KEY TO BRITISH GRAMINICOLOUS RUSTS

1. Teleutospores uniformly 1-celled   **2**
   Teleutospores 2- or more-celled (rarely a proportion of 1-celled spores)   **3**

2. Teleutospores scarce   *U. airae-flexuosae*, p. 359
   Teleutospores abundant   *U. dactylidis*, p. 360

3. Teleutospores with apical processes   4
   Teleutospores lacking apical processes   6

4. Teleutospore pedicels over 8 μ long   *P. festucae*, p. 258
   Teleutospore pedicels less than 8 μ long   5

5. Teleutospore processes less than 11 μ and few   *P. coronata* var. *gibberosa*, p. 254
   Teleutospore processes usually more than 11 μ and more than 4–5 in number
                                                        P. coronata var. *coronata*, p. 251

6. Teleutospores either frequently 1-celled or a proportion with more than 2
   cells (rostrupioid)   7
   Teleutospores almost constantly 2-celled   8

7. Many 1-celled teleutospores present   *P. hordei*, p. 264
   Many 3- or more-celled teleutospores present   *P. elymi*, p. 256

8. Teleutosori erumpent; spores soon exposed   9
   Teleutosori immersed; long-covered by the epidermis   12

9. Teleutospore pedicels usually more than 100 μ long   10
   Teleutospore pedicels less than 100 μ; apex strongly thickened   11

10. Uredospores globose with 3 pores; on *Molinia*   *P. moliniae*, p. 268
    Uredospores obovate with 4 equatorial pores; on *Phragmites*
                                                          P. phragmitis, p. 269

11. Uredospores with scattered pores, capitate uredoparaphyses present
    (Teleutospores verrucose, no paraphyses; *P. pratensis*, p. 275)
                                                          P. magnusiana, p. 266
    Uredospores broadly elliptic with 4 equatorial pores, uredoparaphyses
    absent   *P. graminis*, p. 259

12. At least some uredoparaphyses capitate and intermixed with spores   13
    Uredoparaphyses either absent or not capitate; if present, only clavate and
    peripheral at most   16

13. Uredoparaphyses thick-walled   14
    Uredoparaphyses thin-walled   *P. pygmaea*, p. 276

14. Uredosori pale yellow; arranged in longitudinal rows   *P. brachypodii*, p. 250
    Uredosori not so   15

15. Uredoparaphyses strongly capitate with narrow neck   *P. poae-nemoralis*, p. 271
    Uredoparaphyses absent or at most only clavate   *P. deschampsiae*, p. 255

16. Uredosori yellow or very pale orange   17
    Uredosori orange, brownish-orange or yellowish-brown   18

17. Uredosori in longitudinal rows; teleutospore wall less than 1·5 μ thick in
    upper cell   *P. striiformis*, p. 294
    Uredosori scattered; teleutospore wall more than 1·5 μ thick; on *Vulpia* (and
    other genera?)   *P. schismi*, p. 290

18. Teleutospores usually more than 75 μ long   *P. longissima*, p. 265
    Teleutospores usually less than 75 μ long   19

19. Diplont stages on *Poa* or *Phalaris*; teleutospores abundantly formed   20
    Diplont stages on other graminaceous genera   *P. recondita*, p. 278

20. On *Poa*, aecidia on *Tussilago*   *P. poarum*, p. 274
    On *Phalaris*, aecidia on Monocotyledons   *P. sessilis*, p. 291

Key to predominantly Uredosporic collections on grasses

1. Uredoparaphyses well developed, scattered through the sori   **2**
Uredoparaphyses absent or at most represented by a few marginal thin-walled cells (old uredospore pedicels should not be mistaken for paraphyses)   **6**

2. Uredosori in distinct rows; on *Brachypodium*   *P. brachypodii*, p. 250
Uredosori less regularly arranged   **3**

3. Uredoparaphyses thin-walled, on *Ammophila* and *Calamagrostis*
    P. pygmaea, p. 276
A proportion of uredoparaphyses more or less thick-walled (> 1·5μ)   **4**

4. Uredoparaphyses strongly coloured; on *Phragmites*   *P. magnusiana*, p. 266
Uredoparaphyses almost hyaline; on other hosts   **5**

5. Most uredoparaphyses with a distinct narrow neck; on species of *Poa*, *Puccinellia*, *Glyceria*?, *Festuca*, *Anthoxanthum* and *Arrhenatherum*
    P. poae-nemoralis, p. 271
Uredoparaphyses at most clavate   *P. deschampsiae*, p. 255

6. Uredospores with 3–4 equatorial pores   **7**
Uredospores with scattered pores   **9**

7. Teleutosori always present and teleutospore pedicels over 100μ long; on *Molinia*   *P. moliniae*, p. 268
Teleutosori present or absent, if present pedicels less than 100μ, on other grasses and on cereals   **8**

8. Uredospore wall 3–4μ thick   *P. phragmitis*, p. 269
Uredospore wall 1·5–2μ thick   *P. graminis*, p. 259

9. On host genera list alphabetically below:

Agropyron

 a. Uredosori light chrome yellow, spore wall hyaline   *P. striiformis*, p. 294
Uredosori orange or brownish-orange, spore wall coloured   *b*

 b. Only uredosori present   *P. recondita* f. sp. *agropyrina*, p. 279
Teleutosori usually present   *c*

 c. Teleutospores coronate   *P. coronata*, p. 251
Teleutospores not coronate, sori immersed with brown cylindrical paraphyses
    P. recondita f. sp. *agropyrina* and f. sp. *persistens*, pp. 279 and 286

Agrostis

 a. Only scanty uredosori with immersed teleutosori, teleutospores not coronate   *b*
Either uredosori only or sparse teleutosori with coronate spores
    P. coronata, p. 251

 b. [In mountains with *Thalictrum alpinum*   *P. recondita* f. sp. *borealis*]
In lowlands with aecidia on *Aquilegia*   *P. recondita* f. sp. *agrostidis*, p. 280

Alopecurus

 a. Associated with aecidia on *Ranunculus*: abundant immersed teleutosori usually present   *P. recondita* f. sp. *perplexans*, p. 286
Not so; if teleutospores present they are coronate   *P. coronata*, p. 251

× *Ammocalamagrostis*   *P. elymi*, p. 256

Ammophila   *P. elymi*, p. 256

[*Anthoxanthum*]
  In mountains with *Thalictrum alpinum*   *P. recondita* f. sp. *borealis*, p. 281

Arrhenatherum   *P. coronata*, p. 251

Avena   *P. coronata*, p. 251

Brachypodium   *P. striiformis* (records doubtful), p. 294

Bromus   *P. recondita* f. sp. *bromina*, p. 282

Calamagrostis   *P. coronata*, p. 251

Dactylis
  a. Teleutosori usually absent, uredosori in rows, uredospores light chrome
    in mass   *P. striiformis* var. *dactylidis*, p. 297
    Teleutosori usually present and uredosori orange or brownish-orange   *b*
  b. Teleutospores 1-celled; near aecidia on *Ranunculus*   *Uromyces dactylidis*, p. 360
    Teleutospores 2-celled, coronate   *P. coronata*, p. 251

Deschampsia
  a. On *D. cespitosa*, teleutospores if present coronate   *P. coronata*, p. 251
    On *D. flexuosa*, teleutospores if present unicellular   *Uromyces airae-flexuo-*
           sae, p. 359

Elymus
  a. Uredospores light chrome   *P. striiformis*, p. 294
    Uredospores orange to brownish   *P. elymi*, p. 256

Festuca
  a. Teleutospores usually present, unicellular; on *Festuca rubra* and *F. ovina*
           Uromyces dactylidis, p. 360
    Teleutospores absent or if present coronate   *b*
  b. Teleutospores absent   *Uredo festucae*, p. 259
    Teleutospores usually present, coronate   *c*
  c. Teleutosori dehiscent; on *Festuca ovina*, *F. duriuscula*, (*rubra*?), and *F.*
    rubra var. *arenariae*   *P. festucae*, p. 258
    Teleutosori indehiscent; on *Festuca altissima*, *F. arundinacea*, *F. gigantea*,
    or *F. pratensis*   *P. coronata*, p. 251

Helictotrichon   *P. pratensis*, p. 275

Holcus
  a. Uredosori light chrome, arranged in lines   *P. striiformis*, p. 294
    Uredosori darker, not in distinct lines   *b*
  b. Teleutosori usually present and spores coronate   *P. coronata*, p. 251
    Teleutosori when present with non-coronate spores (iii often present on
    H. mollis but scarcer on *H. lanatus*)   *P. recondita* f. sp. *holcina*, p. 285

Hordeum
  a. Uredosori light chrome; teleutospores if present mostly 2-celled
           P. striiformis, p. 294
    Uredosori darker; many teleutospores 1-celled   *P. hordei*, p. 264

Puccinia brachypodii Otth

Mitt. Naturf. Ges. Bern, **1861**, 81 (1861); Wilson & Bisby, Brit. Ured. no. 106; Gäumann, Rostpilze, p. 514.

[*Epitea baryi* Berk. & Br., Ann. Mag. Nat. Hist. **13**, 461 (1854).]
Puccinia baryi Wint., Rabh. Krypt. Fl. Ed. 2, **1** (1), 178 (1882); Grove, Brit. Rust Fungi, p. 280.

[**Spermogonia** epiphyllous on orange-red spots 2–4 mm., very small, rather numerous, yellow becoming brownish-black. **Aecidia** hypophyllous on yellow spots, peridium pale yellow, rather thick, not revolute, slightly or not torn; aecidiospores yellow in the mass, sub-globoid or ellipsoid, 24–27 × 16μ, wall clear yellow or almost hyaline, of equal thickness throughout, finely verrucose.] **Uredosori** mostly epiphyllous, on linear brown spots, scattered or in groups, often in long, linear series, minute, elongate, reddish-brown, with numerous clavate to capitate paraphyses; uredospores globoid to obovoid, delicately verru-culose, yellow, 20–23 × 17–20μ. Tele-

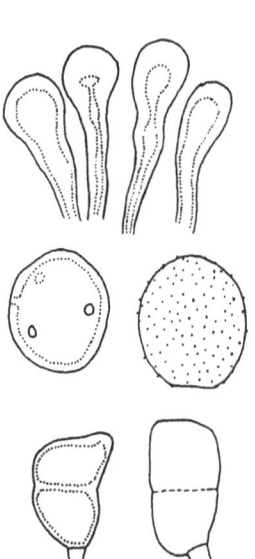

P. brachypodii. Uredo-paraphyses, uredospores and teleutospores.

utosori similar, but long-covered by the epidermis, blackish-brown, teleutospores very irregular, ellipsoid or subclavoid or ovoid, obtuse, or truncate and slightly thickened and undulate above, hardly constricted, somewhat attenuate below, smooth, pale brown, darker above, 25–40×15–25μ; a few mesospores pre-sent; pedicels very short, pale brown, with a dark-brown zone near the base of the spore. Hetereu-form.

[Aecidia on *Berberis vulgaris*]; ure-dospores and teleutospores on living or fading leaves of '*Brachypodium pinnatum*' and *B. sylvaticum*, July–November. Great Britain and Ireland, frequent.

Tranzschel (C. R. Acad. Sc. URSS. **1931**, 45, 1931) suggested that the aecidium would be found on *Berberis vulgaris* on account of the great resemblance between the species and *Puccinia pygmaea* Erikss. Mayor proved that this is true (Bull. Soc. Neuchâtel. Sci. Nat. **58**, 14, 1933) by infection experiments carried out in Switzerland. The aecidia of *P. brachypodii* are similar to those of *P. pygmaea* but are clearly distinguishable from those of *P. graminis* (see p. 259). No deformation of the leaf of the Barberry is caused by *P. brachypodii*; the external walls of the cells of the peridium are smooth while the internal are rather coarsely verrucose, the warts being distinctly larger than those found on the inner wall of the peridial cell of *P. graminis*; also the inner wall of *P. brachypodii* is obviously thicker than that of *P. graminis*. Up to the present the aecidium has not been recorded in this country.

Jørstad (1932, 329) pointed out that the hyaline capitate paraphyses accompanying the uredospores resemble those of *P. poae-nemoralis* (see p. 271) but are shorter; on this account and also because the uredosori and teleutosori of *P. brachypodii* are arranged in longitudinal rows he considers it distinct from *P. poae-nemoralis*.

Grove (*loc. cit.*) stated that the wall at the base of the teleutospore is much thickened, but the wall in this position is of the same thickness as in the other parts of the lower cell. It seems that the appearance is caused by the deposition of a narrow zone of dark-brown material on the wall of the pedicel immediately below the point of attachment. The mesospores which also possess the dark-brown zone at the base, are usually found in the outer part of the sorus.

Puccinia coronata Corda

Icon. Fung. **1**, 6 (1837); Wilson & Bisby, Brit. Ured. no. 141; Gäumann, Rostpilze, p. 572.

Puccinia lolii Niels., Ugeskr. Landm. IV, **9** (1), 549 (1873); Grove, Brit. Rust Fungi, p. 255.

Puccinia coronifera Kleb., Z. Pfl.-Krankh. **4**, 135 (1894).

Spermogonia amphigenous, chiefly epiphyllous and on the petioles, between the aecidia, with projecting paraphyses.

Aecidia amphigenous, chiefly hypophyllous and on the petioles, in rounded groups or irregularly scattered on yellow or yellowish-purple spots, producing some distortion especially on the petioles, cylindrical or cup-shaped with a whitish, laciniate, revolute margin; peridial cells firmly connected, of variable form but often hexagonal not arranged in definite rows, on the outerside projecting for a short distance over the cells below, outer wall strongly thickened up to 7–8 μ, finely punctate when seen from above, inner wall thinner with superficial, rather coarse warts, 20–28 × 18–22 μ; aecidiospores angular-globoid, very finely verruculose, orange, 16–25 × 12–20 μ. **Uredosori** amphigenous, scattered or arranged in rows, sometimes confluent, minute, oblong, pulverulent, orange, usually with a few thin-walled, clavate paraphyses at the periphery; uredospores globoid to ovoid, yellow, variable in size, 14–39 × 10–35 μ; wall finely echinulate with 3 or 4 scattered pores. **Teleutosori** hypophyllous, scattered or sometimes arranged in circles around the uredosori, occasionally confluent, oblong or linear, long-covered by the epidermis, then naked, black, usually with more or less distinct subepidermal paraphyses; teleutospores clavate, apex flattened and crowned with 5–8 digitate, sometimes branched, brownish divergent projections, the projections up to 20 μ long but often shorter, slightly or not constricted, 30–60 × 14–20 μ; tapering towards the base, smooth, brown, pore of upper cell apical, of lower close to the septum; pedicels short, rather thick, very pale brown. Hetereu-form.

Spermogonia and aecidia on *Frangula alnus* and *Rhamnus catharticus* May–June; uredospores and teleutospores on *Agropyron repens, Agrostis canina, A. stolonifera, A. tenuis, Alopecurus pratensis, Arrhenatherum elatius, Avena fatua,*

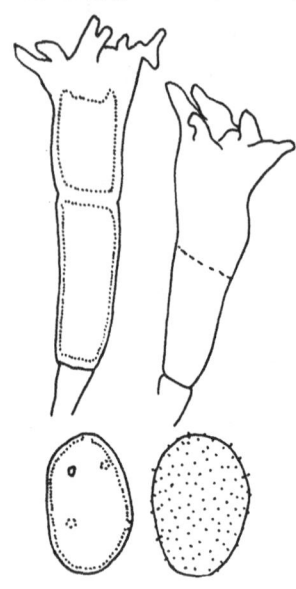

P. coronata. Teleutospores and uredospores.

A. sativa, A. strigosa, Calamagrostis canescens, C. epigejos, Dactylis glomerata, Deschampsia cespitosa, Festuca altissima, F. arundinacea, F. gigantea, F. pratensis, Festuca pratensis × L. multiflorum, Festuca pratensis × L. perenne, Holcus lanatus, 'H. mollis', Lolium multiflorum, L. perenne, Phalaris arundinacea and *Poa pratensis*, August–October. Great Britain and Ireland, common.

The first successful cultures of the crown rust of grasses were made by de Bary (Monatsber. K. Akad. Wiss. Berlin, **1865**, 211, 1866) on detached leaves of *Rhamnus*. He was followed by a number of investigators including Eriksson (Ber. D. Bot. Ges. **12**, 320, 1894; Zbl. Bakt. II, **3**, 294, 1897) whose results were summarised by Klebahn (1904, 254). Among these, Plowright (1889) in this country, found that teleutospores from *Dactylis glomerata* and '*Festuca sylvatica*' (= *F. altissima*?) readily produced the aecidium on *Frangula alnus* but he failed with teleutospores from *Lolium perenne* to obtain a similar result. He noted that the rust on *L. perenne* produced abundant uredospores in the autumn while on *Dactylis* it is an early summer rust with little uredospore production. He thought that two species were confounded under the name *Puccinia coronata*.

Klebahn (Z. Pfl.-Krankh. **2**, 332, 1892) first suggested the division of the rust into two species *P. coronata* Kleb. which had its aecidial stage on *Frangula alnus* and what he later described as *P. coronifera* Kleb. (= *P. lolii* Niels.) with its aecidia on *Rhamnus cathartica*. This division into two species was usually adopted, especially in Europe and in general publications upon the rust fungi, such as those by Plowright, Fischer, Sydow and Grove. As the result of further investigation Treboux (Ann. Myc. **10**, 557, 1912), working in Russia, considered the distinction between the two species was not so clear as was supposed. A similar conclusion was reached by Melhus *et al.* (Res. Bull. Iowa Agric. Exp. Sta. **72**, 211, 1922) and two species are no longer recognised. However, the host specialisation of the rust has been intensively investigated. Gäumann (*loc. cit.*) lists twelve *formae speciales*. These have also been regarded as varieties especially in American publications. These subspecific categories have been subdivided into physiologic races—especially var. *avenae*.

Murphy (U.S. D. A. Tech. Bull. no. 433, 1935) gave a summary of the literature on this subject up to 1935 and Gäumann (*loc. cit.*) lists the major papers by foreign workers on development and host specialisation.

In Britain, Haines (TBMS. **20**, 252, 1936) inoculated oats, wheat, barley, rye, *Lolium perenne* and *L. italicum* with uredospores from oats and obtained infection on oats only, showing that the rust possessed some degree of specialisation. Brown (Ann. Appl. Biol. **24**, 504, 1937, and **25**, 506 1938) published a comprehensive and detailed account of the physiologic specialisation of the uredospore stage of *P. coronata* and its relationship to *Frangula alnus* and *Rhamnus cathartica* which are the aecidial hosts of the fungus in this country. This work is still the most important and most complete published contribution to our knowledge of the rust in this country.

The following seven varieties of *P. coronata* were distinguished in Britain as a result of this work:

var. *alopecuri* infecting *Alopecurus pratensis*
var. *arrhenatheri* infecting *Arrhenatherum elatius*
var. *avenae* infecting *Avena* spp.
var. *calamagrostidis* infecting *Calamagrostis canescens* and *Phalaris arundinacea*
var. *festucae* infecting *Festuca elatior*
var. *lolii* infecting *Lolium perenne*
var. *holci* infecting *Holcus lanatus*

All these varieties had previously been identified in continental Europe.

Four physiologic races were distinguished in the var. *avenae*, race 6, previously identified in North America and Australia and races 42, 43 and 44, hitherto undescribed. Races 6, 42 and 44 were obtained in Great Britain and race 43 in Portugal. According to Brooks (Ann. Appl. Biol. **31**, 364, 1944) a further four races were isolated in later experiments but this number is low compared with 139 races isolated from 144 collecionst on *Avena* by Straib (Arb. Biol. Abt. Berl. **22**, 121, 1937).

There were no clear-cut differences, either in morphological or pathological characters between the varieties assigned by Klebahn to *P. coronata* Kleb. and those assigned to *P. coronifera* Kleb. (*P. lolii* Niels.) but the varieties were found to show a considerable degree of specialisation in their relation to the alternate hosts, vars. *alopecuri, arrhenantheri, avenae, festucae, holci* and *lolii* producing aecidia on *R. cathartica* only and var. *calamagrostidis* on *Frangula alnus* only.

The pathenogenicity of the varieties was not altered appreciably by passage through the alternate host and the varieties did not appear to hybridise readily. Griffiths (TBMS. **41**, 373, 1958) isolated 13 races from 24 collections of var. *avenae* in Wales and south-west England.

Williams & Verma (Ann. Appl. Biol. **44**, 453, 1956) infected many subspecies of *Avena strigosa* and *A. fatua* with British isolates of *P. coronata avenae* in their search for genes conferring resistance to this rust.

A variety of *P. coronata*, now known as var. *gibberosa* (Lagerh.) Jørst. (Jørstad, Skr. Norske Vidensk.-Akad. Oslo, **2**, 9, 1949 and Hylander *et al.* 1953, 46) was described by Lagerheim (Ber. D. Bot. Ges. **6**, 124, 1888) from material found on *Festuca altissima*, near Freiburg, Germany, as *P. gibberosa*. This variety is said to be distinguished by its teleutospores in which the apical projections are few (4–5, rarely up to 7), and wart-like (rarely up to $11 \cdot 5\mu$ long when digitate) and sometimes totally absent; the greatest teleutospore length, in Norwegian material is 61μ. The uredospores in Norwegian material of var. *gibberosa* are approximately $19 \cdot 5$–$32 \times 17 \cdot 5$–25μ whereas those in var. *coronata* are 17–$22 \cdot 5 \times 15 \cdot 5$–$17\mu$ (Jørstad, *loc. cit.*)

The status of this variety in Britain is uncertain. Grove's record on *F. ovina* (J. Bot. Lond. **59**, 314, 1921) is certainly incorrect. One collection on *F. gigantea* from Hamilton, Lanarkshire, has teleutospores with poorly developed apical processes said to be characteristic of the variety. In the only other collection of *F. gigantea* bearing teleutospores, from Horton, Gloucestershire (Herb. Kew), the teleutospores are typically of the var. *coronata* type. The size of the uredospores (all 22–25 × 16–20µ) does not seem to help in the separation of these two and two other collections with uredospores only. The rusts on *F. gigantea* and *F. altissima* require further collection and study.

Puccinia deschampsiae Arth.

Bull. Torr. Bot. Club, 37, 570 (1910).
[*Uredo airae* Lagh., J. Bot. Fr. 2, 432 (1888).]
Puccinia airae Mayor & Cruch., Bull. Soc. Vaud. Sci. Nat. 51, 628 (1917); Wilson & Bisby, Brit. Ured. no. 85; Gäumann, Rostpilze, p. 555.

Spermogonia and **aecidia** unknown. **Uredosori** generally epiphyllous, often on yellowish to violet spots, scattered or in small rows, elliptical to oblong, 0·25–0·75 mm. long, remaining covered for rather a long time, yellow; paraphyses numerous, clavate or capitate, generally hyaline, finally brownish, 60–90 × 10–18µ, wall 2–4µ thick at the apex; uredospores globoid, ovoid or ellipsoid, 24–32 × 18–26µ, wall 1·5µ thick, densely or shortly echinulate or echinulate-verrucose, hyaline, with 8–12 scattered pores. [**Teleutosori** epiphyllous, small, black, subepidermal, paraphysate; teleutospores elongate ellipsoid, rounded, truncate or bluntly pointed at the apex, narrowed at the base, yellowish-brown, darker at the apex, 35–50 × 14–20µ, wall smooth, up to 5µ thick at the apex; pedicel short, brown, persistent.] Hemi-form?
   On *Deschampsia cespitosa.* Great Britain, frequent.

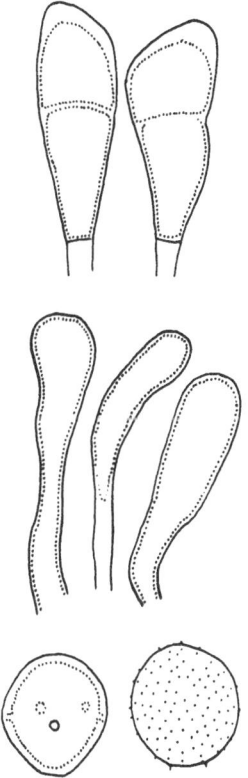

P. deschampsiae. Teleutospores, uredo-paraphyses and uredospores.

Teleutospores have not been found in this country. They were discovered by Mayor & Cruchet in Switzerland in 1917. Treboux (Myk. Zbl. 5, 124, 1914) described the overwintering by means of uredospores in Russia and this evidently takes place here. It is very probable that this is the rust which, according to Plowright (1889, 191) was placed under *Puccinia baryi* (= *P. brachypodii*) by Soppitt.

   Grove (1913, 265) mentioned this rust under *P. dispersa* (*sens. lat.*). Jørstad (1932, 328 and 1950, 9) considered that it is closely related to

P. poae-nemoralis but is distinguished from it by its more clavate uredo-paraphyses which are obviously bent or incurved but are not abruptly narrowed below the head.

It is an arctic rust extending to the extreme north of Norway and is found in Iceland.

Puccinia elymi West.

Bull. Acad. Roy. Belg. **18** (2), 408 (1851); Gäumann, Rostpilze, p. 502.

Rostrupia elymi (West.) Lagh., J. Bot. Fr. **3**, 188 (1889); Wilson & Bisby, Brit. Ured. no. 251.

[*Uredo ammophilae* Syd., Bot. Not. **1900**, 42 (1900).]

Puccinia ammophilae A. L. Guyot, Rev. Path. Vég. **19**, 36 (1932); Gäumann, Rostpilze, p. 550.

[*Rostrupia ammophilae* [Syd.] Wilson, Trans. Bot. Soc. Edin. **33**, iv (1940).]

[**Spermogonia** in small groups, orange. **Aecidia** hypophyllous, on thickened spots which are purple-brown on the upper and yellowish on the under surface with a brownish border, in rounded groups, urn-shaped or more or less cylindrical, yellow with a whitish laciniate margin; aecidiospores angular, subgloboid or ellipsoid, minutely verruculose, orange, 14–28 μ diam.] **Uredosori** epiphyllous, scattered or arranged in lines, often confluent, up to 3 mm. long, surrounded by the ruptured epidermis; uredospores 23–32 × 20–28 μ, wall up to 3·5 μ thick, pores scattered, 7–9. **Teleutosori** hypophyllous, scattered or arranged in lines, often confluent, up to 3 mm. long, surrounded by the ruptured epidermis (or in British collections on *Ammophila* long-covered by the epidermis and sclerenchyma) divided into 2–3 loculi by brown, subepidermal paraphyses, compact, greyish-black; teleutospores mostly 2–3-septate, rarely uniseptate, 45–90 × 10–18 μ, apex rounded and thickened (3–9 μ); pedicel short, brown, persistent. Hetereu-form.

[Spermogonia and aecidia on *Thalictrum minus* associated with *Elymus*];

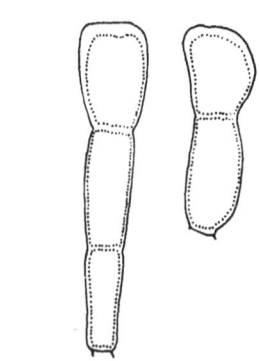

P. elymi. Teleutospores (after Viennot-Bourgin).

uredospores and teleutospores on *Ammophila arenaria*, × *Ammocalamagrostis baltica* and *Elymus arenarius*. England and Scotland, uncommon but probably overlooked.

This rust was first recorded in Britain by Massee (1913) who found it on *Elymus arenarius* at Palling, Norfolk in August. It was later recorded by Ellis (1934) in Norfolk who found it only in the uredospore stage; the uredosori were sometimes accompanied by capitate paraphyses.

Several other collections of uredosori without teleutosori may belong here but they may equally well belong to *Puccinia striiformis* which is also recorded on *Elymus*. Any rust on *Elymus* must develop its sori in rows, due to the longitudinal rows of sclerenchyma and in herbarium material it does not seem possible to distinguish the uredospores of *P. striiformis* and those of the *P. elymi–recondita* group.

Recently (*in. litt.*) an aecidium has been described on *Thalictrum minus* near Aberdeen but this has been shown by culture to form its uredospore and teleutospore stages on *Agropyron junceiforme* and has been assigned to *P. recondita* f. sp. *persistens* (see p. 286); this aecidium appears to be similar to that described for *P. elymi*.

In Denmark Rostrup (Lind, Danish Fungi, 1913, 310) found the aecidium of the *Elymus* race during June and July. This investigator (Overs. K. Danske Vidensk. Selsk. Forh. no. 5, 269, 1898) produced uredospores on *E. arenarius* by sowing aecidiospores. Tranzschel (Myc. Zbl. **4**, 70, 1914) in Russia, obtained aecidia on *T. minus* by sowing teleutospores from *Elymus* sp. Fraser (Mycol. **11**, 129, 1919) and Mains (Pap. Mich. Acad. Sci. **17**, 328, 1933) in North America have shown that aecidiospores from *T. dasycarpum* will infect several species of *Elymus*. Jørstad (1932, 332) has pointed out that in Europe and northern Asia the rust on *E. arenarius* may have 1-, 2- and 3-septate teleutospores but that forms occur in which only 1-septate spores are found.

A small quantity of this rust was discovered on *A. arenaria* growing on the sand dunes on Spurn Head, Yorkshire, in 1936 and was provisionally recorded as *Rostrupia ammophilae* [Syd.] Wilson in 1940 (*loc. cit.*). Later it was noted that A. L. Guyot had found an almost similar rust in 1931, on the same host, in northern France and had recorded it, with a short description, as *Puccinia ammophilae* [Syd.] A. L. Guyot in 1932 (*loc. cit.*). In 1948 Guyot again found the rust in the same locality and gave a fuller account (Uredineana, **3**, 381, footnote, 1951, and **4**, 253, 1953). Guyot in his account of the rust did not mention the occurrence of teleutospores with more than 2 cells but in the material from Yorkshire a number of 3-celled spores were present and few with 4 cells were seen.

In 1900, Sydow in his account of *Uredo ammophilae* recorded the rust on *Ammophila baltica* (= × *Ammocalamagrostis baltica*) as well as on *A. arenariae* and it was found on the former host on the coast of Northumberland in 1949 with both uredospores and teleutospores. The aecidial host is unknown but may well be *Thalictrum minus* subsp. *arenarium* which occurs where the rust has been found.

On *A. arenaria* the teleutosori although hypophyllous are situated more or less in the centre of the leaf and are isolated from the lower surface by the thick-walled epidermis and one or two continuous layers of subepidermal sclerenchymatous cells; there are no stomatal openings on the lower surface of the leaf. There is no information as to how these sori open but if it is by gradual disintegration of the tissues, as would seem to be necessary, it appears that the thin-walled chlorenchyma of the upper surface would decay more readily and sooner than the sclerenchyma of the lower surface. It is possible therefore that the basidiospores escape from the upper side of the leaf.

Puccinia festucae Plowr.

Grevillea, **21**, 109 (1893); Grove, Brit. Rust Fungi, p. 257; Wilson & Bisby, Brit. Ured. no. 263; Gäumann, Rostpilze, p. 580.

Spermogonia in small groups, honey-coloured. **Aecidia** hypophyllous, on round, yellow or brownish spots, in rounded groups, 2–5 mm. diam., shortly cylindrical or cupulate, whitish-yellow, with recurved irregularly laciniate margin, peridial cells not in regular rows, firmly connected, on the outer side shortly projecting over the cell below, outer wall, up to 7 μ thick, finely punctate in plan, inner wall thinner, up to 3–4 μ, tesselate, in surface view finely warted; aecidiospores in distinct chains, globoid, ellipsoid or polyhedroid, finely verruculose, 16–27 μ diam. **Uredosori** epiphyllous, scattered, minute, oblong, yellow, aparaphysate; uredospores globoid to ellipsoid, yellow-brown, 22–30 μ diam., wall 1–2 μ thick, echinulate with spines 2–3 μ apart, with about 6 scattered pores. **Teleutosori** hypophyllous scattered, minute, oblong or sublinear, at first covered, then becoming free by the splitting of the epidermis, blackish-brown, without subepidermal paraphyses; teleutospores clavoid or cylindrical with strongly thickened apex which is sometimes flat and sometimes bluntly pointed, slightly contracted at the septum, tapering towards the base, crowned with 1–6 usually rather elongate, unbranched or bifid, brownish projections, usually directed upwards or occasionally to the side, smooth, light brown, 50–80 μ (including projections) × 16–18 μ, thickness of wall at summit (including projections) 7–20 μ, pores not obvious; pedicels brownish, persistent, 15–25 μ long. Hetereu-form.

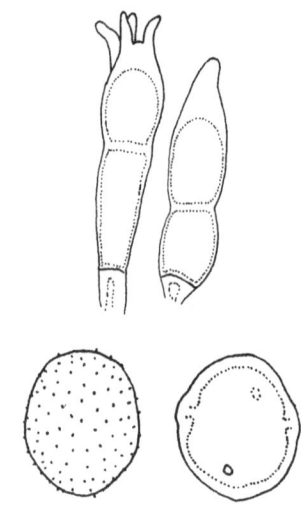

P. festucae. Teleutospores (after Fischer) and uredospores.

Spermogonia and aecidia on *Lonicera periclymenum*, June–August; uredospores and teleutospores on *Festuca longifolia*, *F. ovina*, *F. rubra* and *F. rubra* var. *arenaria*. Great Britain and Ireland, frequent.

The aecidial stage has been recorded from various localities in England, Wales, Scotland and Ireland; records on *Festuca* sp. are less frequent, probably on account of the inconspicuous appearance of the sori. Plowright (*loc. cit.*) first showed that the aecidial stage on *Lonicera periclymenum* was connected with the rust on *Festuca ovina* and this was confirmed repeatedly by Fischer (1904, 377) and Klebahn (1904, 290).

The paraphyses described and figured by Guyot are interpreted by Jørstad (1950, 20) as pedicels of shed uredospores. The drawing appears to be that of a section of an old uredosorus in which one mature uredospore and three young teleutospores are shown, together with a large number of 'paraphyses' which are described as almost linear, thin-walled and colourless, long (up to 80μ) and straight, slightly swollen near the summit where they are 5–8μ wide. A similar section of a mature teleutosorus made from material on *F. ovina* collected near Edinburgh, shows no uredosori and no paraphyses.

[*Uredo festucae* DC.]

Fl. Fr. 6, 82 (1815).

[*Uredo festucae-ovinae* Erikss., Arkiv. Bot. 18 (19), 13 (1923).]

Uredosori epiphyllous usually enclosed by the inrolled host leaves, seated on yellow spots which produce yellow and green banding of the host leaves, 0·5–1 mm. long, subconfluent in heavy infections, aparaphysate; uredospores broadly ellipsoid, 28–32 × 23–25μ, wall 2–2·5μ thick, finely echinulate with 6–8 scattered pores.

Uredosori only, on *Festuca ovina* and *F. rubra*. Common throughout Britain.

The uredosori and uredospores of *P. festucae* very closely resemble those of *Uromyces dactylidis* and it is hardly possible to separate them by morphological characters and it is with *U. festucae* (= *U. dactylidis p.p.*) that Gäumann (1959, 237) associates these uredo collections on fescues. Jørstad (1950, 20) has very convincingly argued that it is unlikely that these uredo infections belong to *U. dactylidis* which appears to be obligatorily heteroecious and usually produces abundant teleutospores. He accordingly associates *U. festucae* with *Puccinia festucae* and this course is followed here.

Puccinia graminis Pers.

Syn. Meth. Fung. p. 228 (1801); Grove, Brit. Rust Fungi, p. 250; Wilson & Bisby, Brit. Ured. no. 160; Gäumann, Rostpilze, p. 715.

Puccinia phlei-pratensis Erikss. & Henn., Z. Pfl.-Krankh. 4, 140 (1894); Grove, Brit. Rust Fungi, p. 283; Wilson & Bisby, Brit. Ured. no. 161; Gäumann, Rostpilze, p. 767.

Puccinia anthoxanthi Fuck., Jahrb. Nass. Ver. Nat. **27-28**, 15 (1873); Grove, Brit. Rust Fungi, p. 269; Wilson & Bisby, Brit. Ured. no. 92.
Puccinia linearis Röhl., Deutschl. Fl. Ed. 2, 3 (3), 132 (1813).

Spermogonia amphigenous mostly epiphyllous, in groups in the middle of thickened coloured spots on the leaf, honey-coloured, flask-shaped, with paraphyses and flexuous hyphae; spermatia, ovoid, 2–4 × 1·3–1·5 μ. **Aecidia** hypophyllous, and on fruits, on thickened, rounded reddish-purple spots with a yellowish margin, 2–5 mm. diam., grouped or scattered, cylindrical or sometimes tubular, white with sub-erect, laciniate margin, peridial cells not in distinct rows, almost rectangular, outer wall 7–8 μ thick, punctate in surface view, inner wall thin, 2·5–5 μ, with closely-set, small warts; aecidiospores angular-globoid, 14–26 μ diam., wall smooth or very finely verruculose, colourless, 1–1·5 μ thick at sides, 5–9 μ above, contents yellow. **Uredosori** amphigenous, often on the sheaths and culms, scattered or arranged in rows, linear, 2–3 mm. long, often confluent and then up to 1 cm. or more long, surrounded by the split epidermis, pulverulent, yellowish-brown; uredospores ellipsoid or ovoid, golden-brown when mature, 21–42 × 16–22 μ, wall echinulate, 1·5–2 μ thick, with 4 equatorial pores. **Teleutosori** amphigenous, in the same or similar sori, soon naked, pulvinate, black; teleutospores broadly ellipsoid or clavoid, rounded or attenuate at the summit, slightly constricted at the septum, attenuate at the base, smooth, chestnut-brown, 35–60 × 12–22 μ, wall smooth, 6–13 μ thick at apex;

pedicel long, persistent. Hetereu-form.

Spermogonia and aecidia on the leaves and fruits of *Berberis vulgaris*, *Mahonia aquifolium* and *M. bealii*; uredospores and teleutospores on *Agropyron repens*, *Agrostis canina*, *A. stolonifera*, *A. tenuis*, *Alopecurus pratensis*, *Anthoxanthum odoratum*, *Arrhenatherum elatius*, *Avena*

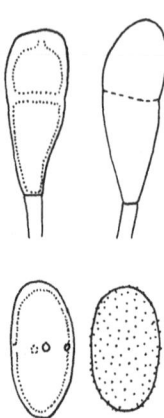

P. graminis. Teleutospores and uredospores.

fatua, *A. sativa*, *A. strigosa*, *Briza media*, *Bromus inermis*, *B. ramosus*, *B. sterilis*, *Cynosurus cristatus*, *Dactylis glomerata*, *Deschampsia cespitosa*, '*Festuca gigantea*' *F. pratensis*, *F. rubra*, '*F. arundinacea*', *Helictotrichon pratense*. '*Hordeum murinum*', *H. vulgare*, '*Lolium perenne*', *Phleum pratense*, *Poa annua*, *P. pratensis*, *P. trivialis*, *Secale cereale*, *Trisetum flavescens*, *Triticum aestivum*.

In Britain *Puccinia graminis* occurs fairly regularly on wheat especially in the south-west of England, but only occasionally is generally epidemic (Moore, 1959, 303). Attack is usually too late to cause serious economic loss. It is less frequent but widespread on oats, and rather uncommon on rye and barley. *P. graminis* f. sp. *tritici* has never been isolated from aecidia on barberry in Britain and there is now good evidence (Ogilvie, *in litt.*) that much infection spreads annually by wind-borne uredospores from western Europe. The appearance of races on wheat in Britain follows closely the pattern of western Europe.

On wild hosts the rust is most frequently found on *Agropyron repens*, *Agrostis canina* and *A. tenuis* and is usually then closely associated with barberries. *Alopecurus pratensis*, *Arrhenatherum elatius*, *Avena fatua*, *Dactylis glomerata* and *Phleum pratense* are less commonly infected. On *Bromus sterilis*, *Anthoxantum odoratum*, *Briza media*, *Bromus asper*, *Cynosurus cristatus*, *Deschampsia cespitosa*, *Festuca rubra*, *Helictotrichon pratense*, *Poa annua*, *P. pratensis*, *P. trivialis* and *Trisetum flavescens* the rust is rare. Records on *Festuca gigantea* (Plowright, 1889), *F. pratensis*, *F. arundinacea*, *Hordeum murinum*, *Lolium perenne* and *Bromus inermis* require confirmation.

Williams & Verma (Ann. Appl. Biol. **44**, 456, 1956) infected several subspecies of *Avena strigosa* and *A. fatua* with British isolates of *P. graminis avenae* in their search for genes for resistance to the rust.

The record of *P. graminis* on *Bromus briziformis* (Wilson & Bisby, 1954) is due to a misidentification. Only uredospores are present but they are not of the *P. graminis* type but belong to *P. recondita* f. sp. *bromina*.

This is the well-known black rust or mildew of cereals about which so much has been written (for a summary up to 1937 see Lehmann, Die Schwartzrost, Munich, 1937). It is now generally known as the Black Rust or Stem Rust since the long linear yellowish-brown uredosori and black teleutosori are found much more frequently on the sheath and culms than on the leaves, a character by which it can generally be distinguished from other rusts occurring on grasses. The fact that the aecidial stage of this rust is found on the barberry is not an absolute distinction from other grass rusts for it is now known that several other species on grasses also form their aecidia on this plant.

The diseases of cereals caused by rust have been known for a very long period. Reference has been made frequently to supposed allusion to it in the Bible but these are very doubtful. There is no doubt that the ancient Greeks and Romans recognised the occurrence of disease in their crops. The latter in their writings used the word Robigo to describe it. It is sometimes translated as 'roughening' or 'rusting' but there is no exact English equivalent. It implied the production of scabs, sores or abrasions of the surface and it is very probable that it was used to describe the condition brought about by attack of the black rust on the culms and sheaths.

On *Anthoxanthum odoratum* this rust was described as a distinct species, *P. anthoxanthi*, by Fuckel; in his description no paraphyses were mentioned in the uredosorus. However, the rust was not properly understood by British authors and Plowright stated that capitate para-physes were present. It was suggested in 1915 (Wilson, J. Bot. Lond. **53**, 47, 1915) that in this country two distinct species existed on this host and in 1921 Grove (J. Bot. Lond. **59**, 311) recorded *P. anthoxanthi* Fuck. without paraphyses and *Uredo anthoxanthina* Bub. with capitate para-physes. This latter rust is included here in *P. poae-nemoralis* Otth (see p. 271). *Puccinia graminis* on *Anthoxanthum* appears to be sparingly present in Britain and is certainly less common than *P. poae-nemoralis* on this host. Many of the early records undoubtedly refer to *P. poae-nemoralis* but authentic records exist from England, Wales (Sampson & Western, 1954, 15) and from most areas of Scotland (Wilson, 1934, 100).

Uredospores are generally abundant on infected plants; the teleuto-spores are small and inconspicuous but are usually present. It appears that in this country the rust is often independant of alternation.

Guyot, Massenot & Saccas (Ann. Ecole Nat. Agric. Grignon, **5**, 92), placed this rust in a small-spored group; they have been unable to obtain infection on *A. odoratum* with rust from wheat, oats, *Arrhena-therum elatus* or *Lolium perenne*. Stackman & Piemeisel (J. Agric. Res. **10**, 429,1917) placed it under *P. graminis avenae*. Jørstad (1950, 26) stated that it is not common in Norway, where he considered it independent of host alternation and noted that the teleutospores are small (22–34 × 13·5–22 μ). Viennot-Bourgin (Rev. Myc. **14**, 19, 1949) stated that it produced aecidia on *Berberis vulgaris* and its variety *atro-purpurea*.

The rust on *Brize media* appears to have been found only twice in this country, near Dunkeld, Perthshire (Dennis & Foister, TBMS. **25**, 277, 1942), in the uredospore condition and on Holy Island, Northumber-land, by Dennis (*in litt.*) in September 1957 in the teleutospore stage. It was placed under *P. graminis avenae* by the Sydows (Monogr. Ured. **1**, 727) and has been recorded by Fischer (1904, 261) in Switzerland on *B. maxima* and by Eriksson in Sweden who infected *Briza media* with *P. graminis avenae*. Jørstad (1950, 27) has described its occurrence in Norway. Guyot, Massenot & Saccas (*loc. cit.*) have found that the teleutospores are small (mean 38·6 × 18·9 μ); they have stated that in Russia *Briza media* has been infected with *P. graminis avenae* by Jaczewski & Vavilov. It appears to be scarce in all these countries.

The rust on *Deschampsia cespitosa* has been found in Kent, Yorkshire (Bramley, *in litt.*), Aberdeenshire (in Herb. Trail collected by Suther-land, and recently near Huntly) and in Midlothian. It has usually been found in the uredospore condition but specimens with numerous teleutosori were collected by Bramley in 1952 (TBMS. **37**, 251, 1954). This was placed in *P. graminis airae* by Eriksson, a variety according to Guyot *et al.* (*loc. cit.* p. 142) limited to *Deschampsia*.

Puccinia graminis has been investigated more intensively than any other rust fungus, Eriksson (1894) and Eriksson & Henning (Z. Pfl.-Krankh. **4**, 66, 1894) early distinguished six *formae speciales* of the fungus based on host specialisation. More intensive work particularly in North America resulted in the recognition (within these forms) of numerous physiologic races variously adapted to host cultivars. These races are accorded numbers and are identified by test on a selected list of cultivars.

Batts (TBMS. **34**, 533, 1951) identified the forms present in Scotland and obtained information on their host range as follows:

1. *Puccinia graminis* f. sp. *tritici* Eriks. & Henn. on *Triticum aestivum*, weakly infects *Hordeum vulgare* and *Secale cereale*.

2. *Puccinia graminis* f. sp. *secalis* Eriks. & Henn. on *Agropyron repens, Bromus sterilis, Hordeum vulgare* and *Secale cereale*.

3. *Puccinia graminis* f. sp. *avenae* Eriks. & Henn. on *Alopecurus pratensis, Arrhenatherum elatius, Avena fatua, A. sativa, Dactylis glomerata* and *Trisetum flavescens*; *Poa pratensis* is easily infected artificially. *Festuca rubra* may also be a host.

4. *Puccinia graminis* f. sp. *agrostidis* Eriks. on *Agrostis stolonifera* and *A. tenuis*.

5. *Puccinia graminis* f. sp. *poae* Eriks. & Henn. on *Poa pratensis* and *P. trivialis; P. annua* is easily infected artificially.

6. *Puccinia graminis* f. sp. *phlei-pratensis* (Eriks. & Henn.) Stak. & Piem. probably this variety on *Phleum pratense*.

The collection on *Deschampsia cespitosa* in Yorkshire (Wilson & Henderson 1954) may belong to a seventh form *P. graminis* f. sp. *airae*. Eriks. & Henn.

Batts found the uredospore lengths of varieties 1–5 above to be significantly different. Uredospore width and teleutospore dimensions could not be used alone to separate the varieties.

Puccinia hordei Otth

Mitt. Naturf. Ges. Bern. **1870**, 114 (1871); Wilson & Bisby, Brit. Ured. no. 167; Gäumann, Rostpilze, p. 449.

Puccinia straminis Fuck., var *simplex* Körn., Thuem. Herb. Myc. Oecon, no. 101 (1873).

Puccinia anomala Rostr., Thuem. Herb. Myc. Oecon. no. 451 (1877).

Puccinia simplex (Körn.) Eriks. & Henn., Landt.-Akad. Handl. Tidskr. **1894**, 175 (1894); Z. Pfl.-Krankh. **4**, 260 (1894); Grove, Brit Rust Fungi, p. 264.

Spermogonia amphigenous, numerous, in groups or widely scattered, at first honey-coloured then blackish, 100–150 μ diam. **Aecidia** amphigenous, scattered, cupulate, yellow, 200–300 μ diam., peridial cells polygonal with external wall smooth, 6–8 μ thick, internal coarsely verrucose, 3–4 μ thick; aecidiospores globoid or ellipsoid, minutely verrucose, subhyaline, 18–26 × 16–20 μ, wall 1·5–2 μ thick. **Uredosori** amphigenous, scattered, minute, cinnamon-brown; uredospores subgloboid or ellipsoid, yellow, 22–27 × 15–20 μ, wall echinulate, with 8–10 pores, scattered, indistinct. **Teleutosori** amphigenous or on the culms, minute, but longer on the culms, oblong, confluent, long-covered by the epidermis, surrounded by a thin layer of subepidermal brownish paraphyses; teleutospores oblong or clavoid, apex truncate or obtuse, slightly constricted, brown, 40–54 × 15–24 μ, wall smooth, 4–8 μ thick at apex, pedicels short, brown; mesospores abundant, usually more numerous than the teleutospores, oblong or clavoid, asymmetrical, 25–45 × 16–24 μ, apex 4–10 μ thick. Hetereuform.

Spermogonia and aecidia on *Ornithogalum pyrenaicum* and on *O. umbellatum*

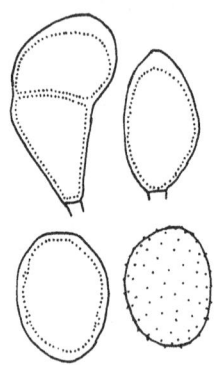

P. hordei. Teleutospore, mesospore and uredospores (after Savulescu).

by culture; uredospores and teleutospores on *Hordeum murinum* and cultivated varieties of *H. vulgare* including *H. distichon*. Spermogonia and aecidia rare; teleutospores, August–September, England, Scotland, frequent.

Puccinia hordei is closely related to *P. recondita* and indeed was included in that aggregate by Grove as *P. dispersa*, however, it is easily differentiated by the high proportion of unicellular teleutospores.

It has been recorded on *Hordeum murinum* by Ellis (1934, 498) in Norfolk. Previously this species had been experimentally infected by Waterhouse (Proc. Linn. Soc. N.S.W. **54**, 615, 1929). Hey (Arb. Biol. Abt. Berl. **19**, 227, 1931; RAM. **11**, 36, 1932) found that in Germany *H. murinum* was strongly resistant.

Successful infection experiments using aecidiospores from *Ornithogalum umbellatum* were first made by Tranzschel (Myk. Zbl. **4**, 70, 1914) in Russia, who also infected *H. vulgare* with aecidiospores. These experiments have been confirmed by Mains & Jackson (J. Agric. Res. **28**, 1124, 1924) in the United States, Mayor (Bull. Soc. Neuchâtel Sci. Nat. **54**, 54, 1920) in Switzerland, by Beck (Ann. Myc. **22**, 291, 1924) in Germany and by Ducomet (Rev. Path. Vég. **13**, 86, 1926) in France. D'Oliviera working in England (Ann. Appl. Biol. **26**, 56, 1939) obtained aecidia on *O. umbellatum* by infection with basidiospores and it appears from these infections that the rust is heterothallic. He later (Agron. Lusit. **13**, 221, 1951) showed that twenty-six Eurasiatic species of *Ornithogalum* were susceptible but that South African species were more or less resistant.

The aecidium on *O. pyrenaicum* was found by Dennis and Sandwith (Nature, **162**, 461, 1948) occurring naturally along the Berkshire and Wiltshire border; they infected *H. vulgare* with aecidiospores from this locality.

D'Oliviera (*loc. cit.*) considered that at Cambridge the rust could overwinter by means of its uredospores and this occurred in Hampshire during the mild winter of 1943. Gassner & Pieschel (Phytopath. Zeitschr. **7**, 355, 1934) described overwintering in Germany by means of mycelium and by uredospores. The teleutospores germinate in the spring.

Investigators in the United States, Argentina, Canada and Germany and D'Oliviera in this country have distinguished about thirty special forms of this rust (see Hey, *loc. cit.*); Stakman *et al.* (Nova Acta. Leop. Carol. **13**, 319, 1935; Straib. Arb. Biol. Abt. Berl. **22**, 43, 1937, reviewed in RAM. **17**, 230, 1938). D'Oliviera has described five from Spain, four from Portugal and four from Great Britain.

Haines (TBMS. **20**, 252, 1936) at Cambridge has shown that *P. hordei* could initiate invasion of wheat, rye and oats, killing the guard cells of the stomata of entry but failing to form mycelia in the tissues.

A discussion on the validity of the specific name of this rust took place in 1939 (TBMS. **23**, 278, 1939) and there has been further discussion by Buchwald (Ann. Myc. **41**, 306, 1943) on this subject.

Puccinia longissima Schroet.

Cohn, Beitr. Biol. Pfl. 3, 70 (1878); Wilson & Bisby, Brit. Ured. no. 179; Gäumann, Rostpilze, p. 724.

[**Spermogonia** rather large, brownish-black, conical, scattered amongst the aecidia. **Aecidia** solitary over the whole leaf, deeply immersed, hemispherical, opening with a minute median pore, yellow; aecidiospores globoid or ellipsoid, mostly angular, verruculose, orange, 22–32 × 18–30 μ.] **Uredosori** epiphyllous, arranged in rows or little groups between the nerves, marked on the lower surface by discoloured spots, oblong, about 0·5 mm. long, surrounded by the ruptured epidermis, deep orange-brown; uredospores ovoid or subgloboid, 25–30μ diam., wall rather thin, finely echinulate, yellowish or pale brown with several, scattered, thickened pores. **Teleutosori** similar but narrower, surrounded and partly covered by the ruptured epidermis, at length naked but not pulverulent, deep chestnut-brown; teleutospores long-ellipsoid or subclavoid, occasionally constricted at the septum, yellowish-brown, very variable in size, 60–120 × 12–20μ, upper cell ellipsoid-cylindroid, rounded or bluntly pointed, dark chestnut-brown, the lower cell usually cylindroid, about one-third longer than the upper cell, paler and narrower, sometimes attenuate downwards, wall thin, smooth, up to 11μ thick at the apex, pedicel thick, persistent, up to 22μ long, colourless. Hetereuform.

[Spermogonia and aecidia on *Sedum*

acre, May]; uredospores and teleutospores on *Koeleria cristata*, June–October. England, Scotland, rare.

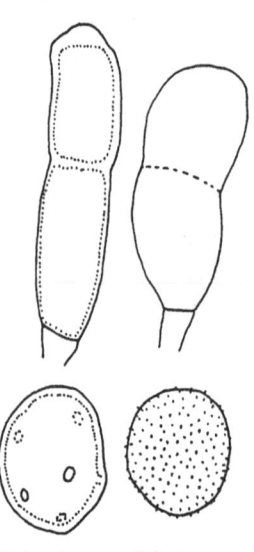

P. longissima. Teleutospores.

The rust on *Koeleria cristata* was previously regarded by Trail (1890, 317 and 367) and Grove (1913, 286) as *Puccinia paliformis* Fuck. but as the result of the discovery of fresh material by the former on Kinnoull Hill, near Perth, in 1916 (see Grove, J. Bot. Lond. **55**, 134, 1917) was later named *P. longissima*. It has also been found in Gloucestershire and in Angus. The aecidium has not been discovered in Britain but has been described on the continent on *Sedum acre* and *S. boloniense*, the mycelium perennating in the latter species according to Bubak (1908, 102).

The connection between the spore stages was proved experimentally by Bubak (Zbl. Bakt. II, **9**, 126 and 922, 1902).

Puccinia magnusiana Körnicke

Hedwigia, **15**, 179 (1876); Grove, Brit. Rust Fungi, p. 271; Wilson & Bisby, Brit. Ured. no. 180; Gäumann, Rostpilze, p. 713.

Puccinia arundinacea DC., Lam. Encycl. Méth. Bot. **8**, 250 (1808), *p.p.* (*nomen ambiguum*).

[*Aecidium ranunculacearum* DC., Fl. Fr. **6**, 97 (1815), *p.p.*]

Puccinia arundinacea β *epicaula* Wallr. Fl. Krypt. Germ. **2**, 225 (1833).

[*Aecidium ranunculacearum* DC. var. *linguae* Grove, Grove, Brit. Rust Fungi, p. 387 (1913).]

Spermogonia epiphyllous in groups, scattered between the aecidia. **Aecidia** hypophyllous in small clusters on yellowish spots, or on the petioles or stems forming elongate groups, cup-shaped, with a laciniate, white margin; aecidiospores densely and finely verruculose, yellowish, 15–25 μ diam. **Uredosori** amphigenous, scattered, rarely confluent, elliptical, 1–2 mm. long, pulverulent, pale yellowish-brown; uredospores mostly ovoid or ellipsoid, pale brownish-yellow, 20–35 × 12–20 μ, wall delicately echinulate, pores indistinct, paraphyses numerous, clavoid, usually pale brownish. **Teleutosori** amphigenous, very numerous, usually scattered over the whole leaf-surface, oblong or sublinear, minute, 1–2 mm. long, or on the culms and leaf-sheaths forming narrow striae several centimetres long, flat, compact, blackish; teleutospores ellipsoid or clavoid, rounded above or rarely conically attenuate or truncate, hardly constricted, attenuate downwards, brown, darker above, 32–55 × 16–26 μ, wall smooth, 5–10 μ thick at apex, pedicels thick, brownish, persistent, as long as or shorter than the spores. Hetereu-form.

Spermogonia and aecidia on *Ranunculus bulbosus*, *R. flammula*, *R. lingua*

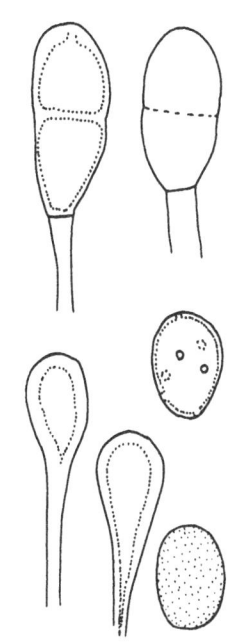

P. magnusiana. Teleutospores, uredo-paraphyses and uredospores.

and *R. repens*, June–August; uredospores and teleutospores on *Phragmites communis*, June–April. Great Britain, scarce.

The connection of the spore forms of this species was first proved by Cornu (C. R. Acad. Sci. Paris, **94**, 1731, 1882) and was confirmed by Klebahn (Z. Pfl.-Krankh. **2**, 332, 1892; Jahrb. Wiss. Bot. **34**, 347, 1900) and Fischer (1898, 50). Plowright (Quart. Journ. Micro. Sci. **25**, 151, 1885) by a long series of cultures proved that the aecidium of this species is produced on *Ranunculus bulbosus* and *R. repens* but not on *R. acris* and other species. In Scotland *Phragmites communis* has been infected by aecidiospores from *R. repens* by Wilson (1934, 402). Further cultures have been carried out in Switzerland by Mayor (Bull. Soc. Neuchâtel. Sci. Nat. **58**, 19, 1933) and Gäumann (Ann. Myc. **34**, 61, 1936), Guyot (Ann. Ecole Nat. Agric. Grignon, ser. 2, **1**, 60, 1937) and in Japan by Hiratsuka (Bot. Mag. Tokyo, **47**, 711, 1933) and Ito (Bot. Mag. Tokyo, **48**, 533, 1934); various species of *Ranunculus* have been infected. Ellis (1934, 498) recorded the occurrence of aecidia on *R. lingua* close to reeds which, later in the year, were infected with *P. magnusiana* and considered

that the aecidium probably belongs to the latter species; the aecidium on
R. lingua had been previously recorded only by Greville (*loc. cit.*) in
1821 at Edinburgh. Bramley (*in litt.*) reported the discovery at Askham
Bog, Yorkshire, of a few aecidia on *R. flammula* in an area where *R.
repens* and *R. bulbosus* were heavily infected close to *P. magnusiana* on
P. communis; it therefore appears probable that *R. flammula* is also a
host for the rust. No infection experiments appear to have been carried
out with *R. lingua* or *R. flammula*.

Dietel (quoted by Andres, Ber. D. Bot. Ges. **52**, 614, 1934) stated that
teleutospores borne on the stems are larger than those found on the
leaves.

The aecidia belonging to *Uromyces dactylidis* occur on the aecidial
hosts of *P. magnusiana* and are morphologically quite indistinguishable,
but begin to appear earlier in the spring.

Puccinia magnusiana is distinguished from *P. phragmitis* by its numerous small teleutosori, and the abundant capitate uredosoral paraphyses;
the teleutospores show hardly any constriction. Both species may occur
together upon the same leaf of *Phragmites communis*.

Puccinia moliniae Tul.

Ann. Sci. Nat. Bot. ser. 4, **2**, 141 (1854); Grove, Brit. Rust Fungi, p. 276; Wilson &
Bisby, Brit. Ured. no. 185; Gäumann, Rostpilze, p. 745.

Puccinia brunellarum-moliniae Cruchet,
Zbl. Bakt. II, **13**, 96 (1904); Gäumann,
Rostpilze, p. 742.
Puccinia aecidii-melampyri Liro, Acta
Soc. Fauna Fl. Fenn. **29**, 55 (1907).

Spermogonia epiphyllous, honey-yellow,
130–150 μ diam. **Aecidia** hypophyllous,
on rounded violet or yellowish-brown
spots, crowded in circular groups, cup-
shaped with coarsely cut, white, slightly
revolute margin, 250–300 μ diam., peridial cells firmly connected, almost
rhombic or quadratic, outer wall smooth,
6–8 μ thick, inner wall verrucose, 4–6 μ
thick; aecidiospores ellipsoid, almost
hyaline, 16–22 × 13–17 μ, wall 1 μ thick,
verruculose. **Uredosori** amphigenous,
generally hypophyllous, often on brown
or purplish spots, scattered or arranged
in lines and confluent, oblong or linear,
brown; uredospores globoid or subgloboid, yellowish-brown, 20–28 × 20–
24 μ, wall 3–6 μ thick, aculeate, with

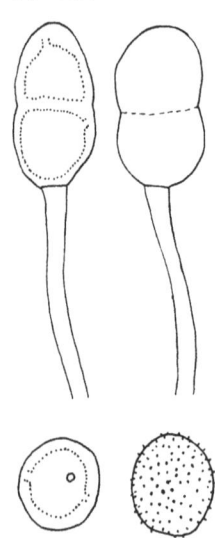

P. moliniae. Teleutospores and uredospores.

3 pores. **Teleutosori** similar, often
confluent and up to 8 mm. long, con-

spicuous, pulvinate, black; teleuto-spores ellipsoid, rounded at both ends, hardly constricted, $32-50 \times 20-30 \mu$, wall smooth, up to 5μ thick at apex, pore of upper cell apical, of lower cell superior, pedicel hyaline, yellowish, apex often brown, persistent, thick, up to 120μ long; mesospores occasionally present. Het-ereu-form.

Spermogonia and aecidia on *Prunella vulgaris*, June-July; uredospores, July–September; teleutospores, July onwards, on *Molinia caerulea*. Scotland, scarce.

Grove (*loc. cit.*) considered that the aecidial stage of this rust might occur on *Melampyrum pratense* but in Britain this has not been established although it occurs on *M. pratense*, *M. sylvaticum* and *Origanum vulgare* in Norway (Jørstad, 1951).

In Britain it is known only in north and west Scotland in localities where the two hosts grow in close proximity. The connection between aecidia on *Prunella* and dicaryont on *Molinia* was proved by Cruchet (*loc. cit.*) and has been recently confirmed by Dupias (Bull Soc. Hist. Nat. Toul. **80**, 33, 1945) in France. The aecidial stage, *Aecidium prunellae* Wint., was first found in Britain by Keith at Forres (Plowright, 1889, 264) and it is known from Killin, West Ross, Eigg and Ben Lui, Argyllshire. The stage on *Molinia* has been recorded from several localities in Perthshire and Argyll. The production of uredospores is scanty and soon ceases; the rust is clearly obligatorily heteroecious.

Puccinia phragmitis (Schum.) Körnicke

Hedwigia, **15**, 179 (1876); Grove, Brit. Rust Fungi, p. 273; Wilson & Bisby, Brit. Ured. no. 195; Gäumann, Rostpilze, p. 746.

Uredo phragmitis Schum. Enum., Pl. Saell, **2**, 231 (1803).
Puccinia arundinacea DC., Lam. Encycl. Méth. Bot. **8**, 250 (1808), *p.p.*
Puccinia trailii Plowr., Monogr. Brit. Ured. Ustil. 176 (1889); Grove, Brit. Rust Fungi, p. 274; Wilson & Bisby, Brit. Ured. no. 233; Gäumann, Rostpilze, p. 750.

Spermogonia amphigenous, in groups, subepidermal, spherical, whitish, 78–90μ diam. **Aecidia** hypophyllous, on circular red or deep purple spots, $0.5-1.5$ cm. diam., in dense clusters, shortly cylindrical or cup-shaped, with a lacini-ate, white, recurved margin; aecidio-spores verruculose, nearly hyaline, 16–26μ diam., spore-mass white. **Uredosori** amphigenous, scattered or subgrega-rious, elliptical, or linear, sometimes

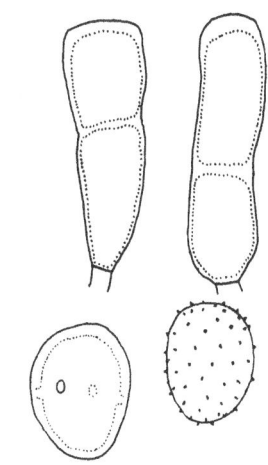

P. phragmitis. Teleutospores and uredospores.

confluent, rather large, convex, pul-verulent, brown, without paraphyses;

uredospores subgloboid or obovoid, brownish, 25–35 μ diam., contents colourless, wall 3–4 μ thick, verruculose, with 4 equatorial pores. **Teleutosori** numerous, similar but larger and thicker, very convex and compact, black; teleutospores oblong, rounded at both ends, constricted, deep yellowish-brown, 45–65 × 16–25 μ, wall 2–3 μ thick at sides, paler and thicker (4–7 μ) above, smooth or sometimes with very fine granulations;

pore of upper cell apical, of lower close to septum; pedicels hyaline or yellowish, thick, persistent, 100–200 μ long. Hetereu-form.

Spermogonia and aecidia on *Rheum rhaponticum, Rumex acetosa, R. 'conglomeratus', R. crispus, R. hydrolapathum* and *R. 'obtusifolius'*, May–June; uredospores and teleutospores on *Phragmites communis*, July–May. Great Britain and Ireland, frequent.

Both Plowright (*loc. cit.*) and Grove (*loc. cit.*) gave *Rheum officinale* as a host; it is possible that the former used this species for inoculations but Grove stated (*in litt.*) that he was mistaken in quoting this as a British host species. Sowerby (Eng. Fungi, 1803, table 398, fig. 6) as early as 1803 published a description and coloured illustration of *A. rhei* on *R. rhaponticum*.

It was Winter (Hedwigia, **14**, 114, 1875) who first showed that *Puccinia phragmitis* has its aecidium on *Rumex* and this was confirmed by Plowright (Proc. Roy. Soc. Lond. no. 228, 47, 1883 and Quart. Journ. Micro. Sci. **25**, 156, 1885) who showed that with his material infection occurred on *Rumex* sp. (not, however, on *R. acetosa*) and on '?*Rheum officinale*'. Further confirmatory experiments have been carried out by Klebahn (1904, 283) and others.

In 1889 Plowright described *P. trailii* as a species closely resembling *P. phragmitis* but differing from it in its teleutospores possessing a granular spore-membrane and being borne on stouter pedicels; he showed by experiment that its aecidial stage developed only on *R. acetosa*, a species not infected by *P. phragmitis sensu stricto*. Klebahn confirmed these results by cultures made in north Germany and Arthur (1934, 155) has done the same for North American forms. In this country the aecidium of *P. phragmitis sensu lato* seems relatively scarce, but it is very conspicuous, and can be found on *Rumex* growing amidst *Phragmites*. Bramley (*in litt.*) produced aecidia by artificial inoculation on *Rheum rhaponticum*.

Puccinia trailii was accepted as a species by Sydow, Fischer & Fragoso (1924, 85) and Savulescu (1953, 70) and doubtfully by Grove and Wilson & Bisby; it is now generally regarded as a biological race and Jørstad (1950, 36) considered it morphologically identical with *P. phragmitis*.

Recently Gäumann & Terrier (Ber. Schweiz. Bot. Ges. **62**, 297, 1952) have shown that in Switzerland teleutospores, produced by infecting

P. communis with aecidiospores from *R. acetosa*, could infect six *Rumex* spp. including *R. acetosa* and *R. obtusifolius*; they suggest, however, that on account of morphological differences *P. trailii* should be retained. Ellis (1934, 497) has observed that aecidia on *R. acetosa* sometimes occur close to those on other species of *Rumex* and a similar observation has been made in Scotland.

The structure of the aecidial peridium was described by Mayus (Zbl. Bakt. II, **10**, 711, 1903). Lamb (Ann. Bot. Lond. **49**, 403, 1935) gave a detailed account of the spermogonium, fertilisation and development of the aecidium and showed that it is heterothallic.

Puccinia poae-nemoralis Otth

Mitt. Naturf. Ges. Bern, **1870**, 113 (1871); Wilson & Bisby, Brit. Ured. no. 198.

[*Uredo poae-sudeticae* West., Bull. Acad. Roy. Belg. II, **11**, 650 (1861).]

Puccinia perplexans Plowr. f. *arrhenatheri* Kleb., Abh. Nat. Ver. Bremen, **12**, 306 (1892).

Puccinia arrhenatheri (Kleb.) Eriks., Cohn, Beitr. Biol. Pfl. **8**, 14 (1898); Grove, Brit. Rust Fungi, p. 284; Wilson & Bisby, Brit. Ured. no. 97; Gäumann, Rostpilze, p. 511.

[*Uredo anthoxanthina* Bub., Ann. Myc. **3**, 223 (1905)]; Wilson, Ured. Scot. p. 438.

[*Uredo glyceriae* Lind (non Opiz), Danish Fungi, 343 (1913)]; Wilson, Ured. Scot. p. 440; Gäumann, Rostpilze, p. 558.

[*Uredo glyceriae-distantis* Eriks., Arkiv. Bot. **18** (19), 18 (1923).]

Puccinia poae-sudeticae Jørst., Nytt Mag. Naturv. **70**, 325 (1931).

Puccinia anthoxanthina Gäum., Ber. Schweiz. Bot. Ges. **55**, 74 (1945); Gäumann, Rostpilze, p. 551.

Puccinia poarum auct. non Niels., Grove, Brit. Rust Fungi, p. 278 *p.p.*

Spermogonia very numerous, minute, covering a great part or the whole of the leaf uniformly. **Aecidia** hypophyllous, generally distributed densely and evenly over the whole leaf, sometimes even on the flowers, deforming the affected branches, between cylindrical and cup-shaped, with a torn whitish revolute margin, peridial cells not in distinct rows, outer wall strongly thickened (7–10μ), punctate in surface view, inner wall thinner but strongly developed,

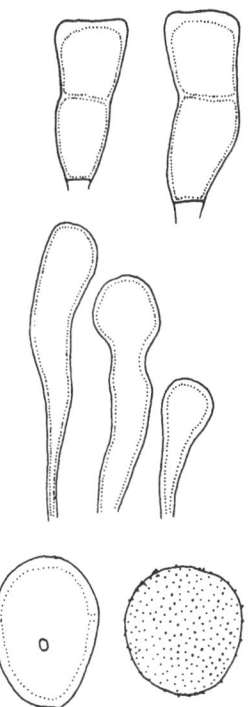

P. poae-nemoralis. Teleutospores, uredoparaphyses and uredospores (after Savulescu).

3–5 μ thick, with tesselate structure and crowded, small warts in surface view; aecidiospores subgloboid or ellipsoid not remaining connected in long chains, verruculose, yellowish, 19–32 × 16–24 μ. Uredosori chiefly epiphyllous, very rarely hypophyllous, on minute yellow spots, scattered, minute, roundish or elliptic, orange-yellow, with numerous paraphyses, generally thick-walled and capitate with a more or less narrow neck, usually bent or incurved, up to 100 μ long with heads up to 21 μ wide, wall up to 4 μ thick; uredospores globoid or ellipsoid, 17–28 × 17–25 μ, wall nearly colourless, 1·5–2 μ thick, densely and minutely verruculose, with 5–6 scattered, rather indistinct pores. Teleutosori hypophyllous, oblong or linear, long-covered by the epidermis, with brownish subepidermal paraphyses, black; teleutospores ellipsoid or clavoid, obtuse or rounded above, somewhat narrowed below, slightly or not constricted, chestnut-brown, becoming paler downwards, 30–45 × 16–22 μ, wall smooth, 5–10 μ thick at apex, pedicels short, brownish, persistent; mesospores occasionally found. Hetereu-form.

Spermogonia and aecidia on *Berberis vulgaris* very rare and only with *Arrhenatherum* race; uredosori and teleutosori on *Arrhenatherum elatius*, *Poa nemoralis* and uredosori only on *Anthoxanthum odoratum*, *Glyceria fluitans*?, *Poa annua*, *Puccinellia maritima* and *P. rupestris*. Great Britain, probably frequent.

Jørstad (*loc. cit.*) regards *Puccinia poae-nemoralis* as a polymorphous species containing various grass rusts characterised by the uredo-paraphyses; with respect to uredospores, teleutosori and teleutospores there exist no essential differences, however, between this rust and other grass rusts with subepidermal stromatic teleutosori such as *P. poarum*. However, *P. poae-nemoralis* commonly does not produce teleutosori and on some hosts they are unknown. The differences between this rust and *P. poarum* Niels. are given under that species (q.v. p. 274).

The aecidium is distinguished from that of *P. graminis* by the larger, distinctly warted spores, the more strongly developed prismatic structure of the inner wall of the peridium and the less distinct spore chains.

There is little known regarding the host specialisation of *P. poae-nemoralis*. The only aecidial stage known is that of the race on *Arrhenatherum elatius* which was proved in 1888 to occur on *Berberis vulgaris* by J. Peyritsch and confirmed by Eriksson, Klebahn and more recently by Gäumann (Ann. Myc. 32, 301, 1934). Guyot & Massenot (Uredineana 4, 338, 1953) infected both *Arrhenatherum elatius* and *A. bulbosum* with aecidiospores from *Berberis vulgaris*. The aecidial stage (*Aecidium graveolens*) covers the under sides of the leaves on the witches brooms. It has been found once in this country, in Northumberland by Dennis in July 1944. An account of the structure and distribution of the mycelium in the stems and buds of the barberry has been given by Magnus (Ann. Bot. Lond. 12, 155, 1898). Hardison (Mycol. 35, 79, 1943) described amphispores on *Festuca pratensis*; these have thicker walls, 1·7–2·6 μ, and germinate only in spring.

The rust on *Anthoxanthum odoratum* is scarce in Great Britain and appears to be more abundant in the north. It was discovered at King's Lynn, Norfolk, in 1884 and was recorded by Plowright as *P. anthoxanthi* 'with paraphyses!' It was found in Scotland in 1915 when it was suggested that two rusts had been confused under the name *P. anthoxanthi*. In 1921 Grove (J. Bot. Lond. **59**, 312) confirmed this and described the occurrence of *U. anthoxanthina* (= *P. poae-nemoralis*) and *P. anthoxanthi* in England and since then the former has been discovered in Tweed, Forth, Tay and Dee areas and in Argyll in Scotland (Wilson, TBMS. **7**, 83, 1921 and **9**, 138, 1924) and from scattered localities in England.

Uredospores capable of germination have been found in March and the rust overwinters in the uredospore condition or as mycelium in the leaf.

A rust bearing teleutospores was found by Gäumann (*loc. cit.*) on a plant bearing *Uredo anthoxanthina* in Switzerland in 1945 and it was named *P. anthoxanthina*. This name is included in a list of Yorkshire fungi by Mason & Grainger (1937, 47) but this appears to have been an error (see Wilson & Bisby, 1954, no. 198).

The rust on *Arrhenatherum elatius* has been frequently recorded from Great Britain and Ireland as *P. arrhenantheri*. Teleutospores are normally present; irregular 3- and 4-celled teleutospores are sometimes found.

The *Uredo* on *Festuca ovina* found by Grove (J. Bot. Lond. **59**, 1921, 314) almost certainly belongs to *P. poae-nemoralis* (see Wilson & Bisby, Brit. Ured. no. 156) which occurs on this host in northern Scandinavia.

The rust on *Glyceria fluitans* recorded as *P. poarum* (Scot. Nat. N.S. 270, 1884), may also belong to this species but the material has not been seen and there are no other records of rusts on *Glyceria fluitans* in Britain.

Uredo glyceriae collected in Scotland in 1925 on *Puccinellia maritima* (TBMS. **12**, 115, 1927) has also been recorded in Norfolk by Ellis (Trans. Norf. Norw. Nat. Hist. Soc. **13**, 499, 1933–4) on this host and also on *Puccinellia rupestris*, all in the uredospore state only. Teleutospores have never been found on this rust on *Glyceria* anywhere throughout its range.

On *Poa*, as Jørstad (1951) reports for Norway, teleutospores never occur on *P. annua*. On *P. pratensis* they have been recorded but not confirmed and sparse development occurs on *P. nemoralis*. The record of this rust (Wilson & Bisby, *loc. cit.*) on *P. trivialis* is unconfirmed. Gäumann does not accept the use of the binomial *P. poae-nemoralis* as used here, particularly as applied to the rusts on *Poa* but presumably places these in *P. petasitis-poarum* Gäum. & Eich. or *P. thalictri-poarum*

Fisch. & Mayor (Gäumann, 1959, 539 and 505). There is no evidence that alternation of *Poa* infecting rusts takes place with either *Thalictrum* or *Petasites* in Britain but these possibilities should be borne in mind.

Puccinia poarum Niels.

Bot. Tidsskr. III, **2**, 34 (1877); Grove, Brit. Rust Fungi, p. 278, *p.p.*; Wilson & Bisby, Brit. Ured. no. 201; Gäumann, Rostpilze, p. 531.

Spermogonia epiphyllous, pale yellow, often very numerous. Aecidia hypophyllous, usually in dense clusters on circular yellowish or reddish thickened spots 1–2 cm. diam., seldom scattered, cup-shaped, with a dentate white revolute margin; aecidiospores verruculose, orange, 18–25 × 16–20 μ. Uredosori on the leaves, sometimes on the culms, minute, roundish or elliptic, soon naked, yellow; uredospores globoid to ellipsoid, yellow, 17–28 × 17–25 μ, wall densely and minutely verruculose, with about 5 indistinct scattered pores. Teleutosori similar, oblong or linear, more or less in short rows, long-covered by the epidermis, surrounded by a small pale area, black, with capitate subepidermal paraphyses; teleutospores oblong-clavoid or cylindroid, rounded, truncate, or rarely conically attenuate above, hardly or not at all constricted, chestnut-brown, becoming gradually paler downwards 30–45 × 16–22 μ, wall smooth 4–8 μ thick at apex, pedicels short, brownish, persistent; an occasional mesospore is found. Hetereu-form.

Spermogonia and aecidia on *Tussilago farfara*, May–June and August–September; uredospores and teleutospores

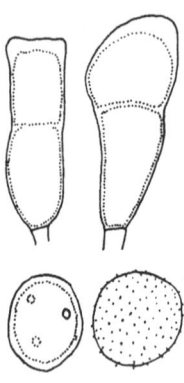

P. poarum. Teleutospores and uredospores (after Savulescu).

on '*Poa annua*', *P. pratensis* and *P. trivialis*, July–August and October–December. Great Britain and Ireland, very common.

The relation between the spore stages was first demonstrated by Plowright (Grevillea, **11**, 56, 1882), and then by many others including Klebahn, Tranzschel, Mayor, Dupias and Guyot (Ann. Ecole Nat. Agric. Grignon, **4**, 122, 1944). This species is unusual in having two aecidial generations in one year. The earlier crop begins to appear in May, and is followed by the uredo- and teleutospores on the surrounding leaves of *Poa*; these germinate quickly and the second crop of aecidia is produced about August, and the second generation of teleutospores may be found on *Poa* from October.

Nielsen (*loc. cit.*) followed by Plowright stated correctly that no paraphyses were present in the uredospores but many authors have stated that the hyaline capitate paraphyses are numerous. These statements have been due to confusion between this species and *P. poae-nemoralis* which possesses paraphyses while *P. poarum* does not; otherwise the uredosori and teleutosori of the two species are identical, and teleutosori alone cannot be distinguished with certainty. The uredosori of *P. poarum* are scanty and easily overlooked whereas teleutosori are generally abundant and soon produced; in *P. poae-nemoralis* the uredosori are abundant while teleutosori are usually scanty and sometimes not formed. *P. poarum* is certainly habitually heteroecious.

A short account of the structure of the aecidium has been given by Sappin-Trouffy (1897, 114) and a description of its development and cytology by Blackman & Fraser (Ann. Bot. Lond. **20**, 35, 1906).

Puccinia pratensis Blytt

Chria. Vidensk. Selsk. Forh. **1896** (6), 52 (1896); Wilson & Bisby, Brit. Ured. no. 204; Gäumann, Rostpilze, p. 870.

[*Uredo avenae pratensis* Eriks., Arkiv. Bot. **18** (19), 17 (1923).]

Spermogonia and **aecidia** unknown. **Uredosori** and **teleutosori** amphigenous, scattered, minute or medium in size, elliptic, oblong or oblong-linear, surrounded by the ruptured epidermis, pulverulent, ferruginous; uredospores globoid or obovoid, yellowish-brown, $27-31 \times 25-27 \mu$, wall pale brown, coarsely echinulate, $2-3 \mu$ thick with 5–7 scattered pores; teleutospores broadly ellipsoid or ovoid, rounded at both ends, scarcely constricted, chestnut-brown, $36-47 \times 25-28 \mu$, wall $3-7 \mu$ thick, densely verrucose, pore of upper cell more or less apical, of lower equatorial or sub-equatorial, pedicels short, thick, fragile, pale yellowish-brown, often attached laterally; mesospores few. Hetereu-form?

Uredospores, August–September, teleutospores, September–November on *Helictotrichon pratense*. England, very rare.

The rust was found in Yorkshire by Bramley in the autumn of 1952 (Naturalist, **1953**, 94). In Norway although the host extends to just north of the Arctic Circle the rust is restricted to the south-eastern coastal lowlands extending only to 50° 59′ N. It overwinters by uredospores (Eichhorn, Denkschr. Bayer. Bot. Ges. **20**, 111, 1936).

This species differs from all other north European grass rusts by its verrucose teleutospores and deciduous pedicels; three species, however, have been described which closely resemble *Puccinia pratensis* and, as they grow on different species of *Helictotrichon*, may possibly be regarded as biological races; these are *P. versicoloris* Semadini (Zbl. Bakt. II, **46**, 451, 1916) from Switzerland and northern Italy, *P. avenastri* Guyot and *P. bromoides* Guyot, from France (Guyot *et al.* Uredineana, 3, 141, 1951). The warts on the teleutospores of *P. pratensis* are relatively large and rounded with small interspaces; approaching the base of the lower cell they become slightly elongated towards the pedicel.

Eriksson (*loc. cit.*) stated that *Helictotrichon sativum* and *Bromus briziformis* are not infected by uredospores from *H. pratense*.

Owing to some resemblance between the teleutospores of *P. bessei* and those of *P. pratensis*, Semadini (*loc. cit.*) has suggested that *Lloydia serotina*, the host of the former, may be the aecidial host of *P. pratense*. Tranzschel (1939, 101) considered that the aecidial host might be a species of *Gagea* or of *Allium* (sect. *Molium*).

Puccinia pygmaea Eriks.

Bot. Zbl. **64**, 381 (1895); Wilson & Bisby, Brit. Ured. no. 209; Gäumann, Rostpilze, p. 515.

[*Uredo ammophilina* Kleb., Krypt. Fl. Mark. Branb. va, 882 (1914).]

Puccinia ammophilina Mains, Bull. Torr. Bot. Club, **66**, 617 (1939); Gäumann, Rostpilze, p. 551.

[**Spermogonia** epiphyllous, in groups. **Aecidia** hypophyllous, in groups; aecidiospores globoid or ellipsoid, finely verrucose, $21–27 \times 16–24\mu$, wall $1·5\mu$ thick, colourless.] **Uredosori** usually epiphyllous on yellow areas, small, oblong, often confluent, in rows between the nerves, orange-yellow, pulverulent with numerous, capitate (apex $10–17\mu$), usually straight, thin-walled, yellowish paraphyses; uredospores globoid to ellipsoid, orange-yellow, $28–36 \times 26–30\mu$, wall finely echinulate with 8–10 scattered, usually indistinct pores. **Teleutosori** amphigenous, compact, long-covered by the epidermis, very small, up to 0·5 mm. long, arranged in rows, black; teleutospores clavoid, apex truncate, rounded or obliquely acuminate, slightly constricted, narrowed below, chestnut-brown, $31–60 \times 15–20\mu$, wall smooth, 1μ thick at sides, $3–5\mu$ at apex, a few mesospores present; pedicels very short, pale brown, with a dark-brown zone at the base of the spore. Hetereuform.

[Aecidia on *Berberis* and *Mahonia* alternating with *Calamagrostis*]; uredospores and teleutospores on *Ammophila arenaria* and *Calamagrostis epigejos*. England, Scotland, Ireland, common on *Ammophila* as uredosori, probably common on *Calamagrostis epigejos*.

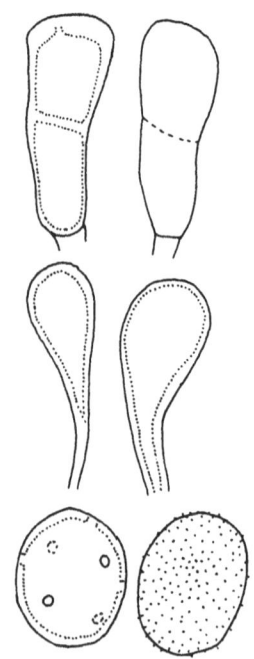

P. pygmaea. Teleutospores, uredo-paraphyses and uredospores.

This rust has been known in the uredospore stage in this country on *Ammophila* since it was recorded by Grove (J. Bot. Lond. **7**, 313, 1921) from Borth in Wales. Since then it has been described from Norfolk (Ellis, 1934, 498) and Suffolk (Mayfield, 1935) and has been found in Devon, Cumberland and Yorkshire. It is widely distributed round the coast of Scotland (Wilson, 1934) and has been found in Ulster.

Teleutospores were first described from Oregon and Michigan in the United States by Mains (*loc. cit.*); in Britain they are known only from Norfolk and Suffolk where they were collected by Ellis in 1950.

The abundant paraphyses in the uredosorus distinguish it from *P. elymi*; these paraphyses appear to be constantly present although Grove (*loc. cit.*) stated that they were sometimes absent—he may have examined some *P. elymi*; it also differs in the larger size of its uredospores which are also thinner-walled than those of *P. elymi*. In the teleutospore stage the two fungi are quite dissimilar.

This rust on *Ammophila* has been referred to *P. ammophilina* Mains. The uredospores on *Ammophila* are slightly larger than those on British *Calamagrostis* collections (32–34 × 26–30μ against 28–32 × 24–27μ) but as regards other characters they are indistinguishable. The aecidial host of the *Ammophila* race is unknown.

On *Calamagrostis* the species was found in Worcestershire by Rhodes in June 1928 (see Grove, J. Bot. Lond. **66**, 211, 1928). Since then it has been found in Yorkshire (Naturalist, **1941**, 3), by Brenan in Buckinghamshire, Perthshire (TBMS. **37**, 248, 1954) and Inverness-shire.

Tranzschel (C. R. Acad. Sci. U.R.S.S. **1931**, 46) made successful cultures in Russia by sowing teleutospores on *Berberis vulgaris* and *B. amurensis* and these results have been confirmed by Mayor (1958, 173) who added *B. polyantha* as an aecidial host and *Mahonia aquifolium* and *B. wilsonae* on which only spermogonia developed. Up to the present the aecidial stage has not been found in this country.

Jørstad (1950, 47) considered that *P. pygmaea* is closely related to *P. poae-nemoralis* but is distinguished by its straight, thin-walled uredoparaphyses. The dark-brown zone on the pedicel of the teleutospore is similar to that found in *P. brachypodii* to which it is also closely related (see p. 251).

Puccinia recondita Rob. & Desm.

Pl. Crypt. Fr. ser. 2, no. 252 (1882) (non *P. recondita* Diet & Holw. 1894); Cummins & Caldwell, Phytopath. **46**, 81 (1956).
[*Uredo rubigo-vera* DC., Fl. Fr. **6**, 83 (1815).]
Puccinia straminis Fuck., Jahrb. Nass. Ver. Nat. **15**, 9 (1860), *nomen ambiguum.*
Puccinia rubigo-vera Wint., Rabh. Krypt. Fl. Ed. 2, **1** (1), 217 (1882); Gäumann, Rostpilze, p. 520.

Spermogonia chiefly epiphyllous, scattered, in small groups. **Aecidia** hypophyllous, in crowded groups, cupulate or rarely cylindric; peridial cells rather variable in size and shape, firmly joined to each other, not in distinct longitudinal rows, usually with a projection of variable length overlapping the upper end of the cell immediately beneath, outer wall up to 10–12 μ thick, punctate in surface view, inner wall up to 6 μ thick with numerous strongly projecting small warts covering the whole surface arranged in irregular groups or lines; aecidiospores globoid or ellipsoid, 13–26 × 19–29 μ, wall colourless, 1–2 μ thick, finely verrucose. **Uredosori** epiphyllous, rarely hypophyllous, scattered, 1–2 mm. long, rarely confluent, oblong or punctiform, cinnamon-brown, aparaphysate; uredospores globoid or broadly ellipsoid, 13–24 × 16–34 μ, wall pale cinnamon-brown or yellowish, 1–2 μ thick, echinulate, pores 6–8, sometimes 4–6, scattered, usually distinct, delimited by a thickened border. **Teleutosori** chiefly hypophyllous or less often on the sheaths, scattered or irregularly aggregated, rarely in distinct lines, small, oblong, long-covered by the epidermis, black, with thin layers of dark-brown subepidermal paraphyses surrounding the sori and dividing the larger sori into loculi; teleutospores mostly 2- occa-

sionally 1- or 3- or more-celled, (the spores with more than 2 cells usually occur at the centre of large sori, the 1-celled spores at the margins of the sori) oblong-clavoid, 36–65 × 13–24 μ, rounded or truncate or obtusely and obliquely pointed above, slightly constricted, wall

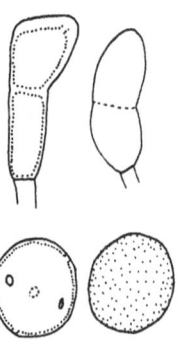

P. recondita. Teleutospores and uredospores.

brown, paler below, 1 μ or less thick at the sides, 3–7 μ above, pedicel short, persistent, pale coloured with usually a chestnut-brown zone just below the spore. Germ tube, basidium and basidiospores colourless. Hetereu-form.

Spermogonia and aecidia on *Ranunculaceae*, *Boraginaceae* and *Crassulaceae*; uredospores and teleutospores on *Gramineae*. Some races very common in Britain.

The brown rusts or leaf rusts of grasses are very frequent in Britain and their discrimination from other graminicolous rusts often presents difficulties. They occur only rarely on the stems and sheaths, thus differing from *Puccinia graminis*. When only uredosori are present they can be distinguished by their lack of well-developed paraphyses and by their dirty-orange colour due to the fact that the walls of the uredospores are brownish, not hyaline; the spore contents are orange in colour. The teleutosori are characteristically subepidermal with peripheral and, in large sori, median paraphyses, partitioning the sori.

Owing to the comparatively small number of races of this species which have been recorded in this country and the very small amount of investigation (especially as regards cultures) which have been carried out on them, any scheme of classification must be regarded as provisional. Arthur (1929, 178) and Jackson (Mem. Torr. Bot. Club, **18**, 1, 1931) have urged the importance of classifying species in the rusts according to the relationship of their hosts and the latter (p. 97) has stated 'In the case of heteroecious species closer relationships would be usually brought out by arranging them according to the host for the haploid phase instead of for the diploid phase.' This suggestion has been attempted here with regard to the races.

Within the complex species *P. recondita*, are included the following *formae speciales* which occur or may occur in Britain and which are mainly distinguishable only by reference to haplont or diplont host relations. The bracketed entries have not been confirmed in Britain. The forms are arranged alphabetically in the following paragraphs.

Aecidial host	Diplont host	Forma specialis
Anchusa and *Lycopsis*	*Secale*	f. sp. *recondita*
Aquilegia	*Agrostis*	f. sp. *agrostidis*
Echium	(*Agropyron*)	f. sp. *echii-agropyrina*
Helleborus	(*Agropyron*)	f. sp. *persistens*
Ranunculus	*Alopecurus*	f. sp. *perplexans*
(*Symphytum*)	*Bromus*	f. sp. *bromina* and (*symphyti-bromorum*)
(*Sedum?*)	*Trisetum* (and *Koeleria*)	f. sp. *triseti*
(*Thalictrum*)	*Triticum*	f. sp. *tritici*
Thalictrum alpinum	(*Agrostis* and *Anthoxanthum*)	f. sp. *borealis*
Thalictrum spp.	*Agropyron*	f. sp. *persistens*
(*Trollius?*)	*Agropyron donianum*	f. sp. *persistens*
None known	*Agropyron*	f. sp. *agropyrina*
None known	*Holcus*	f. sp. *holcina*

f. sp. *agropyrina* Eriks.

Puccinia dispersa f. sp. *agropyri* Eriks., Ber. D. Bot. Ges. **12**, 316 (1894) (*nomen nudum*).

Puccinia agropyrina Eriks., Ann. Sci. Nat. Bot. ser. 8, **9**, 273 (1899); Wilson & Bisby, Brit. Ured. no. 83; Gäumann, Rostpilze, p. 491.

The race on *A. repens* known as *Puccinia agropyrina* does not differ morphologically from *P. persistens* but the aecidial stage is unknown and it is clearly independent of host alteration and overwinters by the uredo stage. The small, scattered, dull-orange uredosori are found on the upper leaf surfaces in July and August; the wall of the uredospores is pale chocolate-brown and the pores (7–8) are surrounded by a slight thickening of the cell wall; paraphyses are few or absent. The teleutosori which are produced from July onwards are mostly hypophyllous or on the sheath; the wall of the teleutospore is slightly thickened at the apex and there is a distinct brown zone at the apex of the pedicel; mesospores are often present at the periphery of the teleutosori. Savulescu (1953) has associated Roumanian collections on *Agropyron cristatum* with aecidia on *Ranunculus sceleratus* and for this reason Gäumann (*loc. cit.*) places this rust in his formenkreis, *P. perplexans*.

The rust is common on *A. repens* in the British Isles; records on *A. caninum* are unconfirmed. For the recent record of *P. agropyrina* on *Agropyron junceiforme* near Bristol (TBMS. **39**, 391, 1956) see under f. sp. *persistens* p. 286.

f. sp. *agrostidis* Oud.

Puccinia agrostidis Plowr., Gard. Chron. III, **8**, 42 and 139 (1890) (*nomen nudum*); Grove, Brit. Rust Fungi, p. 275; Wilson & Bisby, Brit. Ured. no. 84; Gäumann, Rostpilze, p. 477.
Puccinia agrostidis Oud., Rev. Champ. **1**, 528 (1892).
Puccinia rubigo-vera f. sp. *agrostidis* (Plowr.) Mains, Pap. Mich. Acad. Sci. **17**, 352 (1933).

Aecidia on *Aquilegia vulgaris*, uredospores and teleutospores on *Agrostis stolonifera* and *A. tenuis*, Great Britain and Ireland, scarce; it may also occur on *A. gigantea* in southern Ireland (Dennis, in litt.). The aecidial stage is rare in this country; it has been found in Sussex, Worcestershire, Cumberland and Aberdeenshire.

Jørstad (1950, 50) has pointed out that in Norway the uredostage is scanty and is soon replaced by teleutospores; he considered it to be obligatorily heteroecious. He has noted that it is very difficult to distinguish the uredospores from those of *Puccinia coronata*, which also occurs on *Agrostis* sp. but most collections of *P. agrostidis* originated from near *Aquilegia* and collections made away from *Aquilegia* were *P. coronata* whenever teleutospores were present for confirmation.

In this country some of the records of the uredo-stage may refer to *P. coronata*.

The connection between the aecidium on *Aquilegia* and the rust on *Agrostis* was first demonstrated by Plowright (*loc. cit.*); this was confirmed by Jacky (Ber. Schweiz. Bot. Ges. **11**, 18, 1899). Mayor (*ibid.* **47**, 160, 1937) has recorded that some races of *P. actaeae-agropyri* Fisch. on *Agropyron caninum* in Switzerland produce their aecidial stage on *Aquilegia*.

f. sp. *borealis* Juel

Puccinia borealis Juel, Overs. K. Vet. Akad. Förh. **51**, 411 (1894); Wilson & Bisby, Brit. Ured. no. 105; Gäumann, Rostpilze, p. 498.
[*Aecidium thalictri* Grev., Scot. Crypt. Fl. **1**, 4 (1823).]

Spermogonia and aecidia on *Thalictrum alpinum*. Scotland, rare.

Greville's specimen collected on Ben Voirlich (Loch Lomond) in 1823 is in the Edinburgh Herbarium. The aecidium was found on Ben Lui, Perthshire, in 1915 and 1919 (J. Bot. Lond. **53**, 44, 1915, and **57**, 162, 1919) but no British collections of this race on grass hosts are known. Scandinavian collections were proved experimentally to belong to the form known as *Puccinia borealis* on *Agrostis borealis* Hartm. by Juel (Overs. K. Vet. Akad. Förh. no. 8, 411, 1894, and no. 3, 216, 1926). Rostrup (Bot. Tidskkr. **25**, 290, 1903) recorded it from Iceland on *Agrostis* sp. *Calamagrostis stricta, Hierochloë odorata, Anthoxanthum odoratum* and *Deschampsia cespitosa*. Jørstad (1950, 52) reported it from Norway on the same hosts with the addition of *Agrostis borealis* and *A. canina* but omitted *Deschampsia cespitosa*, which he showed to have been misidentified by Rostrup. He described it as an obligatorily host-alternating alpine and northern rust existing in a number of races. In this country the dicaryon host is unknown; it may be *Anthoxanthum odoratum* or *Agrostis* sp. but all the collections from the mountains on these hosts have abundant uredosori, no paraphyses and sparse or no teleutosori, and are certainly uredo-overwintering, non-alternating types belonging to *P. poae-nemoralis*.

The connections with *Agrostis alpina, rupestris* and *borealis* have been confirmed experimentally by Juel (*loc. cit.*) and by Koch & Gäumann (Ber. Schweiz. Bot. Ges. **47**, 448, 1937) but not that with *Anthoxanthum*. Gäumann (1959, 498) maintains that the *Anthoxanthum* rust is not connected with aecidia on *Thalictrum* but is his *P. anthoxanthina* whose aecidia are unknown. Jørstad would attach *P. anthoxanthina* to *P. poae-nemoralis*. It seems most probable that the rusts on *Anthoxanthum* in Norway and Switzerland are quite different. All montane collections on British grasses should be carefully checked for this *Thalictrum* rust.

The aecidial stage of *P. septentrionalis* (see p. 166) which also occurs on *Thalictrum alpinum* is distinguished by the extensive swollen dark-violet spots that it produces on the leaves, petioles, and sometimes on the stem. The two aecidial stages have been found closely associated on Ben Lui in Perthshire.

f. sp. *bromina* Eriks.

Puccinia bromina Eriks., Ann. Sci. Nat. Bot. ser. 8, **9**, 271 (1899); Grove, Brit. Rust
   Fungi, p. 262; Wilson & Bisby, Brit. Ured. no. 107.
Puccinia dispersa f. sp. *bromi* Eriks., Ber. D. Bot. Ges. **12**, 316 (1894) (*nomen nudum*).
Puccinia symphyti-bromorum Müll., Beih. Bot. Zbl. **10**, 201 (1901); Gäumann,
   Rostpilze, p. 523.

[Spermogonia and aecidia on *Ranunculaceae* or *Boraginaceae*]; uredospores and
teleutospores on *Bromus*, common.

Forms of *Puccinia recondita* on *Bromus* (including *Anisantha*, *Cera-
tochloa* and *Zerna*) were first described by Eriksson as *P. dispersa* f. sp.
bromi in 1894 (*loc. cit.*). Eriksson later treated the same rust as a species,
P. bromina Eriks. (*loc. cit.* 1899). He described no aecidial stage in
connection with this species nor has one been certainly found since. In
this work f. sp. *bromina* is used in the broad sense to include all the brome
rusts whose aecidial stages may develop on members of the *Boraginaceae* or *Ranunculaceae*. The races alternating with ranunculaceous
hosts are treated as *P. alternans* by continental workers.

The following *Bromus* species are accepted as naturally infected hosts
in Britain for uredospores, introduced species are bracketed: *Bromus*
(*arduennensis*), (*arvensis*), (*briziformis*), *commutatus*, (*madritensis*), *mollis*,
commutatus, *ramosus*, *secalinus*, *sterilis*. Teleutospores occur on *B.
sterilis* and *mollis*. The following have been artificially infected with
British races (Bean, Brian & Brooks, Ann. Bot. Lond. N.S. **18**, 129,
1954); *B. asper*, *erectus*, *tectorum*, *adoensis*, *arvensis*, *grossus*, *lepidus*,
racemosus and *carinatus*.

The record of *P. bromina* on *Festuca* (*Bromus*) *gigantea* (Wilson &
Bisby, 1954) should be deleted; the specimens are uredosporic and almost certainly *P. coronata*. No aecidial stages are known in Britain
and overwintering is by mycelium or uredospores on the grass host.
Bean *et al.* (*loc. cit.*) failed to infect *Symphytum officinale* and *S. × peregrinum* by inoculation from *Bromus mollis* and *B. sterilis*.

Marshall Ward (Ann. Bot. Lond. **16**, 233, 1902; Ann. Myc. **1**, 132,
1903) and Freeman (Ann. Bot. Lond. **16**, 487, 1902) carried out numerous cultures at Cambridge with races of the brown rust of the bromes
which they assigned to *P. dispersa* Eriks. In the uredospore stage all
these races were morphologically identical; they did not describe
teleutospores or associate the rust with any aecidial stage.

The British races were re-examined by Bean *et al.* (*loc. cit.*) who distinguished four physiologic races; their race 4 is of continental origin.
Race 1 from *B. sterilis* infected only species in the section *Stenobromus*.

Race 2 from *B. arduennensis* infected only species in the section *Serrafalcus*. Race 2*a* from *B. mollis* was similar to race 2 but in addition infected a few species in section *Stenobromus*. Race 3 from *B. madritensis* infected a few species in each section of the genus. Race 4 was less clearly defined but originated from several species of section *Serrafalcus* and infected species in sections *Serrafalcus*, *Stenobromus* and *B. asper* in section *Festucoides*.

Puccinia symphyti-bromorum was described in Switzerland by Müller who carried out infections with teleutospores from *B. erectus* and produced aecidia on *Symphytum officinale* and *Pulmonaria montana*. *Puccinia bromina* and *P. symphyti-bromorum* have been regarded as identical by most workers, although the teleutospores of the two species appear to differ in size and shape. Guyot (Ann. Ecole Nat. Agric. Grignon. II, 1, 67, 1937, and III, 2, 75, 1941) described two rusts on *Bromus* spp. native in Europe which he considered agreed with *P. symphyti-bromorum* and *P. bromina* respectively. He emphasized the differences between the two species; in his *P. symphyti-bromorum* the ratio, length of teleutospores over breadth of lower cell, is more than 3:1—the spores are 'dolichosporous'; whereas in *P. bromina* the ratio is less than 3:1—the spores are 'brachysporous'.

Bean *et al.* on the basis of host range experiments suggested that their British races 1 and 2 were *P. bromina sensu stricto*. They mentioned teleutospores only on *B. mollis* and *B. sterilis* and stated they were of the brachysporous type (*P. bromina*) and this agrees with Jørstad's results from Norway (Jørstad, 1951). All other teleutospores examined by us (all on *B. sterilis*) are also of this type. They are all apparently *P. bromina sensu stricto*.

Further subdivision of *P. symphyti-bromorum* has been proposed; f. sp. *symphyti-bromorum* Müll. forms aecidia on *Pulmonaria* and *Symphytum* species and diplont stages on many bromes; f. sp. *benekeni* Gäum. forms aecidia on *Anchusa*, *Nonnea* and *Pulmonaria* and teleutospores on *Bromus benekeni*; f. sp. *tuberosi-asperi* Dupias forms aecidia on *Symphytum tuberosum* and teleutospores on many *Bromus* species; *P. bromi-maximi* Guyot has been described for the rust on *Bromus maximus* for which an aecidial host is unknown. Mayor (Uredineana, 5, 278, 1958) summarises the information on this rust and gives the results of many of his own experiments.

Species of *Bromus* in Europe are also attacked by rusts of the *P. recondita* type, forming aecidia on *Thalictrum* and *Clematis*. *Puccinia alternans* Arth. infects many species of *Thalictrum* and *Bromus*; one

European race of this predominantly North American species, f. sp. *bromi-erecti* Gäum., has been described. *P. alternans* is said to differ from *P. symphyti-bromorum* in having fewer pores in the uredospores (4–6 as against 7–10 in the latter). *Puccinia madritensis* Maire forms aecidia on *Clematis* spp. and diplont stages on many *Bromus* spp. The status of all these species in Britain is unknown, no aecidial stages have been found and the subject requires intensive research.

Aecidium symphyti Thüm. on *Symphytum officinale* was listed by the Sydows (Monogr. Ured. **1**, 712) and, following them, by Grove (1913, 262) as an aecidial stage of *Puccinia bromina* Eriks.; both indicated that *A. symphyti* occurred in Britain. No native specimen or other record of the aecidium has been found in this country, as pointed out by Wilson & Bisby (1954, 398).

f. sp. *echii-agropyrina* Gäum. & Terr.

Puccinia cerinthes-agropyrina f. sp. *echii-agropyrina* Gäum. & Terr., Ber. Schweiz. Bot. Ges. **57**, 244 (1947); Gäumann, Rostpilze, p. 523.

Aecidium asperifolii Pers. on *Echium vulgare* was mentioned by Plowright (1889, 168) under *Puccinia rubigo-vera* and this appears to be the only record of this rust in Britain but there is none of this material extant. All experimental work points to this rust alternating with species of *Agropyron*, and search should be made on that genus if aecidia are refound in Britain. In the absence of details of aecidial host it is not possible to distinguish this race on *Agropyron* from others of the *P. recondita* group, where they would usually be named *P. agropyrina*.

Tranzschel (Trav. Mus. Bot. Acad. Imp. Sci. St Petersburg, **3**, 52, 1907) described *P. cerinthes-agropyrina* in the Crimea, Russia, and concluded by field observations that its aecidium was borne on *Cerinthe minor* and its uredospores and teleutospores on *Agropyron trichophorum*; he noted that *Echium vulgare* was uninfected. Gäumann & Terrier (*loc. cit.*) found aecidia on *E. vulgare* in Switzerland in 1943 and by field observations concluded that they belonged to a rust on *Agropyron*; they confirmed this by cultures and showed that aecidia occurred on *Echium vulgare* and on *E. lycopsis* but not on various other genera and species of the *Boraginaceae*, including *Cerinthe minor*. The diplont was found to occur on *Agropyron repens* and other *Agropyron* spp.; *A. trichophorum* was not tested. *Puccinia cerinthes-agropyrina* differs from *P. symphyti-bromorum* by not attacking the genera *Symphytum* and *Pulmonaria* as aecidial hosts and from *P. dispersa* by not attacking *Lycopsis*. The Swiss rust is quite similar morphologically to the Russian but differs biologically in attacking different hosts. It was, in consequence, distinguished by the name *P. cerinthes-agropyrina* f. sp. *echii-agropyrina* Gäum. & Terrier and it is to this race that the British collection appears to belong.

Guyot (Uredineana, **3**, 86, 1951) has recorded it in south-eastern France on *Echium vulgare* and with it obtained positive infection on *Agropyron campestre* but failed to infect *A. glaucum* and *A. repens*. Dupias (Bull. Soc. Myc. Fr. **69**, 220, 1953) experimented further with aecidia on *Lithospermum* and *Cynoglossum* and obtained uredosori and teleutosori on *Agropyron repens* and *A. campestre*.

f. sp. *holcina* Eriks.

Puccinia holcina Eriks., Ann. Sci. Nat. Bot. ser. 8, **9**, 274 (1899); Grove, Brit. Rust Fungi, p. 263; Wilson & Bisby, Brit. Ured. no. 166; Gäumann, Rostpilze, p. 558.
Puccinia rubigo-vera f. sp. *holcina* (Eriks.) Mains, Pap. Mich. Acad. Sci. **17**, 381 (1933).

Uredospores and teleutospores on *Holcus lanatus* and *H. mollis*, British Isles, common; uredospores abundant, teleutospores scarce.

Eriksson (*loc. cit.*) showed that this rust is restricted to species of *Holcus*; no other infection experiments appear to have been carried out.

Guyot (Uredineana, **3**, 63, 1951) recognised *Puccinia holcicola* on *Holcus lanata* distinguished by its larger teleutospores, 48–55 × 19–24μ, 40–48 × 22–25μ in *holcina*; and uredospores 24–29 × 22–24μ in *holcicola*, 20·8–24μ diam. in *holcina*. Guyot stated that *P. holcicola* is the common type on *H. lanatus* in central and southern Europe. However, he mentioned only two northern collections which he examined so that the range in spore size in the type area of *P. holcina* has not been adequately sampled. British specimens on *H. lanatus* all agree with his measurements for *P. holcicola*; five uredospore collections have spores 25–25·5 × 20–24·5μ and two teleutospore gatherings, spores 49–61 × 21–26μ. Five uredospore collections on *H. mollis* lie within the range of *P. holcicola*—their maximum range 24·5–28·5 × 21·5–27·5μ but one of these from Hamilton, Lanarkshire, has small teleutospores 42–50 × 22–26μ, within the range of *P. holcina sensu* Guyot. Two collections on *H. mollis* have teleutospores 40–60 × 20–25μ, within the *P. holcicola* range. Obviously, the British rusts on this genus require to be studied carefully. *Puccinia coronata* which also attacks *Holcus* is easily distinguished in the teleutospore stage but appears indistinguishable when only uredosori are present.

f. sp. *perplexans* Plowr.

Puccinia perplexans Plowr., Grevillea, **13**, 53 (1884); Grove, Brit. Rust Fungi, p. 270;
Wilson & Bisby, Brit. Ured. no. 193; Gäumann, Rostpilze, p. 492.
Puccinia rubigo-vera f. sp. *perplexans* (Plowr.) Mains, Pap. Mich. Acad. Sci. **17**, 357
(1933).

Aecidia on *Ranunculus acris*; uredospores and teleutospores on *Alopecurus pratensis*.
Great Britain, scarce.

This race was found at King's Lynn in Norfolk, by Plowright (Trans.
Norf. Norw. Nat. Hist. Soc. **4**, 728, 1884) who carried out many cultures. Later it was discovered by Brebner at Aboyne, Aberdeenshire
(Trail, 1890, 313), by Crossland at Hornsea, Yorkshire on *R. acris*
(Herb. Kew), throughout Yorkshire (Mason & Grainger, 1937, 47) and
at Aberystwyth (Sampson & Western, 1954, 15). It occurs in rather
damp pastures and its scarcity here is difficult to explain as both its
hosts are common.

The connection of the aecidium with the other spore stages was first
demonstrated by Plowright (*loc. cit.*) and has since been confirmed by
Dietel (Hedwigia, **28**, 278, 1889), Klebahn (Z. Pfl.-Krankh. **12**, 145,
1902), Tranzschel (Myc. Zbl. **4**, 70, 1914) in Russia and Fraser (Mycol.
4, 175, 1912) in Canada. Quite similar aecidia on *R. acris* belong to
Uromyces dactylidis.

It appears that there is obligatory alternation of the rust in this
country and Jørstad (1950, 56) has stated that this also takes place in
Norway.

f. sp. *persistens* Plowr.

Puccinia persistens Plowr., Monogr. Brit. Ured. Ustil. 180 (1889); Grove, Brit. Rust
Fungi, p. 282; Wilson & Bisby, Brit. Ured. no. 194; Gäumann, Rostpilze,
p. 495.

Spermogonia and aecidia on *Thalictrum*
minus subsp. *arenarium*, *T. flavum* and
possibly *Helleborus viridis*; uredospores
and teleutospores on *Agropyron juncei-*
forme, *A. repens* and *A. donianum*.
Great Britain, scarce.

In this country aecidia on *Thalictrum flavum* and *T. minus* have been
assigned to *Puccinia persistens* s. str. with the other spore forms on
Agropyron repens and *A. junceiforme*. The aecidium on *T. flavum* has
been recorded from several localities in England extending from Yorkshire, Norfolk and Cambridge to Gloucestershire and Somerset. Plowright showed that when teleutospores on a piece of dead grass were
applied to a plant of *T. flavum* the aecidium appeared in due course. In
July the aecidiospores were applied to a plant of *Agropyron repens* and
gave rise to uredospores in 12 days.

Aecidia have been recorded on *T. minus* by Cooke (1871, 540), Trail (1890, 313) and Grove (*loc. cit.*) but this was regarded as a doubtful host by Plowright (*loc. cit.*). In Hylander *et al.* (1953, 70) this species is said to be the only host in Norway for this rust. Jørstad (1950, 57) considered that in Norway, host alternation is obligatory for this race.

It has been shown recently by Abdell-All (Thesis, University of Aberdeen, 1955) that aecidiospores on a plant of *T. minus* subsp. *arenarium* growing on the sand dunes north of Aberdeen, when applied to a plant of *Agropyron junceiforme* gave rise to uredospores and teleutospores but that *A. repens* when similarly inoculated remained free from infection. Later he produced aecidia on *T. minus* subsp. *arenarium* with the teleutospores produced by infection of *A. junceiforme* and also with teleutospores produced spontaneously on the same plant. He also showed that *Clematis vitalba* was not infected when inoculated with teleutospores from *A. junceiforme*. As the result of further culture experiments with the Aberdeen material and with the infected plants of *T. flavum* from Somerset the existence of two physiologic races was shown; the first on *T. minus* subsp. *arenarium* and on *T. flavum* with its diplont on *A. junceiforme* and not on *A. repens*, and the second on *T. flavum* with its diplont on *A. junceiforme* and on *A. repens*. Neither race would infect wheat.

Uredo agropyri Preuss. recorded on *A. junceiforme* from Norfolk by E. A. Ellis (1934, 499) may belong to the first of these races; his record (1934, 498) on *T. flavum* and *A. repens* evidently belongs to the second.

Wolley-Dod (J. Roy. Hort. Soc. **12**, liii, 1890) recorded an aecidium allied to *Aecidium ficariae* on hellebores in his garden at Malpas, Cheshire. This is probably the aecidial stage which has been recorded as *P. actaeae-agropyri* on several species of *Helleborus* on the continent (Gäumann, 1959, 466). The diplont host is probably *Agropyron caninum* (see Mayor, Ber. Schweiz. Bot. Ges. **47**, 160, 1937).

Collections of uredo- and teleutosori on *Agropyron donianum* from Lawers, Perthshire (TBMS. **37**, 252, 1954), probably belong here. Nothing is known of the aecidial host but it is most probably *Trollius europaeus* which grew in the vicinity and on which Mayor has described a race *P. actaeae-agropyri* f. sp. *trollii-agropyri* alternating with *Agropyron caninum* and a *Trollius* alternating race is common in suitable localities in Scandinavia.

f. sp. *recondita* Rob. & Desm.

Puccinia dispersa Eriks. & Henn., Ber. D. Bot. Ges. **12**, 315 (1894); Wilson & Bisby, Brit. Ured. no. 148; Gäumann, Rostpilze, p. 527.
Puccinia secalina Grove, Brit. Rust Fungi, p. 261 (1913).

Spermogonia and aecidia on *Anchusa officinalis* and *Lycopus arvensis*; uredo- spores and teleutospores on *Secale cereale*. Great Britain, scarce.

Aecidium anchusae Eriks. & Henn. (*Aec. asperifolii* Pers.) on *Lycopsis arvensis* and *Anchusa officinalis* belongs to the race on *Secale cereale* previously known as *P. dispersa* Eriks.; it can be produced only by infection with teleutospores from *Secale cereale* (Eriksson, Ann. Sci. Nat. Bot. ser. 8, **9**, 268, 1899, and **8**, 26, 1898).

The aecidium on *L. arvensis* has been found in Kent, Surrey, Dorset, Worcestershire, Hampshire (Owen, *in litt.*), Norfolk, Yorkshire and Fife and there is a specimen in the Johnston Herbarium from Berwick; it is generally scarce. There is a poor specimen in the Kew Herbarium, on *Anchusa officinalis* from King's Cliff, Northamptonshire, collected by Berkeley in 1848. As the teleutospores are capable of germination as soon as they mature the aecidia are usually found in August or September.

The statement by Plowright (1889, 168) that a bundle of rusted wheat straw, kept in the open, produced aecidia on nearby plants of *L. arvensis* is generally discredited.

Sappin-Trouffy (1897) described the development of the aecidium.

The rust is frequent on *Secale cereale* where the crop is grown; the uredospores usually appear in June or July and are soon followed by the teleutospores. Uredospore overwintering has been confirmed in Russia (Kuprewicz & Kalimonova, abstract in RAM. **18**, 587, 1939).

The alternation was first established by de Bary and numerous experiments have been carried out since; the most comprehensive in Europe, by Eriksson & Henning (Die Getreideroste, 1896).

Hassebrauk (Arb. Biol. Abt. Berl. **20**, 165, 1932) carried out extensive experiments and more recently Dupias (Bull. Soc. Hist. Nat. Toul. **82**, 207, 1948 and Ber. Schweiz. Bot. Ges. **62**, 370, 1952) has confirmed the relation with *Lycopsis arvensis* and showed that *Aegilops ovata* is susceptible to at least some strains. He has noted that aecidia appear on *Lycopsis* in late spring several weeks after development of uredospores and teleutospores on rye.

f. sp. *triseti* Eriks.

Puccinia triseti Eriks., Ann. Sci. Nat. Bot. ser. 8, **9**, 270 (1899); Grove, Brit. Rust Fungi, p. 264; Wilson & Bisby, Brit. Ured. no. 234; Gäumann, Rostpilze, p. 517.
Puccinia rubigo-vera f. sp. *triseti* (Eriks.) Mains, Pap. Mich. Acad. Sci. **17**, 381 (1933).

Uredospores and teleutospores on *Trisetum flavescens*. England and Scotland, scarce.

In south-western France the aecidial hosts are species of *Sedum* section *Eusedum*, *S. reflexum*, *S. rupestre*, *S. nicaeense* (Dupias, Uredineana, **5**, 303, 1958). In the same series of experiments Dupias showed that *Koeleria phleoides* was susceptible when inoculated at the same time as *Trisetum flavescens* with aecidiospores from *Sedum nicaeense*.

f. sp. *tritici* Eriks. & Henn.

Puccinia dispersa f. sp. *tritici* Eriks. & Henn., Z. Pfl.-Krankh. **4**, 259 (1894).
Puccinia triticina Eriks., Ann. Sci. Nat. Bot. ser. 8, **9**, 270 (1899); Grove, Brit. Rust Fungi, p. 262; Wilson & Bisby, Brit. Ured. no. 235; Gäumann, Rostpilze, p. 508.

Uredospores and teleutospores on *Triticum aestivum*. Great Britain, common.

The form on *Triticum* sp. and a few other genera which has its aecidial stage on *Thalictrum* sp., generally known as *Puccinia triticina* Eriks. or *P. rubigo-vera* f. sp. *tritici* (Eriks.) Carleton, closely resembles *P. recondita* f. sp. *persistens* and appears to be separated from it chiefly on account of the disease of wheat, the brown rust or leaf rust, caused by it. The disease is common on wheat especially in southern Britain but is relatively harmless.

The aecidial stage is not known in this country or in Scandinavia but may occur on species of *Thalictrum* in Europe and Asia. According to K. S. Chester (Cereal Rusts, 1946) the only plant spontaneously infected is *Isopyrum fumarioides* in the region of Lake Baikal, Siberia; this has led to the suggestion that Central Asia is the centre of origin of this rust (see also Naumov, Rusts of Cereals in USSR. 1939; see Uredineana, **4**, 460, 1953). Portuguese collections, however, scarcely infect *Isopyrum* according to D'Oliviera (Agron. Lusit. **13**, 223, 1951).

Successful cultures with teleutospores were made in 1919 and 1920 by Jackson & Mains in America (J. Agric. Res. **22**, 154, 1921) on various species of *Thalictrum* and aecidiospores from these cultures were sown on *Triticum* and *Aegilops* with positive results (see Mains, Pap. Mich. Acad. Sci. **17**, 302 and 318, 1933). More recently similar cultures have been carried out in France, Germany, Russia and Japan.

Brown rust is frequent in Britain and ten races have been distinguished

(Roberts, Ann. Appl. Biol. **23**, 271, 1936) but this number is low compared with the one hundred and sixty-three races recognised in a recent revision of the International Index of physiologic races (Pl. Dis. Reporter, Suppl. **233**, 1955). It can overwinter as uredospores on self-sown plants (Mehta, TBMS. **8**, 151, 1923). The uredospore are highly tolerant of frost (Plowright, Gard. Chron. **18**, 234, 1882; Mehta, *loc. cit.*).

The cytology has been studied by Allen (J. Agric. Res. **33**, 201, 1926.) Nelson *et al.* (Phytopath. **45**, 639, 1955) have reported heterocaryosis in brown rust.

The record of aecidia of this group on *Clematis* (as *Aecidium ranunculacearum* var. *clematidis* in Cooke, Handbk. Brit. Fungi, p. 539), copied by subsequent authors, is quite unsubstantiated although Guyot & Massenot (Uredineana, **4**, 281, 1953) have recorded aecidia on several species of *Clematis* in France.

Puccinia schismi Bub.

Ann. K.K. Naturh. Hofmus, Wien, **28**, 193 (1914).
Puccinia fragosoi Bub., Hedwigia, **57**, 2 (1916).
Puccinia vulpiana Guyot, Uredineana, **2**, 53 (1947); Gäumann, Rostpilze, p. 463.
Puccinia vulpiae-myuri Mayor & Vienn.-Bourg., Rev. Myc. **15**, 103 (1950).

[Spermogonia and aecidia not found in Britain]. **Uredosori** epiphyllous, more rarely hypophyllous, elliptic, oblong or linear, covered at first then open, pulverulent, yellowish; uredospores globoid, 24–34μ diam., or ovoid or pyriform with orange contents, 28–32 × 22–26μ, wall hyaline, 2–2·5μ thick, verruculose, with 8–10 scattered pores. **Teleutosori** hypophyllous rarely epiphyllous, scattered, minute, immersed as in *Puccinia* *recondita* with marginal and partitioning, brown, thick-walled, columnar paraphyses; teleutospores 54–60 × 17–21μ, oblong, elliptic or slightly clavate, faintly constricted at septum, apex subtruncate, wall brown, smooth, 2μ thick in upper cell, up to 3–5μ at apex, pedicel up to 10μ long, hyaline. Hetereu-form.
[Aecidia on *Allium* spp.]; uredosori and teleutosori on *Vulpia bromoides*. Wales, rare.

The only evidence for the inclusion of this species is Sampson's record from Aberystwyth and a small sample from the collection then made. The species is distinguishable with difficulty from *Puccinia recondita;* the lateral wall thickness of the upper cell of the teleutospore is 1–1·5μ in *P. recondita* as against 1·5–2·5μ in *P. schismi* (Cummins, 1956); this distinction holds for the British material. The uredosori are light yellow in *P. schismi* but orange to orange-brown in *P. recondita*.

Rusts of the *P. schismi* type have also been recorded abroad (under many names) on the following grasses which occur in Britain: *Koeleria gracilis* (*P. fragosoi* and *P. koeleriae*), *Lolium perenne*, *L. temulentum*,

L. multiflorum (*P. loliina* and *P. loliicola*) and *Vulpia myuros* (*P. vulpiana* and *P. vulpiae-myuri*) (see Jørstad, 1958, 87). Alternation with *Allium* has been established by several authors (Guyot, Ann. Nat. Ecole Agric. Grignon, **1**, 66, 1939), Guyot & Massenot (Uredineana **4**, 338, 1953) Guyot & Malençon (as *P. infraniana* in Urédinées du Maroc, 98, 1957) and Dupias (Bull Soc. Hist. Nat. Toul. **85**, 38, 1950) for various members of this group.

Puccinia sessilis Schroet.

Abh. Schles. Ges. Vaterl. Cult. Nat. Abth. **1869–72**, 19 (1870); Grove, Brit. Rust Fungi, p. 266; Wilson & Bisby, Brit. Ured. no. 221; Gäumann, Rostpilze, p. 444.

Puccinia phalaridis Plowr., J. Linn. Soc. Bot. **24**, 88 (1888); Grove, Brit. Rust Fungi, p. 269; Wilson & Bisby, Brit. Ured. no. 223; Gäumann, Rostpilze, p. 462.

Puccinia orchidearum-phalaridis Kleb., Z. Pfl.-Krankh. **9**, 155 (1889); Grove, Brit. Rust Fungi, p. 268; Wilson, Ured. Scot. p. 399; Wilson & Bisby, Brit. Ured. no. 222; Gäumann, Rostpilze, p. 461.

Puccinia digraphidis Soppitt, J. Bot. Lond. **28**, 215 (1890); Grove, Brit. Rust Fungi, p. 267; Gäumann, Rostpilze, p. 456.

Puccinia paridis Plowr., J. Linn. Soc. Bot. **30**, 43 (1893).

Puccinia winteriana Magn., Hedwigia, p. **33**, 78 (1894) (*nomen nudum*); Grove, Brit. Rust Fungi, p. 268; Wilson & Bisby, Brit. Ured. no. 224.

Puccinia ari-phalaridis Kleb., Jahrb. Wiss. Bot. **34**, 399 (1900).

Puccinia allii-phalaridis Kleb., Jahrb. Wiss. Bot. **34**, 399 (1900); Gäumann, Rostpilze, p. 458.

Spermogonia epiphyllous or hypophyllous amongst the aecidia, reddish-yellow. **Aecidia** hypophyllous, loosely clustered or sometimes circinate, in rounded groups on circular or irregular, large yellow spots, cup-shaped with a laciniate, white, recurved peridium, peridial cells firmly connected, not in distinct rows, with the punctate outer wall strongly thickened, up to $8–10\mu$ thick and the inner wall up to $3–5\mu$, with small warts; aecidiospores globoid or ellipsoid, yellowish, $18–27\mu$ diam., wall thin, very finely verruculose. **Uredosori** amphigenous, scattered, minute, punctiform or shortly linear, soon naked, pulverulent, yellowish-brown, aparaphysate; uredospores globoid to ellipsoid, brownish-yellow, $20–28 \times 18–24\mu$, wall distantly-echinulate, with about 7 scattered pores. **Teleutosori** similar, sometimes confluent, long-covered by the epidermis, pulvinate, black; teleutospores oblong, or oblong-clavoid, round-

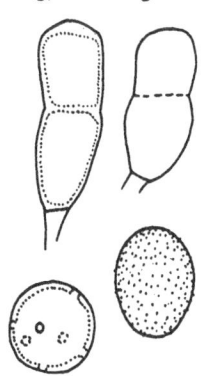

P. sessilis. Teleutospores and uredospores.

ed or truncate and darker above, hardly constricted, somewhat narrowed below, brown, $35–52 \times 15–22\mu$, wall smooth, up to 5μ thick at apex; pedicel brown, short, usually deciduous. Hetereu-form.

Spermogonia and aecidia on *Allium ursinum, Arum maculatum, Convallaria majalis, Dactylorchis fuchsii, D. praeter-* *missa, D. incarnata, Gymnadenia conopsea, Listera ovata, Paris quadrifolia;* uredospores and teleutospores on *Phalaris arundinacea,* July–May. Great Britain and Ireland, scarce.

Four biological races of *Puccinia sessilis,* all with uredospores and teleutospores on *P. arundinacea,* have been recorded in this country; they have often been treated as species in the past and are so treated by Gäumann (1959, 444 *et seq.*).

The race with its aecidia on *Convallaria majalis* and *Paris quadrifolia* (previously known as *P. digraphidis*) is very rare; aecidia on the latter species were found by Plowright (Gard. Chron. III, **12**, 137, 1892) near Carlisle and on the former by Soppitt (Gard. Chron. III, **10**, 643, 1890) on an island in Lake Windermere, Westmorland. The race with aecidia on the Orchidaceae (*P. orchidearum-phalaridis*) is scarce. Aecidia on *Allium ursinum* (*P. winteriana*) are frequently found. The race on *Arum maculatum* (*P. phalaridis*) is uncommon and is recorded from England and Northern Ireland. The distribution of the races is limited by their aecidial hosts, *Convallaria, Paris* and *Arum,* all decrease northwards and scarcely occur beyond the Scottish lowlands; *Convallaria* and *Paris* are not native in Ireland.

Numerous experimental cultures have been carried out with this species, the first in 1874 by Winter (S.B. Naturf. Ges. Leipzig, **41–43**, 1874; Hedwigia, **14**, 113, 1875) who infected *Phalaris arundinacea* from *Allium ursinum* and obtained aecidia and also reinfected the grass host with aecidiospores. Mayor (summarised in Uredineana, **5**, 274, 1958) who carried out series of experiments with this race recorded many *Allium* species as hosts and also *Paris* and *Polygonatum.*

In this country Plowright (*loc. cit.*) using aecidiospores from *Allium, Arum* and *Paris* and Soppitt (*loc. cit.*) with aecidiospores from *Convallaria* obtained successful results on *Phalaris arundinacea.*

Very numerous experiments were carried out by Klebahn (1904) and by others in Germany and by Mayor (Ber. Schweiz. Bot. Ges. **42**, 142, 1933) in Switzerland and Guyot (Ann. École Nat. Agric. Grignon, **5**, 30, 1946) in France. The specialisation in this latter country varies to some extent from that found in Great Britain. They all confirmed that a polyphagous race existed which could infect *Maianthemum, Convallaria, Paris* and *Polygonatum,* whereas in Britain, Soppitt & Plowright experimented with races strictly specialised to the last three host genera in the aecidial phase. The polyphagous race and those closely specialised to *Paris* and *Convallaria* Gäumann (1959, 457) treats as *formae speciales.*

No experiments have been carried out with the orchid races in this country. The relation with *Phalaris* was investigated by Klebahn (summarised in Klebahn, 1904) and more recently by Mayor (*loc. cit.*). With all these races continental workers have evidence of weak specialisation in the aecidial stage. If the results of all these workers is valid it appears that the population of this rust in Britain may differ sharply from those of central Europe.

Puccinia sorghi Schw.

Trans. Amer. Phil. Soc. ii, 4, 295 (1832); Wilson & Bisby, Brit. Ured. no. 228; Gäumann, Rostpilze, p. 727.

[**Spermogonia** amphigenous, in small groups, intermixed with the aecidia, honey-yellow, globoid, 90–110μ diam., paraphyses projecting, short. **Aecidia** amphigenous, mostly hypophyllous, crowded in groups, 2-3 mm. diam., usually circinately arranged around the spermogonia, minute, up to 400μ diam., cupulate, margin of peridium erect, entire or irregularly lacerate, cells of peridium rhomboid, 20–25 × 13–16μ, outer wall faintly striate, 6μ thick, inner verruculose, 3μ thick; aecidiospores globoid or ellipsoid, 18–26 × 14–20μ; wall hyaline, 1–1·5μ thick, very finely and closely verruculose.] **Uredosori** amphigenous, scattered or in elliptical or oblong groups, covered by the bullate epidermis then pulverulent, cinnamon-brown; uredospores ellipsoid to globoid, 32–34 × 20–28μ, wall finely echinulate, cinnamon-brown, 1·5–2 μ thick, with 4 (rarely 3) equatorial pores. **Teleutosori** scattered or in large, compact groups, sometimes confluent, oblong, soon naked, blackish-brown; teleutospores oblong or ellipsoid, rounded or obtuse above and below, slightly constricted at the septum, wall dark chestnut-brown, 1–2μ thick at sides, 5–7μ at apex, smooth, pore in upper cell apical, in lower superior, 28–48 × 13–25μ; pedicel colourless except near the spore, persistent, thick, about equal to the length of the spore. Hetereu-form.

[Aecidia on *Oxalis stricta* and *corniculata*]; uredospores and teleutospores on *Zea mays*, teleutospores in late autumn. England, rare.

This is an American species which was first definitely recorded in Europe, in Holland, in 1837; it was collected in 1822 in Natal. It is now found wherever maize is cultivated but usually does not cause very serious damage.

This rust was found by Ellis (Trans. Norf. Norw. Nat. Hist. Soc. **15**, 371, 1943) near Norwich in November 1941 when teleutospores were present; it appears, however, that it was discovered in a garden in London in 1925. It has since been found occasionally particularly in southeast and south-west England (Moore, 1959, 312).

Puccinia striiformis West.

Bull. Roy. Acad. Belg. **21**, 235 (1854).

[*Uredo glumarum* J. K. Schm., Allg. Oekon.-techn. Fl. **1**, 27 (1827).]
Puccinia straminis Fuck., Jahrb. Nass. Ver. Nat. **15**, 9 (1860) *p.p.*
Puccinia tritici Ørsted, Sygd. Hos. Planterne, p. 95 (1863).
Puccinia glumarum Eriks. & Henn., Landt.-Akad. Handl. o. Tidskr. **33**, 169 (1894);
and Z. Pfl.-Krankh. **4**, 197 (1894); Grove, Brit. Rust Fungi, p. 258; Wilson &
Bisby, Brit. Ured. no. 159; Gäumann, Rostpilze, p. 545.

Spermogonia and **aecidia** unknown. **Uredosori** amphigenous and on the sheaths, culms and inflorescence, oblong 0·5–1 mm. long and 0·3–4 mm. wide, usually not confluent, in long lines on yellowish spots, forming areas of disease which may be up to 70 mm. long, pulverulent, lemon-yellow, uredospores globoid to broadly ellipsoid, 25–30 × 12–24 μ, contents orange, wall hyaline, colourless, with short blunt spines 1·5 μ apart, with 8–10 rather indistinct, scattered pores. **Teleutosori** hypophyllous and culmicolous, in long thin lines or scattered on the inflorescence, oblong, dark brown or black, long-covered by the epidermis, with numerous brown, thick-walled, curved, elongate, subepidermal paraphyses which separate the spores into groups and which by further modification form a loosely agglutinated wall around each group thus producing a compound sorus; teleutospores clavoid, rounded truncate or obliquely conical above, slightly constricted, attenuate below, 30–70 μ long, with upper cell 16–24 μ wide and basal cell 9–12 μ wide; wall of upper cell 1 μ thick at sides, thickened up to 4–6 μ or even 10 μ at the apex, smooth, chocolate-brown, pedicel short, slightly yellowish; mesospores sometimes present 26–32 × 12–16 μ; germ tube of teleutospores

with orange-yellow contents; basidium 4-celled with basidiospores broadly ellipsoid, contents orange-yellow. Hemiform?

Uredospores and teleutospores on *Agropyron caninum*, *A. repens*, *Brachy-*

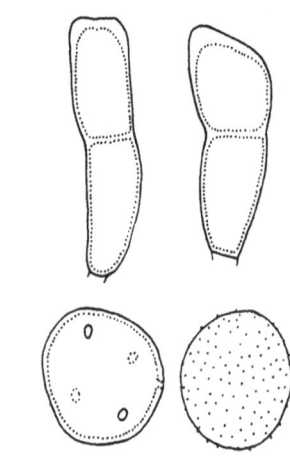

P. striiformis. Teleutospores and uredospores.

podium sylvaticum, *Bromus sterilis*, *Bromus* spp., *Elymus arenarius*, *Hordeum marinum*, *H. murinum* and cultivated *H. vulgare*, *Secale cereale* and *Triticum aestivum*. Great Britain and Ireland, common.

Separation of *Puccinia striiformis* from other rusts on cereals and grasses is difficult. The uredospores, with colourless walls and coloured contents are lighter in colour in the mass than those of the *P. recondita* group with which confusion is most frequent but light-coloured strains have been recorded in the latter (Guyot, Ann. Ecole Nat. Agric. Grignon, III, **2**, 58, 1941). Klebahn (1914, 625) stated that the diameter of the mycelium is greater in *P. glumarum* (up to 11 μ) than in related rusts. It may well be that *P. striiformis* is of polyphyletic origin from a number of races of the *P. recondita* group. The orange contents of the basidium and basidiospores have also been considered diagnostic.

In this country it is usually known as the Yellow Rust on account of its bright lemon-yellow uredosori. In North America it is called the Stripe Rust from the arrangement of the sori on the leaves.

The first report of this rust in Britain was by J. S. Henslow (J. Roy. Agric. Soc. England, **2**, 1, 1841) who observed it on wheat 'White Tunstall' at Hitcham, Suffolk, in June and August 1841; Section v of his report is entitled 'On the Rust Red-rag, Red robin, Red-gum (*Uredo rubigo*)'. He described its occurrence on the stem, leaf, chaff and seed and stated that the orange-yellow stage was followed by a deep-brown stage. There appears to be no doubt that this was *P. striiformis*, although at the time he considered it to be a stage of *Uredo linearis* (= *P. graminis*). Later in the same volume (p. 220) he described the yellow and brown sori of *U. rubigo* on the inner surface of a glume and included drawings of the uredospores and of the teleutospores which he found in their respective sori. According to Eriksson & Henning (Die Getreideroste, 1896, 145) this was the first record of the rust on the glumes.

In spite of much search and investigation by culture in both Europe and America no aecidial stage of this rust has been found. The teleuto-spores germinate as soon as they are mature in the autumn and in the following spring. Infections made by Eriksson & Henning (*loc. cit.* 153) on wheat with basidiospores failed and a similar result was obtained on species of various genera of the *Boraginaceae*; these investigators doubted the existence of an alternate host. Similar cultures have also been made by various investigators but all have given negative results and many consider that the rust has lost its heteroecious character (Eriksson, Arkiv. Bot. **1**, 43, 1903; Klebahn, 1904, 250) but Gäumann (1959, 548) suggests that a monocotyledonous aecidial host may yet be found.

Cereals

It is the most harmful rust on wheat in England and Wales but seldom serious in Scotland; not uncommon on barley, occasionally on rye (Moore 1959, 302).

The rust appears early in the season, even in January (Moore, *loc.*

cit.) and is usually evident by May, it probably persists as mycelium in overwintering host plants. Manners (Ann. Appl. Biol. **37**, 187, 1950) isolated eight physiologic races from wheat and one from barley in Britain.

Agropyron

The rust has been recorded from scattered localities on both *Agropyron caninum* and *A. repens*. It has not been confirmed on the former. Manners (*loc. cit.*) could not infect *A. caninum* and obtained only occasional infection on *A. repens* with isolates from other hosts and from *A. repens*.

Brachypodium sylvaticum

This plant has been recorded several times (TBMS. **33**, 172, 1950, and **21**, 2, 1935; Grove, Brit. Rust Fungi, 259; Wilson & Bisby, Brit. Ured. no. 159; Mayfield, 1935, 1) as a host but none has been confirmed and the fungus may have been *Puccinia brachypodii* which is frequent on this host and also produces sori in lines. *Puccinia brachypodii* has abundant uredoparaphyses in contrast to their scarcity in *P. glumarum*.

Bromus

The records on *Bromus* spp. are unconfirmed except for Manners' report of infection of *Bromus sterilis* by races from wheat, barley and *Hordeum murinum*. Manners failed to isolate any cereal race from wild grasses and considered it unlikely that they are a reservoir for infection. However, cereal races can infect wild grasses, e.g. *B. sterilis* and *Agropyron repens* (Manners, *loc. cit.* 199).

Hordeum marinum

Two collections are known, from Sunk Island, Yorkshire, August 1950, and from Whippingham, Isle of Wight, 1953. Manners (*loc. cit.*) called the rust of this group on *H. marinum* race M and showed that *H. marinum* was uniformly susceptible to it; *H. murinum* and a range of grasses were almost completely resistant. Several wheat, barley and *H. murinum* isolates infected *H. marinum*.

Hordeum murinum

The correct treatment of the rusts on this host in Britain is uncertain. Buchwald investigated collections in Denmark and recognised three rusts, *Puccinia graminis*, *P. hordei-murini* and *P. striiformis*. There is no record of the first named in Britain and it would be readily distinguished by its uredospores from the others. His differential characters for the remaining pair may be summarised as follows: *P. striiformis* about 12 % (2–17 %) mesospores present; *P. hordei-murini* about 28 % (22–38 %) mesospores present. In *P. hordei* which occurs on other species of *Hordeum* the percentage of mesospores was 53–100. The number of pores in the uredospores was 8–9 in *P. hordei* and 10–12 in *P. hordei-murini*.

On *Hordeum murinum* rust has been collected occasionally since it was first recorded near Edinburgh and at Berwick (TBMS. **9**, 137, 1924; Ellis, 1934, 498; Manners, *loc. cit.*, 199; Wilson, 1934, 397). In the Berwick material which bears only teleutospores there are less than 10 % mesospores which suggests that it should be placed in *P. striiformis*.

var. *dactylidis* Manners

Trans. Brit. Myc. Soc. **43**, 65, 1960.

This variety on *Dactylis glomerata* is generally scarce but has been recorded from England by Manners and as *Puccinia glumarum* on *Dactylis* from Wales (Sampson, Agric. Prog. 106, 1924) and from Ireland (TBMS. **41**, 1958, 289).

There is only one race on this host and cultures showed that it does not infect the cereals or wild grasses on which *P. striiformis* is known (Manners *loc. cit.*). Its uredospores are smaller (18·5–25 × 15–20·5 μ). The optimum temperature for infection is considerably higher than that of the other races, 22·5° C. compared with the usual 10–13° C. and it is not a spring but an autumn rust. Batts (*loc. cit.*) reported that it became very abundant in the Lothians, Scotland, in August 1949 and in Co. Kildare, Ireland, its occurrence was noted in the middle of September 1957. Sampson & Western (1954, 12) regard '*P. glumarum*' as the most serious rust of cocksfoot but their measurements of uredospores and presence of teleutospores do not agree with Manner's account of var. *dactylidis*.

Miyagia Miyabe

apud H. & P. Syd. Ann. Myc. **11**, 107 (1913).

Spermogonia subepidermal, flask-shaped with projecting paraphyses and flexuous hyphae. **Aecidia** uredinoid, pustular with a short, delicate, imperfect peridium composed of hyaline pseudo-parenchyma, elongate apically to form club-shaped paraphyses; aecidiospores similar to the uredospores. **Uredosori** subepidermal, peridium similar to that of the aecidium, paraphyses thick-walled, dark brown, sub-opaque, forming a palisade-like layer appearing from above as a black line round the uredosorus; uredospores solitary, borne on pedicels. **Teleutosori** long-covered by the epidermis, surrounded by palisade of paraphyses which also divide the sorus into loculi. Teleutospores 1- or 2-celled, club-shaped, wall thickened at the apex, with persistent pedicels.

All the species of this genus parasitise members of the *Compositae*.

Miyagia pseudosphaeria (Mont.) Jørst.

Nytt Mag. Bot. **9**, 78 (1961).

[*Aecidium sonchi* Johnst., Fl. Berwick, **2**, 205 (1831).]
Puccinia pseudosphaeria Mont., Barker-Webb & Berthelot, Hist. Nat. Iles Canar. III, **2**, 89 (1840); Gäumann, Rostpilze, p. 587.
Puccinia sonchi Desm., Ann. Sci. Nat. Bot. ser. 3, **11**, 274 (1849); Grove, Brit. Rust Fungi, p. 155; Wilson & Bisby, Brit. Ured. no. 227.
Uromyces sonchi Oud., Rabh. Fungi Europ. no. 95 (1872).
Peristemma sonchi (Desm.) Syd., Ann. Myc. **19**, 175 (1921).
Peristemma pseudosphaeria (Mont.) Jørst., Friesia, **5**, 278 (1956).

Spermogonia observed only in culture, chiefly epiphyllous on reddish-brown spots about 1 mm. diam., 72–86μ diam., with paraphyses and flexuous hyphae; spermatia $3 \times 4\mu$. **Aecidia** uredinoid amphigenous, on yellowish spots, scattered or gregarious, minute, pustular, yellowish, surrounded by thin-walled, colourless paraphyses which form a delicate, imperfect peridium; aecidiospores similar to the uredospores. **Uredosori** generally hypophyllous, on the stem collected in small oblong patches 4–5 mm. long, on yellowish spots, scattered or gregarious, minute, at first covered by the epidermis which is raised over them in a hemispherical vesicle and surrounded by a single row of thick-walled, dark-brown, club-shaped paraphyses which form an incomplete peridium, at length pierced at the top but never widely open, in surface view the yellowish spore layer being encircled with a black line made up of the swollen heads of the paraphyses; uredospores ellipsoid or ovoid, contents oily, yellowish, 24–38 × 15–21μ, wall 2–3μ thick, densely verrucose, with 4 pores. **Teleutosori** hypophyllous and on the stems sometimes developing in the emptied uredosori, on irregular brownish areas, scattered or in groups, pulvinate and rounded on the leaves, often confluent on the stems and up to 2 mm. long and 0·25 mm. wide, covered by the epidermis, surrounded by club-shaped, thick-walled, reddish-brown paraphyses

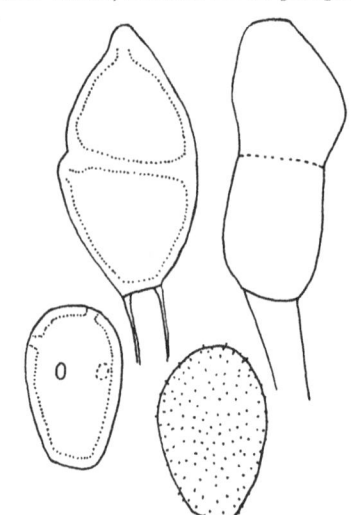

M. pseudosphaeria. Teleutospores and uredospores.

and often compound being divided up by the latter into groups of teleutospores, black; teleutospores ovoid or broadly ellipsoid, rounded or truncate, rarely attenuate above, constricted at the septum, rounded below, 30–60 × 20–30μ, wall smooth, pale brown, 2–2·5μ thick at sides, 3–8μ at apex, pore of upper cell apical, pore in lower cell superior, mesospores often numerous and sometimes exclusively formed, ovoid or clavoid 45–60 × 20–25μ; pedicels long, persistent, pale brown; basidium 4-celled; basidiospores ellipsoid, up to 15μ long. Brachy-form.

On *Sonchus arvensis*, *S. asper*, *S. oleraceus* and *S. palustris*, July–September. Great Britain and Ireland, scarce.

The spermogonia and aecidia have been obtained in culture, by Tranzschel (Ann. Myc. 7, 182, 1909) who sowed teleutospores on *Sonchus arvensis* and by Lamb (Hedwigia, 74, 181, 1934) who infected the same host with mesospores. As described by Lamb, the aecidium generally resembles the uredosorus; in each, the yellow spore-layer is surrounded by a single row of paraphyses. In the uredosorus the paraphyses are dark brown and consequently a black marginal line is seen around the yellow spore mass while in the aecidium they are colourless and the black line is absent. Lamb removed plants of *S. arvensis* bearing uredosori from near Bournemouth and replanted these in Edinburgh; subsequently they produced mesospores only, on the leaves in exhausted uredosori and also in teleutosori; mesospores were also produced on the stems in several superposed layers of sori separated by torn host tissue. An account of the development of the uredosorus has been given by Lamb (*loc. cit.*); the paraphyses arise from the end cells of hyphae which surround the sporogenous hyphae and attain a length of 140μ. Lamb has also described the germination of the mesospores, the basidiospores, the infection process and the development of the spermogonia and aecidia.

Teleutospores have been found in Yorkshire and in Cardiganshire on *S. arvensis* and *S. asper*; on the former host they occurred in abundance on the stems accompanied by a few mesospores; here there was only a single layer of sori. The rust was recorded on *S. palustris* by Ellis from Norfolk (Trans. Norf. Norw. Nat. Hist. Soc. 13, 494, 1933–4) and from Huntingdonshire.

Jørstad (*loc. cit.* 1956, 278) has recently discussed the morphology, relationships and distribution of this rust.

O'Connor (TBMS. 39, 423, 1956) found that in certain areas in Ireland most infections have only 1-celled teleutospores (mesospores), however, he also detected predominantly 2-celled races and admixtures in various proportions of the two teleutospore types. He proposed to recognise the extreme types as two species, *Puccinia sonchi* and *Uromyces sonchi*, but as this is a clear case of the artificiality of division of these rusts by number of teleutospore cells alone they are better regarded as one variable species. Tulasne (Ann. Sci. Nat. Bot. ser. 4, 2, 77, 1854) had long before described similar variability.

Cumminsiella Arth.

Bull. Torr. Bot. Club, **60**, 475 (1933).

Spermogonia subepidermal, punctiform with paraphyses. **Aecidia** subepidermal, cupulate; aecidiospores globose or ellipsoid. **Uredosori** subepidermal, with or without paraphyses; uredospores pedicellate, obovate or fusiform, with yellow, verrucose walls and pores in one or two zones. **Teleutosori** subepidermal, dark brown; teleutospores ellipsoid, of 2 equal cells, with 2 lateral pores in each cell, the wall of 3 layers but not noticeably laminate, verrucose, pedicel long.

A genus of autoecious species on *Mahonia* which were originally included in *Puccinia*. *Cumminsiella mirabilissima*, the only species found in Britain, was removed from *Puccinia* by Magnus and placed in the genus *Uropyxis* Schroet. Arthur established *Cumminsiella* to segregate from *Uropyxis* those species which have subepidermal spermogonia and aecidia. Baxter & Cummins (Mycol. **49**, 864, 1957) recognise six species, all native to the New World.

Cumminsiella mirabilissima (Peck) Nannf.

Lundell & Nannf. Fungi Exsicc. Suec. no. 1507 *a* (1947); Wilson & Bisby, Brit. Ured. no. 2.

[*Uromyces sanguineus* Peck, Bot. Gaz. **4**, 128 (1879).]
Puccinia mirabilissima Peck, Bot. Gaz. **6**, 226 (1881).
Cumminsiella sanguinea Arth., Bull. Torr. Bot. Club, **60**, 475 (1933); Gäumann, Rostpilze, p. 1147.

Spermogonia epiphyllous in groups on dark red spots, dull, orange-coloured, 120 μ high, 100–115 μ wide. **Aecidia** generally hypophyllous, occasionally epiphyllous, on greenish-black, thickened areas, peridium rather shortly and widely cylindrical, with revolute, much cut margin, yellow, peridial cells pitted, 22 × 18 μ, inner walls 2–3 μ thick, outer walls 10–16 μ, prolonged to a long point almost covering the adjacent cell, cell contents yellow; aecidiospores angular-globoid or ovoid, pale orange, 17–24 × 15·5–21 μ, average 19 × 2·5 μ with a very finely echinulate wall uniformly about 1 μ thick. **Uredosori** hypophyllous, seated on small, red or purple spots, scattered or in small, groups, minute, compact, pale brown; uredospores subgloboid, ovoid or pyriform, pale brown, 22–38 × 16–24 μ, wall 2·5–3 μ thick, minutely echinulate with 4 equatorial pores, pedicels long and colourless, joined to the spores by a distinct articu-

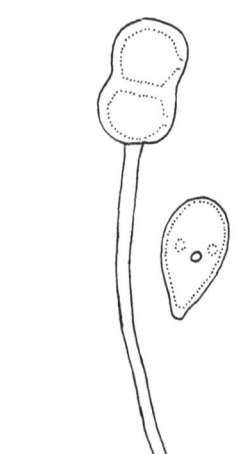

C. mirabilissima. Teleutospore and uredospore.

lation and persistent in the sorus after the spores have fallen. **Teleutosori** occasionally epiphyllous on reddish-brown spots, containing many teleutospores and a few uredospores; usually a few teleutospores are also present in the uredosori, scattered among the uredospores and projecting beyond them; teleutospores ellipsoid or oblong-ellipsoid, rounded at both ends, apex not thickened, constricted at the septum; the wall of 4 layers: (1) the innermost, thin and dark coloured; (2) a thick, brown, warted layer; (3) hyaline layer; (4) the outermost, a very fine cuticularised layer; with 2 equatorial pores in each cell, all in the same vertical plane, $30–36 \times 20–25\,\mu$, pedicels, hyaline, often obliquely inserted, up to $200\,\mu$ long, persistent, very thick-walled and swollen at the base. Auteu-form.

On *Mahonia aquifolium* and *M. japonica* var. *bealii*, aecidia, May–June; uredospores and teleutospores all the year round.

This rust is a North American species found in British Columbia and the western United States. It was discovered in Europe near Edinburgh in 1922 (Trans. Bot. Soc. Edin. **28**, 164, 1923; TBMS. **9**, 135, 1924). In 1923 it was found to be widespread in the Tweed valley and it is possible that it may have been introduced there from North America. Various living plants are known to have been brought into this area in 1915 from Oregon where the rust is native and, although, as far as is known, the importations did not include species of *Mahonia*, spores may have been introduced on leaf fragments. It is now widespread in Scotland extending as far north as Inverness. The first published record from England was in 1930 from Northumberland (Miller, Gard. Chron. **88**, 131, 1930) but there was an earlier record from Denbigh, North Wales, in 1926 (Pethybridge, Gard. Chron. **88**, 312, 1930). It has now been recorded throughout England and Wales, and also from Northern Ireland (Muskett, Cairns & Carruthers, 1934) and from Dublin. The record on *Mahonia japonica* var. *bealii* is from South Wales (*in litt.*). It was found in Denmark in 1925 (Jorgensen, Gartnertidende, November 1925), in Holland in 1926 (Wilson, Gard. Chron. **87**, 132, 1930), in Germany in 1926 (Poeverlein, Ann. Myc. **28**, 421, 1930) and has since been recorded from Sweden (1926), Norway (1927), Finland, Latvia (1928), France (1929), Poland (1930), Czechoslovakia (1930), Switzerland (1930), Italy (1933), Estonia (1936), Roumania (1941).

Accounts of its distribution have been given by Poeverlein (Ann. Myc. **28**, 421, 1930), Klebahn (Z. Pfl.-Krankh. **45**, 529, 1935), Lepik (Ann. Myc. **34**, 439, 1936) and Nicolas (Bull. Soc. Myc. Fr. **52**, 239, 1936).

Aecidia were described almost simultaneously in Scotland (Wilson, Ann. Myc. **28**, 225, 1930) and by Hammarlund in Norway (Bot. Not. **1930**, 380, 1930) and the latter obtained uredospores by sowing the aecidiospores on the older leaves. The aecidia have been found not only on young leaves but also according to Laubert (Mitt. Deutsch. Dend.

Ges. **45**, 273, 1933) on the fruits. There appears, however, to be some doubt regarding their occurrence on the fruits (see Klebahn, *loc. cit.*). They differ from those of *Puccinia graminis* in the yellow peridium, the cells of which possess thicker outer walls and thinner inner walls than those of the grass rust; the wall of the aecidiospore is of uniform thickness, not greatly thickened towards the apex as in *P. graminis*. Hammarlund (Bot. Not. **1932**, 401, 1932) also carried out successful infections on the older leaves with uredospores and on the younger leaves with the basidiospores. The same investigator has made a careful comparison between this rust and *P. graminis* (Arkiv. Bot. **25**, 1, 1933). The thickwalled uredospores can overwinter and germinate readily in the spring; Hammarlund (*loc. cit.*) has shown that they can remain viable when subjected to temperature of $-10°$ and $-20°$ C.; he found that the rust will develop at just above freezing point, its optimum being 8–12° C., and maximum 20° C. Teleutosori containing almost exclusively teleutospores, were found on the upper surface of the leaf in Scotland in the spring of 1939; these were still covered by the epidermis. Similar teleutosori have been recorded by Klebahn (*loc. cit.*).

Tranzschelia Arth.

Res. Sci. Congr. Intern. Bot. Vienne, 340 (1905).

Spermogonia subcuticular, brown or black, conical or hemispherical, without projecting paraphyses. **Aecidia** subepidermal, peridium divided into a few broad lobes; aecidiospores finely verruculose. **Uredosori** subepidermal, erumpent; uredospores intermixed with capitate paraphyses, echinulate, with equatorial pores. **Teleutospores** 2-celled, developing in small groups borne on jointed pedicels, the lower portions being short and united to form compound bases, deeply constricted and readily separating into 2 cells, coarsely verrucose, dark brown with 1 pore in each cell. Heteroecious or microcyclic.

The genus was included in *Puccinia* by Grove and was separated by Arthur chiefly on account of the teleutospore characters. Dietel (Ann. Myc. **20**, 31, 1922) upon biological grounds placed it close to *Ochropsora*. Lindfors (1924, 76) also emphasised the relationship to *Ochropsora* on account of the great resemblance between the developmental stages of the aecidia of *Tranzschelia pruni-spinosae* and *Ochropsora ariae* and the teleutosori of *T. anemones*; he also points out the similarity of the spermogonia in the two genera. Only a few species are known and these occur on the *Ranunculaceae* and *Prunus*.

Tranzschelia anemones (Pers.) Nannf.

Lundell & Nannf. Fungi Exs. Suec. no. 839*a* (1939); Wilson & Bisby, Brit. Ured. no. 256.

Puccinia anemones Pers., Syn. Meth. Fung. p. 226 (1801).

Puccinia thalictri Chev., Fl. Envir. Paris, **1**, 417 (1826); Grove, Brit. Rust Fungi, p. 214.

Puccinia fusca Wint., Rabh. Krypt. Fl. Ed. 2, **1** (1), 199 (1882); Grove, Brit. Rust Fungi, p. 215.

Tranzschelia thalictri (Chev.) Diet., Ann. Myc. **20**, 31 (1922); Wilson & Bisby, Brit. Ured. no. 258; Gäumann, Rostpilze, p. 207.

Tranzschelia fusca Diet., Ann. Myc. **20**, 31 (1922); Gäumann, Rostpilze, p. 205.

[**Spermogonia** hypophyllous, mixed with the teleutosori, prominent, blackish.] **Aecidia** and **uredosori** wanting. **Teleutosori** hypophyllous, rarely epiphyllous, spread uniformly over the whole surface of the leaf, here and there confluent, small, circular, pulverulent, dark brown; teleutospores much constricted, the upper cell almost globoid, the lower globoid, obovoid or clavoid, the cells readily separable, dark brown (the lower cell paler), wall densely covered with large, sometimes pointed warts, 1·5–2·5 μ thick, pore of upper cell apical or subapical, of lower cell, depressed, both covered with a distinct, colourless papilla, 26–55 × 15–30 μ, occasionally 1- or 3-celled spores are intermixed, pedicels hyaline, deciduous, up to 40 μ long. Micro-form.

On *Anemone nemorosa*, Great Britain, common; on *Thalictrum flavum* and *T.*

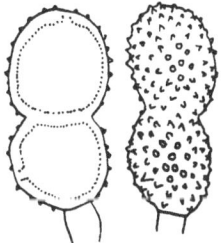

T. anemones. Teleutospores.

minus subspp. *arenarium* and *montanum*, England and Scotland, rare.

This rust is usually separated into two species, *Tranzschelia anemones* and *T. thalictri* (e.g. Gäumann, 1959) but there appears to be no clear morphological distinctions between them.

In the rust on *Anemone* it has been shown by de Bary (Ann. Sci. Nat. Bot. ser. 4, **20**, 95, 1863) and Fischer (1904, 94) that the mycelium is perennial in the rhizome. Dowson (Z. Pfl.-Krankh. **23**, 129, 1913) traced its distribution in the rhizome and leaves and described the structure of the haustoria; the general mycelium is uninucleate. The attacked plants are deformed and rarely flower; they bear paler and narrower leaves with fewer lobes and are much thickened; the structure of these leaves has been described by Maresquelle (Ann. Sci. Nat. Bot. ser. 10, **12**, 42, 1930).

Abnormal flowers on infected plants have been described by Magnin (C. R. Acad. Sci. Paris, **10**, 913, 1890).

Dietel (Ann. Myc. **20**, 177, 1922) has recorded the occurrence of peridial cells in the teleutosorus.

The rust on *Thalictrum* has been found in Perthshire on *T. minus* subsp. *montanum*, on the north coast of Sutherlandshire on *T. minus* subsp. *arenarium* and near Norwich by Ellis (491, 1934) on *T. flavum*; it has also been recorded from Kew Gardens. It appears to possess a perennial mycelium. The same plants are attacked year after year; they are somewhat deformed and taller, with longer internodes, smaller and paler leaves and narrower segments. Spermogonia have not been recorded in British specimens but have been found on the continent and in North America.

On the anemone, early stages in the development of the spermogonium have been described by Kursanov (1922), and Lindfors (1924, 21) has investigated the development of both spermogonia and teleutosorus. The dicaryotic condition arises at the base of the teleutosorus. Walker (Trans. Wis. Acad. Sci. **23**, 567, 1927) has also described the development of the teleutosorus.

Tranzschelia discolor (Fuck.) Tranz. & Litv.

Bot. Zhurn. **24**, 248 (1939).

Puccinia discolor Fuck. Fungi Rhen. no. 2121 (1867).
Tranzschelia pruni-spinosae var. *discolor* (Fuck.) Dunegan, Phytopath. **28**, 424 (1938).
In part included under *Puccinia pruni-spinosae* by Grove, p. 207, and under *Tranzschelia pruni-spinosae* by Wilson & Bisby, Brit. Ured. no. 257, and Gäumann, Rostpilze, p. 201.

Spermogonia amphigenous, scattered, brown or blackish, very shallow, punctiform. **Aecidia** hypophyllous, scattered over the whole surface, flat, with a broad revolute margin which is torn into few (3–5) lobes; aecidiospores roundish 16–24 μ diam., pale yellowish-brown, wall finely verruculose. **Uredosori** hypophyllous, generally on small, yellow or brown spots, scattered but often crowded and confluent, soon naked, pulverulent, cinnamon-brown; uredospores ellipsoid, ovoid or obovoid, 20–40 × 10–19 μ, wall smooth and more or less thickened and dark brown at the conical apex, in lower half paler, sharply verrucose or echinulate, pores 3 or 4, equatorial, mixed with yellowish-brown or pale capitate paraphyses more or less thickened at the apex. **Teleutosori** similar but blackish-brown; teleutospores ellipsoid to oblong, 30–45 × 18–25 μ, deeply constricted at the septum and readily fracturing, the

2 cells dissimilar, the upper cell globoid, wall densely and coarsely verrucose with a basal pore, densely pigmented, the lower oblong-elliptic, wall pale

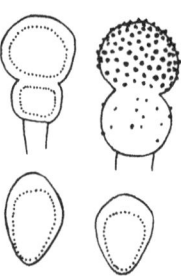

T. discolor. Teleutospores and uredospores.

brown, almost smooth, with a superior pore, pedicels hyaline, short or up to 40 μ long, deciduous, springing in clusters of about 10–20 from a common base. Hetereu-form.

Spermogonia and aecidia on *Anemone*

coronaria and *Anemone × fulgens*, April–
May; uredosori and teleutosori on
Prunus armeniaca, P. domestica and its

subsp. *insititia* and *P. persica* and its
var. *nectarina*, July onwards. Uncommon, but inadequately studied.

The accounts of the two related species of *Tranzschelia* on *Prunus*,
T. pruni-spinosae and *T. discolor*, suffer from failure to recognise these in
the past. In the former, the cells of the teleutospores are similar,
whereas in the latter the lighter coloured, smoother, lower cell contrasts
sharply with the dark, verrucose, upper cell. Most authors are agreed
that *T. pruni-spinosae* attacks wild species of *Prunus*, whereas *T. discolor*
occurs on the cultivated types. As regards cultivated hosts all information on British occurrences confirms this. Most of the British records
are for cultivated *Prunus* species and varieties and all specimens examined on these hosts prove to belong to *T. discolor*. Evidence for *T.
pruni-spinosae* in Britain is scanty on *Prunus* hosts; only a few collections
on *P. spinosa* have been discovered. The two species, at least in northern
Europe, differ in aecidial hosts. Tranzschel (Trav. Mus. Bot. Acad.
Imp. Sci. St Petersburg, **2**, 67, 1905) showed that the species are hetero-
ecious, *T. pruni-spinosae* between *Anemone coronaria* and *P. amygdalus,
P. divaricata* and *P. spinosa* and *T. discolor* between *A. ranunculoides* and
P. spinosa. Arthur (J. Mycol. **12**, 20, 1906, and **13**, 199, 1907) in the United
States confirmed this discovery, proving experimentally that the aecidium
on *Hepatica acutiloba* is a stage in the life cycle of the fungus. However,
Dupias (Bull. Soc. Nat. Hist. Toul. **85**, 37, 1950) obtained infection of
P. domestica and *P. spinosa* following inoculation with aecidiospores
from naturally infected *A. ranunculoides*. The only experiments carried out
in Britain with this group, by Brooks (New Phytol. **10**, 207, 1911) who
infected 'Victoria' plum with spores from *A. coronaria*, agree with these
relations and relate to *P. discolor* rather than to *P. pruni-spinosae*.

The record on *Anemone × fulgens* is from Cornwall in 1937 and 1938
(Bull. Min. Agric. no. 126, 79) and those on peach, nectarine and
apricot by Salmon Ware (Gard. Chron. **94**, 490, 1933).

Grove included *Anemone nemorosa* as a host but there appears to be
no certain evidence for the occurrence of the aecidium on this host and his
record may refer to the aecidium of *Ochropsora ariae* which occurs on
A. nemorosa throughout Europe. Cooke (1871, 536) recorded *Aecidium
quadrifidum* only 'on anemones in gardens' and Plowright *A. punctatum*
only on *A. coronaria*. The record on *A. nemorosa* in Scotland by Wilson
(1934) was an error. Dupias (Bull. Soc. Hist. Nat. Toul. **85**, 33, 1950) in
France noted that neither *A. nemorosa* nor *A. hepatica* was infected
when growing beside severely attacked *A. ranunculoides*. The aecidium

on *Eranthis hyemalis* referred to by Grove (1913) has been proved by Tranzschel (abstract in RAM. **15**, 236, 1936) to be an entirely different species, *Leucotelium cerasi* which has not been found in Britain.

Jacky (Zbl. Bakt. II, **8**, 658, 1901) first observed that the form of the rust attacking certain cultivated hosts in Germany possessed teleutospores that differ from those originally described by Persoon. Fischer 1904 confirmed the observation and treated them as the two forms: f. *typica* in which the two cells of the teleutospore are globose, of the same size and colour and both coarsely verrucose and f. *discolor* in which the upper cell is globose, verrucose and usually thickened above, while the lower is oblong, generally narrowed towards the base, paler in colour and only sparsely, if at all, verrucose. Grove considered that these differences were by no means constant and not of taxonomic importance.

Dunegan (Phytopath. **28**, 411, 1938) and Dunegan & Smith (Phytopath. **31**, 189, 1941) who have investigated *T. discolor* and *T. prunispinosae* find that in both of them typical 4-celled basidia are produced on germination of the overwintered teleutospores. In the former the basidiospores are $8–8\cdot5 \times 6–6\cdot5\,\mu$, whereas in the latter they are $12\cdot5–16 \times 5\cdot5–6\cdot5\,\mu$. These investigators studied the temperature relations for the germination of the uredospores of *T. discolor* and their longevity under different conditions.

The mycelium of the aecidial stage is perennial in the corms of the anemone; it penetrates, in spring, into the growing shoots which become deformed, the affected leaves stand erect, have fewer, smaller lobes and are paler in colour; the flowers are imperfect or altogether wanting. Affected leaves die off soon after the aecidiospores are shed; healthy leaves then develop and persist throughout the following winter. Imperfect flowers of infected *A. ranunculoides* have been described by Magnin (C. R. Acad Sci. Paris, **110**, 913, 1890) and Maresquelle (Ann. Sci. Nat. Bot. ser. 10, **12**, 42, 1930) has studied the deformation of the leaf.

D'Oliveira & Borges (Agron. Lusit. **3**, 71, 1941) have shown that the mycelium in the rhizome of *A. coronaria* perennates in the haploid condition, even in plants with fertile aecidia on the leaves. They have obtained evidence from inoculation tests with basidiospores that the species is heterothallic.

An account of the development of the aecidium on *A. ranunculoides* has been given by Kursanov (1922, 13) but the origin of the dicaryotic cells is not described. In addition to the usual form with dicaryotic aecidiospores Kursanov has also described the occurrence of a form with uninucleate aecidiospores. Rice (Bull. Torr. Bot. Club, **60**, 23, 1933) has investigated the development of the spermogonia and aecidia on *Hepatica acutiloba*; she described nuclear migration in the basal tissue below the aecidium.

Tranzschelia pruni-spinosae (Pers.) Diet.

Ann. Myc. **20**, 31 (1922); Wilson & Bisby, Brit. Ured. no. 257; Gäumann, Rostpilze, p. 201 *p.p.*
Puccinia pruni-spinosae Pers., Syn. Meth. Fung. p. 226 (1801); Grove, Brit. Rust. Fungi, p. 207.
Puccinia pruni DC., Fl. Fr. **2**, 222 (1805).

Spermogonia and **aecidia** as in *Tranzschelia discolor* but not certainly recorded in Britain. **Uredosori** and **teleutosori** as in *T. discolor* but teleutospores with both cells, globose, dark brown, coarsely verrucose, pore of upper cell apical and of lower basal; basidiospores 12–16 × 5·5 –6·5 μ (Dunegan & Smith, Phytopath. **31**, 189, 1941). Hetereu-form.
    [Spermogonia and aecidia on *Anemone*]; uredosori and teleutosori on

T. pruni-spinosae. Teleutospores.

Prunus spinosa and other *Prunus* spp.? Uncommon, perhaps rare.

The distinctions between this species and *Tranzschelia discolor* are discussed under that species. The only British collections which certainly belong to this species are on *Prunus spinosa*. There is no evidence of the occurrence of the aecidia in Britain, which Tranzschel showed may form on *Anemone ranunculoides*.

Endophyllum Lév.

Mem. Soc. Linn. Paris, **4**, 208 (1825).

Spermogonia present, subepidermal. **Teleutosori** similar to aecidia of *Puccinia* or *Uromyces*; teleutospores produced from a fusion cell and germinating as soon as mature by a typical basidium and basidiospores; pores not perceptible; spore-wall coloured, verruculose.

Grove considered the life history of the species included in this genus as primitive. Dietel (in Engler & Prantl, Nat. Pflanzenfam. 1928) considered that they are reduced forms derived from various genera of the Uredinales. A similar opinion was held by Jackson (1931) and Arthur (1925, 1934) who regarded them as microcyclic forms.

Endophyllum is a 'form' genus including species whose teleutospores are similar to aecidiospores. They can only be distinguished from normal aecidiospores on germination.

Endophyllum euphorbiae-sylvaticae (DC.) Wint.

Rabh. Krypt. Fl. Ed. 2, 1 (1), 251 (1882); Grove, Brit. Rust Fungi, p. 333; Wilson & Bisby, Brit. Ured. no. 23; Gäumann, Rostpilze, p. 1221.
Aecidium euphorbiae-sylvaticae DC., Fl. Fr. 2, 241 (1805).
Endophyllum euphorbiae Plowr., Monogr. Brit. Ured. Ustil. p. 228 (1889).

Spermogonia amphigenous, mostly epiphyllous, numerous rounded, 150–190 μ diam., yellow, with projecting hyphae. Aecidia and uredosori absent. Teleutosori aecidioid, amphigenous, mostly hypophyllous, from a systemic mycelium, spread over the whole leaf surface, crowded, sunk in the leaf tissue which is slightly swollen, 200–500 μ diam., with peridium as a shallow cup with revolute, laciniate margin; peridial cells rhomboid, 29–34 μ long, 15–25 μ wide, outer wall 5–7 μ thick, in surface view finely punctate, inner wall 1·5–2 μ thick with tesselate structure, finely warted; teleutospores in evident chains, subgloboid, 17–22 μ diam. or 16–24 × 12–19 μ, wall hyaline, densely verruculose, yellow, about 1 μ thick. Endo-form.

On *Euphorbia amygdaloides*, April–June. England, scarce.

This rust has been recorded from Kent, Norfolk, Gloucestershire and Wiltshire.

The effect on its host has been described by Plowright (1889, 228) and Muller (Zbl. Bakt. II, **20**, 333, 1908). Infection takes place on the buds of the rhizome and on the leaves, the germ-tube boring through the epidermal cell. The mycelium perennates in the plant, especially in the cortex and pith. In the following spring the infected plants develop longer shoots with narrower, paler leaves than the uninfected. In the first year following infection spermogonia and sometimes a few teleutosori are produced on the leaves. Late summer and autumn foliage is almost normal except that the leaves are broader. In the second year abundant teleutosori are produced. Affected plants seldom flower.

This rust, in its mycelium and aecidioid teleutosori and in the deformation which it produces on its host, closely resembles *Aecidium euphorbiae* Pers. (*sens. lat.*) on *Euphorbia cyparissias*, which is known to be the aecidial stage of *Uromyces pisi* (see p. 330).

The spores of *Endophyllum euphorbiae-sylvaticae* germinate freely in water as soon as produced and give rise to a basidium with 3 or 4 basidiospores; these may commence to germinate while still attached to the basidium.

The cytology has been investigated by Sappin-Trouffy (1897, 184) and by Moreau & Moreau (Bull. Soc. Bot. Fr. **66**, 14, 1919). These authors agree that there is no nuclear fusion in the teleutospore; the two nuclei pass into the germ-tube where each divides and in this way the four nuclei of the typical basidium are produced. A form which is uninucleate throughout has been described by Moreau & Moreau (*loc. cit.*). In this, which has been named '*E. uninucleatum*', the uninucleate teleutospore gives rise to a typical basidium which produced four basidiospores.

Endophyllum sempervivi de Bary

Ann. Sci. Nat. Bot. ser. 4, **20**, 86 (1863); Grove, Brit. Rust Fungi, p. 335; Wilson & Bisby, Brit. Ured. no. 24; Gäumann, Rostpilze, p. 1223.

Uredo sempervivi Alb. & Schw., Consp. Fung. Nisk. p. 126 (1805).

Spermogonia amphigenous, mostly epiphyllous, scattered amongst the teleutosori, globoid or conoid, brownish, 120–170 μ diam. **Teleutosori** aecidioid, amphigenous, irregularly scattered, sunken in the leaf tissue, rounded, 0·5–1 mm. diam., peridium well developed, hemispherical, at first closed, then opening by a pore and at length cup-shaped, peridial cells loosely adherent, subgloboid or broadly ellipsoid, minutely verruculose, 24–42 × 20–35 μ, inner wall 3–5 μ, outer 4–7 μ thick; teleutospores bluntly polygonoid or subgloboid, pale yellowish-brown, 18–35 × 18–28 μ; wall 3–4 μ thick, densely and minutely verruculose. Aecidia and uredosori absent. Endo-form.

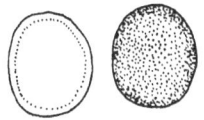

E. sempervivi. Teleutospores.

On *Sempervivum arachnoideum*, *S. calcareum*, *S. globuliferum*, *S. montanum*, *S. tectorum*, '*S. webberi*' (= *S. webbii?*), *S. × fimbriatum*, April–August. England and Scotland, scarce.

It has been proved by de Bary (*loc. cit.*) Hoffmann, (Zbl. Bakt. II, **32**, 137, 1912) and others that the basidiospores produced from the teleutospores infect the leaves and from them arises a mycelium which perennates in the stem. It produces spermogonia and teleutosori in the following spring. The affected leaves are more erect than the normal ones, about twice as long, narrower and yellowish at the base. Infected leaves are only produced in the early summer; the later-formed leaves are normal and do not contain mycelium. Dodge & Reed (J. New York Bot. Gdn. **37**, 54, 1936) traced the parasite from the leaves down to the crown, and to the new runner shoots during the summer, the offshoot plants developing pustules in the following spring. Ramsbottom (J. Bot. Lond. **67**, 237, 1929) noted the very slow spread of the disease and similar observations have been made at Edinburgh and by Dodge & Reed (*loc. cit.*) in the United States.

Cytological investigation of this species was first carried out by Maire (J. Bot. Fr. **14**, 80 and 369, 1900; Bull. Soc. Myc. Fr. **18**, 52, 1902) who described the formation of the basidium and basidiospores; he stated that no nuclear fusion took place in the teleutospore. Hoffman (*loc. cit.*) described the nuclear fusion in the teleutospore and the germination of the latter; he also gave an account of the development of the basidium and basidiospores, the infection and the development of the uninucleate mycelium in the host, and the origin of the binucleate mycelium by the fusion of pairs of cells at the base of the teleutosori. Moreau & Moreau (Bull. Soc. Bot. Fr. **66**, 14, 1919) confirmed Hoffman's conclusions.

Maire (*loc. cit.*) recorded the discovery in France of a form which he called *Endophyllum sempervivi* var. *aecidioides*. This exactly resembles the normal form but the spores on germination do not produce a basidium but only a germ-tube which will not bring about infection of the host. It may be the aecidial stage of some heteroecious species.

Ashworth (TBMS. **19**, 240, 1935) confirmed Hoffman's work and noted the presence of emergent hyphae, protruding through the stomata. She also gave a detailed account of spermogonial development and showed that teleutosori develop even when the spermatia are destroyed.

Stampfli (Hedwigia, **49**, 230, 1910) described the changes in the anatomy of the leaf resulting from infection.

Uromyces Unger

Exanth. Pfl. p. 277 (1833).

Spermogonia deeply embedded in the tissues of the host, flask-shaped with conical mouth and ostiolar filaments, and flexuous hyphae. Aecidia usually with an evident, generally cup-shaped, peridium; aecidiospores with indistinct pores. Uredospores formed singly on their pedicels, with several, usually rather distinct pores, rarely accompanied by paraphyses. Teleutospores 1-celled, on distinct pedicels, almost always with an apical pore. Basidiospores flattened on one side or kidney-shaped. Autoecious or heteroecious.

Uromyces is distinguished from *Puccinia* by its unicellular teleutospores; in all other respects the two genera are similar. The number of cells in the teleutospores in *Puccinia* is, however, not constant and Arthur and others have suggested that the two genera should be combined. The two are here maintained as a matter of convenience.

Uromyces ficariae (Alb. & Schw.) Lév.

Orbigny, Dict. Univ. Hist. Nat. **12**, 786 (1849); Grove, Brit. Rust Fungi, p. 107; Wilson & Bisby, Brit. Ured. no. 288; Gäumann, Rostpilze, p. 309.

Uredo ficariae Schum., Enum. Pl. Saell. **2**, 232 (1803).
Puccinia ficariae DC., Fl. Fr. **2**, 225 (1805).
Uredo ficariae Alb. & Schw., Consp. Fung. Nisk. p. 128 (1805).

Teleutosori amphigenous or on the petioles, about 0·5 mm. diam., rounded frequently in dense, orbicular or elongate clusters, on pale-yellow spots, especially on the petioles, causing distortion, soon naked, pulverulent, chocolate-brown; teleutospores more or less obovoid, often irregular, 22–38 × 18–26 μ, wall smooth, pale brown, apex with conical hyaline papilla, pedicels hyaline, deciduous. Micro-form.

On *Ranunculus ficaria*, March–early June. Great Britain and Ireland, very common.

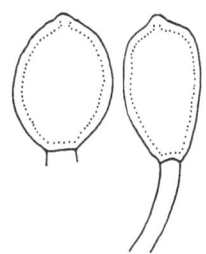

U. ficariae. Teleutospores (after Fischer).

Various continental authors, from Treboux (Ann. Myc. **10**, 74, 1912) onwards, have described uredospores in the teleutosori. British collections are almost exclusively teleutosporic with a few abortive hyaline teleutospores. Grove's description of uredospores is taken from Sydow's Monograph and there is no evidence that he saw them in British collections. The aecidium on the same host belongs to the life cycle of *Uromyces poae* and has no connection with *U. ficariae* though it may be found on the same leaf. K. W. Braid (*in litt.*) has reported the occurrence of a few 2-celled teleutospores in the sorus in material collected near Glasgow in 1948. Klebahn (Jahrb. Hamburg. Wiss. Anst. **20**, Beih. 3, p. 38 extr., 1902) confirmed experimentally that the species is microcyclic.

Several investigators have examined the cytology of this rust. Sappin-Trouffy (1897, 89) figured a rather late stage of development of the sorus and described the mycelium as binucleate. Blackman & Fraser (Ann. Bot. Lond. **20**, 43, 1906) considered the mycelium was generally uninucleate except the mass around the sorus which is dicaryotic and this was confirmed by Moreau (Le Botaniste, **13**, 193, 1914). On the contrary, Kursanov (1922, 53) stated that the mycelium is generally dicaryotic and only found portions of uninucleate hyphae at rare intervals.

Uromyces behenis (DC.) Unger

Einfl. Bodens, p. 216 (1836); Grove, Brit. Ured. no. 109; Wilson & Bisby, Brit. Ured. no. 277; Gäumann, Rostpilze, p. 270.

Uredo behenis DC., Fl. Fr. **6**, 63 (1815).

Spermogonia unknown. **Aecidia** usually hypophyllous, seated on spots that vary both in size and colour (yellow or purple) and are generally very conspicuous, solitary or clustered, cup-shaped, whitish-yellow, with a torn revolute margin; aecidiospores densely and minutely verruculose, yellowish, 15–21 μ

diam. [**Uredosori** aecidioid, similar to
the aecidia but more diffuse.] **Teleutosori** hypophyllous and on the stems, often
surrounding the uredosori, irregularly
scattered, gregarious or circinate, rounded or oblong, long-covered by the lead-
coloured epidermis, rather small and
compact, brownish-black or black;
teleutospores subgloboid or obovoid,
rounded above, $25–35 \times 20–27\mu$, wall
smooth, pale brown, up to 11μ thick at
apex, pedicels persistent, faintly yellow,
thick, up to 75μ long. Auteu-form.

On *Silene maritima* and *S.
vulgaris*, aecidia (uredospores) and teleutospores,
July–October. England, Scotland, Ire-
land, scarce.

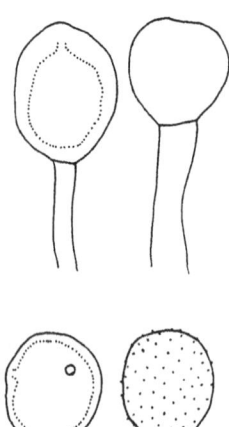

U. behenis. Teleutospores and uredospores.

The spots occupied by the aecidia vary in colour, but are often tinged
or margined with purple. This is one of the species whose aecidiospores
produce aecidioid uredosori as Dietel (Flora, **81**, 394, 1895) has shown.
The aecidia, on the earlier leaves, are in roundish groups on concentric
circles, only a few being scattered. The uredosori on the younger leaves,
stand more often singly and are spread over a larger area; the teleutosori
spring from the same secondary mycelium or are formed separately.

The aecidia arise from infection by comparatively few basidiospores;
the uredosori arise from the more widely dispersed aecidiospores of the
first generation, and their mycelium can produce either uredospores or
teleutospores or both. No spermogonia seem to be known. The uredosori have not been found in Britain.

Kursanov (Bot. Zbl. **135**, 281, 1917) stated that analogous phenomena
to those observed in the primary and secondary generations of *Uromyces
scrophulariae* are found in *U. behenis.*

Uromyces dianthi (Pers.) Niessl

Verh. Nat. Ver. Brunn, **10**, 162 (1872); Wilson & Bisby, Brit. Ured. no. 284;
　　Gäumann, Rostpilze, p. 328.
Uredo dianthi Pers., Syn. Meth. Fung. p. 222 (1801).
Uromyces caryophyllinus Wint., Rabh. Krypt. Fl. Ed. 2, **1** (1), 149 (1882); Grove, Brit.
　　Rust Fungi, p. 108.

[**Spermogonia** amphigenous, globoid, scattered, with exserted paraphyses. **Aecidia** hypophyllous, cupulate; aecidiospores globoid, $16–22\mu$ diam., with a colourless, finely verrucose wall, 1μ thick.] **Uredosori** amphigenous, and on the stems, sometimes on pallid spots, scattered, minute, round or oblong, soon naked, pulverulent, cinnamon; uredospores globoid to ellipsoid, yellowish-brown, $20–35 \times 18–25\,\mu$, wall $2\cdot5–3\,\mu$ thick, sparsely echinulate, with 3–5 pores. **Teleutosori** amphigenous, confluent and large, mostly oblong, surrounded and often covered by the cleft epidermis, subpulverulent, brownish-black; teleutospores globoid to ellipsoid, chestnut-brown, $20–31 \times 18–24\,\mu$, wall $2–3\,\mu$ thick, with an apical, flat, hyaline papilla, densely and minutely echinulate, pedicels short, hyaline, deciduous. Hetereu-form.

[Spermogonia and aecidia on *Euphorbia*], uredospores and teleutospores on *Dianthus barbatus*, *D. caryophyllus* and *D. chinensis*. Great Britain and Ireland,

on cultivated carnations almost all the year round, frequent.

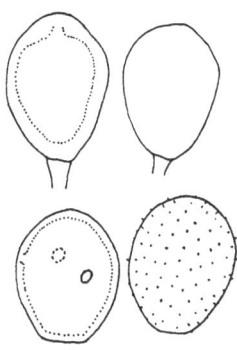

U. dianthi. Teleutospores and uredospores.

The 'Carnation Rust' was introduced into England on imported plants about the year 1890; it sometimes occurs in an epidemic form, causing much injury. The teleutosori are often clustered on the leaves and stems in circinate or elongate swollen patches; uredospores are mixed with them.

Infection experiments with aecidiospores from *Euphorbia gerardiana* were carried out by Fischer (Zbl. Bakt. II, **28**, 141, 1910; Myk. Zbl. **1**, 1, 1912), Treboux (Ann. Myc. **10**, 563, 1912) and Guyot (Ann. Ecole Nat. Agric. Grignon, III, **1**, 61, 1938–9). Guyot (Uredineana, **4**, 333, 1953) has shown that *E. nicaeensis* is also an aecidial host. It is clear from the work of these investigators that specialised forms of the species exist on various species of *Dianthus*, on *Saponaria ocymoides* and on *Tunica prolifera* (see table by Gäumann, 1959, 330). The aecidial host is deformed by the rust, the plants do not flower and the stems and leaves are decreased in size. *Euphorbia gerardiana* does not occur in Britain and the parasite obviously maintains itself here without alternation. The fungus has now spread through the world in greenhouses but only in the sporophytic stage; the aecidium has not been recognised anywhere except in Europe.

Uromyces inaequialtus Lasch

Rabh., Fungi Europ. no. 94 (1859).

[**Spermogonia** honey-coloured. **Aecidia** on yellowish or violet spots, peridium short, cylindrical; aecidiospores roundish or ellipsoid, $15–19 \times 14–16\,\mu$, with orange contents.] **Uredosori** amphi-

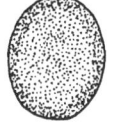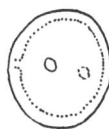

U. inaequialtus. Uredospores.

genous often associated with purplish spots, 0·5 mm. diam., cinnamon in colour; uredospores subgloboid, 21–25 × 18–23 μ, wall 2–3 μ thick, brown, very finely and closely verruculose, with 3 equatorial pores. Teleutosori amongst uredosori, black, about 0·5 mm. diam.; teleutospores subgloboid, 23–30 × 20– 25 μ, wall smooth, 2·5 μ thick, up to 8 μ thick at apex, pore apical, pedicel up to 80 μ long, colourless, persistent. Auteu-form.

[Spermogonia, aecidia], uredospores and teleutospores on *Silene nutans*. Herm, Channel Islands and Kent.

This rust was unknown in Britain until E. A. Ellis collected it in Herm in 1962. It has also been found on herbarium specimens from Kent. Several other species have been listed as hosts for *Uromyces inaequialtus* but none is British; *Silene nutans*, itself, is rare with a scattered discontinuous distribution.

Uromyces sparsus (Schm. & Kze.) Lév.

Ann. Sci. Nat. Bot. ser 3, **8** 371 (1847); Grove, Brit. Rust Fungi, p. 111; Wilson & Bisby, Brit. Ured. no. 311; Gäumann, Rostpilze, p. 271.

Uredo sparsa Schm. & Kze., Deutschl. Schwämme, 7, 5 (1817).

Spermogonia amphigenous, dark red, in little clusters surrounded by the aecidia. **Aecidia** in a contiguous series, more or less forming a single circle, 1–2 mm. diam., around the spermogonial group, bright orange-red, almost scarlet, peridium nearly cylindrical or cup-shaped, with a dentate, erect margin, peridial cells firmly united to one another in a thin membrane, polygonal in outline, not arranged in distinct rows the inner wall 3 μ thick and the outer 4–5 μ, in optical section deeply transversely striate; aecidiospores ovoid, 21–27 × 13–18 μ. **Uredosori** amphigenous and on the stems, small, oblong-oval, 0·5–1 mm. long, dark cinnamon, long-covered by the epidermis which splits and surrounds them, then pulverulent; uredospores globoid to ovoid, brownish-yellow, 22–32 × 14–21 μ, wall faintly echinulate or smooth, 1–1·5 μ equally thick all round with 2–4 almost equatorial pores. **Teleutosori** similar, but darker; teleutospores subgloboid to ovoid, generally tapering downwards, wall smooth, thicker (up to 4 μ) and darker at the apex, brown, 22–32 × 14–21 μ, pedicels persistent, thick, up to 60 μ long, brownish at the apex. Auteu-form.

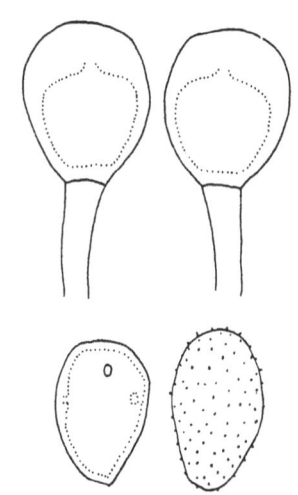

U. sparsus. Teleutospores and uredospores (after Savulescu).

On *Spergularia media, S. rubra* and *S. marina*, aecidia May–June; uredospores and teleutospores, June onwards. England, Scotland, scarce.

Grove in 1913 stated he had seen no British specimens of this rust but since that time it has been found in several localities. Hadden (TBMS.

5, 438, 1916) collected it from Porlock, Wilson (TBMS. **12**, 114, 1927) from East Lothian and Grove & Chesters (TBMS. **18**, 265, 1934) from Norfolk; Grove & Chesters, who also quote a record from the Farne Islands, were the first to record the spermogonia and aecidia; they have published drawings of these stages.

Uromyces betae (Pers.) Tul.

Ann. Sci. Nat. Bot. ser. 4, **2**, 89 (1854); Grove, Brit. Ured. no. 113; Wilson & Bisby, Brit. Ured. no. 278; Gäumann, Rostpilze, p. 310.

Uredo betae Pers., Syn. Meth. Fung. p. 220 (1801).
Uredo betaecola West., Bull. Acad. Roy. Belg. **11**, 650 (1861).

Spermogonia in little clusters, honey-coloured. **Aecidia** amphigenous, often on rounded or irregular, yellowish spots, usually in rather large round or irregular and confluent clusters, cup-shaped, yellowish, with a reflexed, incised margin; aecidiospores delicately verruculose, pale yellowish, 16–24 × 16–20 μ. **Uredosori** amphigenous, scattered, sometimes concentrically arranged, pulvinate, circular, up to 2 mm. diam., covered by the epidermis which at length splits, then pulverulent, cinnamon; uredospores globoid to broadly ellipsoid, 21–32 × 16–26 μ, wall sparsely and minutely echinulate, yellowish, 2·5–3 μ thick, with 2–3 equatorial pores. **Teleutosori** similar, but somewhat compact, dark brown; teleutospores globoid to obovoid, rounded and slightly thickened above, with a minute, hyaline, hemispherical papilla, 22–34 × 18–25 μ, wall smooth,

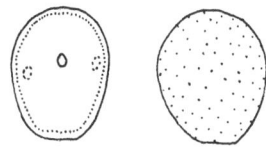

U. betae. Uredospores.

pale brown, pedicels short, hyaline. Auteu-form.

On leaves of several varieties of *Beta vulgaris*, aecidia, April–June; uredospores and teleutospores, May–October. Great Britain, Ireland and Channel Islands, frequent.

This rust is found on cultivated red beet, sugar beet, mangold and spinach beet but rarely causes a serious disease. Aecidia are uncommon in this country and in Europe; they have recently been recorded on sugar beet by Stirrup (Ann. Appl. Biol. **26**, 403, 1939); and at Cambridge and Exeter and on shoots of clamped mangolds (Bull. Min. Agric. **142**, 33, 1960).

The fungus overwinters mainly on seed crop stecklings, clamped mangolds, groundkeeping beets and mangolds. It spreads from there to commercial crops in spring and early summer.

The mycelium and the development of the uredospores and teleutospores has been described by Sappin-Trouffy (1897). Nemec (Bull. Internat. Acad. Sci. Prague, **16**, 118, 1911) has given a detailed account of the formation of the haustoria.

The temperate relations of this rust have been studied by Newton & Peturson (Phytopath. **33**, 10, 1943) who found it very susceptible to high temperatures.

Uromyces chenopodii (Duby) Schroet.

Kunze, Fungi Sel. Exs. no. 214 (1880); Grove, Brit. Rust Fungi, p. 111; Wilson & Bisby, Brit. Ured. no. 279.

Uredo chenopodii Duby, Bot. Gall. 3, 899 (1830).

As *Uromyces giganteus* Speg., Dec. Mycol. Ital. no. 30 (1879); Gäumann, Rostpilze, p. 268.

Spermogonia amphigenous in groups, 100–120 μ diam., with projecting paraphyses. **Aecidia** amphigenous, clustered in circles, 5–10 mm. diam., cylindrical or cup-shaped, bright golden-yellow or whitish, margin deeply laciniate with few lobes, peridial cells almost quadrate, 22 × 19 μ, outer walls transversely striate in section, 11 μ thick, projecting downwards and overlapping the cell below, inner wall up to 3 μ thick, tesselate and warted; aecidiospores rounded or ovoid-polyhedroid, 18 × 20 μ, wall scarcely 1 μ thick, finely verruculose, yellowish. **Uredosori** amphigenous, scattered or gregarious, round or frequently elongate, small, surrounded by the conspicuous, torn epidermis, cinnamon-brown; uredospores globoid to ovoid or oblong, yellowish-brown, 18–25 × 16–21 μ, wall faintly and sparingly echinulate with spines 2 μ apart, about 1·5 μ thick, pores indistinct, apparently several, scattered. **Teleutosori** amphigenous but mostly cauline, on the leaves, rounded and 1–3 mm. diam., on the stems fusoid, up to 3 cm. long, thick, compact, dark brown; teleutospores very variable, ovoid to subpyriform, rounded or subconical at the apex, brown, 24–35 × 18–20 μ, wall smooth, slightly thickened at apex, pedi-

U. chenopodii. Teleutospores and uredospores.

cels pale brown, persistent, up to 80 μ long or more, 5–8 μ wide. Auteu-form.

On the stems and leaves of *Suaeda fruticosa*, *S. maritima* and *S. maritima* var. *flexilis*. England and Scotland, rare.

This rust was first found in Norfolk on *Suaeda maritima* in 1893 by Plowright (Gard. Chron. **18**, 135, 1895) and was later recorded in various localities in Norfolk (TBMS. **1**, 56, 1898; J. Roy. Hort. Soc. **19**, cxxxiii, 1896), Suffolk, Sussex (J. Bot. Lond. **34**, 181, 1898), Durham and Wales (Laundon, *in litt.*) and recently in East Lothian, Scotland, on an upright form of this host; it was found on *S. fruticosa* at Blakeney Point, Norfolk, by C. W. Muirhead in 1959.

There appears to be no record of the occurrence of spermogonia in this country but they have been described by Arthur (1934, 234) and by Savulescu (1953, 586). The teleutospores are very variable, short and broad or long and narrow in the same sorus; the thickening of the apex also varies from 3 to 7μ; the pedicels are often very long and flexuous. No cultures appear to have been performed with the rust but it was noted by Plowright (*loc. cit.*) that *Salicornia* plants occurring in the vicinity of infected *Suaeda maritima*, were not attacked and this has also been observed in Scotland. Jørstad (1958, 110) found no reason for considering *Uromyces salsolae* and *U. salicorniae* specifically different from this rust.

The name of this species is misleading for it has not been found on the genus *Chenopodium* as now understood. Cummins & Stevenson (Pl. Dis. Reporter, Suppl. **240**, 187, 1956) considered that its valid name is *U. giganteus* but this is not accepted by Jørstad (*loc. cit.*).

Uromyces salicorniae de Bary

Rabh. Fungi Europ. no. 1386 (1870); Grove, Brit. Rust Fungi, p. 114; Gäumann, Rostpilze, p. 266.

[*Aecidium salicorniae* DC., Fl. Fr. **6**, 92 (1815).]
Uromyces salicorniae Cooke, Grevillea, **7**, 137 (1879); Wilson & Bisby, Brit. Ured. no. 308.

Spermogonia unknown. Aecidia often on the cotyledons, scattered or in small clusters, at first hemispherical, then cup-shaped, with erect, torn, white margin; aecidiospores finely verruculose, orange-yellow, $17–35\mu$. Uredosori scattered or aggregated, minute, rounded, long-covered by the epidermis, pulverulent, cinnamon; uredospores ovoid, yellow-brown, $24–35 \times 18–25\mu$, wall about $1·5\mu$ thick, very finely echinulate, with 4 pores. Teleutosori similar, but larger and rather compact, dark brown; teleutospores subgloboid to obovoid, $25–35 \times 18–28\mu$, wall smooth, up to 4μ thick about the apical pore, pedicels hyaline, thick, persistent, up to 8μ long. Auteuform.

On leaves and stems of *Salicornia europaea* and *S. ramosissima*, aecidia,

May; uredospores, July–October; teleutospores, August–October. England and Wales, rare.

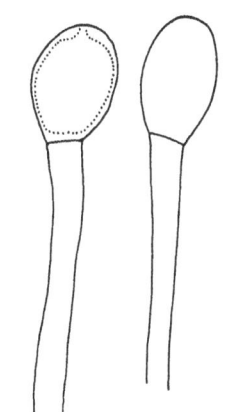

U. salicorniae. Teleutospores.

This rust has been recorded in Norfolk, Suffolk, Sussex, Somerset and Pembrokeshire. The aecidia are found usually on young plants; the teleutosori are chiefly on the stems, as much as 3 mm. long and very pulvinate.

Uromyces geranii (DC.) Lév.

Ann. Sci. Nat. Bot. ser. 3, **8**, 371 (1847); Grove, Brit. Rust Fungi, p. 103; Wilson & Bisby, Brit. Ured. no. 291; Gäumann, Rostpilze, p. 404.
Uredo geranii DC., Syn. Pl. Gall. p. 47 (1806).
Uromyces kabatianus Bub., S.B. Kon. Böhm. Ges. Wiss. II, **46**, 1 (1902); Grove, Brit. Rust Fungi, p. 104; Wilson & Bisby, Brit. Ured. no. 295; Gäumann, Rostpilze, p. 406.

Spermogonia amphigenous, few, large, orange, then darker, 135–150 μ diam. **Aecidia** hypophyllous and on the petioles, on the leaves chiefly in the vicinity of the nerves and there forming large, dense clusters sometimes up to 2 cm. long, on thickened spots and often causing distortion, at first closed and hemispherical, then opening with a circular pore, at length with a slightly revolute, incised margin, orange, peridial cells not firmly connected, overlapping, outer and inner walls of equal thickness (3–4 μ), inner with small warts often joined in small rows; aecidiospores ovoid to ellipsoid, 24–33 × 18–26 μ, wall yellow, 2 μ thick, densely and minutely verruculose, contents orange. **Uredosori** hypophyllous, generally on brownish or reddish-yellow spots, scattered or gregarious, sometimes in circinate groups, minute, rounded, pulverulent, cinnamon, surrounded by the cleft epidermis; uredospores globoid to obovoid, brown, 20–30 × 18–26 μ, wall sparsely echinulate, 2 μ thick, with 1, rarely 2 equatorial pores. **Teleutosori** similar but less pul-

verulent, blackish-brown, teleutospores subgloboid to obovoid, 22–42 × 13–25 μ, wall not thickened above, but with an apical pore covered by a hyaline papilla

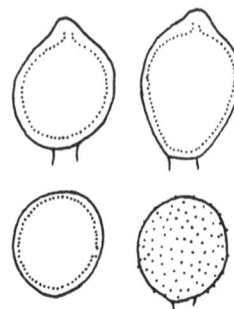

U. geranii. Teleutospores and uredospores (after Fischer).

up to 7 μ high, smooth, brown, pedicels short, hyaline, deciduous. Auteu-form.
On *Geranium dissectum*, *G. molle*, *G. pratense*, *G. pusillum*, *G. pyrenaicum*, *G. robertianum*, *G. rotundifolium* and *G. sylvaticum*, aecidia, March–June; teleutospores, June–October. Great Britain and Ireland, frequent.

This rust was recorded on *Geranium robertianum* by Dennis (Irish Nat. J. **9**, 184, 1948) from Killarney, Ireland; it has been previously recorded on this host from Spain (Fragoso, 1925, 97). The records on *G. pusillum* and *G. rotundifolium* are by Ellis (Trans. Norf. Norw. Nat. Hist. Soc. **15**, 371, 1943.)

Liro (Acta Soc. Fauna Fl. Fenn. **29** (6), 25, 1906) proved that the aecidiospores of this parasite from *G. silvaticum* produced uredo- and teleutospores on the same plant and Bock (Zbl. Bakt. II, **20**, 564, 1908) showed that the uredospores from the same species brought about infection of other hosts of the same genus. Jacob (Zbl. Bakt. II, **44**, 617, 1915) largely confirmed and extended Bock's work.

Uromyces kabatianus was separated from *U. geranii* by Bubak (*loc. cit.*) on the grounds that its sori have a circinate arrangement and that its teleutospores which do not appear until late in October, are paler, larger, more oblong and have more prominent papillae. Jacob generally confirmed these differences but pointed out that the uredosori of *U. geranii* on *G. sylvaticum* are sometimes circinately arranged while those of *U. kabatianus* in heavy infections may be scattered over the whole leaf surface. She pointed out that length and breadth frequency curves for the teleutospores of the two species do not coincide, the modes of *U. geranii* falling at 30μ for length and 22·5μ for breadth, the corresponding figures for *U. kabatianus* being 35μ and 20μ. Lind (Danish Fungi, 1913, 335) succeeded in transferring *U. kabatianus* from *G. pyrenaicum* to *G. molle, G. pusillum, G. rotundifolium, G. dissectum* and other species. Bubak thought that this form occurred only on *G. pyrenaicum* but Bock and Jacob showed that typical *U. geranii* could also be produced on that host. The latter confirmed the infections made by Lind. However, as Guyot (1951, 269) makes clear *U. geranii* is best considered a polymorphic species with several host specialised races, one of which is *U. kabatianus*.

Another aecidium may be found on species of *Geranium*; this is *A. sanguinolentum*, belonging to the heteroecious species *P. polygoni-amphibii* (see p. 164). Liro (Bot. Not. **1900**, 241, 1900) pointed out that the aecidia of the two species differ in the form of the peridial cells; in the peridial cells of *A. sanguinolentum* the outer wall is much thicker than the inner while in the aecidium of *U. geranii* the thickness of the outer and inner wall is approximately equal.

A short description of the binucleate mycelium and of the development of the teleutospore sorus was given by Sappin-Trouffy (1897).

Uromyces anthyllidis Schroet.

Hedwigia, **14**, 162 (1875); Grove, Brit. Rust Fungi, p. 95; Wilson & Bisby, Brit. Ured. no. 273; Gäumann, Rostpilze, p. 358.

[*Uredo anthyllidis* Grev. ex Berk., Smith, Engl. Fl. 5 (2), 383 (1836).]

Uromyces hippocrepidis Mayor, Bull. Soc. Neuchâtel. Sci. Nat. **45**, 40, 1921; Gäumann, Rostpilze, p. 371.

Uromyces jaapianus Kleb., Krypt. Fl. Mark Branb. 5*a*, 239 (1914); Wilson & Bisby, Brit. Ured. no. 293; Gäumann, Rostpilze, p. 389.

[**Spermogonia** and **aecidia** resembling those of *Uromyces pisi* or unknown.] **Uredosori** amphigenous, widely and irregularly scattered, or sometimes with a circle of small ones round a larger one, minute, roundish, black and shining at first, soon naked, then pulverulent, cinnamon-brown; uredospores globoid or subgloboid, yellowish-brown, 18–25 μ, wall 1·5–4 μ thick, sparsely and finely echinulate, with 4–6 pores. **Teleutosori** similar, but darker in colour; teleutospores globoid to ovoid, brown, 16–22 × 15–20 μ, wall 3 μ thick with a minute apical papilla, verrucose, pedicels short, hyaline, deciduous. Hetereu-form.

[Spermogonia and aecidia on *Euphorbia cyparissias*]; uredospores and teleutospores on *Anthyllis vulneraria, Hippocrepis comosa, Lotus hispidus, Trifolium dubium* and *T. campestre*, June–October. Great Britain and Ireland, scarce.

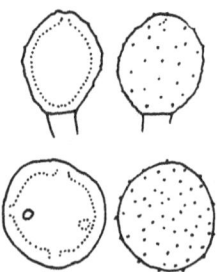

U. anthyllidis. Teleutospores and uredospores (after Savulescu).

This rust has been recorded from the south of England to Orkney and the Hebrides. Teleutospores are usually formed only in small numbers and it appears to overwinter by means of its uredospores.

Gäumann (Ber. Schweiz. Bot. Ges. **55**, 76, 1945) in Switzerland has shown that this rust is heteroecious by producing uredospores on *Anthyllis montana* by infection with aecidiospores from *Euphorbia cyparissias*.

It is distinguished from the races included under *Uromyces pisi* by its slightly larger uredospores with thicker echinulate walls and by its teleutospores with few, strongly developed, scattered warts. Jørstad & Nannfeldt (Bot. Not. **111**, 306, 1958) suggest that these differences are not sufficient to warrant retention of the two species.

The rust on *Hippocrepis comosa* has been recently recorded in the uredospore stage by Garlick (Kew Bull. **1957**, 386, 1959) as *U. hippocrepidis* from Durdham Downs, Bristol. This was regarded as a race of *U. anthyllidis* by Sydow (Monogr. Ured. **2**, 64) and Guyot (1957, 168). According to Fragoso (1925, 73) it differs from *U. anthyllidis* in having uredospores with finer, more clearly spaced warts and more numerous pores (6–7 or 8). It has been recorded from Spain, France and Switzerland.

Uromyces anthyllidis is a collective species and a considerable number of forms or races have been described on various genera of the *Leguminosae*. It has been recorded on species of *Lotus* around the Mediterranean (Guyot, Uredineana, **3**, 84, 1951) and the specimens called *U. euphorbiae-corniculatae* on *L. hispidus* collected by Salwey from Guernsey in 1848 and by Sandwith from the Lizard, Cornwall, should be included here.

The rust on *Trifolium dubium* often referred to *U. jaapianus* also belongs in this aggregate. Its aecidial relations are quite unknown. It should be carefully distinguished from *U. minor* which also infects *T. dubium*.

Grove (Irish Nat. J. **21**, 112, 1912) has recorded *Darluca genistalis* as a parasite on the sori of this rust in Co. Dublin.

Uromyces appendiculatus (Pers.) Unger

Einfl. Bodens, 216 (1836); Wilson & Bisby, Brit. Ured. no. 274.
Uredo appendiculata a *phaseoli* Pers., Syn. Meth. Fung. p. 222 (1801).
Uromyces phaseolorum de Bary, Ann. Sci. Nat. Bot. ser. 4, **20**, 80 (1863); Grove, Brit. Rust Fungi, p. 101.
Uromyces phaseoli (Pers.) Wint., Hedwigia, **19**, 37 (1880); Gäumann, Rostpilze, p. 340.

Spermogonia in little clusters, whitish, then yellowish. **Aecidia** hypophyllous, clustered in little roundish groups, 2–3 mm. wide, on yellowish or brownish spots, cup-shaped, whitish, with a torn revolute margin; aecidiospores polygonoid-globoid, densely and minutely verruculose, colourless, 18–36 × 16–24 µ. **Uredosori** generally hypophyllous, on indistinct spots, scattered or in little clusters, minute, soon naked, surrounded by the cleft epidermis, cinnamon; uredospores subgloboid to ovoid, brownish-yellow, 18–28 × 18–22 µ, wall distantly but sharply echinulate, brownish-yellow, about 1·5 µ thick, with 2 pores, contents colourless. **Teleutosori** similar but confluent, larger, amphigenous and blackish-brown; teleutospores subgloboid to ovoid, rounded above, with a wide pore and a hemispherical, hyaline papilla, chestnut-brown, 24–35 × 18–25 µ, wall up to 3·5 µ thick, smooth or rarely provided, expecially near the apex, with a few hyaline warts, pedicels hyaline, rather thin, about as long as the spore. Auteu-form.

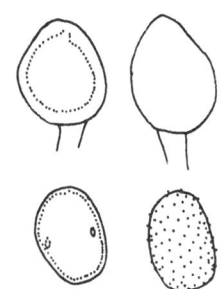

U. appendiculatus. Teleutospores and uredospores.

On leaves of *Phaseolus vulgaris* and *P. coccineus*, May–October. England, until 1952 scarce, since then frequent and damaging in east and south-east England.

This rust has been reported on dwarf bean from several localities in England, including Kent, Cornwall, Somerset, Norfolk, Cambridge and Sussex. It has been found on *Phaseolus coccineus* in Cornwall and Devon. The aecidia occur rarely but have been found in Kent, Somerset, Sussex, Dorset and Glamorgan.

De Bary (*loc. cit.*) proved the genetic connection of the aecidia with the uredospores and teleutosorus. Many specialised forms of the rust have been described (Harter & Zaumeyer, U.S.D.A. Tech. Bull. **868**, 1944). The rust causes a serious disease of dwarf and French beans in many parts of the world.

Andrus (J. Agric. Res. **42**, 559, 1931; J. Wash. Acad. Sci. **23**, 544, 1933) has investigated the cytology of the sex mechanism in this rust and has shown that it is heterothallic and Yarwood (Phytopath. **29**, 933, 1939) has studied the relation of water to infection. Schein (Phytopath. **51**, 674, 1961) related temperature to spore germination and symptom expression.

Uromyces ervi West.

Bull. Acad. Roy. Belg. **21**, 234 (1854); Grove, Brit. Rust Fungi, p. 96; Wilson & Bisby, Brit. Ured. no. 285; Gäumann, Rostpilze, p. 279.

[*Aecidium ervi* Wallr., Fl. Krypt. Germ. p. 247 (1833).]

Spermogonia epiphyllous with ostiolar paraphyses. **Aecidia** amphigenous, or on the petioles, solitary or in small scattered groups, peridium cup-shaped, whitish, faintly revolute, scarcely torn, peridial cells in distinct longitudinal rows, outer wall 5 μ thick, inner wall 3 μ thick, tesselate, with small crowded warts; aecidiospores densely and minutely verruculose, pale yellowish, 16–25 × 14–18 μ. **Uredosori** rarely formed, amphigenous or on the petioles and stems, scattered, minute, oblong, surrounded by the ruptured epidermis, cinnamon; uredospores ovoid or ellipsoid, distantly echinulate, brownish-yellow, 20–30 × 18–22 μ, wall 1·5–2·5 μ thick with 2 (rarely 3) equatorial pores. **Teleutosori** amphigenous, or more frequently on the petioles and stems, scattered, minute, oblong, surrounded by the ruptured epidermis, blackish-brown; teleutospores subgloboid to obovoid, usually darker and rounded above, rounded or attenuate at the base, wall smooth, brown, 20–28 × 14–20 μ up to 8 μ thick at the apex, pedicels brownish,

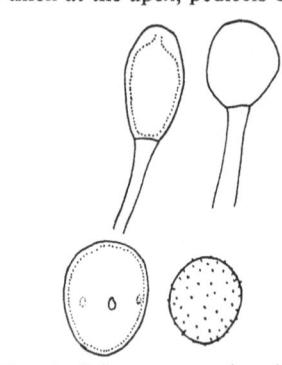

U. ervi. Teleutospores and uredospores.

persistent, as long or twice as long as the spore. Auteu-form.

On leaves, petioles and stems of *Vicia hirsuta*, aecidia May–October; teleutospores from June onwards. England, Scotland, scarce.

Many culture experiments in Europe suggest that Plowright (1889, 140) was correct in his belief that the race on *Vicia hirsuta* is strictly confined to that host, but in Japan it is found on *V. hirsuta*, *V. tetrasperma* and *V. sativa* and Hiratsuka (Jap. J. Bot. **6**, 329, 1933) has infected the two latter species with spores from *V. hirsuta*.

The aecidiospores are capable of reproducing the aecidium and are found throughout the season as Dietel (Z. Pfl.-Krankh. **3**, 258, 1893) and Jordi (Zbl. Bakt. II, **11**, 776, 1904) proved experimentally; the uredospores are, perhaps in consequence, not abundant, only a few being occasionally found and usually intermixed with teleutospores. Schroeter (Cohn, Beitr. Biol. Pfl. **3**, 82, 1883) stated that spermogonia are found with the first-formed aecidia in the spring and not with the aecidia produced later. Hiratsuka also recorded spermogonia in Japan. According to this investigator the spore forms are rather smaller than those of *Uromyces viciae-fabae* and its var. *orobi*. The uredospores of *U. ervi* possess two equatorial pores while in *U. viciae-fabae* and its var. *orobi* 3–5 pores are present.

Uromyces vicial-fabae (Pers.) Schroet.

Hedwigia, **14**, 161 (1875).
Uredo viciae-fabae Pers., Syn. Meth. Fung. p. 221 (1801).
[*Uredo viciae-fabae* Schum., Enum. Pl. Saell. **2**, 232 (1803).]
[*Uredo fusca* Purton, Midland Flora, **2**, 725 (1817), **3**, 507 (1821).]
Puccinia fabae Grev. Scot. Crypt. Flora **1**, tab. 29 (1823).
Uromyces fabae (Grev.) de Bary, Ann. Sci. Nat. Bot. ser. 4, **20**, 80 (1863); Grove, Brit. Rust Fungi, p. 97; Wilson & Bisby, Brit. Ured. no. 287; Gäumann, Rostpilze, p. 275.
Uromyces viciae-fabae (DC.) Otth, Mitt. Naturf. Ges. Bern, **1863**, 86 (1863).

Spermogonia amphigenous, developing amongst the aecidia. **Aecidia** usually hypophyllous, seated on pale-yellow spots which darken with age, solitary or in small, round or elongated clusters, 1–5 mm. long, shortly cup-shaped, with a whitish, torn, revolute margin, peridial cells with outer wall 6–7μ thick, finely punctate in surface view, inner wall thinner with rather densely scattered small warts, aecidiospores densely and minutely verruculose, yellow, 14–22μ diam. **Uredosori** amphigenous, scattered or circinate, surrounded by the ruptured epidermis, minute, pulverulent, pale brown; uredospores globoid to ovoid, distantly echinulate, 20–30 × 18–26μ, wall at length pale brown, 1·5–2·5μ sometimes up to 3·5 or 4μ thick with 3 or 4 pores. **Teleutosori** similar, but darker or blackish-brown; teleutospores subgloboid to obovoid, rounded or truncate at apex, 25–38 × 18–28μ, sometimes with a colourless papilla, wall smooth, brown, up to 7–11μ thick at apex with an apical pore, pedicels brownish, persistent, thick and up to 40–70μ long. Auteu-form.

On leaves and stems of *Lathyrus palustris*, *L. pratensis*, *Pisum sativum*, *Vicia angustifolia*, *V. bithynica*, *V. cracca*, *V. faba*, *V. hirsuta*, *V. lutea*, *V.*

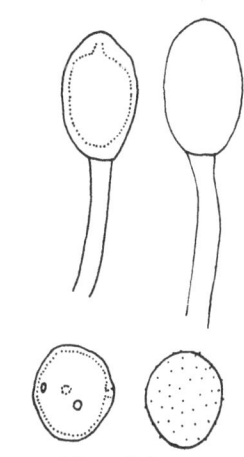

U. viciae-fabae. Teleutospores and uredospores.

sativa, and *V. sepium*. Aecidia, April, May, October and November; uredospores from May; teleutospores from July onwards. Great Britain and Ireland, common.

This species is one of the most widely spread of the Uredinales, reported on many leguminous species and occurring in almost every part of the world. Its life history was accurately described by Tulasne (C. R.

Acad. Sci. Paris, **36**, 109, 1853) and de Bary (*loc. cit.*) about a hundred years ago and the latter performed with it, in 1862, the first successful culture experiment ever made with a fungus.

It appears to have been first recorded in this country by Purton (*loc. cit.*) as *Uredo fusca* on leaves of *Vicia faba*. This was closely followed by Greville (*loc. cit.*) who described *Puccinia fabae* in Latin and in English, with good illustrations of teleutosori on a leaflet of *V. faba*, a dehiscing sorus and isolated teleutospores.

Statements made regarding the thickness of the uredospore wall in the varieties of this rust differ to some extent; the taxonomic importance of this character has been discussed by Gäumann (Ann. Myc. **33**, 464, 1934) who stated that in the rust on *V. sepium*, uredospores developed on the stems possess thicker walls than those developed on the leaves.

In 1862 de Bary (*loc. cit.*) sowed teleutospores from *V. faba* on seedlings of this species and on *P. sativum* and produced aecidia on both these hosts. This experiment has often been repeated both in Europe and America. The specialisation has been studied more recently by Jordi (*loc. cit.*), in Japan by Hiratsuka (Jap. J. Bot. **6**, 329, 1933; Bot. Mag. Tokyo, **48**, 309, 1934) and in Switzerland by Gäumann (*loc. cit.*) Six varieties have been recognised by Gäumann: (1) *viciae-fabae* de Bary on *P. sativum* and *V. faba*; (2) *pisi-sativi* Hiratsuka on *P. sativum*; (3) *craccae* Fischer on *V. cracca*, *P. sativum* and weakly on *V. hirsuta*; (4) *viciae-sepium* Gäumann on *V. sepium* and *V. faba*; (5) and (6) on Japanese species of *Vicia* by Hiratsuka. Kispatic (Phytopath. Zeitschr. **14**, 475, 1943) with the var. *viciae-fabae* from Yugoslavia, has been able to obtain infection on several species of *Vicia* including *V. bithynica* and on *P. arvense* and *P. sativum*.

The only experiments which have been carried out in this country are those by Plowright (1889, 120), evidently with the type race of the species corresponding to Gäumann's *viciae-fabae*; with teleutospores from *V. faba* he infected the same host and *P. sativum* but failed to bring about infection on species of *Lathyrus* and other species of *Vicia*.

As noted by the Sydows (Monogr. Ured. **2**, 103) aecidia are commonly found on some species of *Lathyrus* and *Vicia* and are rare on other hosts such as *V. faba* and *P. sativum*; they were, however, produced in culture on the two latter species by Plowright (*loc. cit.*). They have been found in November 1935 near Harpenden on seedling plants of *V. faba* growing up after harvest and almost simultaneously, under the same condition, at Cambridge by Steven (J. Bot. Lond., **74**, 79, 1936) and again near Harpenden in March 1941 on self-sown plants (Bull. Min. Agric.

126, 32, 1945). The occurrence of aecidia so late in the season suggested that some teleutospores are capable of germinating in the autumn. A small proportion of teleutospores from *V. faba* collected near Edinburgh in September 1939 germinated readily and produced 4-celled basidia and basidiospores in 4 days. Kispatic (RAM. **29**, 5, 1950) in Yugoslavia observed on *V. faba* germination of teleutospores in September and infection of young seedlings from October to the end of December with the formation of aecidia and aecidiospores. He concluded that aecidiospores are unable to survive severe winters in northern regions but probably remain viable in Mediterranean climates. It is possible that records of the occasional appearance of the disease on broad beans in February in south-western England may be due to the survival of aecidiospores on the seedlings plants.

E. A. Ellis (Trans. Norf. Norw. Nat. Hist. Soc. **15**, 271, 1943) has recorded many aecidia and few uredospores in July 1942 on *Lathyrus palustris* in Norfolk.

It was noted by de Bary that uredospores were sometimes produced on the aecidial mycelium and this has recently been confirmed by Brown (Can. J. Res. **18**, 18, 1940) who, as the result of sowing basidiospores from *L. venosus* on *L. venosus* and *L. ochroleucus*, sometimes obtained aecidia and uredosori and sometimes only uredosori.

On the leaves of *V. faba* usually only the uredospores may be found, even as late as the middle of October but on the stems the teleutospores form large black sori from July onwards; the rust has also been found on the pods. On *V. sepium* the uredosori are often darker and covered by the epidermis for a shorter time than on *V. faba*, while the teleutosori occur in greater abundance on the leaves and even on the tendrils. The rust has been recorded by Mayfield (1935, 2) from Suffolk on *V. lutea* and *V. bithynica*, and by Ellis on *V. angustifolia* from Guernsey, Channel Isles (*in litt.*), and on *V. hirsuta* by Bramley from Yorkshire (*in litt.*). The rust on *V. cracca*, although recorded throughout Great Britain and Ireland, appears to be a northern type and has been found in the Shetland Isles by Dennis & Gray (Trans. Bot. Soc. Edin. **36**, p. 220, 1954); according to Rainio (Ann. Soc. zool. bot. Fenn. Vanamo, **3**, 242, 1926) it extends northwards to within the Arctic Circle.

Savile (Amer. J. Bot. **26**, 585, 1939) has investigated nuclear and spore development in this rust; the chromosome number is 4; and Brown (*loc. cit.*) has shown that this species is heterothallic.

Prasada & Verma (Ind. Phytopath. **1**, 142, 1948) have reported the occurrence of repeating aecidia on *Lens esculenta* in India; their development takes place between 17° and 22° C., uredospores being formed above 25° C.

The conditions of humidity and temperature under which the spores germinate have been studied by Hiratsuka (1934) in Japan, Reichert & Palti (RAM. **26**, 523, 1948) in Palestine, Kispatic (*loc. cit.*) in Yugoslavia, Garofalo (RAM. **28**, 45, 1949) in Italy and others.

var. *orobi* (Schum.) Jørst.

K. Norske Vid. Selsk. Skr. **1935** (38), 46 (1936).
Uredo orobi Schum., Enum. Pl. Saell. **2**, 232 (1803).
Uredo orobi DC., Fl. Fr. **6**, 66 (1815).
Uromyces orobi (DC.) Lév., Ann. Sci. Nat. Bot. ser. 3, **8**, 371 and 376 (1847); Grove, Brit. Rust Fungi, p. 99; Wilson & Bisby, Brit. Ured. no. 303; Gäumann, Rostpilze, p. 277.

Uredospores and teleutospores on *Lathyrus montanus*, not common.

The rust on *Lathyrus montanus* was generally regarded as a distinct species until it was transferred to *Uromyces fabae* by Jørstad. It differs morphologically from the rust on *Vicia faba* only in greater average thickness of the wall of the uredospore $1 \cdot 2 – 2 \cdot 5 \mu$ in var. *viciae-fabae*, $3–4\mu$ in var. *orobi*. It would appear to be identical with f. sp. *orobi* Jordi of *U. orobi* (Zbl. Bakt. II, **11**, 763, 1904) which could not be transmitted to *L. luteus*, *L. pratensis*, *V. faba*, *V. hirsuta*, *V. sativa* or *Pisum sativum* or to other species of these three genera.

Uromyces fallens (Desm.) Kern

Phytopath. **1**, 6 (1911).
Uredo fallens Desm., Pl. Crypt. Fr. no. 1325 (1843).
Uromyces trifolii (DC.) Lév., Ann. Sci. Nat. Bot. ser. 3, **8**, 371 (1847) *p.p.*; Grove, Brit. Rust Fungi, p. 90; Wilson & Bisby, Brit. Ured. no. 314; Gäumann, Rostpilze, p. 345.
Trichobasis fallens Cooke, J. Bot. Lond. **4**, 105 (1866).
Uromyces trifolii var. *fallens* (Desm.) Arth., Rusts U.S. Can. p. 305 (1934).

Spermogonia amphigenous, grouped, light yellow turning reddish, then grey, 100μ diam.; spermatia $3–4\mu$; paraphyses $3 \times 40\mu$. **Aecidia** amphigenous, grouped in elongated areas on blades, petioles and stipules, golden turning grey, 190μ diam., 170μ high, peridium yellowish-white, margin recurved, lacerate, peridial cells $12–16 \times 8–10\mu$, outer walls $2 \cdot 5 – 4 \cdot 5\mu$ thick, transversely striate, inner 2μ thick, slightly verrucose; aecidiospores globoid to ellipsoid, light yellow, $20–26 \times 17–22\mu$, wall 1μ thick,

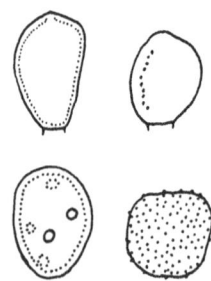

U. fallens. Teleutospores and uredospores (after Guyot).

minutely verrucose. **Uredosori** hypophyllous and on the petioles, scattered or gregarious, small, rarely confluent and larger, pulverulent, soon naked, pale brown; uredospores globoid, ovoid or ellipsoid, yellowish-brown, 18–28 × 20–23 μ, wall about 1·5 μ thick, sparsely echinulate, pores 4–7. **Teleutosori** hypophyllous and on the petioles, scattered or gregarious, small, rounded, sometimes confluent and larger, at first covered by the epidermis, pulverulent, brownish-black; teleutospores globoid to ovoid, rounded at the apex with a minute, hyaline papilla, smooth or with a few scattered or linearly ordered warts, brown, 20–24 × 16–18 μ, wall 1·5–2 μ thick, pedicels short, thin, hyaline, deciduous. Auteu-form.

On *Trifolium pratense*, *T. medium* and *T. incarnatum*, aecidia from April and uredospores from May onwards. Great Britain and Ireland, scarce.

This species is distinguished from *Uromyces trifolii* by the larger number of pores in the uredospores.

The first discovery of aecidia appears to have been by Davis & Johnson (Phytopath. 7, 75, 1917) who obtained them experimentally in 1915 in the United States by infection with teleutospores; they were also observed by Melhus and Diehl (Phytopath. 7, 70, 1917) in 1916. In Europe cultures were made in 1919 by Kobel (Zbl. Bakt. II, **52**, 217, 1921) in which overwintered teleutospores from *Trifolium pratense* produced aecidia on the same host. Davis (Mycol. **16**, 203, 1924) has shown that all the spore forms from *T. pratense* infect the same species and *T. medium* but not *T. repens* or *T. hybridum*. In this country aecidia were discovered in 1931 at Harpenden, Herts. (Bull. Min. Agric. **79**, 49, 1928–32); they have also been found at Cambridge. Grove (*loc. cit.*) recorded the rust on *T. medium* and *T. incarnatum*; Arthur (*loc. cit.*) stated that the latter host is attacked by both *U. fallens* and *U. trifolii-repentis*.

Kobel (*loc. cit.*) has shown that in Switzerland overwintering by means of uredospores can take place. He considered that where uredospores are produced in abundance the dicaryotic mycelium can also live through the winter.

The effect of temperature on spore germination has been described by Davis (*loc. cit.*) and Yarwood (Phytopath. **29**, 933, 1939) has studied the effect of moisture.

Uromyces minor Schroet.

Cohn, Krypt. Fl. Schles. 3, 310 (1887); Gäumann, Rostpilze, p. 350.

Spermogonia absent. **Aecidia** scattered, or in groups of 5 or 6, appearing on young leaves and maturing simultaneously with telia as though from a systemic mycelium, undoubtedly at least

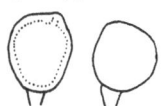

U. minor. Teleutospores.

from the same mycelium, rather deeply sunk and long-covered by the epidermis, peridium slightly toothed at margin but otherwise not protruding beyond epidermis and entire; aecidiospores slightly angular 14–16·5μ diam., wall verruculose. **Uredosori** absent. **Teleutosori** scattered amongst aecidia, hypophyllous, long-covered by silvery epidermis, teleutospores subgloboid to broadly ellipsoid, 20–24 × 14–18μ, with 1 apical or subapical pore with inconspicuous cap, wall smooth, 1–2μ thick, pedicel short. Opsis-form.

Aecidia, uredospores and teleutospores on *Trifolium dubium* and *T. molinierii*. Channel Islands and N. Ireland, rare.

The species is known in Britain only from the collections made by Frost on *Trifolium molinierii* in Jersey in April (TBMS. **43**, 695, 1960) and by McIlwaine on *T. dubium* near Portadown, Northern Ireland. The position of this latter collection has been discussed by Henderson (Notes R.B.G. Edin. **23**, 259, 1961). *Uromyces minor* in Europe is known as a rust principally of *T. montanum* on which the aecidia do not appear to repeat and the teleutosori are early erumpent; whereas, as Dietel (Flora, **81**, 394, 1895) noted, the American rust which occurs on a much wider range of *Trifolium* species including *T. dubium* appears to have repeating aecidia and the teleutosori are long-covered by the epidermis. The Irish collection agrees with this latter type. Moreover the measurements given by Arthur (1934, 305) for the teleutospores of the rust in North America (18–26 × 13–19μ) are closer to those of the Irish collection (20–24 × 14–18μ) than those for *U. minor* in Europe (18–20 × 14–17μ). It has been suggested (Henderson, *loc. cit.*) that the Irish rust is an American element in the Irish flora comparable to those of the higher plants.

Uromyces nerviphilus (Grognot) Hotson

Publ. Puget Sound Biol. Sta. **4**, 368 (1925); Gäumann, Rostpilze, p. 350.
Puccinia nerviphila Grognot, Pl. Crypt. Saône-et-Loire, p. 154 (1863).
Uromyces flectens Lagerh., Sv. Bot. Tidskr. **3**, 36 (1909); Grove, Brit. Rust Fungi, p. 92; Wilson & Bisby, Brit. Ured. no. 289; Gäumann, Rostpilze, p. 351.

Spermogonia unknown. **Aecidia** and **uredosori** wanting. **Teleutosori** hypophyllous, or more often on the nerves and petioles where they cause swellings and distortion, scattered, rather large, 0·5–2 mm. long or confluent and even larger, long-covered by the epidermis, then pulverulent, dark-brown; teleutospores ellipsoid or ovoid, apex rounded with a minute hyaline papilla over the apical pore, brown, 18–30 × 16–25μ, wall 2μ thick, smooth or with a few minute warts arranged more or less in line, pedicels short, thin, hyaline, deciduous. Micro-form.

On *Trifolium repens* and *T. fragiferum*, May–October, Great Britain and Ireland, frequent.

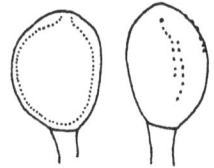

U. nerviphilus. Teleutospores.

This rust known until recently as *Uromyces flectens*, is the most frequent species occurring on *Trifolium repens* in this country and most of the older records on this host referred to it. It has also been recorded on *T. fragiferum* by Ellis in Norfolk and by Sprague in Gloucestershire in 1934.

Dietel (S.B. Naturf. Ges. Leipzig. **15, 16**, 37, 1890) suggested that this rust possessed a perennial mycelium and this was proved to be the case by Kobel (Zbl. Bakt. ii, **52**, 221, 1920) who found mycelium in the rhizome and buds. Kobel has also shown that teleutospores produced during the summer can germinate immediately and bring about infection, whereas those formed in the autumn will germinate in the following spring; as the result of infection by basidiospores, teleutosori are again produced and there is no doubt that the rust is microcyclic.

Schroeter in 1883 (Beitr. Biol. Pfl. **3**, 78, 1883) and Plowright (1889, 125) noted that in some cases plants of *T. repens* infected with *Uromyces* developed teleutospores only and that these were produced on a perennial mycelium throughout the summer. Lagerheim (*loc. cit.*) noticing that the form which produced only teleutospores had sori which were longer, more prominent upon veins and petioles and remained longer covered by the epidermis described this as a distinct species, *U. flectens*. The teleutospores are usually somewhat more verrucose than those of *U. trifolii*. All these investigators were evidently unaware that a similar form had already been described by Grognot in 1863 as *Puccinia nerviphila*; the latter, after its transfer to *Uromyces* by Hotson (*loc. cit.*), must be regarded as the correct name for this microcyclic species.

Although Grognot, in his diagnosis of *P. nerviphila*, did not mention aecidia, Arthur (1934, 305) has reserved the name for an American species which produces spermogonia, aecidia and teleutosori, i.e. an opsis-form; it appears, however, that in addition to the opsis-form a true microcyclic species exists in North America as *U. flectens* has been reported by Mains (Pap. Mich. Acad. Sci. **20**, 87, 1925) from Michigan and he stated that repeated inoculations carried out in 1931 and 1932 with teleutospores on *T. repens* always resulted in the production of teleutosori. Whether the opsis-form occurs in Europe is uncertain although Gäumann (1959, 351) includes it.

Uromyces pisi (DC.) Otth

Mitt. Naturf. Ges. Bern, **1863**, 87 (1963).
Puccinia pisi DC., Fl. Fr. **2**, 224 (1805).

Spermogonia hypophyllous, numerous, scattered amongst the aecidia. **Aecidia** hypophyllous, scattered evenly all over the leaf surface, peridium cupulate, broadly revolute, deeply laciniate, white, peridial cells in radial longitudinal section rhombic, on the outer side projecting slightly over the cell below, outer wall about 7μ thick, finely transversely striated, inner wall about 3μ thick, verrucose; aecidiospores in distinct chains, polygonoid or ellipsoid, $18-23\mu$ diam., wall minutely verrucose, colourless, 1μ thick, contents orange. **Uredosori** usually hypophyllous, on indistinct spots, scattered or in small groups, minute, soon naked, pulverulent, cinnamon yellow; uredospores globoid or subgloboid, $15-32 \times 14-23\mu$, wall minutely verruculose, $1\cdot5-2\cdot5\mu$ thick, yellowish-brown with 3–5 pores. **Teleutosori** similar, amphigenous, but larger, confluent, pale brown; teleutospores subgloboid, ovoid or ellipsoid $20-28 \times 14-24\mu$, wall with a minute, hyaline papilla, 3μ high and 7μ wide, everywhere verrucose with usually elongate warts scattered or more or less in lines,

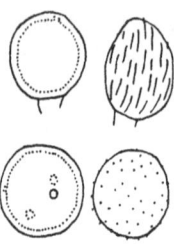

U. pisi. Teleutospores and uredospores.

brown, $1\cdot5-2\mu$ thick, pedicels short, hyaline, deciduous. Hetereu-form.

Spermogonia and aecidia on *Euphorbia cyparissias*, May–June; uredospores and teleutospores on various genera of the *Leguminosae*.

Under the collective name *Uromyces pisi* are here included various closely allied leguminous rust races and race groups often formerly considered distinct species.

The aecidial stage has been found rarely in this country, apparently on a few occasions only, in Kent, West Suffolk and Lindisfarne. The races appear to be independent of host alternation except possibly in certain areas in England where *E. cyparissias* occurs in their vicinity; in general overwintering takes place by means of the uredospores.

Hariot first suggested (Les Urédinées, 1908, 208 and 210) that the type of deformation produced on a common aecidial host differs from one race to another; while some cause elongation of the stems (gigantism) others bring about shortening (nanism). This, to some extent, has been confirmed by Guyot (several papers, well summarised in Guyot, 1957, 582). He has associated certain types of deformation on the aecidial host with the diploid stages of the rusts which have produced them.

Healthy stems may sometimes develop from an infected root-stock and infected shoots may outgrow the mycelium and regain their normal habit in their upper parts.

The subsequent accounts of the *U. pisi* group are arranged alphabetically according to major host genera.

Astragalus

Uromyces punctatus Schroet., Abh. Schles. Ges. Vaterl. Kult. Nat. Abth. **1869–72**, 10 (1870); Wilson & Bisby, Brit. Ured. no. 306; Gäumann, Rostpilze, p. 360.

[*Uredo leguminosarum d. astragali* Opiz, Seznam Kveteny Ceské, 151 (1852), *nomen nudum*.]

Uromyces astragali Sacc., Mycoth. Ven. p. 208 (1873).

Uromyces astragali Schroet., Cohn, Krypt. Fl. Schles. 3, 308 (1887).

Uromyces euphorbiae-astragali Jordi, Zbl. Bakt. II, **11**, 790 (1904).

Uredospores and teleutospores on *Astragalus danicus*. Scotland, rare.

This rust appears to have been first found in this country by U. K. Duncan (*in litt.*) on the sands of Barry, Angus, in 1948 but no record was published. It was recorded by J. A. Macdonald at St Andrews in 1949. It is improbable that the aecidial stage has been found in this country. The uredospore stage is usually dominant but teleutospores have also been collected in Angus and Fife.

Jordi (Zbl. Bakt. II, **10**, 777, 1903; *idem*, **11**, 763, 1904) obtained uredospores on *A. glycyphyllos*, *A. depressus*, on *Oxytropis* sp. and *Lotus* sp. by infection with aecidiospores from *Euphorbia cyparissias*. This was confirmed by Bubak (Zbl. Bakt. II, **12**, 411, 1904) who infected *E. cyparissias* with overwintered teleutospores and obtained aecidial infections; Treboux (Ann. Myc. **10**, 73 and 557, 1912), using aecidiospores from *E. virgata* also infected nine species of *Astragalus*.

Burns (Proc. Roy. Soc. Edin. B, **59**, 212, 1957) using uredospores from *A. danicus* in Scotland infected eleven species of *Astragalus*, including *A. danicus*, *A. glycyphyllos* and *A. alpinus*; infections of *E. cyparissias* with teleutospores gave no result and plants of the spurge planted amongst naturally infected plants of *A. danicus* remained uninfected for 4 years.

Burns has also shown that uredospores collected every month during the year when inoculated onto healthy plants of *A. danicus* in the greenhouse always produced infection; by microscopic examination of infected stems, leaves and petioles throughout the year it was proved that the mycelium is not systemic. As *E. cyparissias* is an uncommon plant in Fife and has never been found infected with aecidia, he concluded that the rust overwinters by its uredospores. The parasite *Darluca filum* has been found constantly in and around the uredosori.

Genista, Laburnum, Sarothamnus and Ulex

Uromyces laburni (DC.) Otth, Mitt. Naturf. Ges. Bern, **1863**, 87 (1863); Wilson & Bisby, Brit. Ured. no. 296; Gäumann, Rostpilze, p. 369.
Puccinia laburni DC., Fl. Fr. **2**, 224 (1805).
Uromyces genistae-tinctoriae (Pers.) Wint., Hedwigia, **19**, 36 (1880).
Uredo appendiculata γ *Uredo genistae-tinctoriae* Per., Syn. Meth. Fung. p. 222 (1801).

Uredospores and teleutospores on *Genista anglica, G. tinctoria, Laburnum anagyroides, Sarothamnus scoparius, Ulex europaeus* and imported *Genista sagittalis*. England and Scotland, scarce.

There is a specimen of rust in Kew Herbarium collected on *Sarothamnus scoparius* in Kent in 1865. It was again recorded in 1934 in the uredospore condition, almost simultaneously by Grove (J. Bot. Lond. **72**, 265, 1934) on *Genista tinctoria* in Worcestershire and by J. A. Macdonald on *G. anglica* near Inverness (Wilson, 1934, 362 and Macdonald, TBMS. **29**, 64, 1946). Ellis (1934) found it on *Sarothamnus scoparius* in Norfolk, and Mayfield (1935, 3) on *G. tinctoria* and *S. scoparius* in Suffolk; teleutospores were found by Ellis and Mayfield. It has also been found on the double garden form of *G. tinctoria* in Sussex (*in litt.*) and from Yorkshire on imported material of *G. sagittalis* (Moore, 1959, 404).

It was found in this country on the laburnum (*Laburnum anagyroides*) by Bramley (*in litt.*) in Yorkshire in 1945; soon afterwards it was discovered on this host from Wingham, Kent by S. P. Wiltshire in 1947; in both localities uredospores and numerous teleutospores were present on the leaves. The second locality is in the region where *Euphorbia cyparissias* appears to be widespread and where it has been twice found infected; here the rusts on *Pisum sativum, Medicago lupulina, M. arabica* and *Onobrychis viciifolia* have also been found.

Kobel (Ann. Myc. **19**, 1, 1921) first proved the connection between the two hosts *L. anagyroides* and *G. sagittalis* by infecting them with aecidiospores from *E. cyparissias*; *G. tinctoria* was not infected; Dietel (Ann. Myc. **17**, 108, 1919) proved the connection with the latter in Germany by infecting with aecidiospores.

Guyot (Ann. Ecole Nat. Agric. Grignon, II, **1**, 45, 1937; Guyot & Massenot, Uredineana, **4**, 323, 1953) has described the distribution of the form on *L. anagyroides* in north-western France and has brought about infection by aecidiospores from *E. cyparissias*; he has shown that it is almost completely restricted to *Laburnum* and to a few closely related species of *Cytisus*. He has also described the nanism of the aecidial host resulting from infection by this form; the stems are short-ened, 20–23 cm. high and are thickened up to 2–3 mm., with leaves 6–8 mm. long and 2·5–3 mm. wide.

The race group included under *U. laburni* is morphologically hetero-geneous and there is considerable variation in the sculpturing of the teleutospore wall; in the form on *G. tinctoria* it is often verrucose-striate, whereas in *U. laburni* the warts are small, rounded, scattered, or arranged in longitudinal rows; these two forms were figured by Kobel. Macdonald (*loc. cit.*) has described three formae speciales of this rust as follows: *U. genistae-tinctoriae* f. sp. *anglicae* on *G. anglica*, f. sp. *scoparii* on *S. scoparius* and f. sp. *europaei* on *Ulex europaeus*. These show slight morphological differences in the uredospores and each is restricted to its host plant.

Lathyrus and Pisum

Uromyces pisi (DC.) Otth, Mitt. Naturf. Ges. Bern, **1863**, 87 (1863), *sens. stricto*; Grove, Brit. Rust Fungi, p. 99; Wilson & Bisby, Brit. Ured. no. 304; Gäumann, Rostpilze, p. 386.

Uredo appendiculata β Uredo pisi-sativi Pers., Syn. Meth. Fung. p. 222 (1801).
Puccinia pisi DC., Fl. Fr. **2**, 224 (1805).

Spermogonia and aecidia on *Euphorbia cyparissias*, May–June; uredospores and teleutospores on *Lathyrus pratensis* and *Pisum sativum*, July–September. Great Britain, rare.

Two collections of an aecidial stage referred to this race but quite unconfirmed have been made in Kent on two occasions: at Dover, May 1909 by T. Taylor (Grove, *loc. cit.*) and by the senior author near Ash-ford in May 1950; the deformation of the aecidial host seems to have been gigantism (see p. 330) on the first discovery and was certainly so on the second; this is characteristic for the race alternating with *Pisum sativum*. At the second locality the specimen was found in a garden where peas had been grown in the previous years, but it is unknown whether they were also infected.

The rust on peas has been found in Kew Gardens, at Faversham, Kent, in 1914 (about 9 miles from where the aecidial stage was found in 1933), in Suffolk (Mayfield, 1935), in Glamorgan (Moore, 1947, 35) and in 1953 at Royston, Herts. (*in litt.*); there are also two records from west and north-west Yorkshire by Mason & Grainger (Cat. Yorks. Fungi, 1937, 45).

Teleutospores have generally been found on *P. sativum* and it is possible that when these occur in localities where *Euphorbia cyparissias* is also found that the rust may be normally or obligatorily heteroecious.

The rust on *Lathyrus pratensis* has been found more frequently and the records are spread over England and Scotland. On this host uredospores are generally abundant and teleutospores are found only rarely on the stems; it overwinters by means of uredospores.

Schroeter (Hedwigia, **14**, 98, 1875) by sowing aecidiospores from *E. cyparissias* obtained uredospores on *P. sativum*, *L. pratensis* and *Vicia cracca* and the connection of the two phases of rust was later confirmed by Rostrup, Fischer and Jordi (Zbl. Bakt. II, **13**, 64, 1904) and others. Jordi also showed that the infection of *E. cyparissias* took place on the underground buds. Buchheim (Zbl. Bakt. II, **55**, 507, 1922, and **60**, 534, 1924) showed that uredospores from *L. pratensis* could infect various species of *Lathyrus* and also *P. sativum*. Grove considered that the rusts on *P. sativum* and on *L. pratensis* were biologically distinct and in France, Guyot (Ann. Ecole Nat. Agric. Grignon, II, **1**, 52, 1937) found an infected stock of *E. cyparissias* the aecidiospores from which infected *L. pratensis* but failed to infect *P. sativum*. Later Guyot (Ann. Ecole Nat. Agric. Grignon, III, **1**, 64, 1939) distinguished three races: (1) infecting *P. arvense* and *P. sativum* but not *Lathyrus* spp.; this is a weak parasite, not forming teleutospores on *P. sativum*; (2) infecting *L. pratensis* but not *P. sativum*; (3) infecting *P. sativum*, *P. arvense* and *L. aphaca* but not other species; this is a vigorous parasite, causing severe deformation on *E. cyparissias* and producing numerous uredospores and teleutospores on *P. sativum*.

Jørstad (Nordisk Jordbruksforskning, **7–8**, 198, 1948) found during 1944–8, the rust on garden peas and on *E. cyparissias* in several localities along the south-eastern coast of Norway. The peas were severely attacked and produced both uredospores and teleutospores. Cultures were carried out with aecidiospores and uredospores developed on the pea plants; he found the race restricted to the pea and concluded that it must be a recent invader.

Rosen (Beitr. Biol. Pfl. **6**, 237, 1892) who was one of the first investigators to study nuclei in the rust fungi, saw them in mycelium, spermatia, uredospores, and teleutospores of this race in 1892 and described some details of their structure.

Stampfli (Hedwigia, **49**, 230, 1910) has shown that various deformations are found in the flowers of infected plants of *E. cyparissias* and has described the occurrence of spermogonia on the surface of the ovary and in the ovarian cavities.

Lotus

Uromyces euphorbiae-corniculati Jordi, Zbl. Bakt. II, **11**, 791 (1904).
Puccinia loti Kirchner, Lotos, **6**, 181 (1856).
[*Uromyces loti* Blytt. Chria. Forh. Vidensk. Selsk. **1896** (6), 37 (1896)]; Grove, Brit.
Rust Fungi, p. 94; Wilson & Bisby, Brit. Ured. no. 300; Gäumann, Rostpilze,
p. 373.

Uredospores and teleutospores on *Lotus*   *tenuis*, July–August. Great Britain and
angustissimus, L. corniculatus, and *L.*   Ireland, scarce.

This rust is widespread on *Lotus corniculatus* but has been rarely found on the other hosts; the records on *L. angustissimus* are from Newquay, Cornwall, by Vigurs (Grove, J. Bot. Lond. **51**, 44, 1913) and by E. A. Ellis in Guernsey (*in litt.*). It was recently discovered on *L. tenuis* in Sussex (Wilson & Henderson, TBMS. **37**, 254, 1954).

A specimen of this rust in the Kew Herbarium stated to be on *L. uliginosus* collected in Gloucestershire (see Wilson & Bisby, 1954, 79) proved on examination to be on *L. corniculatus*; there is no certain record of this rust on *L. uliginosus* in Britain.

There is no definite evidence of the occurrence of the aecidial stage in this country.

Jordi (*loc. cit.*) proved by culture that aecidiospores from *Euphorbia cyparissias* could infect *L. corniculatus* and obtained slight infection on species of *Astragalus* and *Oxytropis* while *Medicago* remained uninfected. Guyot (Ann. Ecole Nat. Agric. Grignon, III, **4**, 147, 1944) confirmed this infection of *L. corniculatus* and found that species of *Lathyrus, Medicago* and *Pisum* together with *L. uliginosus* remained uninfected; later (Uredineana, **4**, 331, 1953) he repeated this culture and proved that, in addition, species of *Cytisus, Trifolium* and *Vicia* were uninfected; he also showed that uredospores obtained from *L. corniculatus* could infect *L. hispidus* and other species of *Lotus*. The stock of *E. cyparissias* from which the aecidiospores for these cultures were obtained showed nanism, infected stem being 24–26 cm. high and 2 mm. diam., with leaves 3–8 mm. long and 3·5 mm. wide.

Medicago

Uromyces striatus Schroet., Abh. Schles. Ges. Vaterl. Cult. Nat. Abth. **1869–72**, 11
(1870); Wilson & Bisby, Brit. Ured. no. 312; Gäumann, Rostpilze, p. 377.

Spermogonia and aecidia on *Euphorbia*   June–September. Uredospores only on
cyparissias; uredospores and teleuto-   *Pisum sativum* after inoculation. Great
spores on *Medicago arabica, M. lupu-*   Britain and Ireland, scarce.
lina, M. polymorpha and *M. sativa,*

The fungus described as *Uromyces striatus* by Grove is not this race but *U. anthyllidis* (Grove & Chesters, TBMS. **18**, 270, 1934 as *U. jaapianus*).

The first discovery of this rust in the British Isles appears to have been made on *Medicago polymorpha* in 1848 by Salwey (Trans. Bot. Soc. Edin. **3**, 78, 1848–50) who recorded it as '*Uredo trifolii* Dec. *apiculosa* Lk. on *M. denticulata*'; the specimen is in Kew Herbarium, and is labelled in Salwey's handwriting with the specific epithet *apiculosa* added later in Berkeley's handwriting; it was found at St Peter's Port, Guernsey, Channel Isles. It was discovered on the same host in Cornwall in 1930 by Arthur (TBMS. **18**, 270, 1933); it has since been recorded on *M. polymorpha* from Somerset (*in litt.*).

It was collected on *M. arabica* near Faversham, Kent, by the senior author in 1903 and was subsequently found in several localities in Somerset (Glasscock & Ware, TBMS. **29**, 167, 1946), Cornwall and the Scilly Isles.

The most frequent host of this rust is *M. lupulina* on wild plants of which it was found in Ireland by O'Connor (1936, 383) and Macdonald in Scotland (Trans. Bot. Soc. Edin. **32**, 557, 1939). It has since been recorded in a few localities in the Midlands and in southern England and in south Wales; it has caused severe damage to crops in Kent (Glasscock & Ware, *loc. cit.*), in Gloucestershire (Moore, 1948, 40; Sampson & Western, 1954, 83) and in Somerset (*in litt.*). *Medicago sativa* was heavily attacked in East Sussex in 1959 (Plant Pathology, **9**, 111, 1960).

The first cultures with this species were made by Schroeter (Pilze Schles. **1**, 306, 1887) with aecidiospores from *E. cyparissias* on *Trifolium agrarium*; similar cultures have also been made by Treboux (Ann. Myc. **10**, 75, 1912), Kobel (Zbl. Bakt. II, **82**, 234, 1920) and Mayor (Bull. Soc. Neuchâtel. Sci. Nat. **51**, 62, 1926). Arthur (Mycol. **4**, 56, 1912) using uredospores from *M. sativa* could infect the same host but not species of *Trifolium*. Recently Guyot (Uredineana, **4**, 323, 1953) has carried out a large number of cultures with aecidiospores in France and has infected with them all the species of *Medicago* which occur as hosts for this rust in Britain.

Butler (TBMS. **41**, 401, 1958) carried out infections with aecidiospores from *Euphorbia cyparissias* collected in West Suffolk and obtained uredospores and teleutospores on *M. lupulina* and uredospores only on *M. sativa* and *Pisum sativum*.

This rust has been recorded from various parts of the world on lucerne (*M. sativa*) on which it produces a serious disease.

Onobrychis

Uromyces onobrychidis Bub., S.B. Kon. Böhm. Ges. Wiss. II, **46**, 7 (1902).
[*Uredo onobrychidis* Desm., Cat. Pl. Omis. p. 25 (1823).]
Uromyces onobrychidis ('Lev.') Thüm., Myc. Univ. 1531 (1879) & Bull. Soc. Imp.
Nat. Moscou, **35**, 211 (1880), *nomen nudum*; Wilson & Bisby, Brit. Ured. no.
302; Gäumann, Rostpilze, p. 381.

Uredospores and teleutospores on *Onobrychis viciifolia*. England and Wales, scarce.

The rust in this country was first found in Surrey by Hadden (TBMS.
5, 438, 1916) and later on cultivated sainfoin in Kent (Wilson, J. Bot.
Lond. **57**, 161, 1919) and has since been recorded from various parts of
England and Wales. The teleutospores of British collections measure
$21-25 \times 14-18\mu$; they are minutely warted with a minute papilla at the
apex.

Gäumann (Ber. Schweiz. Bot. Ges. **55**, 76, 1945) in Switzerland
showed that this rust is heteroecious by producing uredospores on
Onobrychis montana by infection with aecidiospores from *Euphorbia
cyparissias*. There is no evidence that the aecidial stage has occurred in
this country.

Uromyces trifolii (DC.) Lév.

Ann. Sci. Nat. Bot. ser. 3, **8**, 371 (1847).

Puccinia trifolii DC., Fl. Fr. **2**, 225 (1805).
[*Aecidium trifolii-repentis* Cast., Obs. Pl. Acotyl. **1**, 33 (1842).]
[*Uromyces trifolii-repentis* Liro, Acta Soc. Fauna Fl. Fenn. **29**, no. 6, 15 (1906)];
    Grove, Brit. Rust Fungi, p. 91; Wilson & Bisby, Brit. Ured. no. 315;
    Gäumann, Rostpilze, p. 348.

Spermogonia amphigenous, in small groups, on swollen yellow spots, often on veins and petioles, light yellow changing to purple and then grey, 120μ wide, 110μ high; spermatia $3-4\mu$; paraphyses $65 \times 3\mu$. **Aecidia** amphigenous on blades, petioles and stipules, on swollen yellow spots, in small groups rounded on the leaf, elongate on the petioles and leaf veins, up to 5 mm. long, often surrounding the spermogonia, 200μ wide, 200μ high, peridium yellow, slightly incurved, coarsely toothed, peridial cells overlapping, $10-14 \times 8-12\mu$, outer wall 4μ thick, rather punctate in surface view, inner wall $1\cdot6\mu$ thick, verrucose; aecidiospores globoid to ellipsoid-angular, $21-24 \times 14-20\mu$, wall 1μ thick, verrucose. **Uredosori** amphi-genous and on the stems, scattered, rounded or elongate, partly covered by the ruptured epidermis, pulverulent,

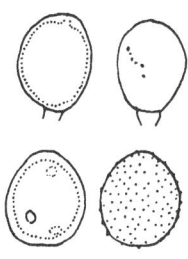

U. trifolii. Teleutospores and uredospores.

cinnamon-brown; uredospores ellipsoid to globoid, $22-24 \times 17-24\mu$, wall $1\cdot5-2\mu$ thick, rather distantly echinulate, pores 2-4, equatorial. **Teleutosori** mostly

W & H

hypophyllous, and on the stems, scattered, circular or elongate, chestnut-brown, covered by the persistent epidermis; teleutospores globoid to ellipsoid, cinnamon-brown, smooth or with a few scattered warts, with a hyaline papilla over the apical pore, $10-28 \times 17-24\mu$,

pedicels colourless, fragile, about 7μ long; basidium 3-4 septate; basidiospores obovoid to globoid, $14 \times 7\mu$. Auteu-form.

On *Trifolium hybridum* and *T. repens*. Great Britain and Ireland, scarce.

As the rusts on *Trifolium hybridum* and *T. repens* are now generally regarded as morphologically similar, following the procedure of Hylander *et al.* (1953, 97) they are both included in *Uromyces trifolii*, although they differ biologically and are almost wholly distinct as to hosts. The earliest cultures with the rusts on *T. repens* were carried out in 1890 by Howell (Bull. Cornell Univ. Agric. Exp. Sta. no. 24, 137; Bot Gaz. **15**, 228, 1890) who, by sowing aecidiospores from *T. repens* on the same host and on *T. pratense*, obtained abundant uredospores on the former and only a slight infection on the latter. In 1906 Liro (Acta Soc. Fauna Fl. Fenn. **29**, 15, 1906) established *U. trifolii-repentis* on the basis of cultures and the morphology of its uredospores.

Aecidia on *T. repens* are rare in Britain but have been found near Perth. The mycelium of the aecidial stage causes long, crooked swellings on the petioles and nerves but not on the leaves. Schroeter (Cohn, Beitr. Biol. Pfl. **3**, 78, 1883) thought that it was perennial in the host but this appears to be doubtful and the statement may have arisen through confusion with *U. nerviphilus* (see p. 328). Dietel (Flora, **81**, 402, 1895) stated that in some localities the aecidiospores can reproduce themselves and gave as evidence the occurrence of aecidia all through the summer. Liro (*loc. cit.*) in 1906 produced aecidia on *T. repens* by infection with aecidiospores from the same host but Davis (Mycol. **16**, 203, 1924) produced only uredospores by infection with aecidiospores. Kobel (Zbl. Bakt. II, **52**, 217, 1921) has denied the occurrence of repeating aecidia and has pointed out that aecidia are always accompanied by spermogonia, a fact which is evidence against repetition. He has explained the occurrence of aecidia throughout the summer by showing that teleutospores formed in summer can germinate immediately and at once bring about infection and the production of aecidia. Teleutospores formed in the autumn last throughout the winter and germinate and cause infection in the following spring. These results have been confirmed by Davis (*loc. cit.*) who has also investigated the optimum temperature for germination of aecidiospores, uredospores and teleutospores.

There has been considerable difference of opinion regarding the status of the rust on *T. hybridum*. It was placed under *U. trifolii* by the Sydows

(Monogr. Ured. **2**, 131) and others, including Grove and Wilson & Bisby and pending further investigation of the rust in Britain that solution is followed here. It has been regarded as a distinct species by Paul, W. H. Davis and Kobel although Paul and Kobel considered it to be morphologically similar to *U. trifolii-repentis*. Arthur (1934, 304) described it as a variety of *U. trifolii*. Kobel showed that it could infect a number of *Trifolium* species including *T. fragiferum* and *T. incarnatum* but not *T. repens* or *T. pratense*. Mains (Pap. Mich. Acad. Sci. **21**, 129, 1936) as the result of cultures on many species found that *T. hybridum* was the only uniformly susceptible host for this rust; it can, however, infect *T. incarnatum*.

As the teleutospores of the rust on *T. hybridum* require a resting period before germination, aecidia appear only in the early summer and are soon followed by abundant uredospore production. Gäumann keeps it as a separate species, *U. trifolii-hybridi*. The aecidial stage has been found by Rostrup (Tidskr. Landöken. 1886, 11) in Denmark and by Fischer (1904) in Switzerland and was first recorded in America by Davis (Proc. Iowa Acad. Sci. **24**, 461, 1917). It has not been found in Britain. Brown (Can. J. Res. **18**, 18, 1940) has produced aecidia by infection with basidiospores and has shown that the rust is heterothallic. Kobel, in Switzerland, has described the occurrence of numerous uredosori on the stems but is doubtful whether the mycelium from which these develop can overwinter; he considered that the persistence of living uredospores throughout the winter is unlikely. The aecidial stage of this species was figured and described in detail by Davis in 1917; he referred to the species as *U. hybridi* but this name was used without proper diagnosis and is consequently not valid.

Uromyces scutellatus (Pers.) Lév.

Ann. Sci. Nat. Bot. ser. 3, **8**, 371 (1847); Gäumann, Rostpilze, p. 315.
Uredo scutellata Pers., Syn. Meth. Fung. p. 220 (1801).

Spermogonia absent in British collection. **Teleutosori** hypophyllous on the deformed host leaves, dark brown, blister-like, surrounded by the white revolute margin of torn host tissue, a few peridial cells occur often mixed with the spores; teleutospores ovoid, 21–29 × 23–27 µ, wall with large longitudinally arranged warts. A few uredospores occur intermixed with the teleutospores, these are ellipsoid, 18–28 × 16–24 µ, wall echinulate with 4 pores. Hemi-form.

Uredospores and teleutospores on *Euphorbia cyparissias*. England, rare.

This species was first definitely recorded in Britain by Butler (TBMS. **41**, 401, 1958) from Tuddenham in West Suffolk. A previous doubtful record by Cooke (Grevillea, **7**, 137, 1878) is unconfirmed.

Butler draws attention to the possibility that this species may not be completely isolated from the heteroecious *Uromyces striatus* which was present in the same host population. The infected shoots are relatively taller, bear shorter, broader and yellower leaves and do not flower. Dietel (Ann. Myc. **34**, 53, 1936) has compared the distribution of size classes of spores in the related species *U. scutellatus*, *U. alpestris*, *U. kalmusii* and *U. striolatus*.

Uromyces tinctoriicola Magn.

Verh. Zool.-Bot. Ges. Wien, **46**, 429 (1896); Gäumann, Rostpilze, p. 318.
Uromyces hybernae Liou, Bull. Soc. Myc. Fr. **45**, 121 (1929).
As *Uromyces sublaevis* Tranz., Wilson & Bisby, Brit. Ured. no. 313.

Spermogonia, aecidia and **uredosori** absent. **Teleutosori** hypophyllous, scattered over the whole leaf, sometimes also epiphyllous, immersed, at first pustular, later surrounded by the ruptured tissue of the leaf, pulverulent, blackish-brown or black; teleutospores chiefly globoid, 22–30 × 20–25 μ, wall 2–3 μ thick, closely and minutely warted, with or without a broad, low, hyaline apical papilla,

pedicel hyaline, fragile, short, deciduous. On *Euphorbia hyberna*. South-west Ireland, quite frequent.

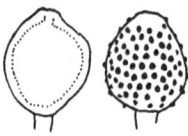

U. tinctoriicola. Teleutospores.

This rust was recorded from the west of Ireland as *Uromyces tuberculatus* by O'Connor (1936, 384), and subsequently by Sandwith from Kerry. Liou (*loc. cit.*) described what is obviously the same fungus on *Euphorbia hyberna* in central and south-west France as a new species, *U. hybernae*. Liou described uredospores as similar to the teleutospore but hyaline; similar spores are present in the Irish collections but we interpret them as imperfect teleutospores. In the Irish collection there is some distortion of the upper leaves on the stems not mentioned by Liou but otherwise there is complete agreement. Dennis & Sandwith (Kew Bull. 1958, 372, 1958) assigned the British collections to *U. tinctoriicola* and very rightly stressed the lack of comparative material for comparison of this group and need for revision. Subsequent collecting in Ireland suggests that the species is quite frequent wherever the host occurs.

Uromyces tinctoriicola has been recorded on a number of species of *Euphorbia* from Spain, including *E. hyberna* (Fragoso, 1925, 53) and occurs throughout central and south Europe and extends eastwards to Persia.

Uromyces sublaevis occurs on almost the same hosts as *U. tinctoriicola* but does not appear to have been recorded on *E. hyberna*; it is described as differing from *U. tinctoriicola* mainly in the more crowded, finer warts of its teleutospores. Savulescu (1953) recognised both species in Roumania but cites them on the same host species and the difference between them appears to be extremely small.

Uromyces tuberculatus Fuck.

Jahrb. Nass. Ver. Nat. 23–24, 64 (1870); Grove, Brit. Rust Fungi, p. 102; Wilson & Bisby, Brit. Ured. no. 316; Gäumann, Rostpilze, p. 323.

Spermogonia hypophyllous, scattered over the whole leaf. Aecidia hypophyllous, scattered amongst the spermogonia, immersed, cup-shaped, with a short denticulate margin; aecidiospores orange, densely verruculose, 17–25 × 14–20μ. Uredosori hypophyllous, scattered, at length naked, cinnamon; uredospores more or less globoid, yellowish-brown, 20–25μ, wall 1·5–2·5μ thick, aculeolate, with 5–7 swollen pores.

Teleutosori amphigenous and on the stems, round, scattered or sometimes arranged in little groups, pulverulent, blackish-brown or black; teleutospores globoid to ellipsoid, chestnut-brown, 20–30 × 18–24 μ, wall 2–2·5μ thick, distantly verrucose with a broad, hyaline, apical papilla, pedicels hyaline, deciduous. Auteu-form.

On *Euphorbia exigua*. England, rare.

This rust has been found in Hampshire (Plowright, 1889, 170), King's Cliff, Northamptonshire (Cooke, Grevillea, 2, 161, 1874), in the Midlands (Purton, Midland Flora, 3, 293, 1821) and recently by Ellis in Norfolk and by Mayfield in Suffolk.

For a long time this species was considered to have only uredo- and teleutospores, but the connection of these with the aecidium occurring on the same species of *Euphorbia* was established by Tranzschel (Ann. Myc. 8, 13, 1910). Berkeley found them together at King's Cliff. The mycelium of the aecidial stage is systemic, but causes little or no deformation; that of the teleutosori is more or less localised.

Uromyces acetosae Schroet.

Rabh. Fungi Europ. no. 2080 (1876); Grove, Brit. Rust Fungi, p. 116; Wilson & Bisby, Brit. Ured. no. 269; Gäumann, Rostpilze, p. 304.

Spermogonia honey-coloured, clustered. Aecidia amphigenous or on the petioles, in dense clusters (up to 1 cm. broad), on reddish purple spots, cup-shaped, whitish-yellow, peridium serrate and revolute, peridial cells on the outer side much overlapping the cell below and much thickened (up to 10μ) and transversely striate in section with finely warted surface, inner wall thinner, 3–4μ thick,

tesselate with coarse warts; aecidiospores nearly smooth or very minutely punctate, pale cream, 18–21 × 12–18μ. Uredosori amphigenous, often seated on red or purple spots, scattered or circinate, minute, pulverulent, cinnamon; uredospores subgloboid to ellipsoid, yellowish or pale brownish-yellow, 18–25 × 17–22μ, wall finely and densely verruculose, about 2·5μ thick, with

3 equatorial pores. **Teleutosori** similar,
but dark brown; spores subgloboid to
ellipsoid, brown, $21-26 \times 20-24 \mu$, wall
rather thick, minutely verrucose, pedicels thin, hyaline, deciduous. Auteu-
form.

On leaves petioles and stems of *Rumex
acetosa* and *R. acetosella*, May–September. Great Britain and Ireland,
frequent.

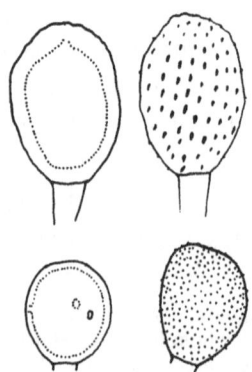

U. acetosae. Teleutospores and uredospores.

This is allied to *U. rumicis* in which the British material was included
by Plowright (1889, 135), but is distinguished by its shorter uredospores
and teleutospores and the lack of hyaline teleutospore papillae. Aecidia
have been found in Scotland (Wilson, 1934, 364), the Isle of Wight
(Grove & Chesters, TBMS. **18**, 269, 1934) and by Ellis in Norfolk (1934,
500). The uredospores and teleutospores are unusually alike, but can be
distinguished by the pores and the fewer warts of the latter; frequently
only uredospores are present.

Puccinia acetosae (see p. 159) occurs on the same host and, in the absence of teleutospores, can be distinguished by the fact that the uredospores have 2 (supra-equatorial) pores and possess few and distant
spines. The aecidium of *P. phragmitis* (see p. 269) is found on *R. acetosa*
but may be distinguished by its white peridium and spores.

Gäumann (Ann. Myc. **29**, 404, 1931) has brought about infection on
Rumex acetosa and on *R. acetosella* with uredospores from *R. arifolius*;
he was unable to find teleutospores on this last species but considered
that the rust on it is undoubtedly *Uromyces acetosae*.

Lindfors (1924, 8) described the cytology of this rust.

Uromyces polygoni-aviculariae (Pers.) Karst.

Bidr. Känned. Finl. Nat. Folk. **4**, 12 (1879); Wilson & Bisby, Brit. Ured. no. 305.

Puccinia polygoni-aviculariae Pers., Syn. Meth. Fung. p. 227 (1801).
Capitularia polygoni Rabh., Bot. Zeit. **9**, 449 (1851).
Uromyces polygoni-avicularis Otth, Mitt. Naturf. Ges. Bern, **1861**, 63 (1861).
Uromyces polygoni (Rabh.) Fuck., Jahrb. Nass. Ver. Nat. **23–24**, 64 (1869); Grove,
   Brit. Rust Fungi, p. 117; Gäumann, Rostpilze, p. 262.

Spermogonia honey-coloured, conical,
only a few together. **Aecidia** hypophyllous, on yellow or violet spots, irregularly aggregated or in circular groups,
cup-shaped, whitish, with a cleft and
revolute peridium, peridial cells not in

distinct rows, outer wall transversely striate in section, up to 8μ thick, projecting downwards and overlapping the cell below, inner 1·5–3μ thick, tesselate and closely warted, warts scarcely 1μ apart; aecidiospores verruculose, yellowish, 15–21 × 14–18μ. **Uredosori** amphigenous or on the stems, scattered or in small clusters, small, round, soon naked, pulverulent, cinnamon; uredospores globoid to ellipsoid, pale brown, 18–26 × 17–24μ, wall 1·5–2·5μ thick, densely and minutely verruculose, with 3 or 4 equatorial pores. **Teleutosori** like the uredosori, but larger and more confluent on the stems, compact, dark brown; teleutospores globoid or obovoid, rounded above, chestnut-brown, 22–38 × 14–22μ, wall smooth, up to 6μ thick at the apex, with an apical pore, pedicels coloured, persistent, thick, up to 90μ long. Auteu-form.

On *Polygonum aviculare* and on *Rumex acetosella*, aecidia, May–June; uredo-

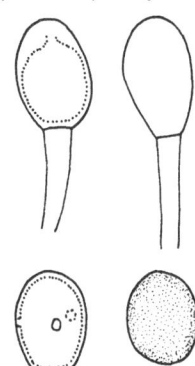

U. polygoni-aviculariae. Teleutospores and uredospores.

spores and teleutospores, July–November. Great Britain and Ireland, common on *Polygonum* but apparently rare on *Rumex*.

The uredospores of *Uromyces polygoni-aviculariae* are similar to those of *U. acetosae* but are more coarsely verrucose.

The aecidium is rare in this country but has been found near Manchester and in Norfolk (Ellis, 1934, 489).

Uromyces rumicis (Schum.) Wint.

Hedwigia, **19**, 37 (1880), Grove, Brit. Rust. Fungi, p. 114; Wilson & Bisby, Brit. Ured. no. 307; Gäumann, Rostpilze, p. 306.

Uredo rumicis Schum., Enum. Pl. Saell. **2**, 231 (1803).
Uredo bifrons DC., Fl. Fr. **2**, 229 (1805).

[**Spermogonia** and **aecidia** as in *Uromyces dactylidis* (p. 361).] **Uredosori** amphigenous, on coloured spots, round, minute, scattered, soon naked, pulverulent, cinnamon; uredospores subgloboid to ellipsoid, pale brown, 20–28 × 18–24μ, wall sparsely echinulate, with 3 equatorial pores. **Teleutosori** similar, but darker; teleutospores subgloboid to ovoid, often narrowed below, brown, 24–35 × 18–24μ, wall rather thick, smooth or nearly so with a hemispherical hyaline papilla, pedicels thin, hyaline, deciduous. Hetereu-form.

Spermogonia and aecidia on *Ranunculus ficaria* not confirmed in Britain.

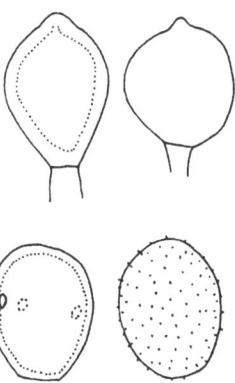

U. rumicis. Teleutospores and uredospores.

Uredospores and teleutospores on *folius*, *R. patientia* and *R. sanguineus*,
Rumex conglomeratus, *R. crispus*, *R.* May–September. Great Britain and
hydrolapathum, *R. maritimus*, *R. obtusi-* Ireland, common.

The spots on the leaves of *Rumex* are small, round, and of various colours, frequently bright red, and the chlorenchyma in the immediate neighbourhood often retains its green colour long after the rest of the leaf has become faded and yellow.

The uredospores and teleutospores of *Uromyces rumicis* are exactly like those of *U. ficariae*, and for this reason Tranzschel (Trav. Mus. Bot. Acad. Imp. Sci. St Petersburg, **2**, 71, 1905) was led to suspect some connection between the two, such as he demonstrated between *Tranzschelia fusca* and *T. pruni-spinosae*, whose teleutospores are equally alike. He reported that he had produced an aecidium on *Ranunculus ficaria* from the spores of *U. rumicis*. Later, he repeated this statement (Trav. Mus. Bot. Acad. Imp. Sci. St Petersburg, **7**, 16, 1910) and added that he had infected *Rumex obtusifolius* with aecidiospores from *R. ficaria*. Other experimenters (Bubak, Zbl. Bakt. II, **16**, 158, 1906; Krieg, Zbl. Bakt. II, **19**, 700, 1907) were unable to repeat the former of these infections; they could produce only the aecidium on *R. ficaria* with the spores of *Uromyces dactylidis*. Grove was unable to infect *Rumex obtusifolius* with aecidiospores from *Ranunculus ficaria*, brought from a place where the aecidium on it and the *Uromyces* on *Rumex obtusifolius* were both very abundant. Gäumann (Ann. Myc. **29**, 399, 1931) in Switzerland has confirmed Tranzschel's discovery; by placing leaves of *R. aquaticus* and *R. aquaticus* × *hydrolapathum* bearing overwintered teleutospores in contact with young plants of *Ranunculus ficaria* he obtained abundant spermogonia and aecidia; *R. bulbosus* was not infected. Gäumann has also carried out infection experiments on various species of *Rumex* with the aecidiospores produced on *R. ficaria* and has concluded that some species including *R. aquaticus*, *R. conglomeratus* and *R. sanguineus* are highly susceptible, others including *R. crispus*, *R. hydrolopathum*, *R. maritimus*, *R. obtusifolius* and *R. pulcher* are slightly susceptible and *R. acetosa*, *R. acetosella* and others are not attacked. As a result of his own experiments and those of Tranzschel, Gäumann concludes that there are two races of *U. rumicis*; f. sp. *aquatici* and f. sp. *obtusifolii*.

Aecidia of *U. rumicis* have not been definitely recorded in Britain. The aecidia of *U. dactylidis* also occur on *R. ficaria* but according to Juel (Sv. Bot. Tidskr. **2**, 169, 1908) may be distinguished from these of *U. rumicis* by their smaller size and lesser number in each aecidial group.

The development of the uredospores and teleutospores is shortly dealt with by Sappin-Trouffy (1896).

Uromyces armeriae Kickx

Fl. Crypt. Flandres, 2, 73 (1867); Grove, Brit. Rust Fungi, p. 89; Wilson & Bisby, Brit. Ured. no. 276; Gäumann, Rostpilze, p. 409.

[*Caeoma armeriae* Schlecht., Fl. Berol. 2, 126 (1824).]

[*Uredo armeriae* [Schlecht.] Duby, Bot. Gall. 2, 899 (1830).]

Spermogonia epiphyllous, honey-coloured, about 40 μ diam.; spermatia 1 × 0·5 μ. Aecidia amphigenous, scattered or in small clusters, at first hemispherical, then cup-shaped, with a whitish, incised peridium; aecidiospores densely and minutely verruculose, yellow, 17–28 × 16–22 μ. Uredosori amphigenous, sometimes on purplish spots, rounded or elongate, surrounded or half-covered by the cleft epidermis, pulverulent, cinnamon; uredospores globoid to ovoid, yellowish-brown, 24–32 × 21–28 μ, wall 2·5–3 μ thick, very densely and minutely verruculose, with 2 or 3 pores. Teleutosori similar, dark brown; teleutospores globoid to ovoid, brown, 24–36 × 21–32 μ, wall smooth, up to 7 μ thick at the apex, pedicel hyaline, nearly as long as the spore, seldom persistent. Auteuform.

On leaves and peduncles of native *Armeria maritima* and its montane var. *planifolia* and on cultivated *A. planta-*

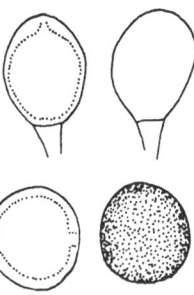

U. armeriae. Teleutospores and uredospores.

ginea, *A. pseudoarmeria* and *A. maritima* var. *elongata*, aecidia in March–June; uredospores from June onwards; teleutospores appear in the uredosori towards the end of July. Common.

This species was united by Plowright with *Uromyces limonii* but is distinguished by the more readily pulverulent sori, the thicker uredospore wall, the shorter and broader teleutospores, and the shorter, hyaline pedicel which is easily detached. Teleutospores are rather scarce but the early production of aecidia suggests that they may germinate and produce infections in the autumn leading to aecidial production in early spring.

This rust has grown on plants in the Royal Botanic Garden, Edinburgh for many years. There are undated specimens of it in the Edinburgh Herbarium collected by Greville in the Garden and labelled *Uredo statices* on *Statice* which appear to be identical with specimens found in 1924. At that time the host was thought to be *Armeria leucocephala* (Wilson, 1934, 359) but has recently been identified (Wilson & Henderson, TBMS. 37, 248, 1954) as *A. plantaginea* var. *plantaginea*.

Spermogonia appear to be rarely produced; they have been found only once on a plant of *A. maritima* var. *planifolia* in cultivation. Aecidia occur abundantly but are overlooked by collectors who do not usually frequent saltmarshes in the early months of the year.

The specimen of the rust on *A. grandiflora* (probably = *A. pseudoarmeria*) was collected by Buchanan White in Perthshire.

Savile & Conners (Mycol. **43**, 186, 1951) have recently divided *U. armeriae* into a number of geographically distinct subspecies; all the British material is referable to their subspecies *U. armeriae* subsp. *armeriae*.

Uromyces limonii (DC.) Lév.

Orbigny, Dict. Univ. Hist. Nat. **12**, 786 (1849); Grove, Brit. Rust. Fungi, p. 88; Wilson & Bisby, Brit. Ured. no. 298; Gäumann, Rostpilze, p. 410.

Puccinia limonii DC., Fl. Fr. **2**, 595 (1805).

Spermogonia honey-coloured in small groups surrounded by the aecidia. **Aecidia** amphigenous, often on red or brownish spots, in roundish clusters or elongated along the nerves, peridium shortly cylindrical, aecidiospores densely and minutely verruculose, yellowish, $21–32 \times 18–26\,\mu$. **Uredosori** amphigenous, scattered, generally roundish or, on the stem, oblong, long-covered by the epidermis, at length naked, pulverulent, cinnamon; uredospores varying from globoid to ellipsoid, densely verruculose, yellowish-brown, $22–32 \times 20–28\,\mu$, wall $1·5–2·5\,\mu$ thick, with 2 or 3 equatorial pores. **Teleutosori** amphigenous or caulicolous, scattered or circinate, roundish or oblong, long-covered by the epidermis, pulvinate, black; teleutospores ellipsoid or clavate, rounded or attenuate at apex, attenuate below, $24–50 \times 14–25\,\mu$, wall smooth, brown, up to $10\,\mu$ thick at apex, pedicels up to $80\,\mu$ long, thick, pale brown, persistent. Auteuform.

On leaves and stems of *Limonium vulgare* and cultivated *L. latifolium* and

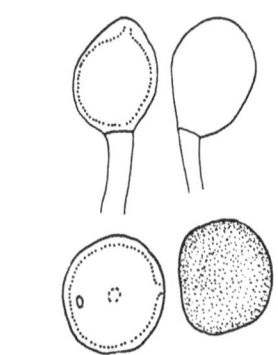

U. limonii. Teleutospores and uredospores.

L. tataricum var. *angustifolium*, aecidia, June–July; teleutospores, July–October. Great Britain, scarce.

The rust has been found on *Limonium latifolium* in Cambridge, Monmouth, Worcestershire, Sussex and Jersey and on *L. tataricum* var. *angustifolium* in Kent (Moore, 1959, 404).

Treboux (Ann. Myc. **12**, 480, 1914) has shown that the teleutospores germinate after the winter's rest. This species was formerly united with *Uromyces armeriae* (see p. 345) but the teleutospores are longer and narrower than in that species.

Uromyces eugentianae Cumm.

Mycol. **48**, 608 (1956).

Uromyces gentianae Arth., Bot. Gaz. **16**, 227 (1891) (non (Str.) Lév., Ann. Sci. Nat. Bot. ser. 3, **8**, 371 (1847)); Wilson & Bisby, Brit. Ured. no. 290 a; Gäumann, Rostpilze, p. 422.

Uredosori amphigenous, without spots, scattered, minute, rounded, about 0·5 mm. diam. surrounded by the ruptured epidermis, pulverulent, pale cinnamon-brown; uredospores globoid, 20–26 × 16–22 μ, wall 1–2 μ thick, echinulate with 3, less commonly 2, equatorial pores. Teleutosori amphigenous, resembling the uredosori, cinnamon-brown; teleutospores globoid, cinnamon-brown, 15–18 × 18–21 μ, wall 1·5 μ thick, finely verrucose, pore with a low, hyaline papilla, pedicel hyaline, short, fragile and deciduous. Hemi-form?

On *Gentianella amarella* agg. England, rare.

This rust has been confused with *Puccinia gentianae* but differs in its smaller, more finely echinulate uredospores with 2 or 3 equatorial pores; the teleutospores are quite different, being 1-celled, globoid and verrucose; the teleutosori are brown, not blackish as in *P. gentianae*. *Uromyces gentianae* (Str.) Lév. refers to *P. gentianae* and pre-dates *U. gentianae* Arth. consequently Cummins had to propose a new name for this *Uromyces* on gentians. In Europe it has been found only on the *Gentianella* sect. *Amarella*.

Specimens collected at Bredon Hill in Worcestershire by Rea in 1928 and identified by Wakefield as *Puccinia gentianae* (TBMS. **14**, 185, 1929) were found by Jørstad (Nytt Mag. Bot. **3**, 103, 1954) to be the *Uromyces*, as were those collected by Ellis at East Cosham, Hampshire. Specimens collected by Clark in Norfolk on plants believed to be *Gentianella campestris* and identified by Plowright (Trans. Norf. Norw. Nat. Hist. Soc. **8**, 680, 1909) as *P. gentianae* are also presumably this species.

Uromyces scrophulariae Fuck.

Fungi Rhen. no. 395 (1863); Jahrb. Nass. Ver. Nat. **23–24**, 63 (1869); Grove, Brit. Rust Fungi, p. 87; Wilson & Bisby, Brit. Ured. no. 310; Gäumann, Rostpilze, p. 419.

[*Aecidium scrophulariae* DC., Fl. Fr. **6**, 91 (1815).]

Uromyces concomitans Berk. & Br., Gard. Chron. N.S. **2**, 238 (1874).

Spermogonia few, single or in little groups, developing simultaneously with the aecidia. **Aecidia** hypophyllous or on the stems, on yellowish spots, in rounded clusters, cup-shaped, yellowish; peridium involute, entire; aecidiospores verruculose, smooth below, yellowish, 18–21 × 14–18 μ. [**Uredosori** similar to

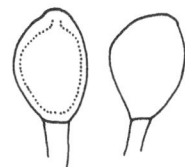

U. scrophulariae. Teleutospores.

aecidia but effuse, uredospores similar to aecidiospores.] **Teleutosori** small and roundish, arranged like the aecidia except that they form more elongated groups (as much as 10 cm. long) on the stems, long-covered by the lead-coloured epidermis, at length naked and pulverulent, dark brown; teleutospores very irregular, obovoid, fusiform, or ellipsoid, angular, rarely subgloboid, 18–35 × 11–18 μ, apex rounded, truncate or slightly pointed, wall smooth, up to 6 μ thick at apex, brown, pedicel persistent, hyaline or yellowish, nearly as long as the spore. Auteu- or autopsis-form.

On leaves, petioles and stems (especially near the soil) of *Scrophularia aquatica, S. nodosa, S. scorodonia* and *S. umbrosa*, July–September. England, Scotland, Ireland, Channel Islands, scarce.

This species appears to be commoner in the western parts of the country. The specimens on *Scrophularia scorodonia* were collected in the Channel Islands by Cooke (Herb. Brit. Mus.); those on *S. umbrosa* by E. A. Ellis in Norfolk in 1959.

The spots on the leaves are pallid, edged with violet-brown. The teleutosori, especially, cause considerable distortion of the leaves and stems. The aecidia and teleutosori can occur simultaneously and intermixed, or the latter surrounding the former. Dietel (Flora, **81**, 394, 1895) showed that the aecidiospores on infection can give rise to aecidioid uredosori which are not accompanied by spermogonia. This stage has not been recorded in Britain.

The development and cytology of the spermogonia and aecidia have been described by Kursanov and discussed by Jackson (Mem. Torrey Bot. Club, **18**, 67, 1931).

Uromyces valerianae (DC.) Lév.

Ann. Sci. Nat. Bot. ser. 3, **8**, 371 (1847); Grove, Brit. Rust Fungi, p. 86; Wilson & Bisby, Brit. Ured. no. 318; Gäumann, Rostpilze, p. 423.
[*Uredo valerianae* Schum., Enum. Pl. Saell. **2**, 233 (1803).]
Uredo valerianae DC., Fl. Fr. **6**, 68 (1815).

Spermogonia epiphyllous, in small clusters, honey-coloured, turning black. **Aecidia** hypophyllous, and often on the nerves, petioles and stalks, seated on pale, thickened spots, densely aggregated or circinate, cup-shaped, whitish-yellow, peridium revolute and torn; aecidiospores covered with minute crowded warts, yellow, 18–25 × 16–20 μ. **Uredosori** amphigenous, usually on indefinite yellow spots, scattered or irregularly grouped, minute, punctiform, pulverulent, brown; uredospores globoid to broadly ellipsoid, yellowish-brown or brown, 21–28 μ diam.; wall 3–5 μ thick, verrucose-echinulate, with

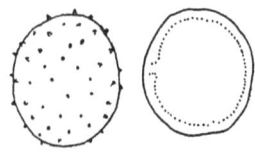

U. valerianae. Uredospores.

2 or 3 pores. **Teleutosori** similar, but longer covered by the epidermis, dark brown; teleutospores ellipsoid or ovoid,

pale brown, 20–30 × 13–21 μ, wall thin, smooth, with an apical pore often with a flat, pale papilla, pedicels short, thin, hyaline, rather deciduous. Auteu-form.

On *Valeriana dioica* and *V. officinalis*, aecidia in May and June; uredospores from June; teleutospores, July–October. Great Britain and Ireland, common.

The uredospores seem to be variable in their markings; some are distinctly verrucose with pointed warts; others are distinctly echinulate. Teleutospores are very uncommon.

Uromyces aloes (Cooke) Magn.

Ber. D. Bot. Ges. **10**, 48 (1892); Wilson & Bisby, Brit. Ured. no. 271.
Uredo aloes Cooke, Grevillea, **20**, 16 (September 1891).
Uromyces aloicola Henn., Engl. Bot. Jb. **14**, 370 (December 1891).

Spermogonia on yellowish spots, sub-epidermal, 150–170 μ diam., orange. **Aecidia** and **uredosori** wanting. **Teleutosori** amphigenous, concentrically arranged in rounded groups about 3 cm. diam., single sori rounded or oblong about 2 mm. long, soon becoming confluent, long-covered by the epidermis, dark brown; teleutospores globoid or subgloboid, ovoid or broadly ellipsoid, often irregular, apex rounded, brown, 28–55 × 20–35 μ, wall verruculose, 4–7 μ thick, pore apical, pedicel hyaline, usually

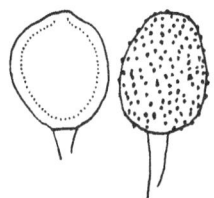

U. aloes. Teleutospores.

persistent, up to 140 μ long. Micro-form.
On *Aloe glauca*, Royal Botanic Garden, Edinburgh, introduced.

The spermogonia are often underdeveloped and abortive.

The rust was found on recently imported plants and, after the removal of the infected leaves, did not appear again.

Thirumalachar (Bot. Gaz. **108**, 245, 1946) has investigated the cytology of the rust; the haploid chromosome number, as determined during the meiotic division of the fusion nucleus in the basidium is six.

Uromyces ambiguus (DC.) Lév.

Ann. Sci. Nat. Bot. ser. 3, **8**, 371 (1847); Grove, Brit. Rust Fungi, p. 121; Wilson & Bisby, Brit. Ured. no. 272; Gäumann, Rostpilze, p. 221.
Uredo ambiguus DC., Fl. Fr. **6**, 64 (1815).

Uredosori amphigenous, long-covered by the conspicuously yellow epidermis; uredospores globoid to ovoid, 20–28 × 17–22 μ, wall 2–3 μ thick with 6 or 7 scattered, indistinct pores. **Teleutosori** amphigenous and on the stems, scattered on the leaves and then small, on the stems confluent and larger, up to 15 mm. long, usually associated with uredosori and covered by a persistent blackish epidermis, teleutospore mass pulverulent; teleutospores subgloboid often rather angular in outline, 20–35 × 17–24 μ, wall smooth, brown, 1–1·5 μ thick

with an indistinct apical pore, pedicel
hyaline, fragile, up to 20μ long. Hemi-
form?
Uredospores and teleutospores on
Allium schoenoprasum, A. scorodoprasum
and *A. babingtonii.* Scarce throughout
Britain.

The relation of this species to
Puccinia allii and what was considered a separate species, *P. porri,* is
considered in the account of the
former (p. 217). It has been listed
as a *Puccinia* by Hylander *et al.*
(1953, 35). While the eventual re-
grouping of the species of *Puccinia*

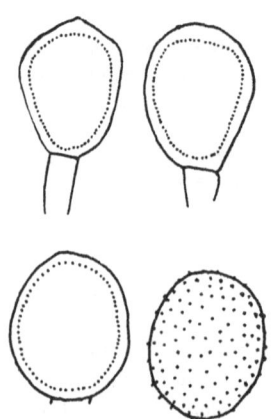

U. ambiguus. Teleutospores and uredospores
(after Savulescu).

and *Uromyces* is desirable according to more natural criteria than the
number of cells in the teleutospores, species as the present one with 100 %
unicellular teleutospores must be kept in *Uromyces* if the genus is recognised at all.

This species is much more restricted in host range than *P. allii* (including *P. porri*) according to von Tavel's experiments with Swiss material
(Ber. Schweiz. Bot. Ges. **41**, 123, 1932), in which he infected only *Allium
sphaerocephalum, fistulosum* and *flavum.* No aecidial stage has been
recorded for *Uromyces ambiguus.*

The record on the endemic *A. babingtonii* is from Ireland (O'Connor,
Sci. Proc. Roy. Dubl. Soc. **25**, 36, 1949).

Uromyces erythronii Pass.

Comm. Soc. Crittog. Ital. **2**, 452 (1867); Wilson & Bisby, Brit. Ured. no. 286;
Gäumann, Rostpilze, p. 286.

Uredo erythronii DC., Fl. Fr. **6**, 67 (1815).
Uromyces erythronii [DC.] Lév., Ann. Sci. Nat. Bot. ser. 3, **8**, 371 (1847) (*nomen
nudum*).

Spermogonia amphigenous, honey-coloured. **Aecidia** mostly hypophyllous or
on the petioles, minute, on indistinct
yellow spots, in rounded or elongate
groups, peridium small, 0·3–0·4 mm.
diam., with incurved margin, at first
covered by the epidermis, opening by a
slightly elongated pore, at length cup-
shaped with coarsely incised, outwardly
turned margin; aecidiospores polyhe-

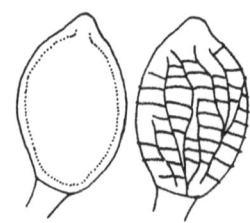

U. erythronii. Teleutospores (after Savulescu).

droid-globoid or ellipsoid, contents orange, $20-30 \times 15-24\,\mu$, wall colourless, densely and minutely verruculose. Teleutosori amphigenous, grouped or spread over the whole leaf, minute, rounded, $0\cdot3-1$ mm. diam. pulverulent, chocolate-brown; teleutospores globoid or ovoid, $22-42 \times 16-25\,\mu$, with a promi-nent apical papilla, wall about $1\cdot5\,\mu$ thick, brown, with longitudinally elongate warts or ridges which are transversely connected, pedicel short hyaline. Opsis-form.

On cultivated *Erythronium dens-canis* in April–May. Rare, introduced.

This rust was introduced into Dorset in 1936 with plants from the French Pyrenees (Moore, 1959, 402).

Uromyces erythronii is closely related with *U. lilii* but is distinguished by the transverse connections between the lines of warts on the teleutospore.

The shape of the peridium was correctly described by Fischer (1904, 6) but has been inaccurately represented by other observers. The outer wall of the peridial cells is thick, the inner thin with small, well-developed warts.

Uromyces gageae Beck

Verh. Zool.-Bot. Ges. Wien, **30**, 26 (1880); Grove, Brit. Rust Fungi, p. 120; Wilson & Bisby, Brit. Ured. no. 290; Gäumann, Rostpilze, p. 289.

Teleutosori amphigenous, scattered, roundish or elliptical, 1–3 mm. long, covered by the lead-coloured epidermis which at length splits longitudinally, then naked, pulverulent, dark brown; teleutospores subgloboid to obovoid, $26-40 \times 18-28\,\mu$, wall smooth, brown, $2-3\,\mu$ thick but usually with a hyaline apiculus, pedicel hyaline, shorter than the spores. Micro-form.

On leaves of *Gagea lutea*, April–May. England, rare.

This rust has been recorded from Yorkshire (Mason & Grainger, 1937) and has been collected again recently by W. G. Bramley (*in litt.*). According to Fischer (1904, 4) the teleutospores germinate in the spring. Plowright (1889, 142, under *Uromyces ornithogali*) described variously shaped pale spots on the affected leaves. Kursanov (1922, 55) was unable to discover uninucleate mycelium in this rust and therefore concluded that the binucleate condition arises very early.

Uromyces holwayi Lagerh.

Hedwigia, **28**, 108 (1889); Wilson & Bisby, Brit. Ured. no. 292.

Uromyces lilii Clint., Ann. Rep. N.Y. State Mus. **27**, 103 (1875) (non *U. lilii* (Link)) Kunze, Rabh. Fungi. Europ. no. 1693 (1873).

[Spermogonia hypophyllous, in groups. Aecidia chiefly hypophyllous, cupulate; aecidiospores globoid, finely verrucose, $16-22 \times 14-19\,\mu$, wall colourless, $1\,\mu$ thick.] Uredosori amphigenous, on pale spots, scattered or in circinate groups about 0·5 mm. diam., pulverulent, cinnamon-brown; uredospores globoid or

obovoid, yellow, 23–30 × 20–26 μ, wall
2·5–3 μ thick, minutely and distantly
echinulate, with 2–3 equatorial pores.
[Teleutosori amphigenous, similar to the
uredosori, chocolate-brown; teleutospores ovoid or ellipsoid, moderately
striately rugose with a hyaline papilla
over the pore, chestnut-brown, 29–39 ×
18–25 μ, wall 2–3·5 μ thick, pedicel
colourless, short, fragile.] Auteu-form.
On *Lilium columbianum*, uredospores
only. Rare, introduced.

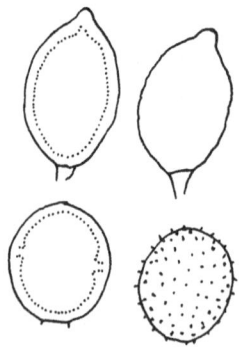

U. holwayi. Teleutospores and uredospores
(after Arthur).

This imported American species has been recorded only by Berkeley
(Gard. Chron. N.S. **11**, 820, 1879), as *Uredo prostii*, from a garden at
Cirencester, Gloucestershire, in 1879. It is distinguished from *U. lilii*
by the presence of uredospores. It is found in north-western America on
Lilium columbianum and other wild and cultivated lilies.

Uromyces aecidiiformis (Str.) Rees

Amer. J. Bot. **4**, 369 (1917).

Uredo aecidiiformis Str., Ann. Wetter. Ges. **2**, 94 (1811).
Caeoma lilii Link, Sp. Pl. Ed. 6, **6**, 8 (1825).
Uromyces lilii (Link) Fuck., Jahrb. Nass. Ver. Nat. **29–30**, Nacht. 3, 16 (1875).
As *Uromyces lilii* (Link) Kunze, Wilson & Bisby, Brit. Ured. no. 297.

Spermogonia punctiform, at first honey-coloured then brownish-black, scattered in centre of aecidial groups. **Aecidia** amphigenous, usually hypophyllous, or on the stems and petioles where the groups may be up to 1 cm. long, hemispherical, at length opening by a central pore; aecidiospores angular-globoid to ovoid, 22–35 × 18–26 μ, wall 3–3·5 μ thick, densely and minutely verrucose, yellowish. **Uredosori** wanting. **Teleutosori** amphigenous, usually hypophyllous on yellowish spots up to 1 cm. wide, rarely elongate on the petioles, pulverulent dark brown; teleutospores irregular in form, globoid, ellipsoid or ovoid, rounded above with a prominent hyaline papilla, brown, 28–44 × 22–30 μ, wall 2–3·5 μ thick with short, anastomosing ridges, pedicels hyaline, 5–8 μ thick at base, short (usually up to 10 μ), persistent. Opsis-form.

On *Lilium candidum*, aecidia April–May; mature teleutospores from June. Rare, introduced.

This introduced species has been recorded from Kew Gardens,
Birmingham (Grove, J. Bot. Lond. **49**, 368, 1911), Middlesex in 1912
and near Norwich (TBMS. **9**, 1, 1923). The specimens from Kew were
mistakenly assigned to *Uromyces erythronii* by Massee (Kew Bull.
1897, 151, 1897).

The Sydows (Monogr. Ured. **2**, 227) and Gäumann (*loc. cit.*) recorded this rust on various species of *Lilium* and *Fritillaria*. Mayor (Bull. Soc. Neuchâtel. Sci. Nat. **42**, 69, 1918) has reported an instance in which it spread from *Lilium candidum* and *L. bulbiferum* on to *L. martagon*. Schneider (Zbl. Bakt. ii, **72**, 260, 1927) showed the connection of the two spore forms and infected *L. martagon* with teleutospores from *Fritillaria meleagris*; he suggested that specialisation is not well developed in this species.

The development of the spermogonia and aecidium on *F. meleagris* has been described by Sappin-Trouffy (1897, 67, under the name *U. erythronii*).

Olive (Ann. Bot. Lond. **22**, 331, 1908) has described the nuclear division in the spermogonium.

Uromyces muscari (Duby) Lév.

Ann. Sci. Nat. Bot. ser. 3, **8**, 371 (1847).
Uredo muscari Duby, Bot. Gall. **2**, 898 (1830).
Uredo scillarum Berk., Smith, Engl. Fl. **5** (2), 376 (1836).
Uromyces concentrica Lév., Ann. Sci. Nat. Bot. ser. 3, **8**, 371 (1847).
Uromyces scillarum (Berk.) Lév., Ann. Sci. Nat. Bot. ser. 3, **8**, 376 (1847); Grove,
    Brit. Rust Fungi, p. 120; Wilson & Bisby, Brit. Ured. no. 309; Gäumann,
    Rostpilze, p. 293.

Spermogonia, aecidia, uredosori unknown. **Teleutosori** amphigenous usually seated on pallid or yellowish spots, small, round or oblong, up to 0·5 mm. diam., in round or oblong clusters often concentrically arranged, sometimes confluent, long-covered by the epidermis which at length splits and surrounds them, pulverulent, dark brown; teleutospores subgloboid to broadly ellipsoid, usually rounded and not thickened above, evenly coloured, brown, 18–32 × 14–22μ, wall uniformly about 1·5μ thick, smooth, occasionally marked with a few, very faint, partly anastomosing ridges, pedicel hyaline, often deciduous, as long as or longer than the spore. Micro-form.

On leaves of *Endymion hispanicus, E. non-scriptus, Muscari polyanthum, Scilla bifolia* and *S. verna*, April–June. Great Britain and Ireland, common on *E. non-scriptus*, less so on other hosts.

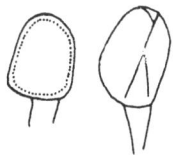

U. muscari. Teleutospores.

The yellow spots and the concentric arrangement of the sori are often very marked. The mycelium is purely local. A few uredospores were found in the young sori on *Scilla obtusifolia* by Juel (Bull. Soc. Myc. Fr. **17**, 259, 1901) and have also been recorded by Schneider (Zbl. Bakt. ii, **72**, 246, 1927) in the sori on *S. bifolia* and other species. Juel describes them as echinulate 27 × 20μ; Schneider as 18–24μ diam., with colourless wall $1\cdot5-2\mu$ thick, warted, with warts $1-2\cdot5\mu$ apart, with 8 pores each covered with a moderate-sized papilla. Schneider has pointed out that there is no pore in the teleutospore and that the germ-tube emerges from a slit in the side of the spore. He has also proved that the teleutospores produced in spring can germinate at once, in the following autumn or in the following spring; infection can take place in the autumn with the production of teleutospores and the latter can germinate at once or in the following spring.

As the result of infection experiments Schneider has shown that the rust on *Muscari racemosum* will not infect other species of *Muscari* or species of *Scilla*; he therefore regards it as a special form. The rust has been recorded on the cultivated hyacinth but not in this country. The record on *M. polyanthum* is from Edinburgh.

Blackman & Fraser (Ann. Bot. Lond. **20**, 35, 1906) found that the general mycelium consists of binucleate cells and concluded that the conjugate condition must arise very early after infection; the absence of spermogonia and the occasional occurrence of uredospores is evidently correlated with this condition. Moreau (Le Botaniste, **13**, 194, 1914) has come to a similar conclusion.

Gräflinger (Ann. Myc. **28**, 321, 1930) has studied the variation in spore size on different hosts.

Uromyces colchici Mass.

Grevillea, **21**, 6 (1892); Grove, Brit. Rust Fungi, p. 122; Wilson & Bisby, Brit. Ured. no. 280; Gäumann, Rostpilze, p. 285.

Spermogonia unknown. **Aecidia** and **uredosori** wanting. **Teleutosori** amphigenous, scattered, rather large, elliptical, sometimes circinate, up to 2 mm. long, covered for some time by the epidermis which at length splits, then subpulverulent, brown; teleutospores subgloboid to ovoid, rounded above, with a broad flat hyaline papilla, smooth, pale brown, $28-40 \times 20-28\mu$, wall $3-3\cdot5\mu$ thick, pedicel hyaline, rather long, but very deciduous. Micro-form.

On leaves of *Colchicum autumnale, C. speciosum, 'C. bavaricum'* (= *C. variegatum*?). England, very rare, introduced.

This rust has been reported from Kew Gardens (Massee, *loc. cit.*) and from Yorkshire (Mason & Grainger, 1937, 45).

Material in Kew Herbarium consists of three fragments of a leaf (or leaves), one of which bears one pustule of *Uromyces*. The spores are more or less as Massee described them. One would suspect some error except that the fungus is stated to have attacked the foliage of *Colchicum speciosum* for three successive seasons, completely destroying it, and although, for the first two seasons it did not attack other species of

Colchicum growing near, during the third season it spread to the other species.

Boerema (Tidskr. Pl.-Ziek. **67**, 1, 1961) has reported considerable subterranean infection of certain varieties of *Colchicum* in Holland. The sori there form on the dry tissues of the bulbs. Leaf infection has not been found in Holland. The rust is not known other than in England and Holland.

Uromyces junci (Desm.) Tul.

Ann. Sci. Nat. Bot. ser. 4, **2**, 146 (1854); Grove, Brit. Rust Fungi, p. 123; Wilson & Bisby, Brit. Ured. no. 294; Gäumann, Rostpilze, p. 253.

Puccinia junci Desm., Pl. Crypt. Fr. no. 81 (1825).
[*Aecidium zonale* Duby, Bot. Gall. **2**, 906 (1830).]

Spermogonia usually epiphyllous. **Aecidia** hypophyllous, in dense circinate clusters, on spots which are zoned with yellow and purple, 2–5 mm. wide, cup-shaped, yellowish-white, with a torn revolute margin; aecidiospores densely and minutely verruculose, yellowish, 17–21 μ diam. **Uredosori** scattered, roundish or oblong, up to 1 mm. long, surrounded by the cleft epidermis, pulverulent, brown; uredospores globoid to ellipsoid, yellowish-brown, 20–28 × 16–22 μ, wall faintly echinulate, with 2 equatorial pores. **Teleutosori** amphigenous or on the culms, scattered or occasionally aggregated, similar to the uredosori, but darker; teleutospores ovoid to clavoid, rounded or conical above, wall smooth, up to 14 μ thick at apex, dark brown, 24–42 × 12–18 μ, pedicel thick, persistent,

brownish, up to 60 μ long. Hetereu-form- Aecidia on *Pulicaria dysenterica*, May–July; uredospores and teleuto.

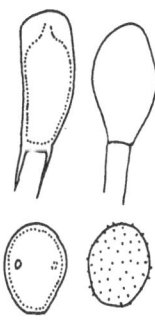

U. junci. Teleutospores and uredospores.

spores on *Juncus effusus*, *J. inflexus* and *J. subnodulosus* from July onwards. England, scarce.

This rust appears to be almost confined to Yorkshire, Cambridgeshire, Norfolk and Suffolk but the aecidial stage had been found at Reading.

The connection of the two forms was proved by Plowright (1889, 132) and has been confirmed by Dupias (Bull. Soc. Hist. Nat. Toul. **80**, 36, 1945) in France. On the continent and in North America it has been found on other species of *Juncus* including most of the species in the British flora; in North America the aecidia have been recorded on a number of genera and species of the *Compositae*. Dietel (Hedwigia, **28**, 23, 1889) has pointed out that near Leipzig this species can overwinter on *J. conglomeratus* by means of uredospores.

Uromyces croci Pass.

Rabh. Fungi. Europ. no. 2078 (1876); Gäumann, Rostpilze, p. 299.

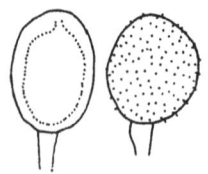

U. croci. Teleutospores.

Spermogonia, aecidia and **uredosori** wanting. **Teleutosori** amphigenous, scattered in rows on each side of the midrib, or in groups, oblong, elongate, 2–5 mm. long, 1 mm. wide, sometimes confluent and longer, covered for a long time by the lead-coloured epidermis, pulverulent, dark brown; teleutospores globoid to broadly ellipsoid, apex rounded, brown, 24–32 × 21–28 μ, wall about 5 μ thick, very delicately verruculose, pore apical; pedicel short, hyaline, up to 10 μ long. Micro-form.

On cultivated *Crocus vernus.* Rare, introduced.

Corms of the crocus were introduced into a garden in Scotland from Como, Italy, in 1955 and the rust developed in abundance on the leaves in 1956. The infected plants have not flowered. Boerema (Versl. Meded. Plantenz. Dienst, Wageningen, **133**, 130, 1959) states that subterranean infection on the fleshy parts of the corms is quite frequent in Holland.

Uromyces ari-triphylli (Schw.) Seeler

Rhodora, **44**, 174 (1942); Wilson & Bisby, Brit. Ured. no. 275.
Puccinia ari-triphylli Schw., Trans. Amer. Phil. Soc. **4**, 297 (1832).
Uromyces ari-virginici Howe, Bull. Torr. Bot. Club **5**, 43 (1874).
[*Aecidium importatum* Henn., Verh. Bot. Ver. Brand. **37**, xxv and 12 (1895).]
[*Aecidium 'dracontii'* Schw.], Grove, Brit. Rust Fungi, p. 387 (1913).

Spermogonia on a systemic mycelium, chiefly hypophyllous, scattered, honey-coloured, becoming brownish; spermatia colourless, ovoid, 4–5 × 3–4 μ. **Aecidia** on a systemic mycelium, chiefly hypophyllous, scattered, reddish, hemispherical before dehiscence, at length cup-shaped with white, recurved peridium; aecidiospores subgloboid or ellipsoid, pale yellow, finely verrucose, 25–30 × 20–27 μ (17–24 × 15–20 μ, Arthur), wall 1·5 μ thick. **Uredosori** and **teleutosori** not found in Europe. Auteuform.

Spermogonia and aecidia on leaves of *Arisaema triphyllum* and on leaves and petioles of *Peltandra virginica.* Rare, introduced.

Aecidium importatum which was regarded by the Sydows (Monogr. Ured. **4**, 358) and Arthur (1934, 214) as the aecidial stage of this species was recorded by Hennings (*loc. cit.*) from the Berlin Botanic Garden in 1894 on plants recently imported from North America. In 1924 it was identified in the Royal Botanic Garden, Edinburgh, where it was said to have been present for some years previously (Wilson & Waldie, TBMS. **12**, 114, 1927). At Edinburgh the spermogonia appear in June on the petioles and on the underside of the midrib of the leaf; the aecidia appear in July. There is abundant colourless mycelium in the midrib and petiole. The mycelium persists in the root-stock and produces aecidia each year in the early summer, the leaves formed later are not attacked. Rapid decay of infected tissue prevents complete maturation of aecidiospores which are not naturally shed.

There is some doubt regarding the identity of the rust with *Uromyces ari-triphylli* as the aecidiospores are larger. The measurements of the aecidiospores given by Hennings (*loc. cit.* p. xxv) in the first note were similar to those found in Edinburgh; in a later note (*loc. cit.* p. 12) he described them as 17–23 × 16–19 μ.

The aecidial stage, *A. dracontionatum* ('*A. dracontii*' in error by Cooke, 1871, 538, and continued by Plowright and Grove) has been found twice in this country on *Arisaema triphyllum*, at Melbury in Dorset by Berkeley in 1863 and at Kew Gardens in 1945, at both places on plants imported from North America.

The aecidia with the spermogonia scattered amongst them are on pallid spots on the leaves, sometimes almost covering them and are limited by the veins; they are arranged without order and are large and elongate; the aecidiospores are orange. Only aecidia and spermogonia develop on plants imported into Europe.

Several investigations have been carried out on this rust on *Arisaema triphyllum* in North America. Dufrenoy (Ann. Inst. Pasteur, **43**, 218, 1929) and Rice (Bull. Torr. Bot. Club, **54**, 63, 1927) have made a detailed examination of the haustoria of the haploid mycelium. Pady (Mycol. **31**, 590, 1939) has shown that the corms are systemically infected with the haploid mycelium; when growth begins in the spring the leaves and flowers become infected. Internal spermogonia are found in various parts of the flowers; the mycelium invades the ovule and young embryo and the disease is possibly transmitted by seed.

Uromyces lineolatus (Desm.) Schroet.

Rabh. Fungi Europ. no. 2077 (1876); Wilson & Bisby, Brit. Ured. no. 299; Gäumann, Rostpilze, p. 222.
[*Uredo scirpi* Cast., Cat. Pl. Marseill. p. 214 (1845).]
Puccinia lineolata Desm., Ann. Sci. Nat. Bot. ser. 3, **11**, 273 (1849).
Uromyces scirpi Burrill, Bot. Gaz. **9**, 188 (1884). Grove, Brit. Rust. Fungi, p. 124.
Uromyces maritimae Plowr., Gard. Chron. III, **7**, 746 (1890).

Spermogonia usually epiphyllous, with projecting paraphyses. **Aecidia** hypophyllous or on the petioles, in rather small clusters, on yellowish spots, cup-shaped, with an incised revolute margin; aecidiospores densely and minutely verruculose, yellowish, 16–24 × 14–20 μ (20–28 × 16–26 μ, on *Oenanthe crocata*).

Uredosori hypophyllous, scattered or arranged in lines, rounded or oblong, up to 1 mm. long, surrounded by the cleft epidermis, pulverulent, cinnamon; uredospores globoid to ovoid, yellowish-brown, 22–35 × 16–25 μ, wall 1·5–2 μ thick, distantly and minutely echinulate, with 3 equatorial pores. Teleutosori amphigenous, on indefinite discoloured spots, scattered or confluent in lines, long-covered by the epidermis, brownish-black; teleutospores clavoid and thickened apically (up to 12 μ), 26–45 × 15–24 μ, wall smooth, pale brown, pedicels brownish, persistent, as long as, or longer than the spore. Hetereu-form.

Aecidia on leaves and petioles of *Berula erecta, Glaux maritima, Oenanthe crocata, O. fistulosa, O. lachenalii*, May–June; uredospores and teleutospores on

Scirpus maritimus. June–September. Great Britain, scarce.

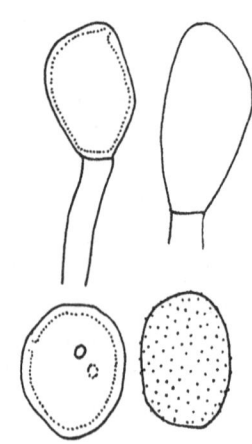

U. lineolatus. Teleutospores and uredospores.

This species has been recorded from various places on the east coast of England and Scotland and from Cornwall. Plowright (*loc. cit.*) with material from near Hull, first proved the connection of the teleutospores on *Scirpus maritimus* with the aecidium on *Glaux maritima*. In 1933 Ellis (1934, 489) found the rust on *S. maritimus* and the aecidia on *G. maritima* in Norfolk and they were recorded later by Mayfield (1935) in Suffolk. Hurst recorded teleutosori on *S. maritimus* in 1931 near Saltash, Cornwall, and in May 1933 found abundant aecidia on *Oenanthe crocata* near the infected *Scirpus* (Grove & Chesters, TBMS. **18**, 265, 1933). Rilstone (J. Bot. Lond. **76**, 353, 1938) at Looe, Cornwall, recorded uredospores on *S. maritimus* and in the following year aecidia on *G. maritima*; although *O. crocata* and *O. lachenalii* were growing close by they were not infected. In 1934 teleutospores were found in East Lothian and in 1959 aecidia on *Glaux* in the same place. Macdonald (Trans. Bot. Soc. Edin. **32**, 556, 1938–9) has recorded the rust on *S. maritimus* and on *O. crocata* near St Andrews. Fort (TBMS. **24**, 98, 1940) has shown that teleutospores from *S. maritimus* collected near St Andrews produce aecidia on *O. crocata* but not on *Glaux maritima*; *S. maritimus* was also infected by aecidiospores from *O. crocata*; it appears that the rust can persist throughout the winter by means of uredospores for these were found in a germinating condition in February on *S. maritimus*. Continuing his investigations at St Andrews, Macdonald (TBMS. **41**, 178, 1958), has isolated two races which correspond to f. sp. *glaucis-scirpi* O. Jaap and f. sp. *scirpi-oenanthe-crocata*

R. Maire. The first has its aecidial stage on *G. maritima* and the second on *O. crocata*; neither is capable of infecting the aecidial host of the other. They are not distinguishable from each other by the symptoms which they produce on *S. maritimus*. Uredospores, formed on *S. maritimus* following infection from either aecidial host, can germinate the following spring and thus perpetuate the rust without the intervention of the alternate hosts. Ellis (*in litt.*) has recently discovered the aecidia on *O. lachenalii* in Norfolk and on *Berula erecta* in Suffolk. Observers on the continent have produced aecidia on other host plants with teleutospores from *S. maritimus*, such as *Hippuris vulgaris* and *Sium latifolium* (Dietel, Hedwigia, **29**, 149, 1890), *Pastinaca sativa*, *O. aquatica* (Klebahn, Z. Pfl.-Krankh. **12**, 141, 1902, and **17**, 136, 1907) and *Berula angustifolia* (Bubak, Zbl. Bakt. II, **9**, 927, 1902). Aecidia of this species have also been found on *Daucus carota*.

In North America (Arthur, 1934, 191) records the rust on various species of *Scirpus* and on hosts in several genera of *Umbelliferae*. It is evident that *Uromyces lineolatus* is, in its aecidial stage, a plurivorous species and the experimental work quoted shows evidence of considerable diplont specialisation.

Olive (Ann. Bot. Lond. **22**, 339, 1908) has given an account of vegetative nuclear division in the gametophytic stage of this species and Fort has described the cytology of the spermogonia and aecidia.

Uromyces airae-flexuosae Ferd. & Winge

Bull. Soc. Myc. Fr. **36**, 162 (1920); Wilson & Bisby, Brit. Ured. no. 270; Gäumann, Rostpilze, p. 235.

[*Uredo airae-flexuosae* Liro, Bidr. Känned. Finl. Nat. Folk, **65**, 573 (1908).]

Spermogonia and **aecidia** not known. **Uredosori** epiphyllous, between the nerves, at first forming conspicuous, orange-yellow spots on the undersurface, occasionally hypophyllous, scattered or arranged in short rows, 0·3–1 mm. long, long-covered by the epidermis, intensely yellow; uredospores subgloboid, ovoid or ellipsoid, $20–27 \times 18–22\,\mu$, wall hyaline, distantly verruculose, up to $3\,\mu$ thick with 6–8 scattered pores; paraphyses absent. **Teleutosori** dark brown, epiphyllous or hypophyllous; teleutospores ovoid or broadly ellipsoid, bright yellowish-brown, $26–37 \times 16–24·5\,\mu$, wall smooth, thickened slightly at base and apex and deeper coloured at the latter, pedicel hyaline, equal in length to the

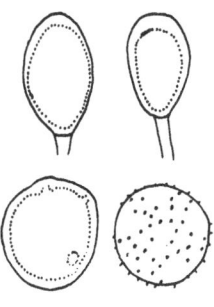

U. airae-flexuosae. Teleutospores and uredospores.

spore, rather thick, persistent. Hemi-
form?
Uredospores and teleutospores on

Deschampsia flexuosa. England, Scotland, scarce, but overlooked.

This species has been recorded from Yorkshire, Norfolk and Surrey, and is widely distributed in Scotland. It was mentioned in a note under *Puccinia dispersa* (*sens. lat.*) by Grove (1913, 265) who associated it with uredosori on *Deschampsia cespitosa* but it is clearly different from the latter which has numerous paraphyses in the uredosorus and, of course, 2-celled teleutospores.

The rust produces bright yellow spots which extend round the whole of the under surface of the leaf, and as these are up to 5 mm. long and often numerous, a very characteristic appearance is produced, the infected leaf showing alternate green and yellow bands. Teleutospores have been rarely found but this may be because they are difficult to discover being generally hidden by the deeply folded leaf. They have been discovered only once in this country, near Peebles, where the rust was abundant, at an altitude of about 1700 ft.

After its discovery by Liro in 1908 the rust was known as *Uredo airae-flexuosae* until the discovery of teleutospores in Denmark by Ferdinandsen and Winge in Denmark in 1920. It is found to the extreme north of Norway (70° 27′ N.) and up to the lichen belt of the mountains. It is clearly not obligatorily alternating.

Uromyces dactylidis Otth

Mitt. Naturf. Ges. Bern. **1861**, 85 (1861); Grove, Brit. Rust Fungi, p. 125; Wilson & Bisby, Brit. Ured. no. 281; Gäumann, Rostpilze, p. 232.

Uromyces poae Rabh., Marcucci, Unio Itineraria Crypt. 38 (1866); Grove, Brit. Rust Fungi, p. 127; Wilson & Bisby, Brit. Ured. no. 283; Gäumann, Rostpilze, p. 239.

Uromyces festucae Syd., Hedwigia, **39**, 117 (1900); Gäumann, Rostpilze, p. 235.

Uromyces ranunculi-festucae Jaap, Verh. Bot. Ver. Brand. **47**, 90 (1905).

Spermogonia epiphyllous, honey-coloured, a few also hypophyllous, among the aecidia. **Aecidia** hypophyllous or on the petioles, in rounded groups on yellow spots often elongate on the petioles, shortly cupulate, with recurved, laciniate margin, yellow; peridial cells not in distinct rows, outer wall thickened up to $14\,\mu$, inner wall up to $7\,\mu$ thick, with numerous small warts and tesselate; aecidiospores globoid or slightly angular, pale yellowish, 17–$25\,\mu$, wall 1–$2\,\mu$ thick, finely verruculose. **Uredosori**

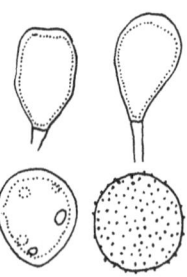

U. dactylidis. Teleutospores and uredospores (after Guyot).

amphigenous, scattered or arranged in rows 0·2–1 mm. long, elliptic or yellowish-brown, without paraphyses; uredospores globoid, subgloboid, or ovoid, yellow or brownish-yellow 20–32 × 18–25 μ, wall 1·5–3 μ thick, finely echinulate with 4–9 scattered pores. Teleutosori generally hypophyllous, subepidermal, scattered or arranged in longitudinal rows, small, often confluent, oblong, compact, shining-black, paraphyses internal, numerous, brown agglutinated, often dividing the sori into compartments; teleutospores at first occurring in the old uredosori, later, in special teleutosori, ovoid or ellipsoid, rounded, bluntly pointed or truncate above where the wall is slightly thickened (up to 4 μ), 1·5 μ thick at the sides, smooth, yellowish-brown, paler at the apex, 18–30 × 14–24 μ, pedicel brownish, as long as, or longer than the spore. Hetereu-form.

Spermogonia and aecidia on *Ranunculus acris*, '*R. auricomus*', *R. bulbosus*, *R. ficaria*, *R. repens* and *R. sceleratus*, March–May; uredospores and teleutospores on *Dactylis glomerata*, '*Festuca ovina*', *F. rubra*, '*Poa annua*', *P. nemoralis*, *P. pratensis* and *P. trivialis*. Great Britain and Ireland, common.

In the uredospore stage of this rust, paraphyses are here stated to be absent although in many previous accounts long, colourless, thin-walled structures are said to be present; Grove (J. Bot. Lond. **72**, 265, 1934) stated that after examination he was convinced that these latter structures were pedicels of uredospores which frequently remain erect and persist for some time after the spores have fallen off and even appear to become longer.

Teleutospores are usually abundant on the grass host and uredosori scanty. Infected grasses grow in the vicinity of ranunculaceous alternate hosts and the rust is obviously obligatorily heteroecious.

This rust was previously regarded as three distinct species, *Uromyces dactylidis*, *U. festucae* and *U. poae*; these were treated by Grove as varieties but their morphological distinctions do not even justify this rank. The differences in effect on their hosts stressed by Grove, distinct yellow banding on the foliage on fescue and larger, more prominent teleutosori on *Dactylis* than on *Poa*, must be interpreted as host reactions.

A large number of physiologic races have also been described.

Recently Guyot (1938, 192) has tabulated the results of infections on the *Ranunculaceae*. In this country cultures appear to have been carried out only by Plowright (Quart. J. Micr. Sci. N.S. **25**, 151, 1885; 1889, 131–2) with the aecidia on *Ranunculus*.

Dactylis

Uredospores 21–32 × 18–25 μ, wall 1·5–2 μ thick. Teleutospores 18–30 × 14–20 μ. Aecidia on *Ranunculus acris*, *R. bulbosus* and *R. repens*, March–May; uredospores and teleutospores on *Dactylis* *glomerata* from July onwards, often covering the leaves, less often the sheaths and culms. Great Britain and Ireland, frequent.

The teleutosori are usually more numerous and more conspicuous than those on *Poa*. Gäumann (1959, 233) recognises six special forms of his *P. dactylidis*.

Poa

Uredospores 14–25 × 14–20 μ, wall 1·5–2 μ thick with 4–9 pores. **Teleutospores** 17–28 × 14–20 μ.

Aecidia on *Ranunculus auricomus*, *R. bulbosus*, *R. ficaria*, *R. repens*, March–May; uredospores and teleutospores on '*Poa annua*', *P. nemoralis*, *P. pratensis*, *P. trivialis*, May–September. Great Britain and Ireland, common.

Grove (1913, 128) stated that the teleutospores on *Poa* are more oblong than those on *Dactylis* and are often provided with shorter pedicels and that the sori are less conspicuous. According to Guyot (1938, 156) the length of the teleutospores is rather greater than those on *Dactylis* (up to 33 μ). They can be most easily found by examining the lower leaves of *Poa* species in June or July, in localities where *Ranunculus repens* is abundant. Gäumann (1959, 239) recognises eight special forms of his *U. poae* on the basis of experiments by a long series of workers which he summarises. It is not possible, however, to relate British material to his information without infection experiments.

Festuca

Uredospores 20–32 × 18–24 μ, wall 1·5–3 μ thick, with 4–8 pores. **Teleutospores** 20–40 × 15–22 μ, thickened up to 2 μ at apex, yellowish-brown, darker towards the apex.

[Aecidia on *Ranunculus bulbosus*]; uredospores and teleutospores on *F. rubra*. Great Britain, scarce.

In this race each group of uredospores is seated on a conspicuous, yellow blotch and a heavily infected leaf shows alternating bands of yellow discoloured tissue and normal green tissue; the yellow bands may be as much as 4 mm. wide. As is explained under *Uredo festucae* DC. (p. 259) in the absence of teleutosori it is not possible to distinguish *Puccinia festucae* from *Uromyces dactylidis* on fescues. We have excluded all these doubtful records based on uredospore collections only from accounts of both species and placed them under *Uredo festucae* which we have associated with *P. festucae* (q.v., p. 258). With this treatment the only certain record of *Uromyces dactylidis* on fescue is Rees' record on *Festuca rubra* at Cardiff in 1933, on plants which had borne only uredospores during the previous years (Grove, 1934, *loc. cit.*).

There is no evidence that the aecidium has been found in this country.

Uromyces ranunculi-festucae Jaap (*loc. cit.*) which occurs on *F. ovina* in continental Europe is only distinguished from *U. dactylidis* on *Festuca* by its slightly longer teleutospores. Klebahn (Z. Pfl.-Krankh. **17**, 135, 1907) showed by culture that its aecidium is borne on *R. bulbosus* also. There is no evidence of this rust in Britain.

The aecidia of *P. magnusiana* which occur on *Ranunculus bulbosus* and *R. repens* are morphologically identical with those of *U. dactylidis* on these hosts but, as pointed out by Grove, those of the former are often developed in July and August while those of *U. dactylidis* begin to appear in the early spring. This statement was partially confirmed near Montrose in central Scotland, when in mid-July 1930 aecidiospores occurring on *R. repens* growing in the vicinity of *Phragmitis communis*, were placed on the leaves of the latter species growing in Edinburgh, uredospores were produced in twelve days and later on teleutosori were formed (see p. 267). The aecidial stage of *U. rumicis* which occurs on *R. ficaria* but has not been confirmed in Britain is similar to that of *U. dactylidis*.

The development of the spermogonia and aecidia on *R. ficaria* has been investigated by Neumann (Hedwigia, **33**, 346, 1894), Blackman & Fraser (Ann. Bot. Lond. **20**, 35, 1906), Pavolini (Bull. Soc. Bot. Ital. **1910**, 83, 1910) and Wang & Martens (La Cellule, **48**, 215, 1939).

As Arthur (1934, 184) pointed out many years ago the correlation of this rust is with the *P. recondita* group; both species possess teleutosori divided into loculi by dark-coloured agglutinated internal paraphyses; in both the haplont may be on the *Ranunculaceae* and the diplont on the *Gramineae*.

Trachyspora Fuck.

Bot. Zeit. **19**, 250 (1861).

Spermogonia unknown. Aecidia uredinoid, without peridia or paraphyses; aecidiospores globoid or ellipsoid, sessile or somewhat catenulate with colourless wall, without pores. Uredosori absent. Teleutosori without paraphyses; teleutospores 1-celled, wall thick and coloured, with large irregular warts, pores absent; pedicel uniseptate, hyaline deciduous. Autoecious, with few species.

The single British species was placed by Grove and others in *Uromyces* but its transfer to a new genus by Fuckel was adopted by Dietel (Engler & Prantl, Nat. Pflanzenfam. Ed. 2, **6**, 57, 1928) and most subsequent authors. The latter considers that the affinities of *Trachyspora* are to *Gymnoconia* and *Phragmidium* rather than to *Uromyces*.

Trachyspora intrusa (Grev.) Arth.

Rusts U.S. Can. p. 97 (1934); Wilson & Bisby, Brit. Ured. no. 255.
[*Uredo alchemillae* Pers., Syn. Meth. Fung. p. 215 (1801).]
Uredo intrusa Grev., Fl. Edin. p. 436 (1824).
Uromyces alchemillae Lév., Ann. Sci. Nat. Bot. ser. 3, **8**, 371 (1847); Grove, Brit.
  Rust Fungi, p. 106.
Trachyspora alchemillae Fuck., Bot. Zeit. **19**, 250 (1861); Gäumann, Rostpilze, p. 215.

Spermogonia unknown, probably not formed. **Aecidia** uredinoid, hypophyllous, crowded, occupying nearly the whole leaf surface, rounded or elongate, confluent and covered by large fragments of the torn epidermis, pulverulent, orange, yellowish or whitish; aecidiospores globoid to ellipsoid, sessile or somewhat catenulate, orange or yellowish, 16–25 × 14–21 μ, wall 1 μ thick, faintly echinulate, pores obscure. **Uredosori** of uncertain occurrence described as small, scattered; uredospores like aecidiospores. **Teleutosori** hypophyllous, at first from a systemic mycelium, replacing the aecidia, later small, scattered, from a limited mycelium; teleutospores globoid to obovoid or oblong 26–40 × 20–30 μ, wall not thickened above, coarsely and irregularly warted especially in the upper part with the lower part almost or quite smooth, brown 3–4 μ thick; pedicel hyaline, about as long as the spore, uniseptate, deciduous. Brachy-form.

T. intrusa. Teleutospores.

On *Alchemilla filicaulis, A. glabra, A. minima, A. monticola, A. vestita* and *A. xanthochlora,* aecidia, April–June; uredosori and teleutosori, June–October. Great Britain and Ireland, frequent but absent in south-east England.

The septum in the pedicel varies in position, it may be close to the base of the spore or up to about 10 μ from it; rupture may occur at the septum or at a variable distance below it.

The systemic mycelium overwinters in the rhizome and grows up with the leaves, causing them to stand more erect and making them paler and conspicuous; they are usually smaller with longer petioles and the laminae are frequently deformed; infected rosettes do not flower. Healthy rosettes may develop from an infected rhizome and bear normal flowers.

The mycelium in the leaves usually produces aecidia first and these are generally soon followed by teleutosori replacing the aecidia with the same distribution (primary teleuto). Secondary teleutosori which are small and scattered, may arise from a limited mycelium later in the season; they may develop on previously infected or on healthy leaves. Secondary uredosori, i.e. uredosori developing on a limited mycelium have been described by some investigators (Kursanov, 1922, 47; Jørstad, 1935, 42); a few uredospores may occur in the secondary teleutosori.

Investigators on the continent have studied the relative abundance of aecidiospores and teleutospores in the primary sori of this rust. They have shown that to the north and at greater altitudes there is an increase in teleutospore production together with a decrease in the number of aecidia until ultimately the latter disappear completely, usually only a few aecidiospores occurring in the teleutosori. The secondary sori may also disappear completely, leaving the equivalent of a microform. Thus Lindfors (1924, 19) found that in Sweden teleutospores are frequently produced in the later stages of the uredinoid aecidia while Rytz (Veröff. Geobotan. Institut. Rubel Zürich, 4, 1, 1927) considered the suppression of the uredo in Scandinavia is due to the arctic-alpine climate; according to Kari (Ann. Soc. Bot. Zool. Fenn. Vanamo, 8 (3), 10, 1936) teleutospores are more abundant in Lapland than they are in southern Finland. Jørstad (1935, 43) has pointed out that in central Norway there is a tendency towards the reduction of the aecidia and that, in extreme cases the aecidiospores and teleutospores develop almost simultaneously in the same sori, and the latter spore form soon dominates completely. In northern Norway the same author (1940, 100) reported that the aecidia are often much suppressed and have been found in abundance only along the sea coast; from northernmost and inland localities at a high altitude they are often represented only by a few aecidiospores among the teleutospores in the young teleutosori. Rytz has shown that often in Scandinavia under these extreme climatic conditions when the rust produces only teleutosori, the infected leaves exhibit no or only slight abnormalities and the rosettes upon which they are borne often bear flowers.

Bubak (Zbl. Bakt. II, 16, 158, 1906) after many trials could not obtain germination of the teleutospores and no successful cultures have yet been made although repeatedly attempted. Infection by aecidiospores with the production of teleutosori has been brought about by Fischer (1898), Klebahn (Z. Pfl.-Krankh. 17, 134, 1907) and by Liro (1908, 58). Klebahn (Z. Pfl.-Krankh. 22, 323, 1912) has described the distribution of the mycelium in rhizomes and buds of infected plants.

Fischer (Mitt. Naturf. Ges. Bern, 1915, 214, 1916) has shown by experiment that the rust can be transmitted by aecidiospores from *Alchemilla*, sect. *Vulgares* to sect. *Splendentes* and *Pubescentes*.

The cytology of this species has been investigated by Kursanov (1922) and Lindfors (1924, 19). Both agree that the aecidiospores are developed from the uninucleate systemic mycelium and that the binucleate condition arises from a fusion of cells or migration of nuclei. Lindfors (*loc. cit.*) believed that teleutospores are also produced on the uninucleate mycelium in a similar manner. Kursanov stated that both a uninucleate and a binucleate mycelium are present and that the latter produces uredospores and teleutospores. The limited mycelium is binucleate.

A rust collected on *Alchemilla alpina* by R. H. Paterson of Glasgow was recorded (Scot. Nat. **4**, 79, 1878) in the following words '*Uredo potentillarum* var. *alchemillae* on the leaves of *Alchemilla vulgaris* in the woods at Innellan and on *Alchemilla alpina* near the summit on Ben Lomond' and 'All the fungi in these communications have been verified by Dr M. C. Cooke, London'. There is no material to substantiate this record which may well be erroneous.

Uredo species on Orchidaceae

A number of collections of solitary uredo infections on orchids have been made in Britain. Except for one on *Spiranthes* mentioned below, the rusts are undoubtedly introduced. These fungi were placed by Grove in three species of *Hemileia* and in *Uredo lynchii*. But for all, the teleutospores appear to be unknown in Britain and they are here treated under their *Uredo* names.

Uredo behnickiana P. Henn.

Hedwigia, **44**, 169 (1905); Grove, Brit. Rust Fungi, p. 383 (under *Hemileia oncidii* Griff. & Maubl.); Wilson & Bisby, Brit. Ured. no. 262.

Uredospores subgloboid 16–18 × 20 μ.
On *Oncidium varicosum* from Brazil and perhaps on other orchids in Glasnevin Botanic Garden. The records of

Hemileia phaji by Grove (1913, 382) on *Phajus wallichii* also probably belong here.

Uredo oncidii P. Henn.

Hedwigia, **41**, 15 (1902); Grove, Brit. Rust Fungi, p. 381 as *Hemileia americana* Mass. and p. 385 as *Uredo lynchii* (Berk.) Plowr.; Wilson & Bisby, Brit. Ured. no. 262 *p.p.* and 265.

Uredospores broadly ellipsoidal 22–30 × 15–22 μ.
On *Oncidium cavendishianum* (Wilson & Henderson, 1954), and imported *Spiranthes* Kew and Glasgow; on *Onci-*

dium and then said to be on *Cattleya dowiana* (Wilson & Bisby, no. 262); on *Oncidium* sp. (Moore, 1959) and on wild *Spiranthes spiralis*.

The rusts here grouped differ only in occurring on various host genera of the Orchidaceae. They all differ from *Uredo behnickiana* described above in having broadly ellipsoid not subgloboid spores. The record of a rust of this type on wild *Spiranthes* rests on a single slide in the British Museum Herbarium of a specimen collected by E. A. Ellis in Hampshire. Only a few sori are present and the spore dimensions agree with those of *U. oncidii.*

Uredo quercus Duby

Bot. Gall, 2, 893 (1830); Wilson & Bisby, Brit. Ured. no. 268.
As *Cronartium quercuum* Miyabe, Bot. Mag. Tokyo, **13**, 74 (1899) (non *C. asclepiadeum* var. *quercium* Berk. Grevillea, 3, 59 (1874)); Grove, Brit. Rust Fungi, p. 315; Gäumann, Rostpilze, p. 84.

Uredosori hypophyllous, thickly scattered, circular, 0·25 mm. diam., hemispherical, dehiscing by an apical pore, at length surrounded by the torn epidermis, yellow, peridium delicate or wanting, uredospores obovoid to broadly ellipsoid, orange-yellow, 15–25 × 10–17 μ, wall colourless, 3 μ thick, evenly echinulate with short, strong points.

On *Quercus robur* and *Q. ilex*, October. England, rare.

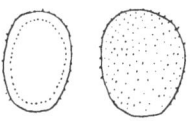

U. quercus. Uredospores.

This rust has been recorded from Norfolk and from several of the southern counties of England on *Quercus robur* and by Hadden (J. Bot. Lond. **54**, 54, 1916) on *Q. ilex* from north Devon; only uredosori have been found. Infection is most frequent on sucker shoots of felled trees. This rust seems to be quite different from the American species on oak which has larger uredospores (20–32 × 15–20μ). Whether it differs from the Japanese rust with only slightly larger spores (24 × 19μ) is more doubtful. Both in Japan (Shirai, Bot. Mag. Tokyo, **13**, 74, 1899; Hiratsuka, Jap. J. Bot. **6**, 25, 1932) and in North America (see Arthur, 1934), the oak rusts have been shown to have aecidial stages on pines on which they form characteristic globoid galls. These are unknown in Europe. Teleutospores have been recorded once in Europe in southern France (Viennot-Bourgin, 1956, 235).

EXCLUDED SPECIES

Uredo plantaginis B. & Br., Ann. Mag. Nat. Hist. ser. 5, 7, 130 (1881). Grove, Brit. Rust Fungi, p. 385; Wilson & Bisby, Brit. Ured. no. 267.

This name has dogged British uredinologists most persistently. Berkeley & Broome's fungus was undoubtedly a *Synchytrium.* The collection from the Isle of Wight mentioned by Grove is *Uromyces betae* on *Beta* sp. The name should be firmly deleted from all lists of rust fungi.

ADDENDUM

Puccinia horiana P. Henn.

Hedwigia **40**, 25 (1901)

Teleutosori 2–4 mm. diam. yollow or greyish, leptosporic; teleutospores slightly constricted, 32–45 × 12–17μ, wall smooth, pale yellow, pedicel hyaline up to 40μ long. Micro-form.

An occurrence of this rust on chrysanthemum cuttings in Essex was recorded by Green & Brooks (Jour. R.H.S. **89**, 127, 1964).

It is a native of China and Japan.

BIBLIOGRAPHY OF MAJOR WORKS CITED IN THE TEXT

Arthur, J. C. (1929). *Plant Rusts.* New York.

Arthur, J. C. (1934). *Manual of the Rusts of the United States and Canada.* Lafayette.

de Bary, A. (1863). 'Récherches sur le développement de quelques champignons parasites.' *Ann. Sci. Nat. Bot.* ser. 4, **20**, 5–148.

Blackman, V. H. (1904). On the fertilisation, alternation of generations and general cytology of the Uredineae. *Ann. Bot. Lond.* **18**, 323–73.

Bubak, F. (1908). 'Die Pilze Bohmens.' *Arch. Naturwiss. Lands. Bohmen.* **13** (5), 1–233.

Cooke, M. C. (1871). *Handbook of British Fungi.* London and New York.

Cummins, G. B. (1956). 'Host index and morphological characterisation of the grass rusts of the world.' *Pl. Dis. Reptr.* Suppl. **240**, 109–93.

Cunningham, G. H. (1931). *The Rust Fungi of New Zealand.* Dunedin.

Ellis, E. A. (1934). 'Flora of Norfolk. Rust Fungi.' *Trans. Norf. Norw. Nat. Hist. Soc.* **13**, 489–505.

Fischer, E. (1898). *Entwicklungsgeschichtliche Untersuchungen uber Rostpilze.* Bern.

Fischer, E. (1904). 'Die Uredineae der Schweiz.' *Beitr. Krypt. Fl. Schweiz,* **2** (2). Bern.

Fragoso, R. G. (1924–5). *Flora Iberica. Uredales.* I, II. Madrid.

Gäumann, E. (1959). *Die Rostpilze Mttteleuropas.* Bern.

Grove, W. B. (1913). *British Rust Fungi.* Cambridge.

Guyot, A. L. (1938). *Le Genre Uromyces.* I. Paris.

Guyot, A. L. (1951). *Le Genre Uromyces.* II. Paris.

Guyot, A. L. (1957). *Le Genre Uromyces.* III. Paris.

Hartig, R. (1894). *Text Book on the Diseases of Trees.* London.

Hiratsuka, N. (1958). *Revision of the Taxonomy of the Pucciniastreae.* Tokyo.

Hunter, L. M. (1936*a*). 'The life histories of *Milesia scolopendrii, M. polypodii, M. vogesiaca* and *M. kriegeriana.*' *J. Arn. Arb.* **17**, 26–37.

Hunter, L. M. (1936*b*). 'Morphology and ontogeny of the spermogonia of the Melampsoraceae'. *J. Arn. Arb.* **17**, 115–52.

Hunter, L. M. (1936*c*). 'The Genus *Milesia* in Great Britain'. *Trans. Brit. Mycol. Soc.* **20**, 116–19.

Hylander, N., Jørstad, I. & Nannfeldt, J. (1953). 'Enumeratio Uredinearum Scandinavicarum.' *Opera Bot. Soc. Bot. Lundensis,* **1** (1), 1–102.

Jørstad, I. (1932). 'Notes on Uredineae.' *Nyt Mag. Naturv.* **70**, 325–405.

Jørstad, I. (1934). 'A study on Kamtchatka Uredinales.' *Skr. Norske Vidensk.-Akad. Mat.-Nat. Kl.* **1933** (9), 1–183.

Jørstad, I. (1935). 'The Uredinales and Ustilaginales of Trøndelag.' *Norske Vid. Selsk. Skrift.* **38**, 1–91.

Jørstad, I. (1937). 'Notes on some heteroecious rust fungi.' *Nytt Mag. Naturv.* **77**, 105–19.

Jørstad, I. (1940). 'Uredinales of northern Norway.' *Skr. Norske Vidensk.-Akad. Mat.-Nat. Kl.* **1940** (6), 1–145.

Jørstad, I. (1950). 'The graminicolous rust fungi of Norway.' *Skr. Norske Vidensk.-Akad. Mat.-Nat. Kl.* **1950** (3), 1–92.

Jørstad, I. (1951). 'Uredinales of Iceland.' *Skr. Norske Vidensk.-Akad. Mat.-Nat. Kl.* **1951** (2), 1–87.

Jørstad, I. (1953*a*). 'Pucciniastreae and Melampsoreae of Norway.' *Uredineana*, **4**, 91–123.

Jørstad, I. (1953*b*). 'Host specialisation within Norwegian blackberry rusts.' *Blyttia*, **11**, 6–15.

Jørstad, I. (1954). 'The rusts of Cyperaceae, Iridaceae and Juncaceae in Norway.' *Skr. Norske Vidensk.-Akad. Mat.-Nat. Kl.* **1954** (3), 1–28.

Jørstad, I. (1958). 'Uredinales of the Canary Islands.' *Skr. Norske Vidensk.-Akad. Mat.-Nat. Kl.* **1958** (2), 1–184.

Jørstad, I. (1960). 'The Norwegian rust species.' *Nytt Mag. Bot.* **8**, 103–46.

Jørstad, I. (1962). 'Distribution of the Uredinales within Norway.' *Nytt Mag. Bot.* **9**, 61–134.

Klebahn, H. (1904). *Die Wirtswechselnden Rostpilze.* Berlin.

Kursanov, L. (1922). 'Récherches morphologiques et cytologiques sur les Urédinées.' *Bull. Soc. Nat. Moscou*, N.S. **31**, 1–129.

Lind, J. (1913). *Danish Fungi as represented in the Herbarium of E. Rostrup.* Copenhagen.

Lindfors, T. (1924). 'Studien uber den Entwicklungsverlauf bei einigen Rostpilzen aus Zytologischen und Anatomischen Gesichtspunkten.' *Sv. Bot. Tidskr.* **18**, 1–84.

Lindroth, J. I. (1902). 'Die Umbelliferen-uredineen.' *Acta Soc. Fauna Fl. Fenn.* **22** (1), 1–223.

Liro, J. I. (1908). 'Uredineae Fennicae.' *Bidr. Känned. Finl. Nat. Folk*, **65**, 642 pp.

Mason, F. A. & Grainger, J. (1937). *A Catalogue of Yorkshire Fungi.* London.

Massee, G. & Crossland, C. (1905). *Fungus Flora of Yorkshire.* London.

Mayfield, A. (1935). 'The Rust-fungi of Suffolk.' *Trans. Suffolk Nat. Soc.* **3**, 1–10.

Mayor, E. (1958). 'Catalogues des Perenosporales, Taphrinales, Erysiphacées, Ustilaginales et Uredinales du canton de Neuchâtel.' *Mém. Soc. Neuchât. Sci. Nat.* **9**, 1–202.

Moore, W. C. (1959). *British Parasitic Fungi.* Cambridge.

Moreau, F. (1914). 'Les Phenomènes de la Sexualité chez les Urédinées.' *Le Botaniste*, **1913–14**, 145–284.

O'Connor, P. (1936). 'A contribution to knowledge of the Irish Fungi.' *Sci. Proc. Roy. Dublin Soc.* N.S. **21**, 381–417.

Plowright, W. B. (1889). *A Monograph of the British Uredineae and Ustilagineae.* London.

Rilstone, F. (1938). 'Cornish micro-fungi.' *J. Bot. Lond.* **76**, 353.

Sampson, K. & Western, J. W. (1954). *Diseases of British Grasses and Herbage Legumes.* Cambridge.

Sappin-Trouffy, M. (1897). 'Récherches histologiques sur la Famille des Urédinées.' *Le Botaniste*, **1896–97**, 59–244.

Savulescu, T. (1953). *Monografia Uredinalior.* II. Bucharest.

Sprague, T. A. (1952). 'The rust fungi of Gloucestershire.' *Cotswold Nat. Field Club*, **31**, 86.

Stevenson, J. (1879). *Mycologia Scotica.* Edinburgh.

Sydow, P. & Sydow, H. (1904–24). *Monographia Uredinearum.* I–IV. Leipzig.

Trail, J. W. H. (1890). 'Revision of the Uredineae and Ustilagineae of Scotland.' *Scott. Nat.* N.S. **4**, 302–27.

Tranzschel, W. (1939). *Conspectus Uredinalium U.R.S.S.* Moscow.

Tubeuf, K. & Smith, W. G. (1897). *Diseases of Plants induced by Cryptogamic Parasites.* London.

Viennot-Bourgin, G. (1956). 'Mildous, Oidiums, Caries, Charbons, Rouilles des Plantes de France.' *Encyclopédie Mycologique*, **26**, 317 pp. Paris.

Wilson, M. (1934). 'The Distribution of the Uredineae in Scotland.' *Trans. Bot. Soc. Edin.* **31**, 345–449.

Wilson, M. & Bisby, G. R. (1954). 'Checklist of British Uredinales.' *Trans. Brit. Mycol. Soc.* **37**, 61–86.

Wilson, M. & Henderson, D. M. (1954). 'Notes on British Uredinales.' *Trans. Brit. Mycol. Soc.* **37**, 248–54.

INDEX

Accepted names of rust taxa and the main page reference to them are printed in bold type; synonyms are in italics. Entries of host plants and reference to the text are in roman type.